Y0-AAY-252

Cardiovascular nuclear medicine

Cardiovascular nuclear medicine

Edited by

H. WILLIAM STRAUSS, M.D.

Associate Professor of Radiology, Harvard Medical School;
Director of Nuclear Medicine Division, Associate Radiologist,
Massachusetts General Hospital,
Boston, Massachusetts

BERTRAM PITT, M.D.

Professor of Medicine, Director of Cardiovascular Division,
University of Michigan School of Medicine,
Ann Arbor, Michigan

SECOND EDITION

with 693 illustrations, including 4 full-color illustrations

The C. V. Mosby Company

ST. LOUIS · TORONTO · LONDON 1979

To

our parents and **teachers**

SECOND EDITION

Printed in the United States of America

The C. V. Mosby Company
11830 Westline Industrial Drive, St. Louis, Missouri 63141

Library of Congress Cataloging in Publication Data

Strauss, Harry William, 1941-
Cardiovascular nuclear medicine.

Bibliography: p.
Includes index.
1. Radioisotopes in cardiology. 2. Heart—Diseases—Diagnosis. I. Pitt, Bertram, 1932- joint author. II. Title. [DNLM: 1. Cardiovascular system—Radionuclide imaging. WG141.5.R2 C267]
RC683.5.R33S77 1979 616.1'07'575 79-18410
ISBN 0-8016-2409-6

CB/CB/B 9 8 7 6 5 4 3 2 1 01/A/028

Contributors

STEPHEN L. BACHRACH, Ph.D.

Associate Professor (Adjunct), Biomedical Engineering, The Catholic University of America; Research Associate, Georgetown University Medical School; Applied Physics Section, Department of Nuclear Medicine, National Institutes of Health, Bethesda, Maryland

GEORGE A. BELLER, M.D.

Head of Division of Cardiology, Professor of Medicine, University of Virginia Medical Center, Charlottesville, Virginia

HARVEY J. BERGER, M.D.

Research Associate in Cardiovascular Nuclear Medicine, Yale University School of Medicine, New Haven, Connecticut

FREDERICK J. BONTE, M.D.

Dean and Professor of Radiology, Southwestern Medical School, The University of Texas at Dallas, Dallas, Texas

EDWARD U. BUDDEMEYER, Sc.D.

Society of Nuclear Medicine; Division of Nuclear Medicine, University of Maryland Hospital, Baltimore, Maryland

L. MAXIMILIAN BUJA, M.D.

Associate Professor of Pathology, Southwestern Medical School, The University of Texas at Dallas Health Science Center, Dallas, Texas

GREGORY D. CURFMAN, M.D.

Instructor in Medicine, Harvard Medical School; Director of Levine Cardiac Unit, Peter Bent Brigham Hospital, Boston, Massachusetts

JOHN T. FALLON, M.D., Ph.D.

Assistant Professor of Pathology, Harvard Medical School; Assistant Professor, Massachusetts Institute of Technology; Assistant in Pathology, Massachusetts General Hospital, Boston, Massachusetts

EDGAR HABER, M.D.

Professor of Medicine, Harvard Medical School; Chief of Cardiac Unit, Massachusetts General Hospital, Boston, Massachusetts

GLEN W. HAMILTON, M.D.

Associate Professor of Medicine, University of Washington School of Medicine; Chief of Nuclear Medicine, Seattle Veterans Administration Hospital, Seattle, Washington

B. LEONARD HOLMAN, M.D.

Associate Professor of Radiology, Harvard Medical School; Chief of Clinical Nuclear Medicine, Peter Bent Brigham Hospital, Boston, Massachusetts

CHARLES J. HOMCY, M.D.

Instructor in Medicine, Harvard Medical School; Assistant in Medicine, Cardiac Unit, Massachusetts General Hospital, Boston, Massachusetts

JOHN W. KEYES, Jr., M.D.

Professor of Internal Medicine and Radiology, University of Michigan Medical School, Ann Arbor, Michigan

DENNIS KIRCH, B.S.E.E.

University of Colorado Medical Center; Denver Veterans Administration Hospital, Denver, Colorado

ANTONIO L'ABBATE, M.D.

Assistant Professor of Medicine, Istituto di Patologia Medica l, University of Pisa; C.N.R. Laboratory of Clinical Physiology, Pisa, Italy

STEVEN M. LARSON, M.D.

Associate Professor of Medicine, Laboratory Medicine, and Radiology; Chief of Radioimmunoassay Section Laboratory Service, Veterans Administration Medical Center, Seattle, Washington

D. E. LIEBERMAN, B.S.E.E., M.S.E.E.

Computer Methods, University of Michigan, Ann Arbor, Michigan

ATTILIO MASERI, M.D.

Sir John McMichael Professor of Cardiovascular Medicine, Postgraduate Medical School, University of London; Hammersmith Hospital, London, England

THOMAS G. MITCHELL, Ph.D.

Associate Professor of Radiology and Environmental Health, Division of Nuclear Medicine, The Johns Hopkins Medical Institutions, Baltimore, Maryland

T. K. NATARAJAN, D.R.E.

Assistant Professor, Department of Radiology and Radiological Sciences, Department of Environmental Health Sciences, The Johns Hopkins Medical Institutions, Baltimore, Maryland

J. A. PARKER, M.D.

Instructor in Radiology, Harvard Medical School; Division of Nuclear Medicine, Department of Radiology, Beth Israel Hospital, Boston, Massachusetts

ROBERT W. PARKEY, M.D.

Professor and Chairman, Department of Radiology, The University of Texas at Dallas Health Science Center, Dallas, Texas

BERTRAM PITT, M.D.

Professor of Medicine, Director of Cardiovascular Division, University of Michigan School of Medicine, Ann Arbor, Michigan

GERALD M. POHOST, M.D.

Assistant Professor of Medicine, Harvard Medical School; Assistant in Medicine, Consultant in Radiology, Massachusetts General Hospital, Boston, Massachusetts

BUCK A. RHODES, Ph.D.

Director of Radiopharmacy, Professor of Pharmacy and Radiology, University of New Mexico College of Pharmacy, Albuquerque, New Mexico

WILLIAM C. ROBERTS, M.D.

Clinical Professor of Pathology and Medicine (Cardiology), Georgetown University, Washington, D.C.; Chief of Pathology Branch, National Heart, Lung and Blood Institute, National Institutes of Health, Bethesda, Maryland

RICHARD S. ROSS, M.D.

Clayton Professor of Cardiovascular Disease, Director of Division of Cardiology, The Johns Hopkins Medical Institutions, Baltimore, Maryland

SOL SHERRY, M.D.

Professor and Chairman of Department of Medicine, Director of Specialized Center for Thrombosis Research, Temple University School of Medicine, Philadelphia, Pennsylvania

MICHAEL E. SIEGEL, M.D.

Associate Professor, Department of Radiology (Nuclear Medicine), University of Southern California School of Medicine; Radiologist, LAC/USC Medical Center, Los Angeles, California

THOMAS W. SMITH, M.D.

Associate Professor of Medicine, Harvard Medical School; Chief of Cardiovascular Division, Peter Bent Brigham Hospital, Boston, Massachusetts

BURTON E. SOBEL, M.D.

Professor of Medicine, Washington University School of Medicine; Director of Cardiovascular Division, Cardiologist-in-Chief, Barnes Hospital, St. Louis, Missouri

PETER STEELE, M.D.

Department of Medicine, University of Colorado Medical Center; Denver Veterans Administration Hospital, Denver, Colorado

ERNEST M. STOKELY, Ph.D.

Assistant Professor, Department of Radiology, The University of Texas at Dallas Health Science Center, Dallas, Texas

H. WILLIAM STRAUSS, M.D.

Associate Professor of Radiology, Harvard Medical School; Director of Nuclear Medicine Division, Associate Radiologist, Massachusetts General Hospital, Boston, Massachusetts

MICHEL M. TER-POGOSSIAN, Ph.D.

Professor of Radiation Sciences, Mallinckrodt Institute of Radiology, Washington University School of Medicine, St. Louis, Missouri

JAN I. THORELL, M.D.

Department of Nuclear Medicine, University of Lund; Malmö General Hospital, Malmö, Sweden

JAMES H. THRALL, M.D.

Associate Professor of Medicine, Director of Nuclear Cardiology Unit, University of Michigan School of Medicine, Ann Arbor, Michigan

SALVADOR TREVES, M.D.

Associate Professor of Radiology, Harvard Medical School; Chief of Pediatric Nuclear Medicine, Divison of Nuclear Medicine, Department of Radiology, Children's Hospital Medical Center, Boston, Massachusetts

ROBERT VOGEL, M.D.

Assistant Professor of Medicine (Cardiology), University of Colorado Medical Center; Director of CCU-MICU, Denver Veterans Administration Hospital, Denver, Colorado

DENNY D. WATSON, Ph.D.

Associate Professor of Radiology, University of Virginia Medical Center, Charlottesville, Virginia

MICHAEL J. WELCH, Ph.D.

Department of Radiology, Mallinckrodt Institute of Radiology, Washington University School of Medicine; Department of Chemistry, Washington University, St. Louis, Missouri

JAMES T. WILLERSON, M.D.

Professor of Medicine, Director of Cardiology Division, Southwestern Medical School, The University of Texas at Dallas, Dallas, Texas

BARRY L. ZARET, M.D.

Associate Professor of Medicine and Diagnostic Radiology, Chief of Cardiology, Yale University School of Medicine, New Haven, Connecticut

Preface

In the five years since the publication of the first edition of *Cardiovascular Nuclear Medicine,* many advances have occurred: (1) the development of a practical means of performing multiple-gated blood pool imaging, (2) the widespread use of ^{201}Tl for myocardial imaging as a replacement for ^{43}K and ^{81}Rb, (3) an understanding of some of the mechanisms of thallium redistribution—enough that imaging following injection of a single dose of thallium is now the norm for detecting myocardial ischemia—and finally, (4) increasing clinical acceptance of the results of nuclear imaging methods in the care of cardiac patients. These new techniques and clinical applications suggest that there is a sufficient change in the body of knowledge and enough new developments on the horizon to justify a new edition of *Cardiovascular Nuclear Medicine*. More than 90% of the material in this text has been rewritten, attesting to the dynamic developments in the field.

Based on current experience, there appear to be several indications for specific cardiovascular nuclear medical procedures that are likely to stand the test of time. The clinical problems, specific tests, and diagnostic points from their outcome are listed in the boxed material.

We anticipate that during the useful life of this edition several technical developments will become widely used in community hospitals, including (1) tomographic imaging to assist in the detection of myocardial perfusion abnormalities and to quantify regional wall motion, (2) ultra short-lived tracers to perform repetitive studies of ventricular function while limiting the radiation burden to the patient, (3) techniques employing platelets, clotting factors, or labeled fats to indicate the growth or regression of atheromatous lesions, (4) simple metabolic substrates as sensitive indicators of cells afflicted with ischemia or the early phases of cardiomyopathy, and finally, (5) equipment to permit the ambulatory monitoring of ventricular function in conjunction with electrocardiographic monitoring while patients go about their daily chores.

In preparing this edition we have attempted to provide sufficient material for the cardiologist intent on learning tracer procedures and for the nuclear physician intent on understanding the physiology and clinical rationale for applying these procedures. To meet these goals, there are chapters on both the technical and clinical aspects of the procedures.

H. William Strauss
Bertram Pitt

Clinical problem	Technique	Comment
Suspected ischemic heart disease		
Atypical chest pain, positive ECG without pain, high risk factors—no symptoms	Exercise thallium study with redistribution	Normal at >85% heart rate—Ischemia unlikely although coronary disease may be present Abnormal—Ischemia likely
	Exercise ventricular function measurements	Normal—Increases of 5% or greater in ejection fraction, no new regional wall motion abnormalities Adequate left ventricular reserve Abnormal—Inadequate left ventricular function reserve
	Ergonovine thallium scan	Determine if spasm-related ischemia is present
Documented ischemic heart disease		
Precatheterization	Exercise thallium or exercise ventricular function	Extent of myocardium at risk of ischemia Impairment of left ventricular function produced by ischemia
Unstable angina	Rest and redistribution thallium	Differentiation of scar from ischemia
	Pyrophosphate	Detect acute necrosis
	Ventricular function at rest	Extent of left ventricular function impairment
Postoperative	Exercise thallium Exercise ventricular function Infarct avid	Document method of pain relief: New infarction Improved perfusion
Myocardial infarction		
"Rule out"	Rest and redistribution ^{201}Tl Infarct avid	Triage
Documented infarction (CCU phase)	Rest and redistribution thallium	Site and extent of infarction and ischemia
	Infarct avid	Site and extent of necrosis
	Ventricular function	Site and extent of left ventricular dysfunction
Convalescent phase	Limited exercise ^{201}Tl Ventricular function	Identify peri-infarct ischemia Measure ventricular function reserve for exercise program
Heart failure		
Recent or chronic congestive heart failure	Rest and limited exercise ventricular function	Separate left ventricular aneurysm from diffuse hypokinesis
	Exercise thallium	Separate ischemic from nonischemic myopathy
Acute heart failure	Rest ventricular function	Evaluate right and left ventricular function Define effects of therapy
Valvular and congenital heart disease		
Valvular heart disease (asymptomatic, preoperative)	Exercise ventricular function	Measure functional reserve to identify patient for early valve replacement
Valve replacement (postoperative)	Rest and exercise ventricular function	Detect perioperative changes in function Separate cardiomyopathy from paravalvular leak
Congenital disease	Rest thallium Radionuclide angiogram	Separate ischemia from congenital myopathy of newborn Detect and quantify shunts

Contents

SECTION THREE

Infarct-avid imaging

SECTION FOUR

Cardiomyopathies

SECTION FIVE

Peripheral vascular disease

SECTION SIX

Clinical applications in vitro

SECTION ONE

PRINCIPLES FOR CARDIOVASCULAR NUCLEAR MEDICINE

1 ☐ Instrumentation in nuclear cardiology

Edward U. Buddemeyer
Stephen L. Bachrach
Thomas G. Mitchell

The instruments employed by the nuclear cardiogist range from the simplest survey meters through more complex beta- and gamma-counting equipment up to the most sophisticated nuclear imaging systems available. Some knowledge of the operation, capabilities, and limitations of this equipment is required if the physician is to properly assess the data and/or the images obtained from nuclear radiation detection devices. This chapter is organized roughly in order of increasing complexity so that the reader may enter it at the beginning or at some intermediate point, according to the level of prior knowledge of the subject matter. The first section deals with the basic physics of radioactive decay and with the properties of the products of decay, especially those that furnish a basis for detection. The second section covers the several kinds of counting systems suitable for the radioassay of in vitro specimens and discusses the concepts of resolution in energy and in time. The third section describes instruments for in vivo radioassay, primarily camera imaging systems, and presents the concept of resolution in space together with a discussion of the problem of statistical "noise" in images. The last section is a presentation of the general considerations and specific limitations and advantages associated with the use of computers in the construction and processing of images, with special emphasis on nuclear cineangiography at equilibrium. First-pass studies are mentioned only briefly out of deference to the more extensive discussion of that technique elsewhere in this text.

PRODUCTS OF RADIOACTIVE DECAY

Radionuclides may emit a variety of radiations, depending on their mode of decay. A given radionuclide may emit only particulate radiation (e.g., ^{14}C), particulate and photon radiations together (e.g., ^{131}I), or photons alone (e.g., ^{99m}Tc). If these emissions are to be detected, they must first undergo some kind of an interaction within the detector. The type of interaction depends on the kind of radiation, particulate or photon, and it is therefore convenient to group the products of decay into these two categories for the purpose of this discussion.

Charged particles

Charged particles are the primary products of two modes of decay, beta (or negatron) decay and positron decay. A beta particle is an electron that has been emitted from the decaying nucleus. An essentially undetectable particle, an antineutrino, is emitted at the same time, carrying with it a variable fraction of the overall energy of decay, $E_{\beta\,max}$. The beta particles may be left with all, some, or practically none of the decay energy; thus their emission spectrum

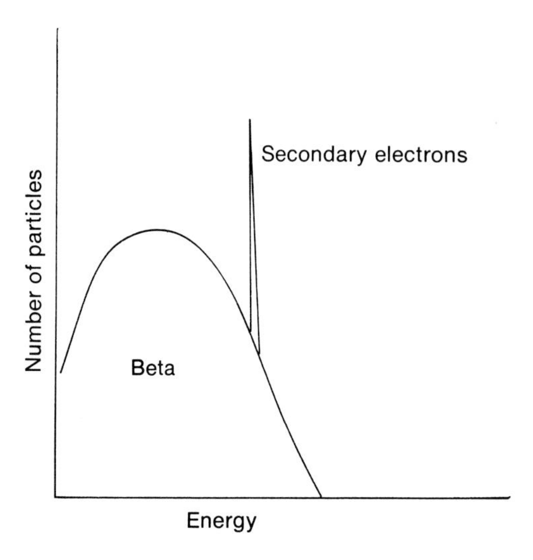

Fig. 1-1. Typical particle spectrum from radionuclide. Beta particles exhibit continuous spectrum of energy from slightly above zero to a maximum E_β, which is characteristic for that radionuclide. Sharp spike shown for the secondary electrons corresponds to those which might have been produced by internal conversion of gamma rays, when conversion electrons would have energy equal to $E\gamma - \phi$, where ϕ is binding energy of electron.

follows a continuum from near zero to $E_{\beta\,max}$ (Fig. 1-1). Average beta energy, $\overline{E}_{\beta-}$, is typically 30% to 40% of $E_{\beta\,max}$.

Positrons are identical to electrons in mass and in the absolute magnitude of their charge, but the charge is positive in sign. They are also similarly distributed in energy.

The other particulate radiations of interest are secondary electrons, formed as alternatives to gamma emission (conversion electrons), and x-ray emissions (Auger electrons). The secondary electrons are characterized by discrete energies; the energy depends on the energy the gamma rays and x rays would have if they were not converted as well as on the atomic shell from which the electrons emanate. Thus:

$$E_{ce} = E_\gamma - \phi$$

$$E_{aug} = E_{x\,ray} - \phi$$

where ϕ is the electron-binding energy (Fig. 1-1).

Beta particles and secondary electrons interact with matter in several ways, some of which provide a means of detection. The particle may interact with one of the native electrons of the atoms of the absorber in such a way as to transfer sufficient energy to the target electron to liberate it from its parent atom. This process produces an ion pair consisting of a relatively massive, positively charged atom (minus one electron) and the negatively charged free electron. If these ions are produced in a gas in which they are free to move, they may be collected and measured. Ions formed in semiconducting solids may be collected with great speed and measured with exquisite precision. Both semiconducting and gas-filled detectors are in use.

Beta particles may also transfer to a target electron an amount of energy sufficient only to raise the electron to an excited state. If the target material is fluorescent, the excited electron will promptly return to its ground state, emitting the difference in energy (around 3 eV) as a photon of visible or near-ultraviolet light. Bursts of light produced in this fashion are called "scintillations," and scintillation detectors are available in a variety of forms and configurations.

The decay time of the excited state is a property of the fluorescent material. If the fluor is to be useful as a scintillation counter, it is required that its decay time be very short, on the order of a microsecond or less, so that the light pulse will be of short duration. The shorter the decay time of the fluor, the briefer the pulse and the greater the counting rate that can be accommodated without mutual interference between pulses closely spaced in time.

Beta particles and secondary electrons may produce other effects in matter, such as chemical bond breaks and molecular vibrations, which become manifest in the forms of chemical changes and heat production, respectively. The chemical changes in biologic targets are responsible for many of the deleterious effects of radiation but are not useful for detection at the levels of activity employed in diagnostic nuclear medicine. Heat production is similarly immeasurably small except at extremely high dose rates.

Beta and betalike radiations are categorized as "nonpenetrating" because their very low mass (0.0005 amu) and consequent lack of momentum makes them relatively easy to scatter and, ultimately, to bring to a halt. The range of a 1 meV beta particle in tissue is about 4 mm. If such particles are emitted within the body, they are totally absorbed therein and thus cannot be detected by external counting devices. Radionuclides that decay by beta emission are therefore useless for in vivo imaging procedures except in the case in which the primary negatron decay events are accompanied by associated photon emission.

Photons

Photons are discrete packets of electromagnetic energy having neither charge nor mass and traveling at the velocity of light. Gamma photons (or rays) are emitted when excited nuclei decay to their ground state. The energy of the photon is precisely equal to the difference in energy between the excited and ground states of the nucleus. For any given nucleus the energy difference between any two states is a fixed quantity; thus all gamma rays emitted from a particular nuclear transition have exactly the same energy. To a detector with perfect energy resolution they would thus appear to be monoenergetic.

Nuclei may be left in an excited state as a consequence of any decay mode. Gamma emission may follow negatron decay, positron decay, electron capture decay, alpha decay, or spontaneous fission. The transition to the ground state may be immediate—the gamma ray is emitted at the same time as the primary decay products (negatrons, positrons, etc.), as in the case of ^{131}I, in which the beta particle and gamma ray are emitted in temporal coincidence. In certain nuclei the excited state may persist for a finite length of time, that is, it may be "metastable." The decay of metastable nuclei such as those of ^{99m}Tc produces gamma

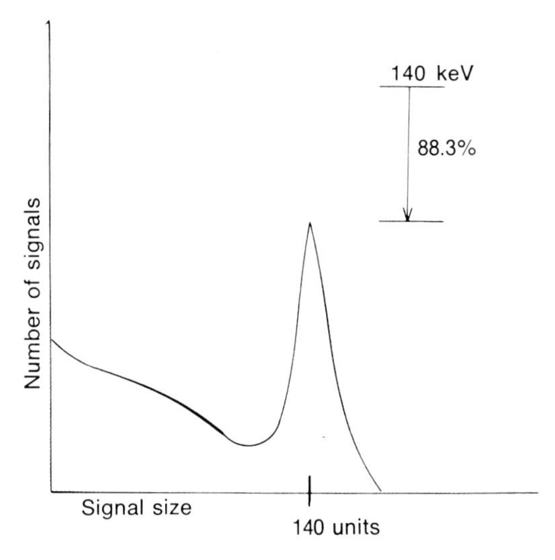

Fig. 1-2. Typical signal spectrum generated by gamma emitter (^{99m}Tc) whose rays are incident on NaI-crystal detector. Sharp spike to right corresponds to gamma rays that have interacted in crystal and deposited all their energy in crystal (photopeak); broad sloping area to left corresponds to gamma rays that have deposited only part of their energy in crystal (Compton continuum).

photons only (neglecting the possibility of internal conversion). Since gamma rays are highly penetrating and experience little difficulty in escaping from the body, metastable isomers in a suitable chemical form can serve as ideal scanning agents.

The transition to the ground state may occur in a single step, as in the example of ^{99m}Tc, in which case only a single photon will be produced, yielding a simple, monoenergetic gamma spectrum. In other cases (e.g., ^{75}Se) the route to ground proceeds through a number of discrete intermediate steps with a gamma photon being emitted at each transition. The several transitions are usually coincident in time, with the result that the nucleus emits several gamma photons "in cascade" (i.e., at the same time), thus producing a more complex gamma spectrum.

The electron capture mode of decay results in an orbital vacancy, usually in the K shell of the electrons surrounding the nucleus. The vacancy is immediately filled, often by an electron from the nearest L shell, but in any case by one that loses energy when it becomes more tightly bound. The lost energy is emitted in the form of an x ray.

X rays emitted after electronic vacancies are refilled are characteristic of the element from which they come, the shell being filled, and the shell from which the replacement electron comes. Characteristic x rays are represented by Roman capital letters, which indicate the shell being refilled, with subscript alpha, beta, gamma, and so on, indicating the shell from which the replacement comes. For example, $K_{\alpha 1}$ indicates the refill of a K shell vacancy with an electron from the L_1 shell, and $L_{\alpha 1}$ represents the refill of an L_1 shell vacancy with an electron from the M shell.

X rays may also be produced by deceleration of high-speed secondary electrons, beta particles, or positrons in the vicinity of nuclei of high atomic number. The x rays produced in this fashion (by deceleration of particulate radiation) are called bremsstrahlung (that is, braking radiation). Bremsstrahlung are characterized by a maximum energy equivalent to the maximum energy of the electrons. However, most of these x rays are of lesser energy. The bremsstrahlung process is the principal mode of production of x rays in fluoroscopic and radiographic units. Bremsstrahlung may also be used to detect the presence of high-energy beta particles (for example, $^{32}P{:}E_{\beta\,max}$ = 1.7 mev) in tissues of high atomic number such as bone.

Gamma rays and x rays, because of their electromagnetic nature, have a low probability for interacting with matter, the probability decreasing in general with increasing radiant energy. Source self-absorption of gamma rays and x rays is small compared to particulate radiation; therefore external counting techniques are feasible for x-ray and gamma-ray emitters. Self-absorption within small samples of blood or other tissue specimens of 3 to 5 ml is usually small enough to be disregarded in counting systems in vitro for x rays or gamma rays above about 25 keV. For gamma rays or x rays originating from within the body of a patient, self-absorption of the radiation may or may not be a problem, depending on the energy of the radiation and the size, shape, and depth of the structure containing the x-ray or gamma-ray emitters. Fig. 1-3 indicates self-absorbed fractions for uniformly distributed and central point sources in objects of various sizes.

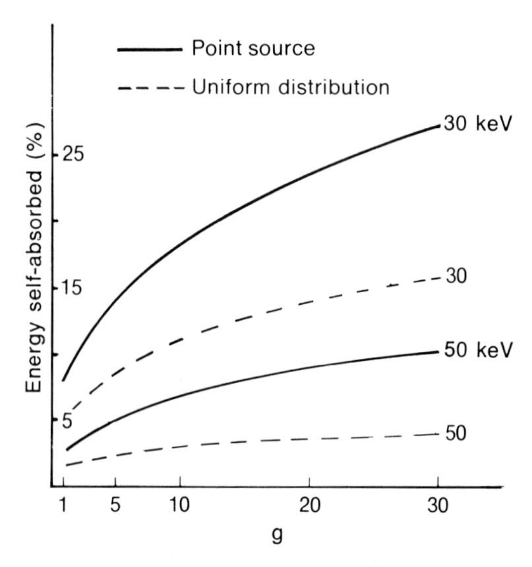

Fig. 1-3. For low-energy gamma rays, self-absorption within the source can be a problem. Shows effect of increasing mass of spherical sources on self-absorption of gamma-ray energy for sources that are uniformly distributed and those in which source is centrally located.

Photons are not themselves detectable. In their interactions with matter, however, detectable charged particles are produced. Within the range of photon energies employed in nuclear medicine, the two most likely interactions are photoelectric absorption and Compton scattering.

In photoelectric interaction, the photon disappears, transferring all its energy to an orbital electron of an atom of the absorber. Photoelectric absorption is not possible unless the energy of the photon exceeds the binding energy of the target electron. The electron leaves its parent atom with the full energy of the photon less the characteristic binding energy of the shell from which the electron was ejected. The ejected "photoelectron" may be detected as a result of any of the interactions (ionization, excitation, etc.) previously described for charged particles. The orbital vacancy in the target atom is filled immediately, yielding a characteristic x-ray photon. If this X photon is also absorbed, which is likely if the absorber is dense, the full energy of the original incident photon is deposited in the absorber and is available for detection. If the photoelectric interaction occurs near the surface of the absorber, it is possible for the characteristic x ray to escape absorption. In that case, the energy deposited in the absorber will be that of the photoelectron only, and if the absorber were also a detector, an "escape peak" might be detected at an energy corresponding to the original photon energy minus the binding energy of the electrons of the detector atoms.

The occurrence of photoelectric absorption is favored in the vicinity of strong electromagnetic fields, such as those adjacent to atomic nuclei. Therefore provided that the energy of the incident photon exceeds the binding energy of the innermost electron shell (i.e., the K shell), it is most likely to interact with K electrons. When photon energy is insufficient for K-shell interaction, photoelectric absorption will occur with less tightly bound electrons, but always preferring inner- to outer-shell electrons.

Since photoelectric absorption requires a strong electromagnetic field, its occurrence depends strongly on the number of positive charges in the nucleus (i.e., on Z, the atomic number). The probability of photoelectric absorption increases approximately as the fourth power of the atomic number of the absorber. For this reason, detectors constructed of high Z materials are especially useful for photon detection.

Photoelectric absorption is most likely when the energy of the photon just exceeds the binding energy of the absorber electrons. Above the critical binding energy, therefore, the probability of photoelectric interaction diminishes as photon energy increases. Higher energy photons thus are more likely to interact with an absorber by alternative mechanisms such as Compton scattering.

In the Compton process, the incident photon again interacts with an electron of the absorber, but, rather than being completely absorbed, the photon is scattered through an angle, losing only part of its energy. The maximum energy loss occurs at the maximum angle of scatter (i.e., 180 degrees backscatter), lesser amounts being lost at lesser angles of scatter. The residual energy of the scattered photon (E_S) depends strictly on the initial energy of the incident photon (E_ϕ) and the angle of scatter (θ); thus:

$$E_S = \frac{E_\phi}{1 + \frac{E_\phi}{0.511}(1 - \cos\theta)}$$

where

E_S = Energy of scattered photon (meV)
E_ϕ = Energy of incident photon (meV)
θ = Angle of photon scatter (degrees)

The energy lost by the incident photon is transferred to the target electron. This electron, a "Compton" electron, may be detected as a result of any of the interactions described previously for charged particles. It is evident from the equation that 180-degree scatter represents the absolute limit of the energy that can be transferred to a Compton electron. The energy transferred is always less than the full energy of the incident photon. Since (1) it is the Compton electron rather than the photon that actually is detected, and (2) the Compton electron can never obtain all the energy of the incident photon, there evidently will be a sharp cutoff, corresponding to 180-degree scatter, in the detected energy spectrum, and this cutoff will occur at an energy below the full energy of the photon. The cutoff point is called the "Compton" edge.

Photons may be totally absorbed in matter by a series of Compton scattering events culminating in photoelectric absorption. Part of the incident photon energy is deposited at each scattering site and the last of its energy at the site of photoelectric absorption. Such a series of events will deposit all the energy of the incident photon within the absorber at essentially the same

Table 1-1. Factors affecting counting rate of source at various depths in water

Depth (cm)	Factor due to inverse square law	Factor due to gamma-ray absorption of energy		
		30 keV	100 keV	300 keV
0	1	1	1	1
1	0.055	0.69	0.84	0.89
2	0.012	0.48	0.71	0.79
3	0.006	0.33	0.60	0.70
5	0.002	0.16	0.43	0.55
10	0.0005	0.025	0.18	0.30

instant. If the absorber is a detector, it will detect all the energy of the incident photon.

Detector geometry

It is evident from a consideration of the detectability of gamma radiation or x radiation in a narrow beam arising from within an object that the beam of radiation will be diminished in intensity according to the source's distance below the surface. With a detector of 1 cm^2 cross-sectional area and an intrinsic efficiency of 100%, using a point source at the surface as a reference, the apparent counting rate would be reduced with increasing depth, according to the schedule given in Table 1-1.

Two factors are operative in decreasing the number of photons reaching the detector. As the point source is moved farther away, the solid angle subtended by the detector becomes smaller, and the geometric factor (G) decreases as the inverse square of the distance (r) (Fig. 1-4). Thus:

$$G = \frac{\text{Area of cap of sphere of radius r}}{\text{Area of sphere of radius r}}$$

$$= \frac{2\pi r^2 (1 - \cos a)}{4\pi r^2}$$

$$= \frac{1 - \cos a}{2}$$

If a is 180 degrees, G is 1 (100% or 4π geometric factor); if a is 90 degrees, G would be 0.5 (50% or 2π geometric factor). An approximation can be made as follows:

$$G \approx \frac{\text{Area of detector face}}{\text{Area of sphere of radius r}} = \frac{\pi r_d^2}{4\pi r^2}$$

$$G \approx \frac{1}{4}\left(\frac{r_d}{r}\right)^2$$

Thus the geometric factor for a point source seen by a detector of 1-inch radius at a distance of 6 inches is as follows:

$$G = \frac{1}{4}\left(\frac{1}{6}\right)^2 = \frac{1}{144}, \text{ or considerably less than 1\%}$$

For a crystal and multiaperture collimator with N holes, the geometric factor for a point at the focal spot is as follows:

$$G = N \times \frac{1}{4}\left(\frac{R}{r}\right)^2$$

where

R = Radius of each inlet aperture in collimator

r = Average distance from point source to collimator face

The geometric falloff per unit increase in distance becomes progressively less as the source and detector are moved farther apart. This is one reason why most detection systems are set up in such a way that the detector is at some distance from the skin overlying the tissue or organ containing the radionuclide. The minimum distance approachable is usually determined by the thickness of the collimator in front of the detector. When the detector face is moved back to about 10 cm from the surface of the object being scanned, the falloff due to the inverse square law is minimized (Fig. 1-5).

By increasing the distance between source and detector, an attempt is made to decrease the variability in the counting geometric factor due to inverse square law effects alone. A source at 9 cm depth (19 cm total distance) would have a geometric factor of $\frac{20^2}{19^2}$, or 1.1 times higher than a source of 10 cm depth, and a point source at 11 cm depth (21 cm total distance) would have a geometric factor of $\frac{20^2}{21^2}$, or 90% of that at 10 cm depth.

Absorption effects appear at first to be independent of the distance from source to detector. However, in reality, this is not so. In a consideration of only photoelectric absorption in which the gamma-ray energy is completely absorbed, distance from source to detector is unimportant. However, in the range of gamma-ray energy, most useful in nuclear imaging, the principal mode of interaction in tissue is the Compton effect. The result is that part of the gamma-ray energy is given to a secondary electron, and the residual gamma ray travels on in a different direction at reduced energy. At distances far from the detector, gamma rays that undergo even small angle scattering are deflected away from the detector face; this decreases the number of gamma rays detected (Fig. 1-6). The effect of source-to-detector distance on apparent attenuation of a gamma ray depends on the ability of the detection system to separate the unscattered rays from those gamma rays that have been scattered before reaching the detector (Table 1-2).

The combined effect of the inverse square law and the attenuation within the overlying absorbing medium tends to reduce drastically the detectable radiation coming from a patient. The inverse square law is important at short distances and is independent of gamma-ray energy. Fractional attenuation per unit path length, although slightly dependent on depth, is strongly dependent on gamma-ray energy, being greater for gamma rays of lower energy.

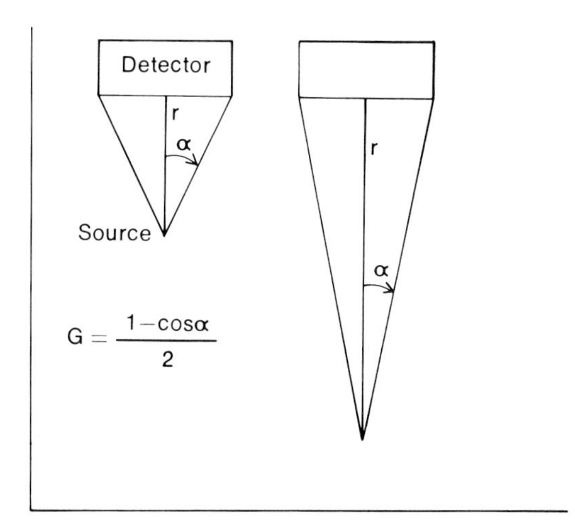

Fig. 1-4. Solid angle subtended from point source of electromagnetic radiation to detector's face decreases with increasing distance between source and detector. Fraction of rays given off that are incident on the detector face is expressed in terms of geometric factor G, which varies from 0 to 1.

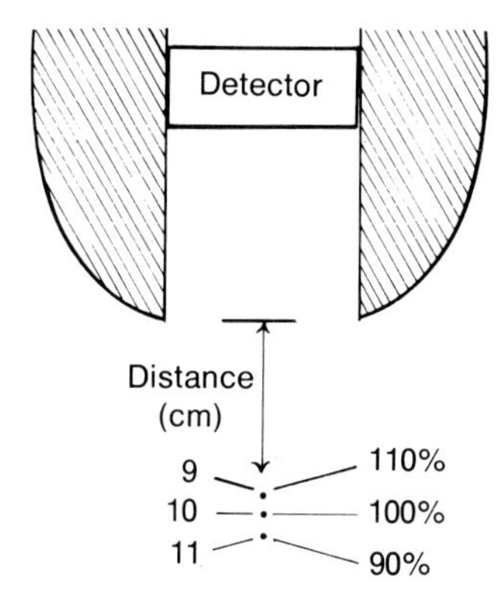

Fig. 1-5. Effect of even slight displacement of source of radiation along the axis of detector. At distance shown, 1 cm movement of source in either direction leads to a 10% change in counting rate. These differences are accentuated for sources closer to detector and minimized at greater distances.

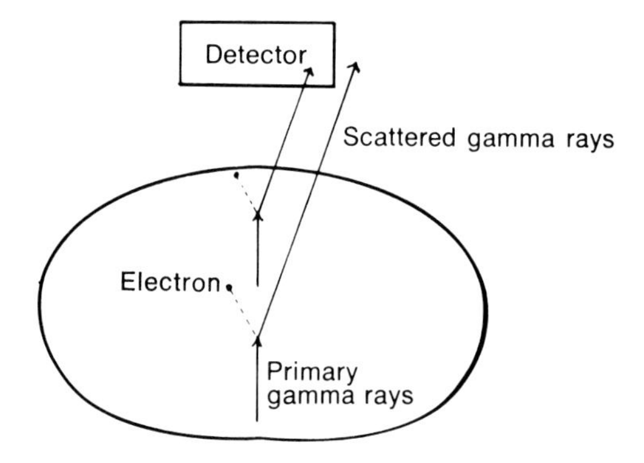

Fig. 1-6. Gamma rays scattered through small angles may be detected by counter if scattering occurs close to crystal. Gamma rays scattered through small angles may not be separable from primary, unscattered gamma rays (Table 1-2).

Table 1-2. Scattering angles necessary to produce a 10% difference in gamma-ray energy

Nuclide	Principal gamma-ray energy (keV)	Angle (degrees)	Scattered gamma-ray energy (keV)
^{125}I	35	152	32
^{133}Xe	81	68	73
^{99m}Tc	140	39	126
^{131}I	364	31	330
^{137m}Ba	662	22	600

Optimum half-life

The chart of the nuclides offers a virtually limitless choice of physical half-lives. For any given administered activity it is generally true that the shorter the half-life, the lower the absorbed radiation dose. On first consideration it would thus appear that the optimum half-life would be as short as possible, but that is a naive conclusion.

During in vivo tracing studies there is always some lapse of time, established by the kinetics of the traced system, between the time of administration of the radiopharmaceutical, t_a, and the time at which the examination (scan, uptake estimation, etc.) is made, t_x. In the case of a 24 hr thyroid iodine uptake study, where $t_x - t_a = 24$ hr, the use of an extremely short-lived radionuclide would require an enormous initial administered dose so that sufficient activity would remain at the time of examination.

It has been shown that, on the assumptions that mixing is instantaneous and elimination is the result of physical decay alone, the optimum half-life for dose reduction, $t_{1/2}$(opt), is as follows:

$$t_{1/2}(\text{opt}) = 0.6931\ (t_x - t_a)$$

For a 24 hr study the optimum half-life is $0.6931(24) \cong 17$ hr. It is for this reason among others that ^{123}I ($t_{1/2} = 13$ hr) is so very nearly ideal in this application.

The time span in nuclear cardiologic tracing is apt to be very short, a first-pass study, for example, being substantially complete within 1 min of injection. In that circumstance the optimum half-life is also very short; thus isotopes like 13 sec ^{81m}Kr become extremely attractive choices. Such short-lived nuclides could be safely administered in multiple curie quantities so as to provide an exceedingly high counting rate, which could be translated into either spatial or temporal resolution far superior to that provided by 6 hr ^{99m}Tc. Such advances, exciting in prospect, will have to await the development of much faster counting systems, since none of the currently available instruments can tolerate counting rates much higher than those already obtained from, say a 20 mCi bolus of ^{99m}Tc. If ultrafast, solid-state counting systems become available, then these devices, together with ultrashort-lived radionuclides, may be expected to provide whole orders of magnitude improvements in resolution, to the end that the spatial resolution of isotope scans might approach that of good-quality radiographs.

SYSTEMS FOR IN VITRO ASSAY

Instruments for dose calibration

Almost any instrument in the laboratory can be used for calibration of a radionuclide provided that it is sensitive to the radiation emitted by the radionuclide. The activity of an unknown sample can be determined directly from a comparison with a counting standard of known activity under identical geometric conditions, thus:

$$\frac{\text{Counts/unit time of unknown sample}}{\text{Counts/unit time of standard}} = \frac{\mu\text{Ci in sample}}{\mu\text{Ci in standard}}$$

The sensitivity of the usual well scintillation counting system is so high that it permits introduction of only a small quantity

have shorter decay times (e.g., CsF, 0.005 μsec), are available, but none of these have as high a light output as does sodium iodide.

Sodium iodide is hydroscopic. If permitted to remain in contact with atmospheric moisture, the crystal will quickly lose the optical clarity that is a prerequisite for efficient light collection. For this reason the crystals must be hermetically sealed ("canned") at the time of manufacture. The can ordinarily supplied is rather robust to maintain the integrity of the seal despite the mechanical stresses and strains anticipated in routine use. Typically, the can is constructed of aluminum 2 mm or more in thickness. Such thicknesses present an all but impenetrable barrier to particulate radiations so that sodium iodide crystals are not sensitive to incident negative beta particles, positrons, or secondary electrons. Aluminum 2 mm thick is also sufficient to seriously attenuate the very low energy x and gamma photons of ^{125}I. When choosing a detector for radioimmunoassay with ^{125}I, it is well to select one with the special thin can that is offered as an option by most vendors.

Detector assembly. A crystal scintillation detector assembly consists of a solid, transparent crystal of sodium iodide optically coupled to a photomultiplier (PM) tube. The output of the PM tube is an electric voltage pulse of a size that depends on the amount of energy deposited in the fluor by an incident photon. The pulses may be counted by means of digital scalers or digital or analog count rate meters. The scintillation counter acts in the following manner:

1. A photon of x or gamma radiation penetrates the can and interacts within the crystal to produce a secondary electron (Compton electron, photoelectron) having all or part of the initial energy of the photon. The energy of the secondary electron is totally absorbed within the crystal. A portion (~10%) of the absorbed energy produces excitation, which leads to the emission of light photons in the visible and near-ultraviolet regions of the spectrum.
2. Photons of light that escape from the fluor may penetrate the glass envelope of the PM tube and release photoelectrons from its photocathode. The photoelectrons released at the photocathode are drawn by a positive potential difference of 50 to 150 V to an electrode that can act both as a cathode and as an anode and hence is called a dynode. The kinetic energy of the photoelectrons may be sufficient to release three to five electrons from this first dynode.
3. Secondary electrons released from the first dynode are accelerated by a positive electric field of 50 to 150 V to the second dynode, where they may release additional electrons. This cascading effect takes place at ten to twelve dynode stages, depending on the PM tube. If every electron striking a dynode results in the emission of three electrons from that dynode, then for every photoelectron striking the first dynode, approximately 3^{10}, or 60,000, electrons will be emitted from the tenth dynode. This occurs when a total of about 1000 V is placed across the PM tube.
4. A positive electric field at the anode is responsible for collection of electrons given off by the last dynode. If the interelectrode and lead capacity of the anode is 10 picofarads, the 60,000 electrons produced in the previous case will produce a momentary voltage change on the anode, thus:

$$\frac{60{,}000 \text{ electrons} \times 1.6 \times 10^{19} \text{ coulombs/electron}}{10^{-11} \text{ farads}} = 0.96 \text{ mV}$$

This pulse may then be amplified to drive a scaler, count rate meter, or oscilloscope or to be sorted according to its size in a pulse-height analyzer.

Because of variations in conversion efficiencies along the line, the output pulses from the PM tube will vary. Part of this stems from variations in conversion of absorbed energy into photons of light and conversion of light into photons at the photocathode and variations in electronic amplification along the dynodes of the PM tube. The net result is a distribution of output pulses from the PM tube for events of identical initial energy absorption in the scintillator. As long as the voltage difference between dynodes does not materially change because of a previous high pulse of current, the output of the scintillation counter is approximately proportional to the energy absorbed in the initial ionizing event. The scintillation counter is therefore considered to be a proportional counter; the output signal is proportional to the input signal and is also proportional to the energy of incident radiation, if the incident photon has been totally absorbed.

Energy resolution. The proportionality between the energy absorbed in the crystal and the output signal can remain perfectly constant only if all the components that make up the signal are perfectly precise; but that is not generally the case. The causes of imprecision include the following:

1. A dependence between the amount of energy deposited and the light output of the crystal, negligible at energies above kiloelectron volts but considerable at lower energies
2. A dependence between the location of the scintillation in the crystal and the efficiency of light collection, in that light originating from a point at some distance from the photocathode is likely to be partially self-absorbed in the crystal
3. A lack of perfect reproducibility in the response of the photomultiplier tube

Additional uncertainty is introduced according to the complexity of the light piping, which may intervene between the crystal and the phototube, and when the signal is derived from the output of more than one phototube, each contributing its own independent error.

Energy resolution is measured by determining the relative spreading (in percent) of the detected photopeak at its half-maximum points. Thus, if the 662 keV photopeak of $Cs^{137} \rightarrow Ba^{137m}$ passed through the half maximum at 637 keV on the low side of 662 keV and 687 keV on the high side, then the full width half-maximum (FWHM) resolution would be as follows:

$$\frac{\Delta E_{\gamma}}{E_{\gamma}}(100) = \frac{687 - 637}{662}(100) = \frac{50}{662}(100) = 7.55\% = \text{FWHM}$$

The best energy resolution is achieved with the simplest design. A single, small crystal directly coupled to a single phototube may exhibit a resolution as good as 7%. When several phototubes are involved in the production of the signal (as in the Anger camera), resolution is typically degraded to a value of 15%, and when lengthy and complex light piping is a necessary part of the design (as in the Bender and Blau camera), energy resolution is a very poor 50%.

In a nuclear imaging device a consequence of poor energy resolution is a loss in the ability to reject off-focus, scattered-in photons by means of pulse-height analysis. Including scattered photons in the image reduces contrast and makes it more difficult to discern lesions, especially cold lesions in a hot field, for example, as in thallium scanning.

Resolving time and coincidence losses. As mentioned earlier in this chapter, the scintillation light pulse produced in sodium iodide is of brief but finite duration. If two such pulses occur close enough together in time to be counted as one, the phototube(s) will produce a single output signal equivalent to their sum. Failure to resolve the pulses thus leads to two errors: (1) the false indication that one rather than two events has occurred and (2) an overestimation of pulse height, since the energy of both has been attributed to one.

The minimum time that must elapse between successive pulses for them to be distinguished as two separate events is called the "pulse-pair resolving time" of the counting system. If the system is to be used at high counting rates, then a short resolving time is required to avoid intolerable coincidence losses.

The resolving time of a counter can be determined by exposing it to a series of sources with activities that increase in a predictable fashion. At an activity sufficient to produce significant coincidence losses the observed disparity between expected and indicated counting rates provides an estimate of resolving time according to the following expression:

$$R - r = Rr\upsilon$$

where

R = Predicted counting rate (cps)
r = Observed counting rate (cps)
υ = Resolving time (sec)

When the resolving time is known, this equation can be used to predict the coincidence loss ($R - r$) that will occur at any given counting rate.

Coincidence losses lead to an underestimation of the activity within the field of veiw of the detector, the greatest losses occurring at the highest counting rates. As a consequence, errors may be introduced into comparisons drawn from two separate counts or image frames, one of which contains higher activity than the other. The error will be in the direction of underestimating the difference between the two counts or frames. When comparing end-diastolic and end-systolic images, for example, coincidence losses, if any, will be more severe in the more active end-diastolic image, thus yielding an erroneously low estimate of ejection fraction. In addition to the effect on indicated counting rate, unresolved coincident pulses confound the logic of the position-sensing circuits in Anger-type cameras, leading to a degradation in spatial resolution.

The resolving time of simple crystal scintillation counters is of the order of 1 to 2 μsec. In the Anger camera system the coincident pulses from several phototubes must be processed to synthesize position signals, which takes additional time; therefore the overall resolving time of the camera is (typically) 5 or 6 μsec.

Liquid scintillation counters

Gamma-emitting radioisotopes of the elements of which biochemicals are chiefly composed (carbon, hydrogen, oxygen, nitrogen) either do not exist or have half-lives that are too short to be practical in routine laboratory use. One can, of course, use a "foreign" gamma-emitting label (i.e., a radioisotope of an element not present in the native molecule) to tag biochemicals of interest (e.g., ^{131}I in albumin), but such compounds are never chemically identical to the native substance, and in some cases the difference is so marked as to preclude their use in tracing the behavior of sensitive systems.

There are, however, readily available, beta-emitting isotopes of the "native" elements of biochemical compounds, especially ^{14}C and ^{3}H (tritium). These two radionuclides have half-lives long enough to be convenient to use, yet not so long as to severely limit the maximum attainable specific activity. Compounds synthesized from these radionuclides are exact chemical duplicates of the native substance and are indistinguishable from it, even in the most fastidious systems.*

The counting of beta emissions is complicated by the fact that they are readily scattered and ultimately stopped even in thin absorbers. A beta particle has difficulty escaping from the sample in which it is contained and penetrating the sample cover, air space, and entrance "window" that may intervene between the sample and sensitive volume of the detector. For these reasons, in the most sensitive counting system now in common use, the sample and detector are in one and the same solution. This obviates losses due to absorption outside the detec-

*There is a slight "isotopic effect" arising from the difference in molecular weight that results from the substitution of the heavier ^{14}C for native ^{12}C or ^{3}H for ^{1}H. The isotopic effect is usually small enough to be neglected.

tor and, at the same time, provides a geometry factor of 1 in that all emanations from the source, regardless of their direction of propagation, must enter the sensitive volume of the detector. The system is a liquid scintillation counter.

Except in regard to the fluor, a liquid scintillation counting system resembles the crystal scintillation system previously described. The systems have in common a PM tube with associated power supply, a variable-gain linear amplifier, a pulse-height analyzer, and a scaler-timer readout device.

Peculiar to the liquid scintillation system is the detector fluor, which usually consists of an alkyl benzene solvent such as toluene, in which is dissolved a small amount (~5 g/L) of an organic scintillator with the beta-emitting sample. In the operation of this system, a specific sequence of events leads to detection.

STEP 1. Emission of a beta particle by a sample molecule dissolved in the solution. Most biochemical specimens are hydrophilic substances with only limited solubility in toluene. Although there are exceptions (for example, fine suspensions can be counted), it is generally true that the lack of sample solubility is one of the primary limitations of the overall sensitivity of detection.

STEP 2. Excitation of solvent molecules by the beta particle. Since the solvent molecules are the most numerous constituents of the solution, they are also the most likely recipients of the energy lost by the beta particle.

STEP 3. Resonant transfer of the solvent molecule's excitation energy to a scintillator molecule. Electrons involved in the chemical bonds of aromatic rings have the capacity to momentarily absorb energy, that is, to become excited, without disrupting the structure of the ring. The energy is promptly reemitted and excites a neighboring solvent molecule, which in turn excites its neighbor; this process continues until the excitation energy reaches one of the relatively rare scintillator molecules in the solution. The process is resonant in that the transfer takes place without a loss of energy.

Anything that interferes with resonant transfer quenches, that is, diminishes the intensity of, the scintillation signal by intercepting the excitation energy before it can cause excitation of a scintillator molecule. Any substance that decomposes when excited, either irreversibly or reversibly but endothermally, acts as a chemical energy "sink" into which energy may be lost. Such substances are called chemical quenching agents. The organic compound to be assayed may be a quencher itself, and quenching agents, such as trichloroacetic acid, chloroform, and oxygen, may be introduced in varying quantities in admixture with the sample during preparation. Quenching by such agents is called solute quenching.

STEP 4. Emission of light by the scintillator. Excitation of a scintillator molecule results in the emission of a photon of visible light (wavelength, ~ 4000 Å; energy, ~ 3 eV). Light emission from organic scintillators is more prompt than that from inorganic crystals. (Decay time, $\tau_{1/2}$, for NaI is ~ ¼ μsec; whereas $\tau_{1/2}$ for a typical liquid scintillator is ~ 1 nsec.)

STEP 5. Transmission of light photon(s) to the photocathode of the PM tube. The solution should be both transparent and colorless so that absorption of light photons is minimized. The biologic sample may, however, contain chromophores that strongly absorb light in the blue end of the spectrum; hemoglobin is a common example. Such substances diminish the scintillation signal by absorbing light photons (color quenching) before they can reach the PM tube(s). Yellow solutions, such as urine and plasma, also strongly absorb the light emitted by most scintillators.

The light photons must next penetrate the sample counting vial. It seems to make little difference whether the vial is made of glass, which is completely transparent, or of polyethylene, which is only translucent; transmission of light signals is approximately the same for both materials.

Finally, to reach the photocathode, the

light photons must penetrate the glass envelope of the PM tube. Older PM tubes* were constructed of a glass that was only semitransparent to the emissions of relatively short wavelength of the primary scintillator. For this reason, scintillation cocktails formerly contained small quantities (0.3 to 0.5 g/L) of a secondary scintillator, the function of which was to absorb the emissions of short wavelength and reemit them as light photons of a slightly longer wavelength, which were better able to penetrate the window of the PM tube. The secondary scintillator was called a wavelength shifter. PM tubes of recent manufacture have glass windows that are transparent to the emissions of the primary scintillator. Secondary scintillators are thus no longer required with modern counting equipment. Most proprietary scintillation media continue to include a secondary scintillator to be compatible with all equipment, new and old.

STEPS 6 AND FOLLOWING. Release of photoelectrons, photomultiplication, amplification, pulse-height analysis, and so on. Succeeding steps in the detection process are, with the exceptions to be noted, similar to crystal scintillation systems.

Problem of noise. The two most commonly used beta-emitting radionuclides for labeling of organic compounds are ^{14}C and ^{3}H. ^{14}C has an average beta-particle energy of 45 keV and a maximum of 155 keV. The beta particles from ^{3}H have a maximum energy of only 18 keV and an average energy of about 6 keV (6000 eV): if an ^{3}H atom decays within the scintillator solution (volume of 20 ml), all the beta-particle kinetic energy is deposited within the solution, but only about 10% of it leads to light production, the remainder being lost to other competitive mechanisms of absorption, such as heat production. Thus approximately 600 eV of the 6000 eV is converted to light. Each light photon has about 3 eV of energy; thus there are $\frac{600 \text{ eV}}{3 \text{ eV/photon}} = 200$ light photons produced by an ^{3}H beta particle of average energy. Because of the geometry of the PM tube, the counting-vial arrangement, and the probability of some loss in light transmission, the efficiency of light collection is not likely to be much greater than 10%; therefore, of the 200 photons produced, only twenty may be incident on the photocathodes of the PM tubes. The quantum efficiency (photoelectron per incident light photon) of the photocathode is approximately 10%; therefore a beta particle of average energy from ^{3}H can be expected to produce only two photoelectrons at the photocathodes of the PM tubes. Certain of the efficiency estimates used in this example are susceptible to modest improvement.

Nevertheless, it is evident that in assay of ^{3}H, we are dealing with a system that produces very weak signals. In some cases, these signals are indistinguishable from random noise signals generated in the PM tube. At normal room temperature, some of the surface electrons at the photocathode of the PM tube may momentarily acquire sufficient thermal energy (approximately 3 eV) to become liberated, generating a noise pulse. The low-strength signals from ^{3}H beta particles are difficult to distinguish from the noise signals generated by thermionic emission.

One effective method of suppressing noise is to cool the PM tubes to between 0° and −8° C; this lowers the average thermal energy of the photocathode electrons so that fewer of them acquire sufficient energy to be released. For practical reasons, cooling of the PM tubes requires that the sample vials be cooled also. Cooling of the samples raises the possibility of phase separation in the sample solution, which at normal room temperature might already be nearly saturated with water. Phase separation effectively isolates the aqueous sample from the organic scintillator, dramatically lowering counting efficiency. For this reason, scintillation media must form single clear homogeneous solutions, not only at the room temperature at which they are prepared under direct observation but also at

*RCA 5819 and Dumont 6292.

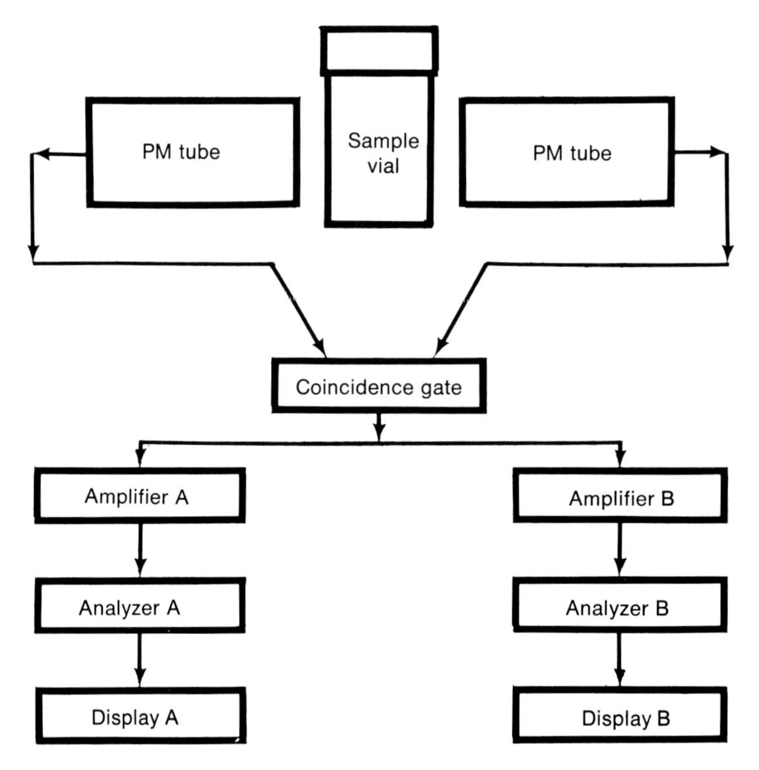

Fig. 1-7. Components of liquid scintillation counter.

the colder temperature at which they will be assayed when they are hidden from view.

Improvements in PM tube design have made it possible to produce systems that do not require cooling. Despite the improved characteristics, these systems generally have a somewhat higher background (noise) level than those that are refrigerated. They are, however, less expensive (by at least the cost of a refrigeration unit) and, in certain circumstances, more convenient to use.

Another method of dealing with random PM tube noise is to use two tubes arranged as in Fig. 1-7. The output of each tube is led through a coincidence gate that opens only if pulses occur simultaneously in both tubes. A scintillation occurring anywhere in the solution or sample vial is necessarily seen simultaneously by both tubes; thus coincident signals are produced that pass the gate and go on to be analyzed and displayed. In contrast, thermal noise pulses occur randomly and independently in each tube and are only very rarely in accidental coincidence. Most noise pulses are blocked at the coincidence gate; thus noise background is reduced to a tolerable level. All liquid scintillation counters in current production employ this technique for reduction of random noise. This coincidence method has no effect on reduction of background due to local sources of radiation.

Problem of quenching. Certain components of the sample may quench the scintillations produced by beta-particle interactions in the sample. Quenching agents can be removed by procedures such as combustion of samples to $^{14}CO_2$ and $^{3}H_2O$, but such methods can be tedious and slow and not conducive to the counting of large numbers of samples. The ordinary user is generally obliged to count samples that contain variable and unknown amounts of quenching agents even when all the samples are similar in nature and especially when they are not. Since quenching causes a decrease in

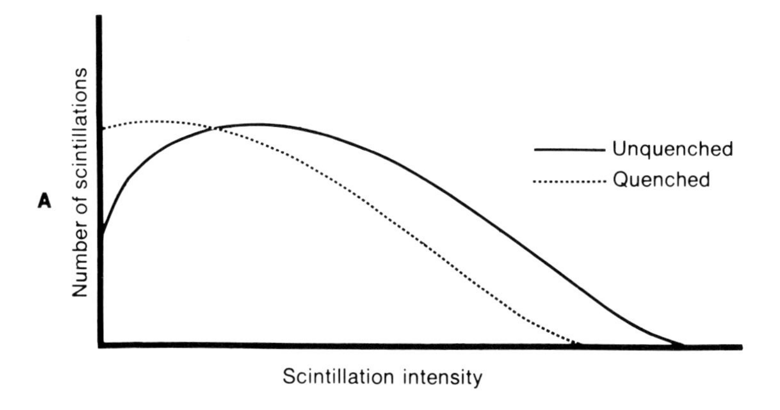

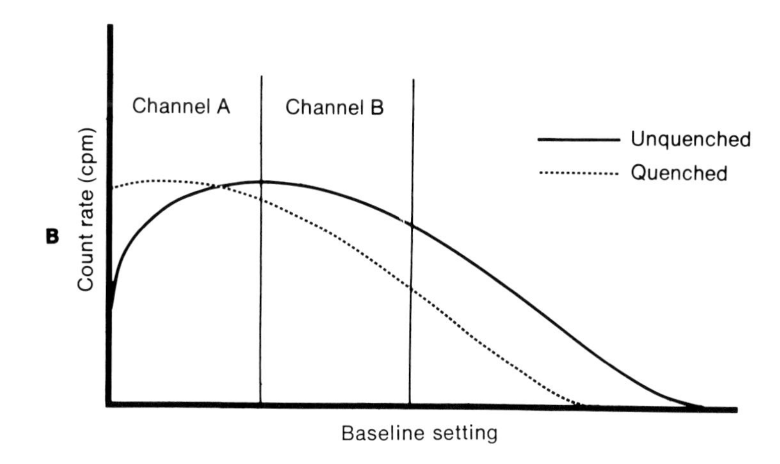

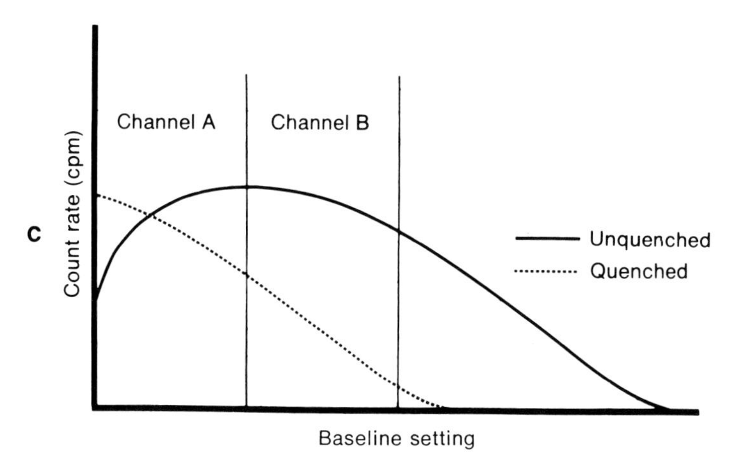

Fig. 1-8. A, Spectral shift in quenching. **B,** Channel ratio shift in moderate quenching. **C,** Channel ratio shift in severe quenching.

counting efficiency, some method must be employed to counteract its effect so that errors can be avoided in the estimation of sample activity.

The effect of quenching on the spectrum of detected scintillations is shown in Fig. 1-8, *A*. Quenching reduces the apparent intensity of all the scintillations; therefore the spectrum of the quenched sample shifts in relation to the spectrum obtained with an unquenched sample. The magnitude of the shift is a direct function of the severity of the quenching.

Correction of quenching. By monitoring the amount of spectral shift, we can estimate the degree of quenching and apply an appropriate quenching correction factor.

CHANNEL RATIO METHOD. This monitoring can be done conveniently with a dual-channel system by the channel ratio method. We can count the lower-energy portion of the spectrum in one channel (channel A) and the higher-energy portion in another channel (channel B). The occurrence of quenching causes an increase in the ratio of counts in channel A to those in channel B (Fig. 1-8, *B*). More severe quenching causes further increase in the A/B ratio (Fig. 1-8, *C*). By careful execution of pilot experiments in which standards of constant total activity containing various amounts of quenching agent are counted in both channels, we can obtain sufficient data to establish a smooth relationship between channel ratio and counting efficiency with sufficient precision to construct a quenching correction curve. Thus to use the channel ratio method, it is necessary to perform a quantitative preliminary experiment to establish the correction curve. Furthermore, it is necessary that the data be obtained from samples with a constitution similar to those with which the curve is to be used. Changing to a different radionuclide or to samples of a different nature necessitates construction of a new curve appropriate to the new circumstances (Fig. 1-9).

Many clinical procedures will continue to generate large numbers of similar samples for as long as they remain in use. In that case, the initial investment of effort re-

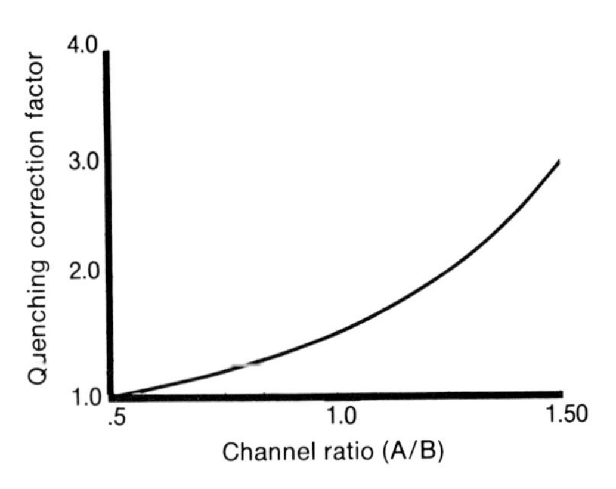

Fig. 1-9. Quenching correction factor versus channel ratio.

quired for the establishment of the channel ratio method is time well spent, since thereafter it is a simple matter to obtain the quenching correction by looking up the appropriate value for the ratio obtained. With dual-channel systems, counting time is not prolonged by the need to count each sample in two channels; this factor accounts for the popularity of such machines. Some of the more recent instruments equipped with small dedicated computers carry out the correction automatically, confidently printing out the results as disintegrations per minute (dpm). Since the correction itself is subject to a number of possibilities for error, the prudent user ought to regard such an implicit claim to infallibility with a certain degree of skepticism.

INTERNAL STANDARD METHOD. An alternative method of determining the quenching correction is the internal standard method. It is probably the most accurate and broadly applicable of the available methods, but it is also the most tedious to carry out. The basic idea is that if one can measure counting efficiency independently in each and every sample, then the measured efficiency can be used to correct each detected count rate to its equivalent absolute decay rate. Counting efficiency is assessed by the process of adding to each sample a standard of known activity that is the same radionuclide as is contained within the sample and noting the increase in counting rate produced by this. The standard is usually added in the same chemical form as the solvent. The same sample can be counted twice, before and after the addition of the standard, but it is usally more convenient to divide equal amounts of the sample between two counting vials and then to add the standard to one of the pair. Both vials can then be counted in sequence in an automatic counter, as in the following example:

	Observed count rate
10,000 dpm ^{14}C standard alone	5000 cpm
^{14}C sample and standard	19,000 cpm
^{14}C sample alone	16,000 cpm

Note that in the presumably unquenched ^{14}C standard, the counting efficiency is only 50%.

In the sample, the addition of ^{14}C of 10,000 dpm activity results in an increase in counting rate of 19,000 − 16,000 = 3000 cpm. The ^{14}C counting efficiency in that sample thus is $\frac{3000\text{ cpm}}{10{,}000\text{ dpm}} = 0.30$ cpm/dpm. The diminished efficiency can be attributed to the presence of quenching in the sample. The same quenching was present before addition of the standard; therefore the decay rate in the original sample must have been $\frac{16{,}000}{0.300} = 53{,}333$ dpm. This particular sample, therefore, is $\frac{53{,}333\text{ dpm}}{10{,}000\text{ dpm}} =$ 5.33 times as active as the standard. The remaining samples in the series can be similarly and individually corrected to an absolute decay rate.

The occasions when it is actually necessary to know the absolute decay rate of a sample are rather few. It is usually sufficient to know the relative activity among the various samples and some standard. In those cases, an absolutely calibrated standard is not necessary; it is sufficient to correct all results to some common relative efficiency, such as the relative efficiency of the standard. The desired result can be obtained as follows (the data used are from the previous example): The count rate of the standard is 5000 cpm. Adding this activity to that of the unknown sample results in an increase of 19,000 − 16,000 = 3000 cpm. The relative counting efficiency in the sample thus is $\frac{3000}{5000} = 0.60$. The relative sample count rate corrected for quenching is $\frac{16{,}000}{0.60} = 24{,}000$ cpm. Note that the relative result obtained by this method is the same as that obtained previously. The sample is, as before, $\frac{26{,}667\text{ cpm}}{5000\text{ cpm}} = 5.33$ times as active as the standard.

The internal standard method is tedious in that three quantitative pipettings (two aliquots of sample solution and one of standard) are required for each sample and the counting time is doubled because the two sample vials must be counted sequentially. The method is precise, however, since sample counting efficiency is measured directly with the radionuclide being counted. The method will work even with dissimilar samples provided only that quenching is not so severe that it prevents a statistically significant increase in counting rate when the standard is added.

EXTERNAL STANDARD METHOD. The external standard method is a quenching correction method that is considerably easier to apply (it can be done automatically). It is the approximate equivalent of the internal standard method. In this method, the samples are first counted alone and then recounted while being bombarded by gamma rays from a standard external to the solution that can be moved into an exactly reproducible position for irradiation of the vial. Gamma-ray photons interacting in the fluor and vial produce a population of Compton electrons of varying energies. The Compton electrons mimic (with something less than 100% fidelity) the behavior of beta particles similarly situated. By noting the increased count rate due to the external standard, one can obtain an estimate of relative counting efficiency in a manner analogous to that used with an internal standard. Slight alterations in sample volume, height, and reproducibility of gamma-ray source lead to variability of count rate with this method.

To minimize variability from these sources, the external standard may also be used to generate correction factors by the channel ratio method. The samples are counted with and without the external standards in place. Counts added by the presence of the external standards are recorded in two sample channels, one of high energy and one of low energy. Net standard counts are obtained by subtracting the previously recorded counts of the sample alone in each channel. Counts for the determination of gross sample activity (obtained when the sample is counted alone) can be recorded in a third independent channel or in one or both of the channels used for the standard. Shifts in the ratio of net added counts in the standard channels are correlated with sample quenching by a pilot experiment in which variously quenched samples of known activity are used. Quenching correction factors are then established from the standard channel ratio.

The advantage in using the external standard for determining the channel ratio is that the counting rates of the standard are consistently high (the standard is always hot), and the ratio can thus be determined with precision. Correction factors can therefore be determined even for low-activity samples. This would not be the case if the factor had to be computed from sample activity.

The relationship between external standard counting efficiency and sample counting efficiency is by nature inferential and not very secure. The constancy of this tenuous relationship depends heavily on the maintenance of identical conditions in the pilot experiment and in actual use with unknown samples. Even relatively minor variations in sample makeup may necessitate a repeat pilot experiment. For this reason, it is best to postpone the pilot experiment until all the "bugs" have been worked out of the sample preparation procedure.

The external standard method, even though convenient, cannot match the inherent accuracy of the internal standard method. Nevertheless, the method works well enough to persuade most instrument manufacturers to provide the external standard feature, if not routinely, at least as an option.

Finally, some samples quench so severely that they cannot be reproducibly counted at all, no matter what method of quenching correction is used. Highly colored specimens are not amenable to the liquid scintillation counting method. Such samples can be counted only if they are first subjected to chemical or physical procedures such as combustion to remove quenching agents.

Summary of in vitro counting methods

In vitro radiometric procedures are coming into increasing use, not only in cardiology but also in clinical medicine in general. If the in vitro source is a gamma emitter, its activity can be sensitively and reproducibly measured with a crystal scintillation detector. In all but a few instances, gamma-emitting radionuclides are a foreign element in native biochemical compounds and may fail to trace the behavior of a given compound in a discriminating system. Native labels of ^{14}C and ^{3}H are available, but both are beta emitters.

Beta emitters can be sensitively detected by liquid scintillation counting, but the method is technically difficult. The primary problem is the occurrence of varying degrees of quenching that can cause large errors even in relative radioassay. Quenching correction methods must therefore be applied to all samples. The simples quenching correction methods are the least precise, but fortunately, the precision required in clinical work is usually such that one of the more convenient methods, such as the channel ratio or external standard method, will suffice.

IMAGING DEVICES IN CARDIOVASCULAR NUCLEAR MEDICINE

Characteristics of images

Images produced in nuclear medicine have a number of characteristics in common with other kinds of images. A good deal of research has been conducted on photographic and television imaging systems that is applicable to nuclear medicine. The principles thus revealed are helpful in the assessment of the current and anticipated performance of imaging devices in nuclear cardiology.

Such an evaluation is conveniently begun with the defining of a "perfect" imaging system so that the imperfections of existing systems can be more readily appreciated. A "perfect" image for human observers is one that the human eye cannot distinguish from the real object at the same viewing distance. An object containing a radionuclide emits gamma radiation of one or, at most, a few discrete wavelengths, and the image formed that we ultimately perceive has a variable density of shades of gray, varying from white to black at the extremes.

A perfect, monochromatic imaging system can be defined as one in which (1) real contrasts in object density are faithfully reproduced in image contrast with neither amplification nor attentuation; (2) all detail in the object, down to the finest that can be resolved by the viewer's eye, is reproduced in the image; and (3) variations in image contrast are exclusively the product of variations in object contrast and not the result of some extraneous variable (that is, the image is free from noise). Nuclear imaging systems fall short of perfection on two of these counts. Following is a discussion of these imperfections.

Image versus object contrast in nuclear medicine

The contrast in the object to be reproduced by a nuclear medical imaging device is the contrast between the level of radioactivity in a diseased area, that is, "lesion," and the average activity in surrounding normal regions, thus:

$$C_0 = \frac{A_{lesion} - A_{normal}}{A_{lesion} + A_{normal}}$$

where

C_0 = Contrast of object
A = Activity

Lesion detectability is evidently a function of object contrast, and the object contrast, in turn, is a complex function of the properties of the diagnostic radiopharmaceutical. For the limited purpose of assessing the performance of the imaging system, it is legitimate to neglect the factors that affect object contrast and to be satisfied with a system that faithfully reproduces that contrast, whatever it may be.

Reproduction of object contrast in the image is related more closely to the contrast modulation available in the display system than it is to the inherent response of the detector system. The linear response range of NaI crystal scintillation detectors extends from the activity of background radiation to the maximum count rate that can be recorded without significant coincidence losses; this range is about 100 counts/min (cpm) to 100,000 cpm (some would claim 1 million cpm), or three or four orders of magnitude. Within this range, the detector can distinguish differences as small as plus or minus one count. It could therefore provide 99,900 degrees of shading for the gray scale of the image.

If the display system were linear (this is a large assumption), how many gradations of density could it reproduce? Of these, how many could be perceived? More than likely, the answer to both questions would be, fewer than the detector can provide.

The film used in the majority of display systems has a dynamic range of 0.2 to 3.0 optical density units. When the film is viewed by the transmitted light provided by an x-ray view box (this is the usual circumstance), film densities in excess of 2 optical density units appear, to the human eye, to be completely opaque. The useful range for most film is limited to optical densities in the range of 0.3 to 1.8 density units. In the lower portion of this range, when viewing conditions are ideal, the human observer can distinguish light-intensity differences of 2% (corresponding to a density difference of 0.0086 unit) if there is a sharp line of demarcation. It follows that a generous estimate of the number of perceptible density gradations would be 100 shades of gray. The presence of random noise, eye fatigue, unsharpness of borders, and space between image elements reduces this to a great extent.

It is within the inherent capabilities of the detector system to provide a greater range of image contrast differences and finer shading than can be perceived. In this respect, nuclear medical imaging devices are already perfect. The imperfections in the images produced in the clinic are caused by defects in one or both of the two remaining criteria of image perfection, that is, lack of spatial resolution and presence of noise in the image.

Information density and statistical fluctuation noise in images

The images with which we are most familiar, photographs, are produced by processes in which millions to billions of photons are concentrated on the film or paper. They thus have a very high information density, that is, a high number of photons per unit area, and in such images it is possible to resolve very fine detail. Sharp photographs commonly contain more detail than can be resolved with the unaided eye.

The information in nuclear medical images is produced by gamma-ray photons emanating from the source. Dosimetric considerations limit the strength of the source when that source is a radiopharmaceutical contained within a patient. For this reason, images in nuclear medicine tend to be relatively low in information content. In this context, the phrase *information content* is synonomous with *count density,* which refers to the number of counts per square centimeter of image. Count densities employed in current practice are low enough to limit spatial resolution to such a degree that "cold" lesions must be at least 1 cm (and preferably more) in diameter before they can be visualized. Observers are thus confronted with an image that contains much less detail than the eye is capable of resolv-

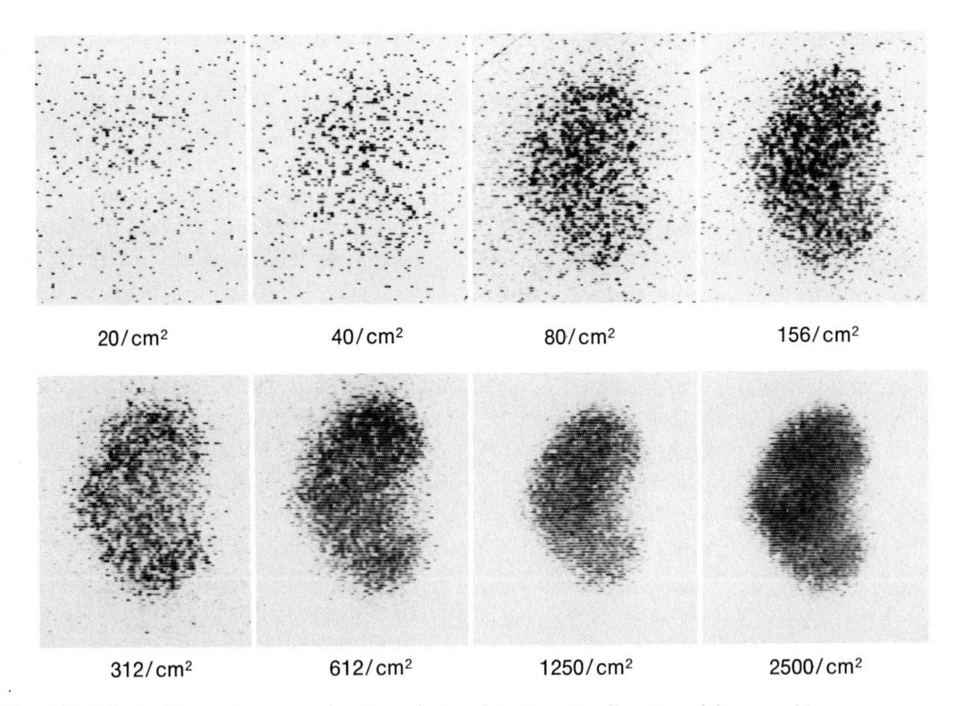

Fig. 1-10. Effect of increasing count density on lesion detection. Small region of decreased tracer accumulation in the midportion of the lateral aspect of kidney can best be seen on image with count density of 2500/cm².

ing and that is a visibly imperfect reproduction of an object that may be difficult or impossible to interpret when the lesion is small.

A consequence of low count density is a coarseness in the image that can be readily appreciated by simple experiment. Try, for example, to draw an image of an object by using 100 randomly placed pencil dots. Then try again using 1000, 10,000, 100,000, and 1 million dots. Reproductions of such an object at various information densities are shown in Fig. 1-10.

In images of distributions of radioactivity, the effect of count density can be directly related to Poisson statistics. Random variations in image density, arising from the random nature of radioactive decay, introduce statistical fluctuation "noise" into the image. The amount of such noise in a unit area of image depends on the square root of the number of counts in that area and hence on the (square root of) count density. If, for example, the image contains 100 counts/cm^2, then any given square centimeter will exhibit a purely statistical relative standard deviation of $\pm 10\%$ from the mean. At a count density of 10,000 counts/cm^2, the statistical deviation of the same square centimeter will be only $\pm 1\%$ from the mean. The image with the higher count density will thus appear to be much more uniform (in normal areas), and abnormalities within it will be more readily perceived. If we establish plus or minus two standard deviations as the limits of normality, then a 1 cm^2 lesion could be called abnormal at a count density of 100 counts/cm^2 only if it differed by 20% or more from the mean. At 10,000 counts/cm^2, the same lesion could be detected with equal confidence if it differed by only as much as 2% from the mean.

Clearly, lesion detectability is related to count density, as is minimum detectable lesion size. If it were possible to produce images of 10,000 counts/mm^2, then theoretically we could resolve millimeter-sized lesions with the same confidence that we now resolve centimeter-sized ones. However, the ability of an observer to perceive lesions in an image depends not only on statistical considerations but also on the inherent spatial resolution of the system used to produce the images. If the system can only resolve relatively large lesions (e.g., a centimeter or more in diameter), then the count density required to permit their recognition is not so great. Objective tests of the per-

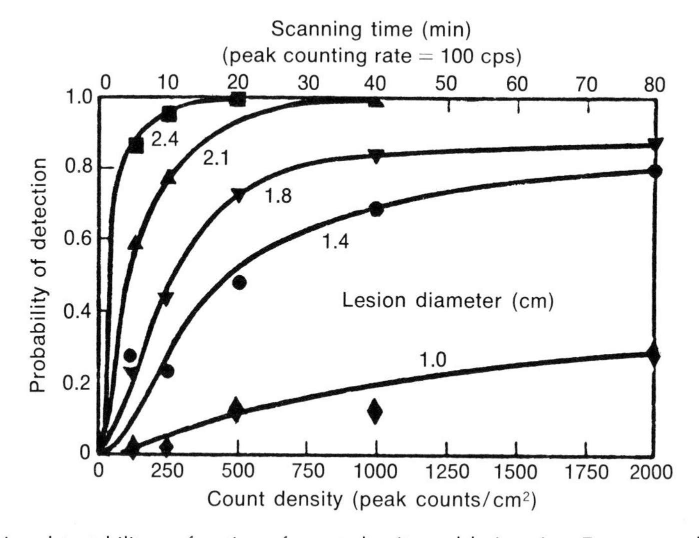

Fig. 1-11. Lesion detectability as function of count density and lesion size. Data were obtained with collimator with ⅜-inch (0.95 cm) resolution. (From Schulz, A. G., Knowles, L. G., and Kohlenstein, L. C.: APL Tech. Dig. **10**:2, 1971.)

formance of observers viewing images produced by a system with a (FWHM) resolution of 1 cm[1] indicated that their perception was not susceptible to much further improvement by the use of count densities greater than 1000 counts/cm², a density that is readily attainable within reasonable dosimetric limits (Fig. 1-11). At that count density, the noise variability in the image was already low enough to permit recognition of the smallest lesion that the system could resolve. The most recent imaging systems have considerably finer resolution; thus it would now be advantageous to employ much higher count densities, but progress in that direction is limited by dosimetric considerations.

Spatial resolution

Spatial resolution is that properly of an imaging system which describes its ability to portray fine detail in the image. Spatial resolution can be measured in a variety of ways. One of the simplest methods is to measure the minimum distance by which two points (or lines) must be separated in space to be separately portrayed in the image.

Measurement of minimum separation distance can be done using a source in which a series of radioactive points or lines are placed closer and closer together. Minimum separation distance could then be found by noting the point at which they first appeared to merge in the image. Such a source could be made from two infinitely thin lines of radioactivity that converge into a point in a predictable fashion. The source can be constructed of capillary tubing that contains a radioactive solution. Capillary tubing is useful because it can be filled with any desired radionuclide. Since system resolution depends in a complex manner on the energy of the photons, it will prove helpful if the source can be used with a variety of radionuclides. The bore of the capillary tube and hence the width of the line source cannot, of course, be "infinitely" thin. As a practical matter, however, a bore diameter of 1 mm or less is sufficient to serve the purpose, given the limited resolution in the imaging systems currently employed in nuclear medicine.

Fig. 1-10 is an image of such a radioactive source made with a rectilinear scanner. Note the apparent spreading (widening) of the lines, the result of the less-than-perfect spatial resolution of the scanner. This spreading makes it difficult to say precisely just where it is that the lines in the image do, in fact, "merge." In practice, therefore, this method is limited to yielding a quick but more or less subjective estimate of system resolution. For precise work, a more objective estimate is required.

Despite the subjectivity of this method, it is so readily accomplished that measurement of spatial resolution should be part of the routine, daily quality control on every imaging system. It is most conveniently done using a bar phantom and inspecting the image to determine the finest bars that are just resolved.

The spreading of the line in Fig. 1-10, evident to the eye, is susceptible to more precise measurement than that provided by "eyeballing." The degree of spreading is inversely related to the spatial resolution of the system in that the greater the spreading in the image, the poorer the resolution, and the converse. Measurement of the spreading would thus provide an indication of the resolution of the system.

With a rectilinear scanner the measurement may be accomplished by slowly scanning across a line (or point) source and accurately determining the counting rate at each point in space along the scan line. The scan must be made "slowly" to obtain a large number of counts at each point, so as to reduce statisitcal error to insignificance. With camera systems interfaced to a computer, equivalent data can be obtained from the digitized image of such a course after correction for whatever scaling factors may exist. In addition, the image should be stored in and the data retrieved from a matrix, the elements of which represent very small regions of space of, say, a millimeter or less.

In either case, a plot of detected activity versus distance is made centered on the position of the line (or point) and extending on either side to distances at which background counting rates are observed. The mathematic expression that describes the curve is called the "point-spread response function" (PSRF) or the "line-spread response function" (LSRF), according to whether a point or line source was used. In either case the spreading of the curve is an inverse function of the spatial resolution of the system. The spreading is conveniently measured at the half-maximum points of the curve to yield the "full width at half maximum" (FWHM) in centimeters.

This is the most commonly used estimate of spatial resolution. If, for example, a vendor specifies that an imaging system (or collimator or whatever) has a spatial resolution of 1 cm, what is most probably meant is that the FWHM of the system (or collimator) is 1 cm. The FWHM thus fulfills the need for an objective measurement of resolution, but even so, it is not a totally satisfactory estimate. The lesions that we seek to portray are not lines or points; they exist, rather, in a very wide variety of irregular shapes and sizes. A more complete description of spatial resolution would be one that predicted the contrast in the images of lesions of any size. A complex mathematic function called the "modulation transfer function" (MTF) has been developed, which permits this prediction to be made.

Derivation of the MTF of a system is a formidable mathematic problem. Suffice it to say that if the LSRF or PSRF of the system is known or can be determined, then it is possible to calculate the MTF.

The MTF is a prediction of the contrast that will be found in images of varying sized objects. The hypothetical "object" on which the prediction is based has a contrast that varies sinusoidally. Cyclic variations in regions of space are described by a "spatial frequency." If the cycle takes 2 cm to complete, it thus has a spatial frequency of $\frac{1 \text{ cycle}}{2 \text{ cm}} = 0.5$ cycle/cm.

The use of spatial frequency to specify object size may at first seem odd, but it is directly analogous to optical systems, in which resolution is given as the number of lines per millimeter that the system can resolve. We have no difficulty understanding that an optical system that can resolve 400 lines/mm is better than one that can only resolve 200 lines/mm. The resolution of nuclear medical imaging systems is not nearly that good, but it is apparent that a scanning system that can resolve 1 cycle/cm is superior to one that resolves only 0.5 cycle/cm.

The MTF predicts the contrast in the image relative to that in the object. At any given spatial frequency, therefore, $\frac{\text{Image contrast}}{\text{Object contrast}} = \text{MTF}$. If the imaging system is perfect at that frequency, then image contrast will equal object contrast, and the MTF takes the value 1.0. At some higher spatial frequency, the distance between succeeding cycles will become too small for the system to fully resolve, image contrast will diminish, and the MTF will have a value less than 1.0. Ultimately, at some still higher spatial frequency, resolution and image contrast will disappear altogether, and the value of the MTF will fall to zero. A curve showing the MTF as a function of increasing spatial frequency (the usual form of presentation) will thus start at unity and diminish to zero.

The MTF curves of superior imaging systems will extend to higher spatial frequencies. They are therefore useful in the comparison of rival imaging systems. They can also be used to compare various components within a given imaging system. Thus, in a scintillation camera, one could separately determine the MTF of the collimator, of the positioning circuits, of the display oscilloscope, of the recording film, and of any other component of the system that might affect spatial resolution. Such as analysis would be useful in attempting to improve the design of the device, since it would reveal which of the several components was most responsible for the limited resolution of the system. The analysis would not only indicate which component would benefit most from improved design, it would also give an estimate of the degree of improvement that might be expected to result from the perfection of that component.

At the present time (and for the foreseeable future) the imaging system of choice in nuclear cardiology is the isotope camera in one of its forms. These systems have undergone a steady improvement in their spatial resolution during their relatively brief history. Earlier models exhibited a resolution of a centimeter or more. In the better models of recent manufacture the resolution has been sharpened to the order of 6 mm. Despite this improvement, the ability of the physician to perceive millimeter-sized lesions in cardiac scintigraphs is not much increased because of the interference caused by the continued presence of statistical "noise" in the image.

Camera imaging systems

Gamma-ray cameras are stationary-detector imaging devices. In most cases, they have a large-diameter parallel-hole collimator and an equally large crystal scintillation detector with the capability of seeing the entire area of interest at once, which is often a decided advantage.

Anger camera. The Anger camera detector has a crystal of NaI that is either 1.25 or 2.5 cm thick and 27.5 to 40 cm in diameter. A hexagonal array of nineteen to ninety-seven PM tubes views the back surface of the crystal through a light pipe at a small standoff distance. Scintillations produced within the crystal are transmitted not only to the nearest PM tube, but also to neighboring tubes because of the arrangement of light piping and distance. Coincident signals are generated from each PM tube that receives light after a gamma-ray interaction in the crystal. These signals are sampled by an x-y analyzer, which contains an array of resistors, two for each tube. The resistor values are determined according to the spatial coordinates (x and y) of the PM tube to which they are connected. The sum of the coincident signals for all PM tubes thus generates a signal with an x and y address of amplitudes proportional to the spatial coordinates of the original scintillation. In addition, the total coincident output of all PM tubes is summed to form a pulse called the z signal, which is subjected to pulse-height analysis. Each scintillation in the crystal thus results in the production of three simultaneous analog signals: x and y signals dependent on the position of the scintillation in the crystal, and a z signal dependent on the overall intensity of the scintillation.

The z signal is synthesized from the coincident outputs of a number of phototubes, and it therefore contains a considerably larger sampling error than that which would be found in a similar signal from a single-phototube crystal scintillation probe system. The FWHM of the Anger camera z signal is thus broadened to a value of, typically, 15% or more. When this relatively wide photopeak is subjected to pulse-height analysis, a correspondingly wide counting window (20% to 30%) is required to include the bulk of the photopeak events. A 20% window centered on the 140 keV photopeak of ^{99m}Tc would have a width of roughly 30 keV, encompassing apparent energies from 125 to 155 keV. In choosing to accept such a wide range of z signals, one necessarily agrees to accept an equally wide range of x and y position signals as well. Using the raw x and y signals would therefore lead to an intolerable positioning error with a consequent loss of spatial resolution. This problem is solved by the following technique.

In the case, say, of an especially bright scintillation (i.e., one at the upper end of the counting window), the x and y signal pulses will be atypically high. The z signal, however, will also be atypically high and to the same extent; thus the ratios x/z and y/z will be nearly constant over a considerable range of energies. For these reasons, the positioning signals transmitted to the display unit are not the raw x and y signals but rather x/z and y/z, thus removing most of the interdependence between (apparent) position and (apparent) energy.

The display unit of the camera is a

cathode ray tube. The electron beam within the tube is normally blocked from the tube face by a biased grid. The x-y signal is used for deflecting the beam. If the coincident z signal is of the proper height (it passes the pulse-height analyzer), the bias is momentarily removed, producing a brief flash on the phosphorescent tube face at a position corresponding to that of the original scintillation. An image is produced by the integration of a number of flashes, either with a scope that has a persistent phosphor or, more commonly, with a time-exposure photograph of a short persistent phosphor.

In any crystal scintillator, a gamma ray produces a scintillation at the site of initial scattering. In a thick crystal, the scattered ray can also be absorbed; therefore a second coincident scintillation is produced at its site of absorption. The two scintillation events are indistinguishable from a single photoelectric interaction if the z signal is sufficiently great to pass the pulse-height analyzer. The x-y analyzer, however, mislocates these scintillation events; thus a single spot is produced on the readout scope at a position corresponding to the center of luminosity of the two events. If this process occurs frequently, spatial resolution is lost; hence it is necessary to have a thin crystal, in which multiple interactions are less likely.

A more pressing reason for the use of a thin crystal arises from the following circumstances. Suppose that a 5 cm diameter phototube is viewing the rear surface of a 2.5 cm thick crystal from a standoff distance of 2.5 cm. Suppose further that a photon entered the crystal directly under the phototube. The signal produced by that phototube could vary as much as 64% due to the geometry of light collection alone, depending on whether the scintillation occurred at the distant front surface of the crystal, the nearby rear surface, or somewhere in between. Even worse, the signals from neighboring phototubes would also vary, but not in the same proportion. The disproportionate response between the phototubes introduces uncertainty into the position signals, which results in a loss of spatial resolution. The thicker the crystal, the greater the loss of resolution.

Thin crystals have a low intrinsic photopeak efficiency for higher energy photons, which results in a loss of counting efficiency. This limitation is not of great practical consequence, since nuclides that emit high-energy gamma rays are avoided for imaging with isotope cameras because of practical constraints on collimator design for high-energy imaging. Anger had foreseen this problem from the inception of his design and had experimentally demonstrated the relationship between crystal thickness, sensitivity, and resolution (Fig. 1-12). Most manufacturers of commercially available instruments have followed the inventor's original practice of using a 2.5 cm thick crystal, this being a reasonable compromise between sensitivity on the one hand and resolution on the other. In a recently introduced camera the compromise has been deliberately shifted in favor of enhanced resolution by the use of a crystal only 1.25 cm thick. For the nuclear cardiologist, who is likely to be using relatively low-energy gamma emitters (e.g., ^{99m}Tc-labeled blood pool agents, ^{201}Tl), the cost in lost sensitivity is not very great; therefore the improved resolution may represent an attractive bargain trade-off. However, to date, the improved resolution resulting from 1.25 cm thick detectors has been marginal.

The original commerical version of the Anger camera employed nineteen phototubes. More recent models have appeared with thirty-seven and even ninety-seven tubes. The additional light-collecting power provided by the more numerous tubes can be exploited in one of two ways.

1. By maintaining the same between-tube separation distance as in the original nineteen-tube models, the thirty-seven tubes could be distributed over a wider area so as to achieve a wider field of view (ca 37.5 to 44 cm) with approximately the same spatial resolution.
2. The thirty-seven tube array could be more tightly packed into the standard

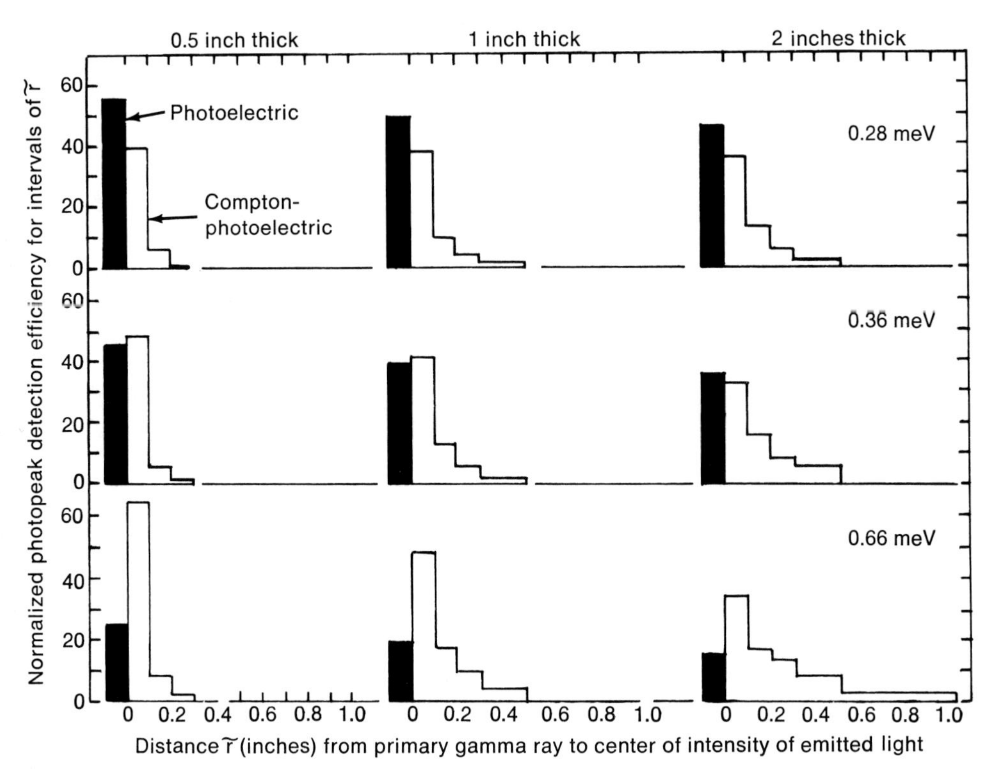

Fig. 1-12. Calculated loss of position resolution in sodium iodide due to multiple Compton-photoelectric interactions. Curves show relative number of events occurring within intervals of r̄, where r̄ is distance from incident gamma ray to center of intensity of light produced. Curves for 0.5-, 1-, and 2-inch thick crystals at three gamma-ray energies.

field of view (ca 25 mm) to measure light output and its distribution more sensitively and acutely, yielding a distinct improvement in intrinsic spatial resolution.

The standard field of view is more than sufficient to encompass the heart. Enlarged fields of view, appreciated in other applications, are less useful in nuclear cardiology because such cameras spend the majority of their counting time acquiring, processing, sorting, and displaying events of extracardiac origin. Extracardiac activity may very well saturate the counting circuitry when enough of it is included in the field. Anger cameras are already hard pressed to accommodate the very high counting rates resulting from a bolus injection in a first-pass study without the added burden of excessive extracardiac activity. For this reason, a thirty-seven–tube Anger camera selected for cardiology should be one in which the additional tubes provide a benefit useful to the cardiologist, that is, a superior spatial resolution within the standard, 25 mm field of view.

The maintenance of uniform response within the field of an Anger camera depends on the preservation of uniform gain among the PM tubes in the array. PM tube gain is subject to both long- and short-term drift; therefore field uniformity should be checked daily. Even in a camera that is performing to manufacturer's specifications there may be as much as a ± 20% variation in sensitivity across the field. This degree of variation, distributed gradually over the whole field, is unlikely to be apparent even to an astute observer and is therefore no impediment to the visual interpretation of images. When such images are subjected to computer analysis, however, the nonuniformity will not be overlooked by the computer. Such errors will appear in any data extracted from designated portions of the image(s). It is therefore imperative that acquired images be corrected for residual nonuniformities before drawing conclusions based on data obtained by region-of-interest analysis.

Time constraint in nuclear cardiology

Pulse-pair resolving time now runs around 5 μsec in the better Anger camera models. With this resolving time, a state-of-the-art camera will exhibit a 10% coincidence loss at a detected counting rate of 20,000 cps. In ordinary nuclear medical procedures this counting rate is not likely to be exceeded. In contrast, the techniques of nuclear cardiology, especially nuclear cineangiography, are apt to lead to extremely high counting rates.

Unlike livers, brains, bones, and most other objects of nuclear imaging procedures, the heart is in constant, rapid motion no matter how closely the patient is confined. Thus, while 5- or 10-minute camera exposures taken of, say, the motionless brain are perfectly acceptable, such lengthy exposures of the heart would reveal only the cardiac outline during diastole. For most diagnostic purposes, a diastolic image alone is not sufficient. To reveal wall motion defects or to determine ejection fraction from images, at least two image frames are required, one sharply focused on the fleeting instant of end of systole and the other on end of diastole. To portray the movement of the heart with a motion picture, a whole series of such frames are required, each covering a discrete portion of the cardiac cycle.

It has been established that no less than 25 frames/beat are needed for optimum portrayal and analysis of cardiac motion. If the heart rate were as slow as one beat/sec, then individual frames could be no longer than $\frac{1 \text{ sec/beat}}{25 \text{ frames/beat}} = 0.04$ sec/frame. If the physician were to insist on a count density sufficient to take full advantage of the spatial resolution of the camera, for example, 5×10^5 counts/frame, then the required counting rate to record the data in one cycle would be as follows:

$$\frac{5 \times 10^5 \text{ counts/frame}}{0.04 \text{ sec/frame}} = 1.25 \times 10^7 \text{ cps}$$

This rate is much too high to be accommodated by the camera. At the more realistic counting rate previously mentioned (20,000 cps), which already admits a 10% coincidence loss, the number of counts per frame could not be greater than 0.04 sec/frame (2×10^4 counts/sec) = 800 counts/frame. Image frames consisting of only 800 counts are very difficult to interpret due to the random spatial destruction that is likely to be encountered when so few events are recorded. Frames of acceptable count density can be built up by accumulating counts from similar portions of a number of successive beats in the memory of a digital computer coupled to the camera through an ECG gating system. The mean counting rate during acquisition thus need not be so high. Even so, it is likely that some means of correcting for coincidence losses will be required, for instance, to avoid underestimation of ejection fraction in an equilibrium study. The problem is especially severe in a first-pass procedure, in which the entire injectate will appear in the field of view during the early frames.

For the foregoing reasons there is a special premium on short resolving time for imaging devices that are to be employed in nuclear cardiology. Although Anger-type, single-crystal, multiple-phototube cameras have undergone steady improvement in this regard, the commercially available imaging device that currently exhibits the shortest resolving time is the latest production model of the multicrystal system of Bender and Blau.

Multicrystal camera

The multicrystal detector in the Bender and Blau system consists of 294 optically isolated crystals arranged in a fourteen by twenty-one rectangular array. The individual crystals are square in cross section, approximately 1 cm to a side and 3.8 cm long. The whole array measures 15 by 23 cm (6 by 9 inches). In front of the array is a 294-hole collimator, in which each hole is aligned with a specific crystal.

Spatial resolution is achieved in an entirely different manner than that used in Anger cameras. In the multicrystal system, each crystal is optically coupled through plastic light pipes to two PM tubes, one corresponding to the row and the other to the column in the array where the crystal is located. A total of thirty-five PM tubes is required for the fourteen rows and twenty-one columns. Time-coincident pulses in both a row and column PM tube uniquely identify the crystal in which the scintillation occurred. The localization can be made quickly because all that is required is fast coincidence circuitry, which is readily available. The pulse-pair resolving time of the most recent (1978) commercial version of the Bender and Blau system is only about 2 μsec.

With a 2 μsec resolving time, a 10% coincidence loss will occur at an indicated counting rate of 50,000 cps, which is two and a half times greater than the equivalent rate for the Anger camera. Moreover, the digital position-sensing system continues to function properly even at saturation counting rates, although coincidence losses are then high (66%). The imaging properties of the multicrystal system, therefore, remain unchanged up to the saturation rate of 235,000 indicated cps. At that counting rate, nearly 10,000 counts/frame could be correctly portrayed at the previously mentioned framing rate of 25/second.

Neighboring crystals are optically isolated, but the between-crystal separation is not large enough to admit a thickness of shielding sufficient to be radiopaque. It is therefore possible that a gamma photon scattered within the detector array could produce scintillations in more than one crystal, thus confounding the logic on which spatial resolution is based. Accordingly, the circuits are programed in such a way that any events characterized by coincident signals from more than two PM tubes are rejected. By this simple expedient, the problem of within-detector scattering is circumvented without the need to use thin crystals. The 3.8 cm deep crystals have a considerably higher intrinsic photoelectric efficiency for the more energetic gamma rays than do the 2.5 cm crystals used in the Anger system. This potential advantage is of limited practical importance, however, because high-energy photon emitters are not chosen for imaging procedures because of difficulties in collimation.

The geometry of the detector array and PM tube arrangement is such that the average length of the plastic light pipes is approximately 50 cm (20 inches). Fully 75% of the light emitted by the crystals is lost in transmission. The effect of these transmission losses is to increase the spreading of the photopeak to an FWHM of 50%. With a photopeak this wide, pulse-height analysis is not very effective at blocking scattered-in photons. A pulse-height window set to include the FWHM of the 140 keV photopeak of ^{99m}Tc would be 70 keV wide (i.e., 50% of 140 keV) with a base level at 105 keV. Technetium photons scattered through as much as 100 degrees would still have enough energy (106 keV) to enter this window.

While sharp energy resolution is always to be preferred, other things being equal, its importance in imaging can be exaggerated. By far, the greatest amount of scatter rejection takes place in the collimator; only a relatively small fraction of photons is scattered at exactly the right angle to become aligned with the long axis of the holes in the collimator. In addition, the positioning signals in the multicrystal system are all but totally independent of (apparent) light intensity; thus intrinsic spatial resolution is not degraded, despite the less than ideal energy resolution.

Using a matrix of 1 cm^2 detectors, one would initially suppose that the ultimate spatial resolution could not be any finer than the size of an individual matrix element, that is, 1 cm. It is true, of course, that a point source could be located almost anywhere within the 1 cm^2 field of view (assuming straight-bore collimation) of any given detector without affecting the response of that particular crystal. The response of neighboring crystals in the array, however, would vary according to whether the source was located closer to the one or to the other, and this information can be used to refine the estimate of source position by a process of matrix analysis extended over an interval of time sufficient to permit the accumulation of statisitically valid data from adjacent matrix elements. Matrix analysis of the stored data is accomplished retrospectively by the digital computer supplied with the commercial version of the Bender and Blau system. The process is more sophisticated than simple spatial averaging in that it takes into account the known response functions of the elements in the array. The result is a display that exhibits a real spatial resolution finer than the 1 cm dimensions of the array elements.

Thus two fundamentally different camera systems are available to the nuclear cardiologist. The principal performance differences between the two are that the Anger camera has much better energy resolution, whereas the Bender and Blau system has a higher count rate capability. Other differences are less significant in cardiology. The smaller field size of the multicrystal camera is not a disadvantage in cardiology. On the other hand, its higher intrinsic detection efficiency for high-energy photons is not much of an advantage because photons of relatively low energy are the preferred emissions in imaging procedures. The choice, therefore, depends on whether one places greater value on resolution in energy or in time.

GATED CARDIAC STUDIES AT EQUILIBRIUM

The performance of gated cardiac studies requires an understanding of the basic physical assumptions on which the technique is based. Additionally, such studies put special demands on the instrumentation and computer software. In this section, a typical study will be described to illustrate the assumptions and instrumentation requirements. The validity of some of these assumptions will then be examined in detail and various instrumentation requirements discussed. Several methods of computer cardiac analysis and data collection are described to present the advantages and limitations of each.

A typical study

From 10 to 20 mCi of a blood pool label (^{99m}Tc-labeled human serum albumin, or red blood cells) is administered intravenously to the patient. The injection need not be a bolus. After 3 to 5 min, it is assumed that the ^{99m}Tc-labeled substance is in equilibrium in the blood pool, that is, the number of millicuries per milliliters of blood is the same throughout the body. The gamma camera (with parallel-hole collimation) is then positioned over the patient. Many different camera orientations are possible, but for determining left ventricular function a 30- to 40-degree left anterior oblique (LAO) view is often used. In this orientation the approximate plane of the septum lies perpendicular to the face of the camera, resulting in good left-right ventricular separation. Occasionally, a 15-degree caudal modification is also applied. This has the effect of better separating the left atrium from left ventricle and also causes the left ventricle to be viewed more nearly perpendicular to its long axis.

An ECG machine is connected to the patient. The signal from this ECG is fed into a trigger device, which produces a pulse at each R-to-S transition. This trigger device (called the gate) may be an electronic device specifically designed for this purpose or, equally satisfactorily, may be the "synch" output or QRS beeper output of a standard ECG monitor. This gate pulse, along with the signals from the gamma camera, is fed into a computer. It is possible to use a multiformat photographic device instead of the computer, but, as will become clear, clinically significant gains are best achieved with the use of the computer.

The purpose of the computer (or multiformat programer) is to divide the cardiac cycle into many segments of time and to create a separate image during each of these time segments. Suppose, for example, it was desired to divide the cardiac cycle into twenty time segments, giving twenty images throughout a single cardiac cycle. Suppose also that the patient being studied had a heart rate of exactly 60 beats/min, corresponding to an R-to-R time of 1 sec (1000 msec). Each of the twenty time segments would therefore be 50 msec long. The ECG-triggered pulse is used to begin the picture-making process. At the first ECG-triggered pulse, when the heart is at end of diastole, all camera data are put into the first picture for a period of 50 msec. At the end of 50 msec all data are put into the second picture for the next 50 msec, and so on until all twenty pictures are formed. At the next ECG-triggered pulse, gamma camera data are again added to the first picture for 50 msec, then data are added to the second picture, and so on. At the end of, for example, 100 beats, the first image is a composite image of the blood in the heart during the 0 to 49 msec interval after the R wave, the second from 50 to 99 msec and so on. The reason it is necessary to create a composite image made up of many beats is that, with 10 or 20 mCi of ^{99m}Tc at equilibrium in the blood, the count rate is too low to give an image of reasonable quality in any single 50 msec time interval. Hence data from many beats are summarized. If the cardiac function during one beat is identical to the function during all other beats making up the composite, the 100-beat composite image should appear the same as the image that would be obtained from a single beat using 100 times the dosage of ^{99m}Tc. The series of images, then, span not a single cardiac cycle but a single average cardiac cycle.

Two forms of analysis are generally used to interpret the data obtained from a gated equilibrium study. These are (1) visual analyses of the image sequence and (2) quantification of the image sequence to obtain parameters such as ejection fraction, ejection rate, etc. The optimum method for visual analysis is to display the sequence of images as an endless loop movie, replaying the images from a single average beat over and over again. If each image of the movie is displayed sequentially, with little time lost between images, the resulting movie appears as a smooth contraction of the cardiac chambers. If fewer then ten images span the cycle, the movie begins to appear discontinous in time. If more than about thirty images are used to span the cycle, additional imaging time is required because each image may have too few counts. Low-count images result in poor contrast between the left ventricle and surrounding tissue, and a statistical "snow" appears in the movie sequence. A study with 50 to 100 images spanning the cycle (usually done for research purposes) can generally be compressed by summing consecutive images into a satisfactory movie sequence. A study with too few images cannot be similarly recovered. It would be difficult to overemphasize the increase in clinical utility gained by displaying the image sequence as a movie. Anatomic and functional features difficult or impossible to visualize from gated static images (subtle wall motion defects, alterations in global and regional rates of ventricular contraction, volume variations and motions of the aorta, etc.) are readily apparent in the movie display. The human eye and brain are well suited to the interpretation of such spatially and temporally varying information when presented in movie format. The movie display not only provides temporal information, but also allows the eye to average out some of the noise in the spatial data due to limited counting statistics. The result is a display format from which visually subtle features of cardiac anatomy and function can be discerned.

Quantitative analysis of the data is based on the assumption that the counts over any region of an image are proportional to the

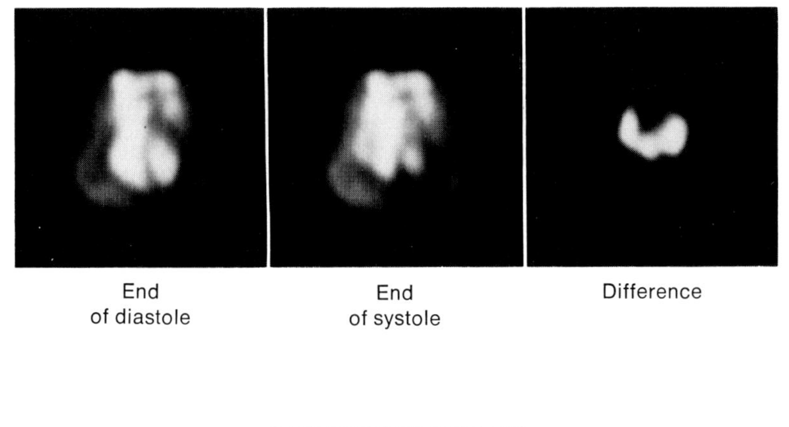

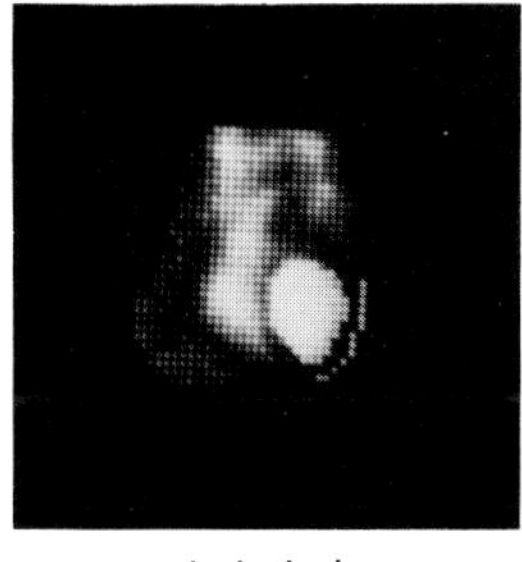

Fig. 1-13. Selection of regions of interest from end-diastolic, end-systolic, and difference images.

blood volume in that region. If the left ventricle is identified by the observer and the counts over that structure plotted on the ordinate and image number on the abscissa, a plot proportional to blood volume versus time is obtained. As the volume of blood in the left ventricle changes with time, decreasing after the R wave pulse during systole and increasing during diastole, so will the counts over the region of the left ventricle. Unfortunately, the counts over the region of the left ventricle are not due solely to the blood volume within the ventricular chamber. Also contributing to the counts are photons emitted from the blood in the myocardium and from the overlying and underlying tissue (such as chest wall and lungs). It is necessary to correct this background of counts due to extraventricular blood. Ideally, one would like to know the background count rate over the region of the left ventricle with the heart removed. Since this is impossible, the count rate from tissue immediately adjacent to the left ventricle is frequently used to estimate the background. Fig. 1-13 shows one popular method for estimating background. Care must be taken not to include structures such as liver or spleen in the background region of interest. To correct for background, the counts from the background region of interest must be normalized to the same area as the left ventricular region of interest. After the normalization is performed, the background value is subtracted from each point of the curve of Fig. 1-14 to give a plot of left ventricular volume versus time over the single average cardiac cycle (Fig. 1-15). From this background-corrected curve, many quantities can be calculated, the most clinically important being ejection fraction. Ejection fraction is defined as the counts at

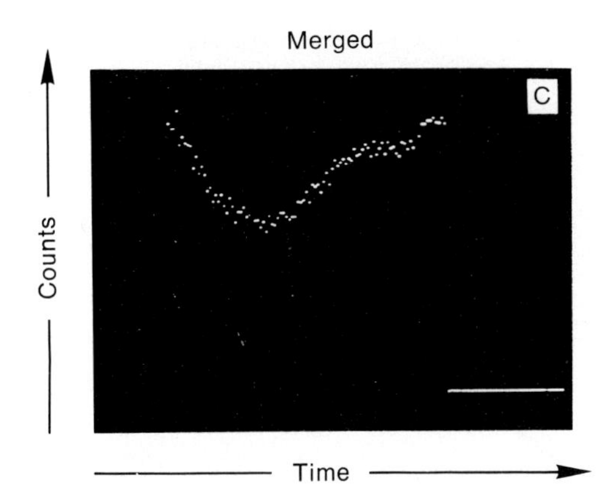

Fig. 1-14. Counts over left ventricle as function of time from R wave. Each point represents 10 msec.

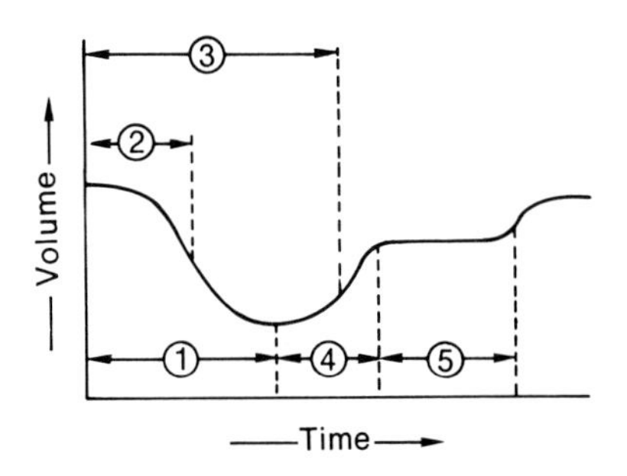

Fig. 1-15. Typical timing parameters derived from LV volume curve. *1,* Time to end of systole; *2,* time to peak ejection rate; *3,* time to peak filling rate; *4,* filling time; *5,* diastasis period.

end of diastole (ED) minus the counts at end of systole (ES) divided by the counts at end of diastole, thus:

$$EF = \frac{ED - ES}{ED}$$

Care must be taken in defining the time of end of systole. It is frequently taken as the lowest point on the volume curve. However, when counts in each point of the systolic portion of the curve have less than 500, the statistical uncertainty of the point exceeds 10%, and the procedure tends to consistently underestimate the end-systolic count rate, thus to overestimate ejection fraction. Curve-fitting or averaging procedures are occasionally used to reduce this small source of error.

Other quantities of possible interest are various timing parameters and maximum ejection rates and filling rates. One of the current limitations of blood pool scintigraphy is that, although counts may indeed be proportional to blood volume, the proportionality factor is not known. This prevents calculation of absolute and stroke volumes from the curve of activity versus time. However, volumes can be calculated from outlines of the left ventricle made directly from the images.

The gated image sequence obtained from scintigraphic studies is often compared to similar image sequences obtained during contrast ventriculography. When used to quantitate left ventricular function by calculation of ejection fraction, the two techniques have little in common. The contrast ventriculography technique usually assumes the ventricle to be a perfect prolate spheroid. The scintigraphic technique makes no such geometric assumption. Volume changes in contrast ventriculography are determined by observation of edge motion in one or two views. If a defect does not occur at the edge in the view used, its contribution is not considered. The scintigraphic technique, on the other hand, calculates relative volume directly from counts. Aside from attenuation effects (discussed on pp. 8 and 9) all regions of the ventricle are observed simultaneously in the calculation of volume changes. In contrast ventriculography, very high resolution images are necessary to accurately define ventricular borders for calculation of the projected ventricular area. Interobserver errors in defining these edges are known to be large. In the scintigraphic technique, resolution need not be high because area tracings need not be used to calculate volume. Resolution need only be high enough to separate the left ventricle from other structures. If background is adequately compensated for, variations in the definition of the ventricular region of interest should produce only small errors in ejection fraction.

Validity of assumptions

There are several assumptions that must be made in the analysis of ECG-gated equilibrium blood pool studies. In the previous

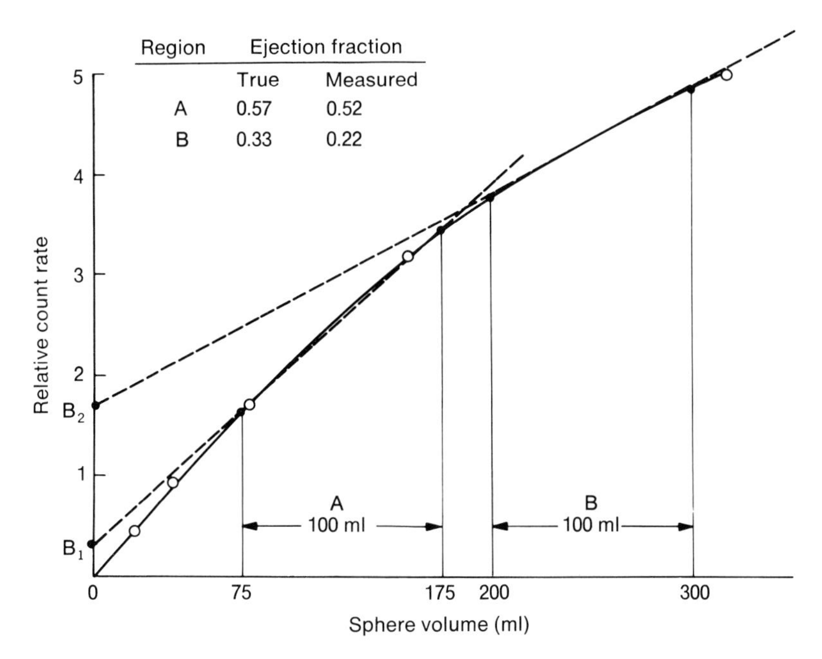

Fig. 1-16. Count rate as a function of volume for ^{99m}Tc-filled spheres viewed by parallel-hole collimator.

description of a typical study, most of the assumptions seem plausible. Certain assumptions, however, deviate slightly from reality; the deviations can occasionally affect the clinical interpretation of the data. Following is a discussion of the validity of each of the assumptions.

Count-volume linearity. One of the most basic assumptions of cardiac blood pool imaging is that the count rate from a region is proportional to the blood volume of that region. There are several reasons why this assumption is not strictly true. First, the pharmaceutical (labeled albumin or red blood cells) may break down and distribute itself outside the blood pool. The effects this has on the quantitative results are discussed in the section on sources of background error (p. 39). Second, photons from the heart may be attenuated (absorbed or scattered) by blood from the heart itself. As the volume of blood in the heart changes with time, the amount of attenuation also changes with time. In a patient with a large stroke volume, the counts at end of diastole may be reduced by this attenuation more than will the counts at end of systole. Fig. 1-16 illustrates the magnitude of the effect for a gamma camera with parallel-hole collimation. The slight deviation of this curve from a straight line could cause small errors in the measured value of ejection fraction. A typical stroke volume in a normal individual is about 70 ml. Over this small range of volumes the curve is quite linear. Extrapolating this linear portion of the curve back to zero volume, however, results in a nonzero intercept on the count rate axis (Fig. 1-16). Thus there would be an apparent count rate even at zero blood volume. This produces the same effect as would additional background counts, except that true background could presumably be measured and corrected for, whereas this "virtual" background cannot. Moreover, the magnitude of the virtual background differs for various end-diastolic volumes, even if the stroke volume is constant. Subjects with large ventricles might therefore tend to have a larger virtual background than would subjects with small ventricles. This in turn would cause measured ejection fractions in subjects with large ventricles to be systematically lowered as much as 5% or 10%. Whereas the count versus volume nonlinearity can cause errors in the scintigraphic technique, these errors are probably much smaller than the errors arising from the geometric assumptions necessary for planimetric calculation of ventricular ejection fraction.

Average versus single beat. In a gated equilibrium study, the data from many beats are summed to form a single, composite beat. In a normal individual, even at rest, not all beats are identical. The most notable beat-to-beat variation is the variation in the length of each beat. In a normal individual the length of each beat may fluctuate by as much as several hundred milliseconds. A plot of number of beats versus length of each beat (called the R-R distribution) might appear as in Fig. 1-17. How will these fluctuations affect the results of an equilibrium study? The most obvious effect will be a loss of counts from the images near the end of the cycle. Consider, for example, the individual whose beat length distribution is shown in Fig. 1-17. Assume that the computer sorts data from this individual into twenty images of 50 msec each. The last image, then, comes from data occurring during the 950 to 1000 msec interval following the R wave. Some beats, however, are shorter than 950 msec. Data from these beats thus are not included in the last image. The last image will contain data from a smaller number of beats than will images from earlier in the study; a similar phenomenon will occur with the next-to-last image,

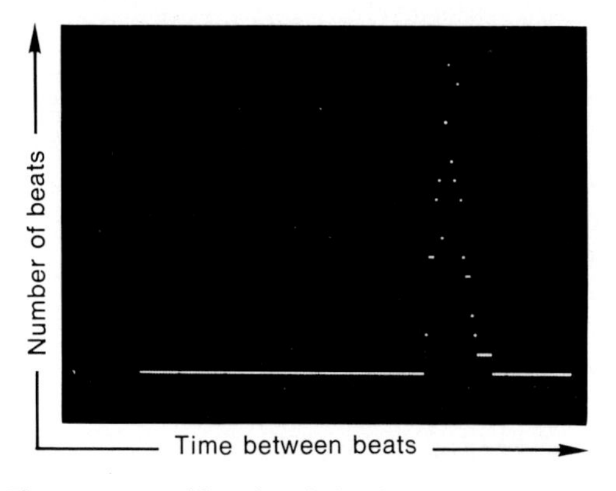

Fig. 1-17. Typical beat-length distribution. Peak at 1000 msec.

and so on. If the subject's shortest beat is 880 msec long, then only the first seventeen images (0 to 850 msec) will contain data from all the beats. The eighteenth, nineteenth, and twentieth images will each contain data from fewer and fewer beats. In general, then, the last few images of the sequence must be ignored for both qualitative and quantitative analysis. (A method of data collection, which avoids this problem is discussed in Chapter 6.) The distortion in the last few images usually causes no significant loss of clinical utility. This is because the clinical significance of events during diastasis and atrial contraction is not yet clear.

A more important concern is whether the normal fluctuations in beat length cause a blurring of events during systole or early diastole. Such blurring, if it occurred, could alter the value of quantities calculated from the volume curve, such as ejection fraction, ejection rate. Fortunately, it has been shown that these fluctuations primarily cause shortening or lengthening of the diastasis period without significantly altering the ejection or active filling phases of the cycle. At higher heartbeat rates (for example, during exercise) when the diastasis period begins to disappear, the R-R distribution becomes narrower in both normal subjects and in subjects with a variety of heart diseases (excluding subjects with arrhythmias).

Summarizing, although the normal fluctuations in beat length may distort the periods of diastasis and atrial contraction, the more clinically important systolic and active filling periods remain relatively precise. In some collection systems even the late portions of the cardiac cycle can be recovered.

Premature ventricular contractions, or wide fluctuations in beat length due to arrhythmias, can adversely affect the data. Ideally, these beats (and the beat that follows them) should be eliminated from the data. In practice, the quality of the movie image sequence is not significantly degraded even when a small percentage of the total beats are premature contractions. The quantitative data, however, are affected. Unless these beats are removed from the data, they can cause errors in the calculation of ejection fraction and other quantitative parameters of ventricular function.

Background determination. One of the largest sources of potential error in the determination of ejection fraction is variability in defining the background region of interest. Background usually accounts for approximately 50% of the total counts over the left ventricle at end of diastole. A fractional uncertainty of 10% in the value of background (because of counting statistics or interobserver variability in background region-of-interest selection) will cause a fractional uncertainty of about 10% in ejection fraction. Thus when background constitutes about 50% of the counts over the left ventricle at end of diastole, a given fractional error in background causes an equal fractional error in ejection fraction. By applying the equation of error propagation to the equation for ejection fraction it is possible to estimate the effect a fractional error in background will have on the fractional error in ejection fraction. As the fraction of background increases (for example, due to a poor pharmaceutical preparation or loss of binding with time), the same uncertainty in background can cause a much larger error in ejection fraction.

Left ventricular attenuation effects may also cause some error in background. In correcting for background the assumption is made that counts from a region adjacent to the heart approximate counts from cardiac muscle and tissue lying in front of and behind the heart. For large stroke volumes, the true background may vary with time because the ventricular volume alternately shields and does not shield the underlying tissue. Since it is likely that the majority of the background counts come from structures in front of the heart, the magnitude of this effect is probably small.

There is no universally accepted method for selecting a background region of interest. It is probably more important that the user be consistent with the method than that the background obtained be accurate in absolute terms. The requirement for accurate absolute background correction is especially relaxed for sequential intervention studies performed with a single dose of isotope, such as rest/exercise or drug intervention studies. In studies of this sort, the critical parameter is usually a change in ejection fraction. (For example, did ejection fraction increase or decrease with exercise?) In measuring changes in ejection fraction from a single administration of isotope, it is no longer as important that background be correct in absolute terms. Rather, it is more important that the background method be applied consistently from rest to exercise.

Instrumentation

Performing gated cardiac studies often puts special requirements on the instrumentation, both the gamma camera and the computer acquisition system. In this section some of these requirements will be discussed. In addition, the general features of several different real time methods of acquiring and processing cardiac data will be described.

Resolving time. The ability to perform studies during exercise is proving to be of great clinical importance. This ability requires that relatively high count rates (20,000 to 30,000 counts/sec) be tolerated by the system with little or no data losses so that the study may be completed in as little time as possible. The primary sources of data losses have in the past been dead time in the gamma camera, in the analog-to-digital converter, and in the system software. The latter two are easily modified to cause only negligible losses. Some of the newer gamma cameras also have acceptable losses at these count rates. The so-called "maximum counting rate" of a camera is not an indication of its losses at lower count rates. Actual losses of less than 2% to 3% at the maximum counting rate expected are acceptable. Typically, with a high-efficiency collimator in a 35-degree LAO with 15 degree caudal modification, about 1200 to 1500 counts/sec/mCi may be expected from a ^{99m}Tc-labeled blood pool agent. At dosages of ^{99m}Tc of 10 mCi or more, many gamma cameras will have unacceptable losses. Dead time of 6 μsec or greater is typical for all but the newest gamma cameras. Dead times of this magnitude, even if nonparalyzable, will result in count losses of about 10% at 10 mCi dosages and considerably higher at 15 to 20 mCi.

One method for dealing with this problem is to use a lead mask to reduce the field of view of the gamma camera to 15 to 20 cm (6 or 8 inches) in diameter. This is an adequate field of view for most cardiac blood pool studies. Most computer systems allow the user to "zoom" in on the heart, eliminating structures outside the user's area of interest. This, of course, does not achieve the same effect as masking the camera because zooming does not reduce the count rate to the camera, only the count rate to the computer.

There are two potentially deleterious effects that can arise from camera dead time: an increase in the length of time necessary to perform the study and a distortion of the quantitative data. The latter effect can occur if the total count rate seen by the camera at end of diastole is greater than that seen at end of systole. In an unmasked camera, most of the counts within the field of view are from structures extraneous to the heart, and hence the whole-field count rate does not change significantly during the cardiac cycle. If the camera is masked, this is not longer the case. In new generation cameras, count losses at 20,000 to 30,000 cps are negligible. It still may be desirable to mask the field of view so that the computer does not spend unnecessary time handling data from irrelevant structures. As previously mentioned, often the zoom feature performs this function. However, limiting the field of view is a poor approach, since changes in pulmonary blood volume may occur, which can be of diagnostic importance. Therefore the camera and computer field should encompass a 25 cm diameter.

Computer requirements. (See also Chapter 4.) There are three basic functions that an equilibrium cardiac computer system must perform: (1) creation of the image file and concurrent acquisition (i.e., in real time) of the gated image sequence, (2) display of the image sequence in endless loop movie format, either concurrently with acquisition or immediately following a pause in data collection, and (3) provision of a reasonably rapid quantitative analysis following completion of data collection. Without these capabilities, it is at best awkward and inconvenient, if not impossible, to apply the system to before- and after-intervention studies.

Gated equilibrium cardiac studies have been performed for many years. Only recently, however, have they been considered to be of general clinical utility. The reason for the current popularity of the technique is probably directly traceable to two simple innovations: making the computer produce its results in real time and applying the techniques to intervention studies. The two ideas are obviously related. Intervention studies become clinically useful when the physician is able to actively participate, basing the type and extent of further interventions on the status of the patient as evidence by the computer display. The system becomes truly clinically useful when the physician is able, for example, to perform a study at rest, observe the movie (and perhaps get an ejection fraction), and then proceed with appropriate diagnostic or therapeutic interventions based on these results. If no wall motion defect is present at rest, perhaps exercise will elicit one. If a wall motion defect is elicited during exercise, it might be of value to determine whether or not it improves after administration of nitroglycerin.

The ability of the physician to observe the display (or data derived therefrom), intervene with the appropriate diagnostic or therapeutic maneuvers, and quickly observe the effect of the intervention adds a new dimension of clinical utility to cardiac imaging. It is for these reasons that the computer system must not only acquire and sort the data in real time, but also be able to quickly display the data in movie format. Ideally, the movie display would be continuous throughout data collection, updated as frequently as new data came in. Alternatively, the movie display should at least be accessible at any time during a few seconds pause in data collection. This capability is useful not only to assess the patient's clinical status, but also to allow the user to decide when to terminate the study on the basis of visual image quality rather than on less reliable indicators, such as time elapsed or number of counts in the whole field. In addition, viewing the movie as it is acquired is the simplest, most reliable way to ensure proper camera positioning, an important consideration if the camera is masked or the field of view expanded by the computer. As gated equilibrium studies begin to be performed more frequently as a method of continuous monitoring of cardiac function in the intensive care unit, the real time capabilities become even more crucial.

Quantitative evaluation. Calculation of ejection fraction from the volume curve is often merely a quantitative refinement of a previously determined qualitative assessment made from observation of the movie sequence. This is especially true in the case of intervention studies. By observing both a rest movie and an exercise movie displayed simultaneously side by side, the physician can almost always assess the direction of change in the ejection fraction, whether it increases, decreases, or remains approximately the same with exercise. This being the case, it is probably adequate for the derivation of quantitative results to be postponed until completion of the entire sequence of studies.

Computer methods

There are many different approaches to the problem of constructing a useful gated cardiac computer system. The descriptions that follow are meant to illustrate the basic approaches used in many systems.

The gamma camera x-y coordinates are digitized by the analog-to-digital converter (ADC). Typically, the ADC digitizes the gamma camera field of view into an array of 32 dots by 32 dots (called "pixels" for picture elements), a 64 by 64 array of pixels, or a 128 by 128 array of pixels. A single image with these digital resolutions would take up 1024 words, 4096 words, or 16,384 words of computer memory, respectively. (There are techniques that permit the use of half words instead of full words, effectively halving the memory needed for an image.) In gated cardiac studies, twenty or more images per cardiac cycle are generally desired, and these images usually must be resident in the computer memory at the same time. For just twenty images this would require 24,080, 81,920, or 327,680 words of memory, respectively, for image storage. Many existing minicomputers cannot handle more than approximately 32,000 words of memory. Some of the newer generation minicomputers can be equipped with as large as a million words, but only at fairly high cost and often only with considerable extra bulk. (Very recent advances are being implemented in a few machines that eliminate these disadvantages, as will be mentioned in Chapter 4.) For acquiring data from a gated cardiac study performed with a high-efficiency collimator, a resolution in excess of 64 by 64 pixels is probably not justified. It is wasteful to digitize to millimeters an image whose inherent resolution may be in excess of a centimeter. Even a 64 by 64 pixel image sequence acquisition requires a large amount of memory. To economize on memory, many ADCs have the ability to amplify the analog x-y signals from the camera. Thus if normally the ADC digitizes the full field of the gamma camera into a 64 by 64 pixel array, after amplification, only a small portion of the field will be digitized into the same 64 by 64 pixel array. This effectively increases the digital resolution (but not the spatial resolution) of the image.

This analog amplification (or zoom) feature, in addition to eliminating extraneous data, can allow an amplified 32 by 32 pixel cardiac image to have nearly the same pixel resolution as an unamplified conventional 64 by 64 cardiac image. Consider a typical 35-degree LAO view digitized as 64 by 64 pixels. In some subjects, the cardiac chambers would probably occupy only a quarter of the image. Thus of the total of 4096 pixels, the cardiac chambers would occupy only about a fourth of these, or 1024 pixels. In this case digitizing the full field as 64 by 64 pixels would be identical to blowing up the quadrant containing the cardiac chambers by a factor of 4 (2 amplification in x and 2 amplification in y) and digitizing the result as 32 by 32 pixels. Thus a 32 by 32 pixel image of a "blown-up" object can have the same resolution as would be obtained in a conventional 64 by 64 pixel image. Typical amplification factors that are satisfactory for most subjects are on the order of 1.5 to 1.7. Thus a 32 by 32 pixel image sequence can possess an effective resolution of 45 by 45 pixels or 50 by 50 pixels. This is quite satisfactory for a gated cardiac acquisition. (Acquisition pixel density requirements are not related to the pixel density at which the images should be displayed.)

An alternative to analog amplification is to digitize the unamplified gamma camera data, for example, to 64 by 64 pixels and then digitally select (e.g., by software) that region which contains the cardiac chambers. Both methods are capable of producing nearly identical results, although the analog method may be somewhat easier to implement and be capable of operating at higher data rates.

All ADCs possess some dead time. A 1 or 2 μsec dead time is readily achievable. Some ADCs have a dead time that increases with increasing digital resolution, 64 by 64 pixel resolution requiring twice the time as 32 by 32 pixel digitization. As long as the ADC dead time is less than the camera dead time, the ADC contribution to the overall dead time usually can be ignored, since closely spaced pulse pairs will have already been lost in the camera or its associated electronics.

Image sorting. The computer can be made to sort the data in the following manner. If there are thirty 32 by 32 pixel images available to the user, thirty 1024 arrays are set aside in the computer memory. The user is then asked what temporal width is wished for each image to span. That is, should each image include 50 msec of data, 20 msec of data, or what? Some systems use a fixed temporal width for each image or fix the number of images that must span the entire cycle. It is usually more flexible to allow the user to set the temporal width of each image, depending on the heart rate and information desired from the study. In subjects with very long R-R intervals (e.g., subjects on drugs such as propranolol [Inderal]) whose R to R times might be as long as 1.2 sec, a system that forced the thirty available images to span the entire 1.2 sec time period would result in 40 msec per image—a temporal width that is probably too coarse for accurate ejection fraction calculations. Instead, it might be more desirable to span only the first portion of the cycle with the thirty images, choosing perhaps 20 to 30 msec per image. This would prevent visualization of the long period of diastasis in such a subject, but would permit better resolution of the more interesting systolic and active filling periods.

The basic sorting algorithm is usually quite simple. A program "pointer" can be made to point to the origin of any picture in the computer memory. Each digitized x-y signal from the gamma camera causes the appropriate pixel of the image "pointed to" to be incremented by one. The ECG-triggered signal causes the pointer to be set to the first image. The elapsing of a certain amount of time on a clock within the computer causes the pointer to move to the next image in the sequence. Thus the pointer progresses from image to image within the gated sequence and returns to the first image at each ECG.

The computer clock generally works by generating pulses at fixed time intervals, perhaps a clock pulse every 10 msec. These pulses are then counted as they occur to keep track of elapsed time. The patient's ECG will rarely coincide with one of these clock pulses. Thus the pointer will be returned to the first image at some arbitrary point within a 10 msec interval. This will cause the first image to consist of data from a nonintegral number of clock pulses. After the occurrence of an ECG, it is best to wait until the next clock pulse arrives (an average wait of 5 msec for a 10 msec clock) before beginning the sorting process at the first image. This is one reason why it is desirable to have a clock that generates pulses at a rate somewhat faster than the smallest temporal image width desired. Another reason for a fast clock rate (perhaps several hundred Hertz) is to allow the user more flexibility in choosing how to divide up the cardiac cycle.

The time interval for each image beyond which quantitative calculations become blurred is not precisely defined. For a rest study, 30 msec is probably satisfactory, whereas during exercise a 20 msec/image may be required.

With this method premature beats or other beats outside of the usual beat-length distribution cannot be easily eliminated. By the time the length of the beat is known, the data from that beat have already been added to the images. Although such beats usually do not affect the visual quality of the movie, they can, as was mentioned previously, affect the quantitative information derived from the images. The normal fluctuations of beat length will also distort the last portion of the image sequence for the reasons mentioned previously. Because of these facts, it is very desirable to know what the distribution of beat lengths that made up a particular study looked like. This plot of number of beats versus beat length is easy for the computer to construct and display and ideally should be part of the cardiac analysis system. Although there is little one can do about the fluctuations in beat length or about PVCs in the system just discussed, at least the user should be aware of what the magnitude of these fluctuations was and whether or not there were a significant number of beats of abnormal length included in the data.

There is another method of accumulating the data that avoids many of the previously mentioned problems. With this method it is possible to eliminate beats of undesirable length and to correct for fluctuations in beat length. In this approach, the system sorts the data into movie images, as was described earlier, and also simultaneously writes list-mode data (that is, the event-by-event list of data, just as is comes from the gamma camera, ECG gate, and timer) onto a high-density disk. The movie sequence is available for display at any instant during the study. The quantitative data are obtained either directly from the list-mode data on the disk or by reformatting the disk into images. In either case, 1 to 2 min are required to read through the disk after the study. Only one pass through the data is necessary if a list of the length of each beat is created in real time during the study. Thus prior to rereading the disk, the user can select not only what length beats should be excluded from analysis, but also when, temporally, the beats occurred. This latter facility, employed in a system at the National Institutes of Health (NIH), has proved convenient for analyzing exercise studies at gradually increasing levels of load. The principal disadvantage of storing the entire study in list mode is the cost of the disk. The time necessary to reformat the study need not be a significant factor if a suitably fast algorithm is used.

By storing the entire study onto a disk in list mode, at least for a few minutes, it is quite easy to eliminate beats lying outside a user-selected window of values (called beat-length windowing or beat-length discrimination). Data from beats one wishes to ignore are simply skipped over. By storing in sequence the length of each beat, those blocks of data that are to be skipped can be identified in advance. It is equally simple to compensate for the effect of beat-length fluctuations. The data are "gated" in two directions, forward in time from each R wave, as has already been described, and backward in time from each R wave. The forward data fall at the end of the cycle due to beat-length fluctuations, whereas the backward-gated data fall off in time near the beginning of the cycle. By merging the first two thirds of the forward data with the end-diastolic third of the backward data, a complete, average cardiac cycle is approximated, even showing the effects of the volume increment due to atrial contraction. Again, this is, at least conceptually, easy to do from the list data, since the length of each beat is known in advance.

Recent advances in the cost (roughly $100 per 1024 16-bit words and dropping) and density (64,000 16-bit words per small circuit board) have permitted a third approach to be investigated at NIH. This approach retains most of the advantages of storing list-mode data on the disk but does not actually require a disk. Instead, approximately 40,000 to 60,000 words are set aside for temporary storage of list-mode data within the computer's memory. If two beats worth of data can be stored in list mode within the computer's memory, the data from the current beat can be acquired while the computer is concurrently sorting the data from the previous beat. As the data from the entire previous beat are within memory, its length is known prior to sorting the data, just as with the disk. Thus the user can exclude beats lying outside the user-selected beat-length window as well as do the "backward" gating previously described without the use of a disk. Obviously, there is no time required to "reread" the data as there is with a disk. With the NIH system, the cost of the additional memory is presently between a fourth and a third of the cost of the disk.

This approach, using very large amounts of memory, would have been inconceivable, both economically and for reasons of insufficient physical space, only a very short time ago. Now minicomputer systems are available with more than 500,000 words of 16-bit memory, which can fit into the mainframe of the computer. Thus it is possible to do a 10 msec/image study, to eliminate undesired beats, to correct for beat-length fluctuations, and to display the image sequence, all in real time, all in memory. No disk or tape is required. The small size and potentially lower cost of such a system may have important ramifications for cardiac nuclear medicine in the future.

Most computer systems employed for gated scan acquisition in nuclear cardiology do not use these elegant approaches because of their expense or the time involved for analysis. Usually the data from a given cardiac cycle are only analyzed and sorted into an image file during acquisition, without concomitant list-mode recording. The images are displayed in live time from the image file memory concurrent with data acquisition. If beat-length windowing is desired, it is done by observing the length of each beat in comparison to a series of reference beats collected at the beginning of the acquisition. If the observed beat is within ±10% of the reference R-R interval, the collection is allowed to proceed. If the observed beat is not in this window, the acquisition is placed on hold, while a series of five to ten beats are sampled to determine when the heart rate has returned to the reference rate; then the acquisition is allowed to proceed. Analysis of the data from these collections for ejection fraction and ejection and filling rates from the right and left ventricles is performed with semiautomated techniques using the data from the image file only. Although these collections may not be as precise as those recorded with the combined list- and picture-mode methods, they have shown a high correlation against results obtained in the catheterization laboratory and appear to supply data of sufficient precision for use in clinical cardiology.

First-pass studies

First-pass studies, in which bolus injections of isotope are followed as they pass through the cardiac chambers, are fully described in later chapters. Most of the physicial limitations and sources of error inherent in gated studies are also present in first-pass studies. Most notably, all the comments concerning the effects of attenuation are applicable in the case of first-pass techniques. If an RAO orientation is used, the self-attenuation effects and virtual background effects may be more severe than in the LAO orientation, due to the superposition of both chamber volumes in the RAO orientation. Similarly, background correction is equally important in gated equilibrium and first-pass studies. Additionally, background varies with time in a first-pass study, as the isotope distributes itself within the body. Count rate requirements are more severe for first-pass studies because of the bolus nature of the injection. There are, however, some measurements (for example, transit time studies and shunt detection) for which first-pass studies give information not readily attainable from equilibrium studies.

REFERENCE

1. Schulz, A. G., Knowles, L. G., and Kohlenstein, L. C.: Simulation studies of nuclear medicine instrumentation, Appl. Tech. Dig. **10**(2), 1971.

BIBLIOGRAPHY

Alpert, N. M., and Burnham, C. A.: A new data processing system for gated cardiac studies. In IEEE Computer Society Proceedings: Symposium on computer applications in medical care, Nov. 1978.

Alpert, N. M., McKusick, K. A., Pohost, G. M., et al.: Non-invasive nuclear kinecardiography, J. Nucl. Med. **13**:1182, 1974.

Bacharach, S. L., Green, M. V., Borer, J. S., Douglas, M. A., Ostrow, H. G., and Johnston, G. S.: Real-time scintigraphic cineangiography. In IEEE Computer Society Proceedings: Computers in cardiology, St. Louis, 1976, pp. 45-48.

Bacharach, S. L., Green, M. V., Borer, J. S., Douglas, M. A., Ostrow, H. G., and Johnston, G. S.: A real-time system for multi-image gated cardiac studies, J. Nucl. Med. **18**:79, 1977.

Bacharach, S. L., Green, M. V., Borer, J. S., Ostrow, H. G., and Johnston, G. S.: A computer system for clinical nuclear cardiology. In IEEE Computer Society Proceedings: Symposium on computer applications in medical care, Nov., 1978.

Borer, J. S., Bacharach, S. L., Green, M. V., et al.: N. Engl. J. Med. **296**:839, 1977.

Borer, J. S., Kent, K. M., Bacharach, S. L., et al.: Circulation **56**(4):III, 1977.

Cohn, P. F., Levine, J. A., Bergeron, G. A., et al.: Am. Heart J. **88**:713, 1974.

Douglas, M. A., Ostrow, H. G., Green, M. V., Bailey, J. J., and Johnston, G. S.: A computer processing system for ECG-gated radioisotope angiography in the human heart, Comp. Biomed. Res. **9**:133, 1976.

Green, M. V., Brody, W. R., Douglas, M. A., Borer, J. S., Ostrow, H. G., Line, B. R., Bacharach, S. L., and Johnston, G. S.: Ejection fraction by count rate from gated images, J. Nucl. Med. **19**:880, 1978.

Green, M. V., Ostrow, H. G., Douglas, M. A., Myers, R. W., Bailey, J. J., and Johnston, G. S.: Scintigraphic cineangiography of the heart, Medinfo 74, Amsterdam, 1974, North-Holland Publishing Co., pp. 827-830.

Green, M. V., Ostrow, H. G., Douglas, M. A., Myers, R. W., Scott, R. N., Bailey, J. J., and Johnston, G. S.: High temporal resolution ECG-gated scintigraphic angiocardiography, J. Nucl. Med. **16**:95, 1975.

Parker, J. A., Secker-Walker, R., Hill, R., Siegel, B. A., and Potchen, E. J.: A new technique for the calculation of left ventricular ejection fraction, J. Nucl. Med. **13**:649, 1972.

Zaret, B. L., Strauss, H. W., Hurley, P. J., Natakajan, T. K., and Pitt, B. A.: A noninvasive scintiphotographic method for detecting regional ventricular dysfunction in man, N. Engl. J. Med. **284**:1165, 1970.

Zir, L. M., Millar, S., Dinsmore, R. E., Gilbert, J. P., and Hawthorne, J. W.: Intraobserver variability in coronary angiography, Circulation **53**:627, 1976.

2 □ Emission computed tomography of the myocardium

John W. Keyes, Jr.

TECHNIQUE OF COMPUTED TOMOGRAPHY

Tomography is a general term covering any imaging technique that produces a picture of a slice or section through the object of interest. Conventional tomographic techniques, used in x-ray studies for many years and later adapted to nuclear medicine, produce tomograms that are parallel to the plane of the detector. These tomograms are usually produced by moving the imaging system in relation to the patient in such a way that a single plane passing through the patient remains stationary (as seen by the detector), while all other planes in the patient appear to move and thus become blurred and out of focus. The presence of this out-of-focus data causes these images to have very low contrast and distorts quantitative relationships within the image.

Computed tomography (CT), on the other hand, produces tomograms that are at right angles to the plane of the detector. The detector system is again moved in relation to the patient, but a computer is used to calculate an in-focus image from data acquired at each of the imaging positions. Computed tomograms are superior to conventional tomograms both because of the way the data is collected and the way the final image is reconstructed. The net result of these differences is that computed tomograms contain no out-of-focus data, and if the process has been carried out properly, the spatial and quantitative relationships of the structures within the field of the tomogram are undistorted in the final image. In addition, computed tomograms are by their very nature extremely high in contrast, and consequently, density or activity differences, which are invisible in conventional projection images or tomograms, are readily discernible in computed tomograms.

There is complete analogy between x-ray or transmission computed tomograms and radionuclide or emission computed tomograms. The basic data-gathering apparatus for an x-ray CT system consists of a source of x rays and a detector, which are so constructed that a narrow beam of x rays can be scanned across a single section of the body and the resulting attenuation of the x-ray beam recorded. The set of attenuation measurements recorded from one such scan is called a projection. The apparatus is then rotated around the patient and another projection recorded until a series of such scans have been obtained through the same section from a number of different viewing angles, resulting in a set of projections, one for each angle. The resulting projections are then fed into a computer, which reconstructs the distribution of x-ray attenuation coefficients throughout the section of the body that is under study. A number of different techniques or algorithms have been developed to accomplish these reconstructions. Some common ones are backprojection, convolution, and a number of iterative algorithms such as ART and SIRT. Interested readers are referred to the reference list for a more thorough discussion of these techniques.[2,5,6,8,10]

Emission computed tomography (ECT) relies on the same basic concept except that the patient is the radiating source and the detector produces a series of projections that measure the radioactivity across a section of the patient as seen from different angles rather than recording the attenuation of an x-ray beam as it passes through the patient.

The only difference between these data sets is the units that are applied to the measured data. As far as the computer is concerned, it makes no difference what units are applied to such sets of numbers, and the same reconstruction algorithms that are used for x-ray tomography can also be used to reconstruct tomograms of the distribution of radioactivity in the body.

One major difference between x-ray CT and ECT relates to the fact that emission CT projection data are distorted by absorption losses occurring within the patient before the emitted radiation is detected. Consequently, in addition to simply reconstructing the distribution of radioactivity within the body, most ECT systems also incorporate provisions to correct for these internal attenuation losses. The final resulting image produced by an ECT system is thus a high-contrast, quantitatively accurate representation of the distribution of radioactivity within a cross section of the body. If a series of contiguous, two-dimensional emission tomograms is obtained, the resulting set of images is, in effect, a complete three-dimensional reconstruction of the distribution of radioactivity within a known volume of the body.

In a recent review, Ter-Pogossian[9] pointed out several major drawbacks to conventional imaging of the heart with the scintillation camera. These include the following:

1. The scintillation camera projects a three-dimensional object onto a two-dimensional plane, introducing in the process noise and artifacts and reducing image contrast.
2. The changing resolution and sensitivity of the scintillation camera with depth complicates quantitative measurements.
3. The internal attenuation of the emitted gamma radiation distorts the quantitative relationships in the image.

It should be apparent that the innate high-contrast and three-dimensional character of ECT images offers a solution to these problems. In an attempt to make use of these advantages of ECT in overcoming the limitations of conventional nuclear imaging techniques, diverse approaches have been tried.

APPROACHES TO EMISSION COMPUTED TOMOGRAPHY

Two basic approaches to ECT imaging have developed in recent years. These may be broadly grouped as (1) positron coincidence ECT techniques and (2) single photon ECT techniques.

Positron CT imaging systems rely on a special property of positron-emitting radionuclides, namely, the simultaneous production of a pair of 511 keV gamma rays, which are emitted 180 degrees from each other on annihilation of the positron particle. Because of this special property, positron emission systems can be designed to utilize what has been called "electronic collimation" to achieve very high sensitivity and uniform spatial resolution and sensitivity across the tomographic reconstruction plane. In addition, this same property allows for a highly exact and technically fairly easy correction for the internal radiation attenuation that occurs when any three-dimensional emissive source is imaged by external means.

The major advantages to positron tomography include the following:

1. A great potential for the development of new radiopharmaceuticals incorporating positron-emitting nuclides. Several of the common organic building-block atoms, including car-

bon, oxygen, and nitrogen, have positron-emitting isotopes that could be incorporated into a vast array of biologically active molecules.
2. The extremely short half-life of many positron-emitting nuclides, which makes it possible to use large (millicurie) amounts to obtain good statistical quality at much lower radiation doses than comparable images obtained with the radiopharmaceuticals in current use. The short half-life also makes it possible to perform serial studies at short intervals.
3. The possibility of accurate attenuation correction inherent in the positron system, which allows the production of images of very high quantitative accuracy.
4. A high detector efficiency that is made possible by the elimination of mechanical collimators. This also contributes to images with good statistics at minimal radiation doses.

To obtain these advantages, however, positron systems pay a fairly high price. Following are significant disadvantages to the positron approach:

1. The potentially useful isotopes of carbon, oxygen, and nitrogen are cyclotron produced and have very short half-lives. Thus a cyclotron is a necessary adjunct to any positron ECT system if its full potential is to be realized.
2. Positron systems require unconventional and highly specialized detector systems that are useful only for positron imaging.
3. The extremely short half-life of many of the positron-emitting isotopes makes them extremely difficult to work with if elaborate biochemical syntheses are required.
4. The high energy of positron annihilation gamma rays increases problems of radiation safety and requires special shielding compared to conventional radionuclides.
5. The realization of the full potential of radiopharmaceuticals based on positron emitters will require the development of an entirely new radiopharmaceutical technology.

The alternative to positron ECT is so-called "single photon" ECT. The term "single photon" has been coined to signify radionuclides that do not emit the paired, opposed gamma rays characteristic of positron emitters. Although the term is slightly misleading because many of these nuclides do give off more than one photon per disintegration, the term is in fairly wide use and will be used throughout this discussion to indicate all nonpositron-emitting radionuclides. In this sense, single photon emitters include virtually all radionuclides used in medical diagnosis at present, including ^{99m}Tc and ^{201}Tl.

Single photon tomographic systems are advantageous from a number of standpoints. The major advantages include the following:

1. These systems are typically based on conventional imaging instrumentation such as scintillation cameras. The high state of development of such instrumentation and its wide availability represents a significant advantage in terms of the rapid development and dissemination of this technique.
2. The basic radiopharmaceuticals already in use in nuclear medicine today can be utilized without further radiopharmaceutical development.
3. A properly designed tomographic system could make use of points 1 and 2 to efficiently perform all conventional imaging procedures in nuclear medicine as well as to obtain tomographic images.

Following are potential disadvantages to the single photon approach:

1. There is a lack of a completely accurate method of correcting for internal attenuation of the emitted radiation. Consequently, this approach cannot achieve quite the same quantitative accuracy as the positron approach.

2. The range of potential radiopharmaceuticals appears to be less than with positron emitters. However, the immediately available range of radiopharmaceuticals is much larger, and these pharmaceuticals are likely to be accessible to a far larger group of users than will positron agents.

The significant differences between these approaches has produced a fair amount of controversy as to which method is really best. Recent evaluations by myself[6] and by Budinger and his co-workers[3] would suggest that in terms of image accuracy and quality at a practical, applied clinical level, there is probably no significant performance difference between the positron approach and the single photon approach.

Positron ECT imaging of the myocardium and the use of this technique for sizing myocardial infarcts are discussed in another chapter of this book and in several recent publications.[1,11,12] The remainder of this chapter will present the results of preliminary studies of myocardial imaging and infarct sizing using several different approaches to single-photon emission tomography.

MYOCARDIAL IMAGING AND ESTIMATION OF INFARCT SIZE BY SINGLE PHOTON EMISSION COMPUTED TOMOGRAPHY

We have developed in our laboratory a prototype single photon emission transaxial tomograph based on a scintillation camera.[7] This unit, the Humongotron, has been specifically designed to realize all the advantages previously mentioned that are inherent in the single photon approach to CT. This unit is illustrated in Fig. 2-1. Using this device, we have been investigating in a canine model the comparative advantages of CT imaging of the myocardium compared to conventional scintillation camera imaging. As a further extension of the CT work, we have been evaluating the quantification of myocardial infarct size.

Acute myocardial infarctions were produced in sixteen dogs. These included eight anterior myocardial infarctions, which were produced by a technique of induced partial ischemia followed by total occlusion, and eight posteroinferior myocardial infarctions, which were produced by temporary occlusion of the left circumflex artery for 60 minutes, followed by reperfusion. The reperfusion technique was used for the posteroinferior infarct because most dogs will not survive permanent complete occlusion of the left circumflex artery.

Forty-eight hours after the induction of the acute myocardial infarctions, the animals were given 15 mCi of [^{99m}Tc] pyrophosphate. One hour later the animals were sacrificed by an overdose of pentobarbital. Both tomographic and conventional images were obtained immediately on all animals. The hearts were then removed and sectioned at right angles to the base-apex axis into slices approximately 1 cm thick. These slices were then stained with nitro blue tetrazolium, which causes a deep blue color to develop wherever dehydrogenases are present in the tissue. The areas of infarction, which are depleted of dehydrogenase, appear as easily delineated pale zones. The areas of infarction were separated from areas of viable myocardium by careful gross dissection, and the actual weights of the infarcted and viable portions of the left ventricular and septal myocardium were obtained.

Computed tomograms representing contiguous sections approximately 8 mm thick, which included the entire area of the myocardium, were produced for each dog. The total volume of myocardial infarct was then calculated from the tomograms and multiplied by the average density of myocardium; the resulting figure was compared with the measured infarct size obtained from the gross specimens. In one dog with an anterior myocardial infarction the infarct could not be separated from the image of the sternum, and an estimate of infarct size could not be obtained. Size comparisons that follow are based on the fifteen studies in which infarct size was calculable.

Excellent tomographic reconstructions

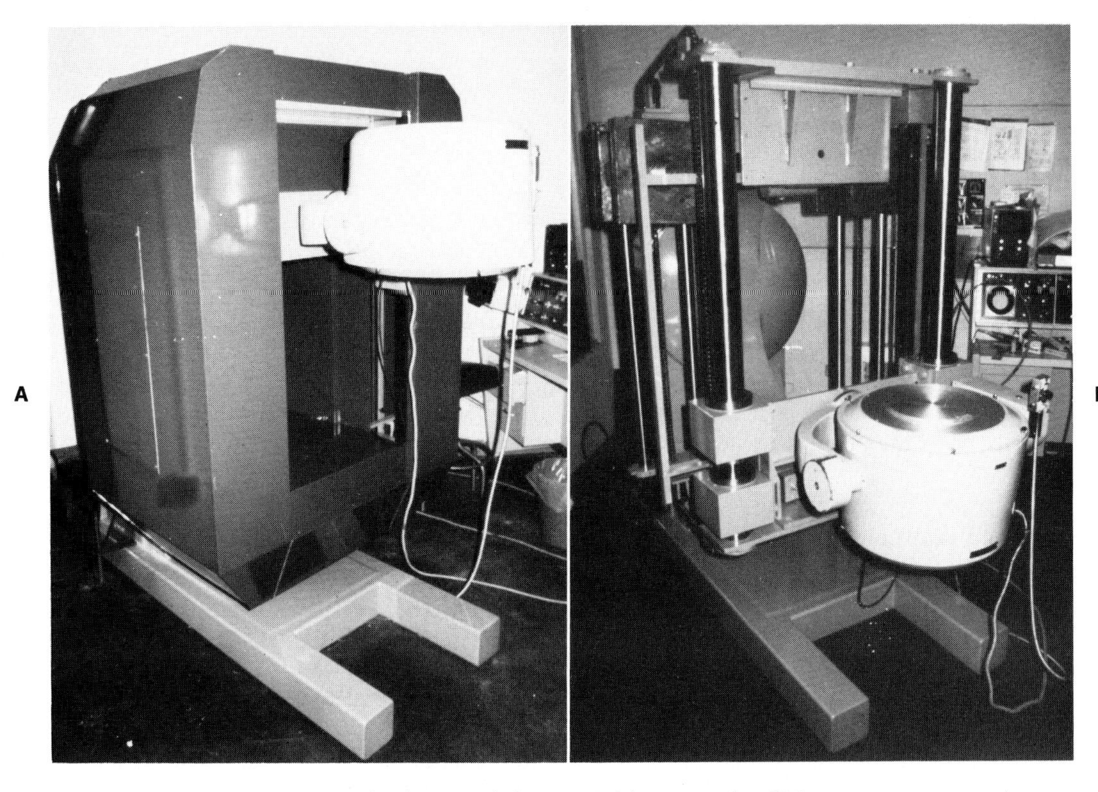

Fig. 2-1. Humongotron, a single-photon emission transaxial tomograph utilizing a gamma-camera detector. **A,** In operational form. **B,** Sheet metal covers removed to show details of mounting, counterweights (for balance), and C arm with bearing, which permits detector to rotate full 360 degrees in either direction.

were obtained in all cases. Figs. 2-2 and 2-3 show tomographic reconstructions of anterior and posterior myocardial infarctions compared with representative conventional scintillation camera views.

The calculated infarct sizes showed a good correlation with the corresponding measured infarct sizes. Fig. 2-4 illustrates the relationship between the calculated and measured infarct size in the fifteen studies in which a satisfactory estimate of infarct size could be obtained.

For best image quality it seemed that gating of some sort to eliminate myocardial movement would be necessary. Since our present tomographic system does not incorporate the facility for gating, all imaging was accomplished in dead animals. It was felt that this would represent a "physiologic" equivalent of gating and give the best assessment of the overall accuracy of this approach for estimation of myocardial infarct size.

A few of the dogs were studied alive and ungated to test the need for gating. A somewhat surprising finding was that there was no discernible difference in studies obtained in the living animals without gating when these were compared with repeat studies obtained after the animals were sacrificed. Fig. 2-4 shows a comparison of tomograms in a living animal compared to a repeat series with the heart not beating. In nine dogs the correlation between the measured infarct size in beating versus nonbeating hearts was 0.94.

EMISSION TOMOGRAPHY BY MEANS OF MULTIPLE-PINHOLE COLLIMATORS

Several approaches to emission tomography have been developed that utilize specially designed collimators containing a

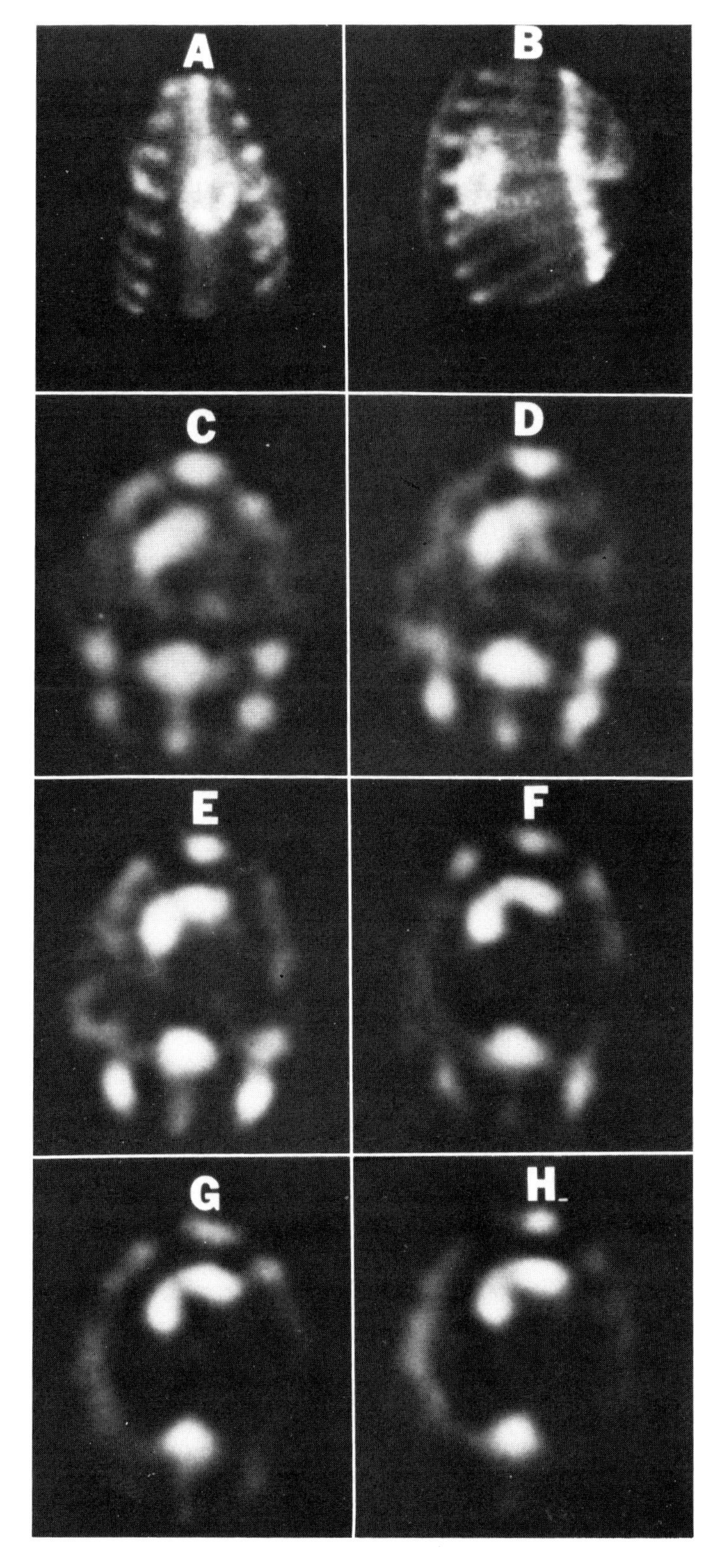

Fig. 2-2. Anterior myocardial infarct. **A** and **B,** Conventional anterior and left lateral views with gamma camera. **C** to **H,** Tomographic sections viewed from above with anterior surface of chest up. Sections progress from cardiac base toward apex.

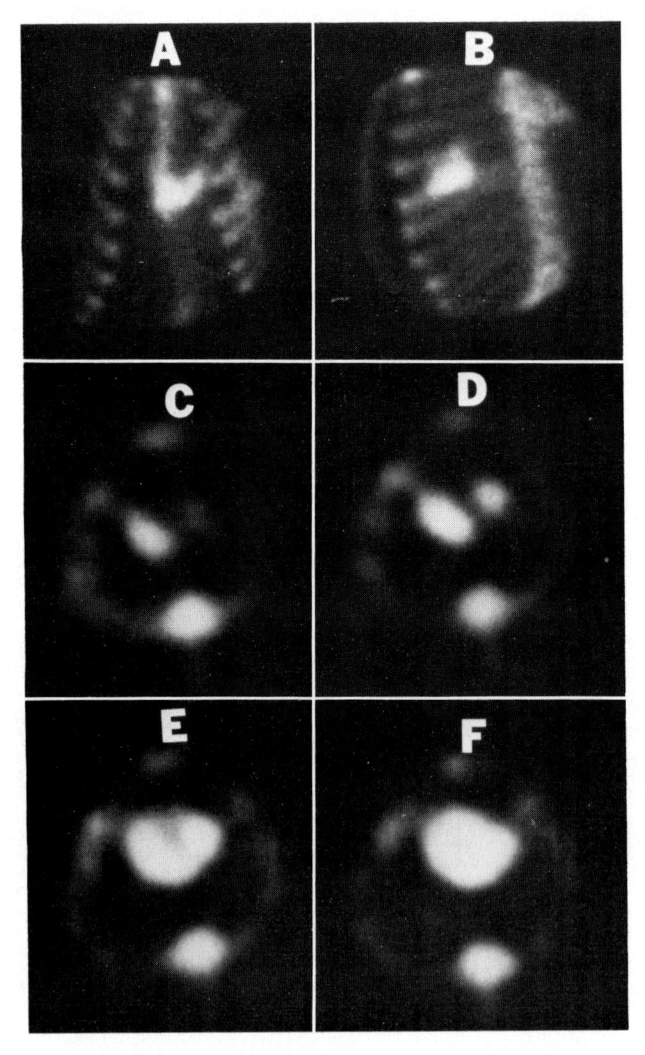

Fig. 2-3. Posterior myocardial infarct. **A** and **B,** Conventional anterior and left lateral views. **C** to **F,** Tomographic sections viewed from above with anterior surface of chest up.

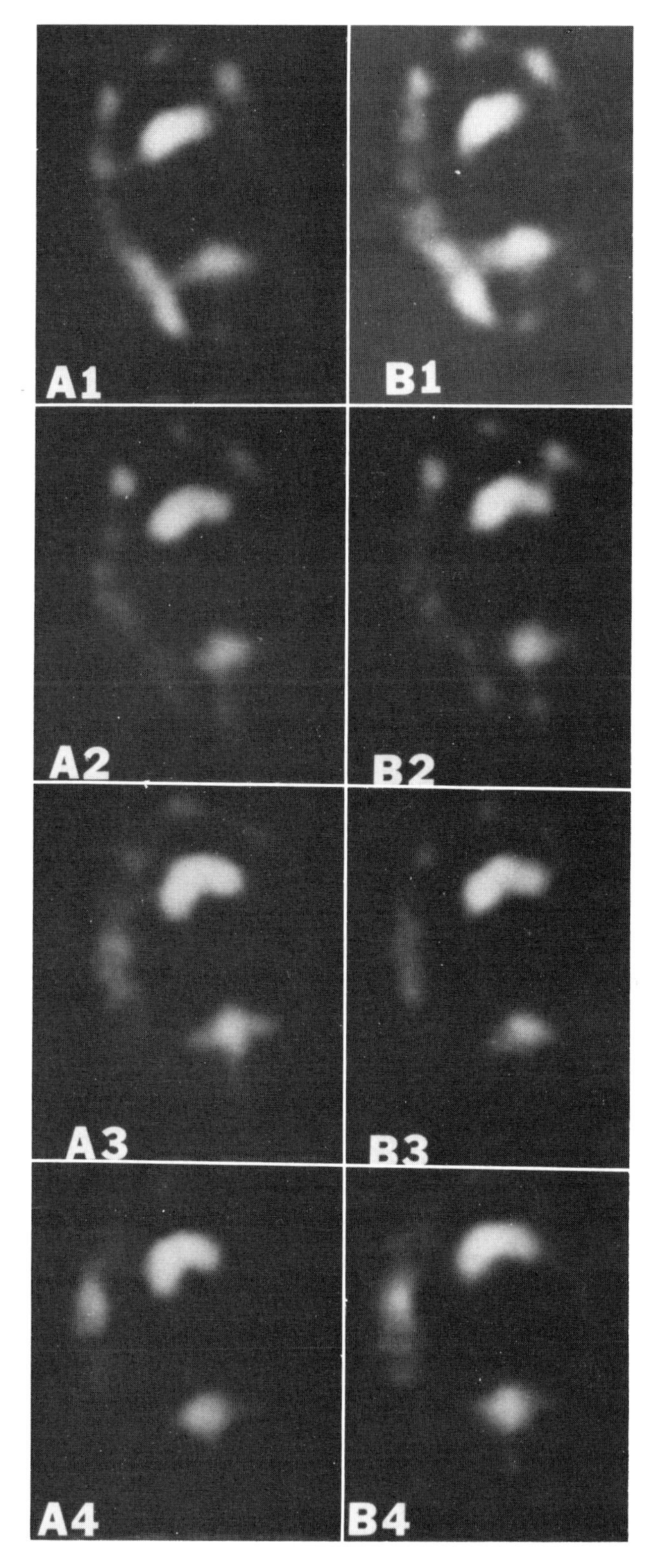

Fig. 2-4. Anterior myocardial infarct. Comparison of sections obtained after death, **A** series, with sections obtained ungated during life, **B** series.

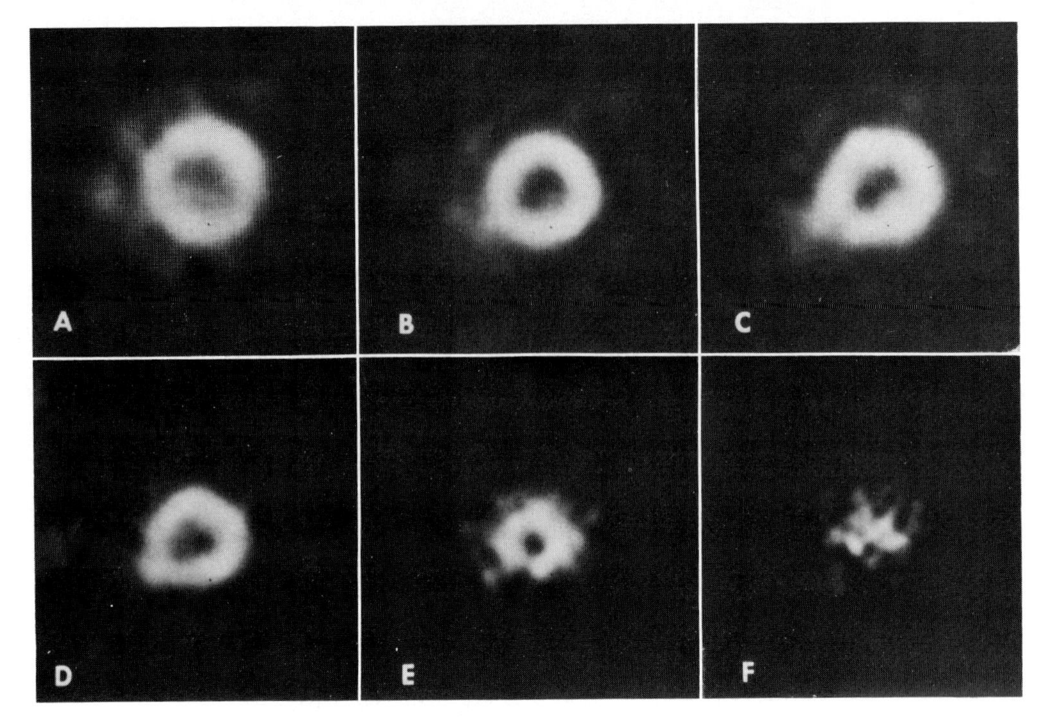

Fig. 2-5. Tomographic images of thallium distribution in normal individual obtained 1.5 hr after exercise using wide field-of-view camera and coded aperture. Images were acquired in 30-degree LAO projection and are viewed from apex. Sections begin near base of heart, 13.5 cm from aperture (**A**) and proceed toward apex in approximately 1 cm steps (**B** to **F**).

number of pinholes. These can have either a small number of fixed pinholes as in the seven-pinhole collimator of Vogel and associates[10a] or a larger number, which are opened and closed in a predetermined sequence, the so-called "coded aperture" as described by Rogers and co-workers.[7a] Although the multiple-pinhole methods use a computer to reconstruct the tomographic images, they are otherwise quite different and generally simpler than the other approaches to ECT that have been discussed. In use, a specially designed collimator is attached to an otherwise standard scintillation camera. Since these special collimators contain several pinholes, each pinhole views the organ of interest from a different angle. The disparity in images produced by each pinhole provides the information necessary to reconstruct tomographic slices corresponding to different depths in the organ in much the same way that the disparity between the images seen by each of our two eyes allows us to perceive depth.

These multiple-pinhole techniques have several advantages for myocardial imaging. They provide high sensitivity because several pinholes view the organ simultaneously. For small organs such as the heart, pinhole imaging provides significantly better resolution than can be achieved with other types of collimators, including those used in other types of ECT. Finally, the compact size and relative simplicity of the multiple-pinhole systems facilitates their use in a clinical environment. The coded aperture systems in particular are readily adaptable to portable gamma cameras and can be used at the patient's bedside in the coronary care unit.

The disadvantages of the multiple-pinhole systems appear to be minimal in clinical use. The separation of tomographic planes is less distinct than in other com-

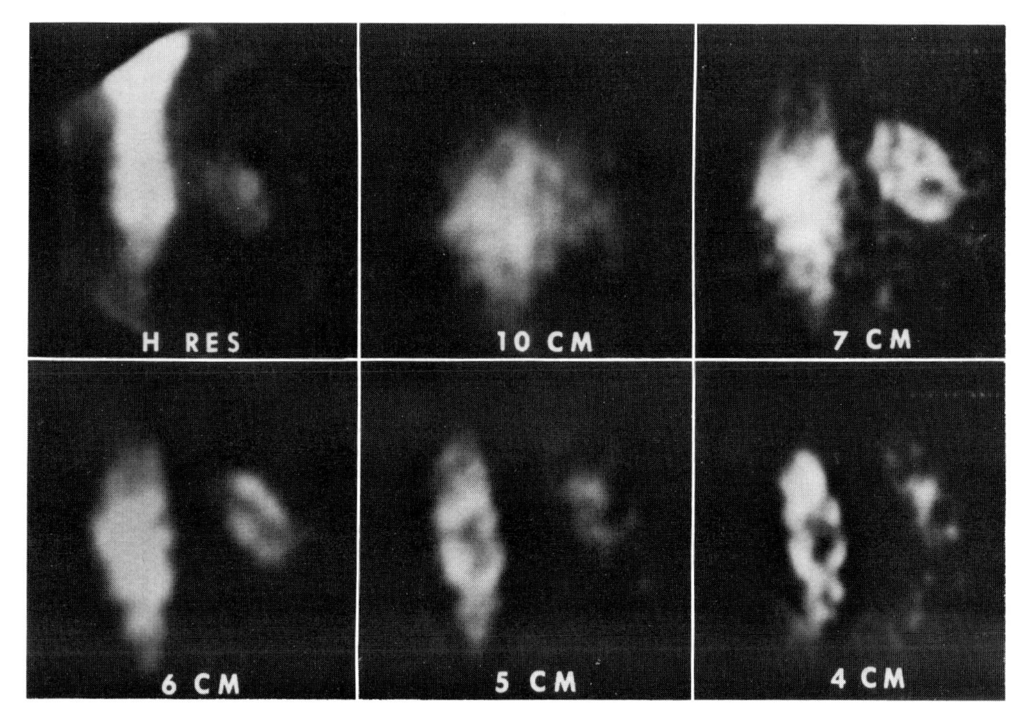

Fig. 2-6. Coded aperture tomograms obtained with portable gamma camera. These images of [^{99m}Tc] pyrophosphate uptake in acute anterolateral myocardial infarction were obtained in anterior projection. *H Res:* Conventional anterior scintigram of same patient for comparison.

puted tomography systems, and image contrast is somewhat less. The utility of the pinhole techniques for quantifying the true amount of radioactivity in an organ remains to be determined. Finally, because of geometric considerations, certain views or projections may not be feasible with these systems.

Figs. 2-5 and 2-6 illustrate representative results used in a coded aperture system in my laboratory. Tomograms of similar quality and character have been demonstrated by Vogel and colleagues[10a] and others with the seven-pinhole collimator.

The multiple-pinhole techniques lend themselves well to several problem areas in cardiac images. As can be seen from Fig. 2-5, high-quality images of a ^{201}Tl distribution are readily obtainable. Work done to date suggests that the high-resolution, three-dimensional nature of these images will materially improve the sensitivity and accuracy of ^{201}Tl studies in the evaluation of ischemic myocardial disease.

Blood pool imaging is another area in which multiple-pinhole tomography seems to hold great promise. The sensitivity of the method allows gated images to be readily obtained, and the tomographic nature of the study should permit greater sensitivity in detecting small areas of abnormal wall motion, better localization of wall motion abnormalities, and perhaps true regional ejection fraction measurements.

SUMMARY AND CONCLUSIONS

Emission computed tomography is the nuclear medicine analog of computed tomography in radiology. It is a technique that can produce a high-quality, high-contrast, spatially and quantitatively accurate image of the distribution of a radionuclide within a section of the living body. A contiguous set of emission computed tomograms represents, in effect, a complete three-dimensional reconstruction of the distribution of radioactivity within a volume of the body.

There are two major approaches to emission computed tomography: techniques employing specialized instrumentation for the coincidence detection of the emitted annihilation photons from positron-emitting radionuclide and single photon systems, which utilize conventional gamma-emitting radionuclides. Both approaches have individual advantages and disadvantages, but at an applied level, both produce essentially equivalent images and both probably have equal capability for estimating myocardial infarct size.

A third approach, based on special collimators containing multiple pinholes, shows great promise for myocardial tomography. These systems produce high-resolution images and are simple enough for bedside use.

The single photon approach is capable of producing high-quality renditions of the distribution of a variety of conventional radionuclides within the myocardium. Tomographic images of the distribution of [^{99m}Tc] pyrophosphate within acute myocardial infarctions give an accurate depiction of the location and configuration of the infarct and can be used to accurately estimate infarct size.

REFERENCES

1. Alton, W. J., Beller, G. A., Gold, H. K., et al.: Three-dimensional myocardial imaging after intracoronary injection of ^{68}Ga-labeled albumin microspheres, Circulation **53, 54** (suppl. II):II-81, 1976.
2. Brooks, R. A., and DiChiro, G.: Principles of computer assisted tomography (CAT) in radiographic and radioisotopic imaging, Phys. Med. Biol. **21:** 689, 1976.
3. Budinger, T. F., Derenzo, S. E., Gullberg, G. T., et al.: Emission computer assisted tomography with single-photon and positron annihilation photon emitters, J. Comp. Assisted Tomo. **1**:131, 1977.
4. Budinger, T. F., and Gullberg, G. T.: Three-dimensional reconstruction in nuclear medicine emission imaging, IEEE Trans. Nucl. Sci. NS-21 **3**:2, 1974.
5. Kay, D. B., Keyes, J. W., Jr., and Simon, W.: Radionuclide tomographic image reconstruction using Fourier transform techniques, J. Nucl. Med. **15**:981, 1974.
6. Keyes, J. W., Jr.: Emission CT—whence, what and where. In Proceedings of Seventh Symposium on Sharing of Computer Programs and Assisted Data Processing, Springfield, Va., 1977, National Technical Information Service, CONF-770101, 278.
7. Keyes, J. W., Jr., Orlandea, N., Heetderks, W. J., et al.: The Humongotron—a scintillation camera transaxial tomograph, J. Nucl. Med. **18**:381, 1977.

7a. Rogers, W. L., Koral, K. F., Mayans, R., et al.: Coded aperture imaging of the heart, J. Nucl. Med. (in press).

8. Shepp, L. A., and Logan, B. F.: The Fourier reconstruction of a head section, IEEE Trans. Nucl. Sci. NS-21 **3**:21, 1974.
9. Ter-Pogossian, M. M.: Limitations of present radionuclide methods in the evaluation of myocardial ischemia and infarction, Circulation **53**(suppl. I):I-119, 1976.
10. Ter-Pogossian, M. M., Phelps, M. E., Brownell, G. L., et al., editors: Reconstruction tomography in diagnostic radiology and nuclear medicine, Baltimore, 1977, University Park Press.

10a. Vogel, R. A., Kirch, D., LeFree, M., et al.: A new method of multiplanar emission tomography using a seven pinhole collimator and an Anger scintillation camera, J. Nucl. Med. **19**:648, 1978.

11. Weiss, E. S., Ahmed, S. A., Welch, M. J., et al.: Quantification of infarction in cross sections of canine myocardium in vivo with positron emission transaxial tomography and ^{11}C-palmitate, Circulation **55**:66, 1977.
12. Weiss, E. S., Hoffman, E. J., Phelps, M. E., et al.: External detection and visualization of myocardial ischemia with ^{11}C-substrates in vitro and in vivo, Circ. Res. **39**:24, 1976.

3 □ Radiopharmaceuticals

Buck A. Rhodes

The radiopharmaceuticals applicable to cardiovascular nuclear medicine can be broadly classified as follows: (1) blood volume indicators; (2) blood flow indicators; (3) blood clot indicators; and (4) infarct indicators, which are preferentially accumulated in ischemic or infarcted muscle. Both volume and flow indicators have been available for some time. Indicators of blood clots and myocardial ischemia or infarction are more recently available. For example, ^{125}I-labeled fibrinogen for clot detection was approved for use in the United States in 1976; [^{99m}Tc] pyrophosphate and other ^{99m}Tc tracers have more recently been approved for infarct scanning.

BLOOD VOLUME INDICATORS

Plasma proteins and blood cells labeled without damage with radioactive tracers are used as blood volume indicators. If damaged by the labeling, plasma proteins are sequestered by the liver and red blood cells by the spleen or liver or both. Serum albumin can be labeled with radioiodine (^{123}I, ^{125}I, or usually ^{131}I), radiochromium (^{51}Cr), or technetium (^{99m}Tc). Fibrinogen can be labeled with radioiodine (^{123}I, ^{131}I, or usually ^{125}I) or ^{99m}Tc. Transferrin can be labeled with radioiron or an isotope of indium (^{111}In or ^{113m}In). Red blood cells can be separated from plasma, labeled with ^{51}Cr, ^{99m}Tc, or ^{203}Pb, and then resuspended and injected. O-negative donor cells have at times been labeled and used as the tracer. Usually, however, the patient's own cells are removed, labeled, and injected. A recent innovation is to tag the cells in vivo with ^{99m}Tc. With these approaches to labeling, cells of various ages are tagged; hence this is referred to as random labeling. When the cells are labeled during their formation, all the tagged cells are the same age; this is referred to as cohort labeling and is accomplished with the use of radioiron (^{52}Fe or ^{59}Fe) or labeled amino acids such as [^{75}Se] selenomethionine. At present, the most commonly used volume indicators are albumin or red blood cells labeled with ^{99m}Tc, transferrin labeled with ^{113m}In, and radioiodinated serum albumin or ^{51}Cr-labeled red blood cells. ^{99m}Tc and ^{113m}In are preferred because they provide high photon yield and low radiation exposure to the patient.

Following is an outline of the uses made of the various blood volume indicators:

I. Measurement of pool size
II. Measurement of pool turnover rates
III. Measurement of hemodynamics
   A. Physiologic studies
      1. Transit times
      2. Cardiac output
      3. Right and left ventricular ejection fraction and velocity
      4. Stroke and end-diastolic volumes
      5. Shunt quantification
      6. Peripheral volume changes with exercise or gravity
   B. Morphologic studies
      1. Cardiac flow patterns to localize extracardiac shunts
      2. Cardiac chamber size and shape
      3. Motion of each chamber
      4. Peripheral venous flow patterns

^{99m}Tc-labeled tracers

Although ^{99m}Tc-labeled human serum albumin was introduced in 1964 by McAfee, Stern, and associates[69] and ^{99m}Tc-labeled erythrocytes in 1967 by Fisher and associates[29], it took several years for these tracers to be developed to the point at which they could be prepared using kits. However, the clinical use of these agents often still requires specialized procedures designed to label the proteins or the cells under sterile, pyrogen-free conditions immediately prior to injection. The adaption of in vivo tagging techniques has simplified the ^{99m}Tc-labeled radiopharmaceutical supply problem.

General principles of ^{99m}Tc labeling. ^{99m}Tc is usually obtained as a solution of sodium pertechnetate ($NaTcO_4$) in 0.09% saline solution, either directly from a commercial supplier or by elution of it from a ^{99m}Mo-^{99m}Tc generator. As the pertechnetate ion, ^{99m}Tc is in its highest and most stable oxidation state (VII). To use ^{99m}Tc as a label, it is first necessary to reduce it to a lower redox state. Redox states varying from III to VI result from reduction reactions; often multiple redox states are produced at the same time. The reaction products depend on the strength of the reducing agent and the affinity and concentration of available complexing groups. ^{99m}Tc, when complexed, can be maintained in these lower redox states. ^{99m}Tc in the intermediate redox states VI and V tends to disproportionate into mixtures of ^{99m}Tc (VII) and ^{99m}Tc (IV). ^{99m}Tc labeling occurs when an intermediate ^{99m}Tc complex undergoes ligand exchange to become bound to a molecule such as albumin. The chemical reactions and formulation steps for preparing ^{99m}Tc-labeled tracers are (1) reduction, (2) complex formation, (3) labeling, (4) purification, and (5) formulation.

Complexes of reduced ^{99m}Tc tend either to hydrolyze, forming insoluble compounds such as technetium dioxide (TcO_2), or to oxidize, reforming pertechnetate ions (TcO_4^-). These two forms of technetium are the most common radiochemical impurities in ^{99m}Tc-labeled radiopharmaceuticals. The insoluble oxides are concentrated in the liver, spleen, and reticuloendothelial system of the bone marrow, whereas the highly soluble pertechnetate ion is concentrated in the thyroid and stomach.

^{99m}Tc-labeled albumin. Steigman and associates[103] have demonstrated that free sulfhydryl groups are essential for the labeling of technetium to albumin. Excess carrier technetium (^{99}Tc) can decrease labeling yields, probably by competing for the sulfhydryl labeling sites. A 25%, salt-poor solution of albumin usually gives the best labeling. Dimerization or denaturation of the albumin either before or during labeling causes blood clearance to be faster. This is undesirable because it prolongs imaging times and increases background radioactivity in the body.

At least four methods are available for preparing ^{99m}Tc-labeled human serum albumin. Most of these have been adapted to labeling kit procedures so that reproducible radiopharmaceutical quality products can be prepared daily for routine clinical use. These methods and their literature sources are listed in Table 3-1. Williams and Deegan have shown that their hydrochloric acid method yields ^{99m}Tc-labeled albumin with a slower blood clearance time than that prepared by the iron–ascorbic acid method.[119]

It has also been shown that albumin tagged by the stannous chloride method has a different tissue distribution than that prepared by the iron–ascorbic acid method (Table 3-2). Electrolytically labeled albumin has been shown to be satisfactory for nuclear cardiac ventriculographic imaging in humans; however, the method is subject to variables that are difficult to control even when using kits. Thus up to 20% of the preparations may not pass quality control. More recently the simpler $SnCl_2$ procedure has been improved, making this the method of choice for routine clinical studies.

Quality control tests for ^{99m}Tc-labeled albumin usually include (1) visual inspection for particulates or cloudiness (indicating denatured protein), (2) pH measurement,

Table 3-1. Methods for preparing ^{99m}Tc-labeled albumin

Method of reduction	Basic references	References to kit procedure
Iron–ascorbic acid	Stern, H. S., McAfee, J. G., and Zolle, I.: In Technetium-99m albumin. In Andrews, G. A., et al., editors: Radioactive pharmaceuticals, United States Atomic Energy Commission Conference 651111, 1966, p. 359.	Cooper, J. F., Stern, H. S., and DeLand, F. H.: A "kit" for preparation of high specific-activity ^{99m}Tc albumin for cisternography and blood pool imaging, Radiology **95:**533, 1970.
Hydrochloric acid	Williams, M. J., and Deegan, T.: ^{99m}Tc-albumin: an assessment of physical methods for measuring labeling efficiency, Int. J. Appl. Radiat. Isot. **22:** 775, 1971.	
Stannous chloride	Eckelman, W. C., Meinken, G., and Richards, P.: ^{99m}Tc–human serum albumin, J. Nucl. Med. **12:**707, 1971.	Eckelman, W. C., Meinken, G., and Richards, P.: High-specific-activity ^{99m}Tc human serum albumin, Radiology **102:**185, 1972.
Electrolytic	Benjamin, P.: A rapid and efficient method of preparing ^{99m}Tc–human serum albumin: its clinical applications, Int. J. Appl. Radiat. Isot. **20:**187, 1965.	Dworkin, H. J., and Gutkowski, R. F.: Rapid closed-system production of ^{99m}Tc-albumin using electrolysis, J. Nucl. Med. **12:**562, 1971.

Table 3-2. Blood clearance of labeled human serum albumin (HSA) in humans and animals

	Percentage of injected dose in blood			
Preparation	Mice (30 min)	Rabbits (60 min)	Dogs (60 min)	Humans (60 min)
^{131}I HSA	63.1*			
^{99m}Tc HSA (iron–ascorbic acid)		52.6†		
^{99m}Tc HSA (electrolytic)	42.4	59.0 ± 6.4		46.0 ± 10.5‖
^{99m}Tc HSA ($SnCl_2$)‡	33.3			
^{99m}Tc HSA ($SnCl_2$) NEN	(43.4)§		58.4 ± 5.3	

*38 at 60 min.
†32 at 60 min.
‡See Rhodes, B. A.: Semin. Nucl. Med. **4:**281, 1974 for more detailed list of values.
§In rats, assuming blood volume is 5% body weight. For mice a blood volume of 7% body weight was assumed, and for dogs, 7.9% was assumed.
‖From Callahan, R. J., McKusick, K. A., Lamson, M., III, et al.: J. Nucl. Med. **17:**47, 1976.

and (3) chromatography for determining free (TcO_4^-) and unbound, reduced (TcO_2) radioactivity. A variety of chromatographic methods have been reported. We prefer to use two silica gel–impregnated glass fiber sheets (0.635 by 6.35 cm). One is developed in acetone to separate TcO_4^- and the other in 0.9% saline solution to separate TcO_2. To detect polymerized proteins or soluble Tc complexes, a column chromatographic method would have to be used. ^{99m}Tc-labeled albumin with greater than 10% free TcO_4^- or less than 85% ^{99m}Tc tagged to albumin can be expected to result in some apparent image degradation.

A test of efficacy is to determine the percentage of injected doses in mice at 30 min. This value approaches 60% in the best preparations; 40% is usually taken as a lower limit for the results of this test.

Results from our laboratory have demonstrated that albumin tagged by the stannous chloride method has a different tissue distribution than that prepared by the iron–ascorbic acid method (Table 3-2). Preliminary results suggest that electrolytically labeled albumin may have the greatest in vivo stability; however, none of the ^{99m}Tc methods have produced a tagged albumin that is as stable as radioiodinated serum albumin.[119]

^{99m}Tc-labeled red blood cells. Although the chemical mechanism of the binding of ^{99m}Tc to cells has not been fully described, it probably parallels that of the binding of ^{51}Cr.[5,92] Gray and Sterling in their studies of chromium labeling demonstrated that the oxidized, anionic chromium as the chromate ion (CrO_4^{-2}) diffuses freely across cell membranes, whereas reduced cationic chromium as the chromic ion (Cr^{+3}) is both nondiffusible and reactive with protein molecules.[33] Thus to be labeled with ^{51}Cr, the cells are incubated with chromate ions, which diffuse into the cells. Once inside, the ions become reduced by intracellular reactions to chromic ions, which then bind to cellular components. In this chemical state, they do not readily diffuse back out of the cells. ^{99m}Tc as anionic pertechnetate ions also diffuses into cells. Pertechnetate, however, is apparently less amenable to intracellular reduction than is chromate; thus the original labeling methods have been difficult to reproduce. Eckelman and associates demonstrated that the method could be made reproducible by increasing the redox potential of the reaction media with small amounts of stannous chloride.[25] Korubin and co-workers showed that the cells labeled by this method are satisfactory for determinations of red blood cell volume in humans.[59] Cooper found that five washes of the ^{99m}Tc-labeled cells were necessary for the method to give results comparable to those obtained with ^{51}Cr-tagged cells (Table 3-3).

The recommended methods using $SnCl_2$ are listed in Table 3-4. More rapid methods are available by employing reagent kits.

Table 3-3. Comparison of red blood cell mass measurements of two methods in ten patients*

Method	Sampling time after injection (min)	Percent ± standard deviation compared to 15 min ^{51}Cr method
^{51}Cr	15	100
^{51}Cr	180	102.4 ± 5.7
^{99m}Tc	15	101.7 ± 2.5
^{99m}Tc	180	105.4 ± 2.4

*Data provided by Dr. Malcom Cooper, University of Maryland Hospital, Baltimore, Md.

Table 3-4. ^{99m}Tc labeling of red blood cells in vitro

For blood volume studies	For blood flow studies
1. Draw 8.5 ml of blood from patient into a 10 ml syringe containing 1.5 ml of ACD solution* and mix.	
2. Transfer blood into a 20 ml sterile serum vial and centrifuge 10 min at 2500 rpm.	
3. Remove plasma and top layer of cells, leaving 4 ml of RBC.	3. Remove plasma and top layer of cells, leaving 2 ml of RBC.
4. Add 100 μCi of ^{99m}Tc in 0.1 to 2.0 ml of saline solution and mix gently.	4. Add 100 mCi of ^{99m}Tc in 0.1 to 2.0 ml of saline solution and mix gently.
5. Add 0.1 ml of 1 mg/ml of $SnCl_2 \cdot 2H_2O$ in saline solution (freshly prepared and passed through a 0.22μg Millipore filter); mix gently and let stand 5 min.	
6. Add approximately 15 ml of saline solution, mix, centrifuge, and remove saline solution.	
7. Wash cells 2 times with 15 ml or 4 times with 5 ml of saline solution.	7. Wash cells 1 time with 15 ml of saline solution.
8. Suspend cells in 10 ml of saline solution.	8. Suspend cells in 2 ml of saline solution.

*Catalog no. 6761, Abbott Laboratories, North Chicago, Ill.

Smith and Richards[99] have developed a kit using a stannous citrate; Gutkowski and coworkers[37] use stannous glucoheptonate; Ducassau and colleagues[21] use stannous pyrophosphate. The cells can be labeled in vivo by administering stannous salts intravenously prior to the administration of pertechnetate.[386]

By far, the most common method of labeling the blood for studies of cardiac function at this time utilizes in vivo labeling of red blood cells. This method was developed from the serendipitous observation that, following a bone scan, a brain scan performed with ^{99m}Tc demonstrated the vascular anatomy with unusual clarity. Further evaluation of this finding led to the conclusion that the red blood cells were labeled.[78a] Studies performed by Thrall and his associates[108a] indicate that the quality of images recorded with in vivo labeled red cells as determined by blood pool/lung background equals or exceeds that from ^{99m}Tc-labeled albumin.

Following is the method for ^{99m}Tc labeling of red blood cells in vivo:

1. Dissolve contents of stannous pyrophosphate of diphosphonate in 5 ml of 0.9% saline solution.
2. Inject 3 to 5 ml intravenously* (sufficient to deliver 3 to 5 mg stannous ions).
3. Wait 15 to 30 min.
4. Inject $^{99m}TcO_4^-$ intravenously.

^{113m}In-labeled transferrin

Shortly after ^{113m}In was introduced as a short-lived ($t_{1/2}$ = 1.7 hr) generator-produced isotope for tracer studies,[105] Stern and associates demonstrated that when carrier-free solutions of ^{113m}In at pH less than 3.5 were injected intravenously, the radioactivity remained in the bloodstream.[104] Subsequently, it was shown that the ionic ^{113m}In was bound to plasma transferrin,[47] producing a tag sufficiently stable to permit plasma volume measurements.[48,122] The low radiation dosage and the lack of placental transfer of indium make this tracer particularly suitable for placental localization.[49,122,123]

Ionic ^{113m}In solutions suitable for intravenous injections in small volumes are obtained directly from sterile, pyrogen-free ^{113}Sn-^{113m}In generators. Since the eluting solution is slightly acidic (0.05 NHCl), the tracer is injected with caution to avoid a localized burning sensation if some of the solution is extravasated. Usually blood is pulled into the syringe, allowed to mix with the solution, and then reinjected. This permits neutralization of the acid by the blood and labeling of the plasma transferrin immediately prior to injection. The solution can also be partially neutralized with buffers provided that the pH is not allowed to exceed 3.5, the point at which insoluble indium hydroxide begins to form. Several methods for preparing partially neutralized ionic indium radiopharmaceuticals have been reported;[1,11,18,104] however, these are rarely used, since the generator eluate can be injected directly without chemical alteration.

Choosing the indicator

For a simple red blood cell volume determination, 30 μCi of ^{51}Cr provides adequate radioactivity for the measurement of blood volume with 1 to 10 ml samples.[113] This tracer has been used routinely for a number of years because of its long shelf life, its low cost, and tradition. Since the introduction of reliable methods for ^{99m}Tc labeling of red blood cells, this tracer may replace ^{51}Cr for routine use. Some further advantages of ^{99m}Tc are that larger doses of radioactivity can be used and that blood pool scanning is readily accomplished with it.

For cardiac volume studies, ^{99m}Tc-labeled red blood cells are probably used more than any other tracer.

BLOOD FLOW INDICATORS

A useful classification of blood flow indicators is based on the amount of the tracer

*This injection must be made directly into a vein. Many drugs or IV solutions appear to interfere with binding when the solution is injected through IV tubing, but they do not interfere with a direct IV injection.

Table 3-5. Blood flow tracers

Extent of extraction during first transit	Type of radiopharmaceutical
Complete	1. Particles removed by capillary blockade
	a. Microspheres
	b. Macroaggregated albumin
	c. Inorganic flocs
	2. Myocardial scanning agents
	3. Diffusible ions such as pertechnetate
	4. Chelates, etc.
	5. Noble gases
↓	6. Blood proteins or cells
None	7. Particles too small to be removed by capillary blockade

extracted from the bloodstream during the first transit of the tracer through a capillary bed. On one extreme are tracers such as radioactive microspheres greater than 15 μm in diameter that are completely extracted from the blood as it flows into the capillaries. On the other extreme are tracers such as tagged red blood cells in which approximately 100% of the radioactivity flows through the capillaries. Intermediate between these two extremes are various tracers that are partially extracted. Alkaline metals such as K^+ have high extraction efficiencies as the tracers pass through muscle capillaries, whereas labeled proteins have low extraction efficiencies. Anions such as iodide and pertechnetate and noble gases such as radioxenon have intermediate extraction efficiencies (Table 3-5). When a radiation detector is placed over a capillary bed and a bolus of a tracer is injected upstream, radioactive time curves can be generated. Curves for each type of tracer are shown in Fig. 3-1. Recirculation of nonextracted and poorly extracted tracers gives secondary and sometimes tertiary peaks on the curve. When a tracer is extracted from the blood by some other mechanism, such as the removal of colloids by phagocytosis, recirculation is minimized, as illustrated by the colloid curve in Fig. 3-1. This phenomenon can be used to advantage when repeat studies are required.

The uses of radioactive blood flow indicators range from the gross visualization of patency of large vessels in nuclear angiography to the quantitative determination of regional blood flow. The blood flow studies in which these indicators are used are as follows: (1) nuclear angiography, for visualiza ion of flow channels and patterns; (2) myocardial scanning, for visualization of relative regional perfusion; and (3) clearance determination, for quantification of nutrition flow. If the time-activity data are collected as a series of gamma camera images and stored as a series of numerical matrices, then the data can be reduced by a digital computer and displayed as a functional image of the particular flow parameters that are diagnostically useful. This principle was first illustrated by Kaihara and associates.[53]

A further use of radioactive blood flow indicators is in the separation of nutritional flow from total flow through capillary beds. One way of accomplishing this is to use two tracers with different extraction (or diffusion) coefficients and different gamma-ray energies. The tracers are mixed and injected together as a bolus. The flow curve for each tracer is determined. Then from the solution of a set of simultaneous equations, the two components of flow can be determined. Chien described this method and included the use of a third nonextractable tracer to permit correction for the distribution of the intravascular transits.[15] Another way of accomplishing this is to separate arteriovenous shunt flow from total flow by

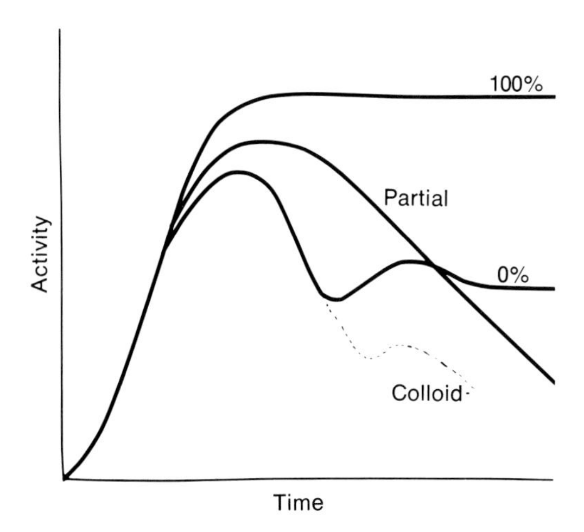

Fig. 3-1. Radioactivity from capillary bed as function of time after the upstream injection of bolus of radioactive blood flow tracer. The 100% curve results from complete trapping of tracer in capillary and precapillary vessels, as in case of radioactive microspheres. The partial curve results from incomplete removal of tracer by capillary bed, as in case of radioactive chelates. The 0% curve results from no removal of tracer as bolus passes capillary bed, as in case of labeled red blood cells. The colloid curve *(dashed line)* shows that recirculation is decreased when another mechanism such as phagocytosis removes tracer from bloodstream.

determining the fraction of a dose of radioactive microspheres that is not extracted during the first transit through a capillary bed.[84] Because of their size (15 to 30 μm) relative to that of capillaries (less than 10 μm), the microspheres can only escape entrapment by being shunted through arteriovenous anastomoses, whose diameters (estimated to be 60 to 200 μm) are large enough to permit the microspheres to flow through.[88,114]

Albumin microspheres

Albumin microspheres labeled with short-lived tracers were developed to meet the specialized demands of cardiovascular nuclear medicine.[114] The use of microspheres permits control of particle size, number, and rate of metabolism.[6,90,125] Furthermore, because the particles are first prepared, sized, tested for quality, and then labeled as needed just before use, they are a highly reproducible tracer. Since the same

Table 3-6. Radionuclides for tagging albumin microspheres

Tracer	Gamma-ray energy (keV)	$t_{1/2}$
^{99m}Tc	140	6 hr
^{203}Pb	279	52 hr
^{113m}In	392	1.7 hr

particles can be labeled with different gamma-emitting radionuclides, particle-size distribution can be held constant, with gamma-photon energy being the only significant variable in the different microsphere preparations. This permits dual-tracer studies for the evaluation of myocardial[52] and extremity arterial perfusion patterns.[97] Table 3-6 lists the three principle gamma-emitting nuclides that have been used for labeling the microspheres. Albumin microspheres are preferred over ^{131}I- or ^{99m}Tc-labeled macroaggregated albumin because of the inherent variability of macroaggregates. This variability was carefully documented in 1973 by the Isotope Pharmacy of the National Health Service of Denmark.[75] A disadvantage to the use of albumin microspheres is the slightly higher probability of adverse reactions.[30a]

Labeling of albumin microspheres with ^{99m}Tc is usually performed with commercially available labeling kits. These kits provide 5 mg of 10 to 35 μm microspheres that can be labeled with unlimited amounts of carrier-free ^{99m}Tc (approximately 750,000 particles). Depending on the number of patients to be studied per day and the types of studies, 5 to 250 mCi of radioactivity is used for labeling. The tagged particles are finally suspended in 10 ml of saline solution to give a concentration of 0.5 mg (75,000 particles)/ml with a specific activity of 1 to 25 mCi/mg. For lung perfusion studies in adults, the administered dose is kept above 0.5 ml (37,500 particles) to avoid image degradation caused by too few particles. Satisfactory myocardial images have been obtained with less than one tenth this number of particles injected directly into one of the coronary arteries. The ^{99m}Tc label is slowly eluted from the suspended microspheres; thus after 4 to 6 hours at room temperature, as much as 10% nonbound radioactivity may be present in the suspending solution. This is usually not a problem in lung scanning, but it is a particular disadvantage in quantitative estimations of shunting. Therefore, immediately prior to such studies, the suspending solution is replaced. The ^{99m}Tc that leaches from the microspheres during standing behaves biologically as a blood pool tracer, that is, its distribution and blood clearance approximate those of ^{99m}Tc-labeled human serum albumin.

Labeling kits for tagging microspheres with radioindium or radiolead are not generally available; hence, for dual microsphere tracer studies, it is necessary to carry out the tagging reactions in the radiopharmacy. Several methods for indium labeling have been reported.[9,10,83] The method currently used to prepare high-specific activity ^{113m}In-labeled microspheres for myocardial scanning is as follows[38a]:

1. Add 1 ml of 5% sodium acetate to a vial containing 5 mg of 3M instant microspheres.
2. Add 3 to 8 ml of ^{113m}In in 0.05N HCl (as it comes from the generator).
3. Sonicate 15 to 30 sec in an ultrasonic bath.
4. Heat for 5 min at 100° C.
5. Sonicate for 5 min in an ultrasonic bath.

The ^{203}Pb method is as follows:

1. Add 3 to 10 ml of ^{203}Pb to an empty vial.
2. Add 0.5 ml (1 mg) of microspheres suspended in acetate buffer.*
3. Add 0.3 to 1.0 ml of 1N NaOH to exactly neutralize the acetic acid added in step 1; that is, if 5 ml of ^{203}Pb solution was used, add 0.5 ml of NaOH.
4. Incubate 10 min in a boiling water bath.
5. Remove tagging solution after transferring reaction mixture to a Hopkins tagging vial.* Wash microspheres once with saline–Tween 80 solution† and then resuspend the tagged microspheres in the desired volume of saline–Tween 80 solution.

Problems have been encountered with the lead-labeling procedure when chemical impurities in the original lead solution were high enough to precipitate in basic solution.

Myocardial scanning agents

The myocardium is readily visualized by scanning or imaging devices when a particulate tracer such as ^{131}I-labeled macroaggregated human serum albumin (MAA) or ^{99m}Tc-labeled microspheres is injected directly into the coronary arteries or into the left side of the heart. Intra-arterial injections are usually avoided except when they can be carried out in conjunction with other invasive techniques such as contrast angiography.

The myocardium can be visualized after the intravenous administration of tracers that have a relatively high extraction efficiency during their first transit through the muscle. For example, an intravenous dose of radiopotassium or radiothallium is extracted by muscle, liver, and kidney tissues during its transit in proportion to the distribution of cardiac output to these tissues.[96] Once intracellular, the tracer becomes diluted by the high concentration of intracellular potassium; hence the clearance rate of tracer potassium or thallium from the myocardium is much slower. The uptake is approximately 70% of the amount of tracer in a single transit through the capillary bed

*Suspend 10 mg of 15 to 30 μg albumin microspheres in a 0.1 M acetate buffer. To prepare buffer, adjust 0.1 M sodium acetate solution to pH 10.0 with 0.1 M NaOH.

*Wheaton-Hopkins tagging vial, catalog no. 655725, Wheaton Glass Company, Millville, N.J.

†To prepare saline–Tween 80 solution, add 0.1 ml of Tween 80 to 100 ml of 0.9% saline solution and adjust pH to 10.0 with 0.1 M NaOH.

for potassium and 88% for thallium. The efflux rate is sufficiently slow that a period of 6 to 18 hr is required before the tracer potassium eventually comes into equilibrium with total body potassium and far longer with thallium.[76] Thus, although total cellular influx and efflux rates of potassium are equal, the relative dilution effect of the tracer potassium results in its being moved intracellularly much faster than it is discharged from the cells. For imaging, 2 mCi of $^{201}TlCl$ in 0.9% saline solution is administered. Imaging can begin almost immediately.[51]

Cations of other radioactive alkaline-earth metals such as Rb^+ and Cs^+ and their analog $^{13}NH_4^+$ are also being evaluated as possible myocardial scanning agents (Chapter 9). Tracer rubidium closely parallels radiopotassium in its initial tissue distribution and its rate of redistribution.[66]

Table 3-7. Myocardium-specific agents

Agent	$t_{1/2}$	Gamma-ray energy (keV) and abundance (%)
$^{13}NH^+$	10 min	510 (200%)
$^{43}K^+$	22 hr	371 (100%)
		614 (91%)
		Others
^{81}Rb	4.7 hr	450 (13%)
^{82m}Rb	6.3 hr	775 (76%)
^{127}Cs	6.2 hr	410 (50%)
		Others
^{128}Cs	32 hr	372 (48%)
		411 (25%)
		548 (5%)
^{201}Tl	73 hr	70-80 (Hg-K_α x-rays 98%)
		135 + 167 (10%)
^{123}I fatty acid	13.3 hr	159 (83%)
^{77}Br fatty acid	57 hr	240 (30%)
		Others

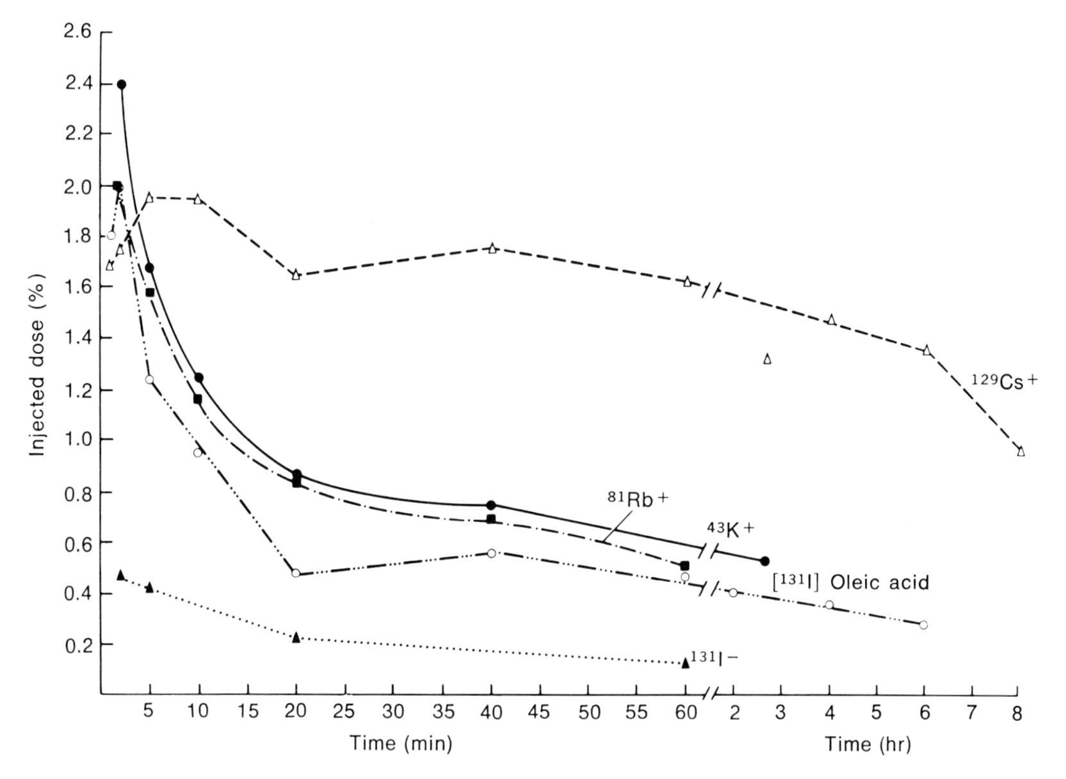

Fig. 3-2. Concentration of various myocardial scanning agents in mouse heart as function of time after intravenous injection of dose. [^{131}I] oleic acid was bound to human serum albumin. [^{131}I] iodide is included for reference. Each point is average of determinations for three animals.

Radiocesium, on the other hand, is probably partially discriminated against by the potassium transmembrane transport system; thus it moves more slowly into the cells, and, once intracellular, it tends to remain there for a longer period of time. Poe documented this difference quantitatively in both dogs and rats.[79] This is illustrated in Fig. 3-2. ^{201}Tl, originally proposed by Lebowitz and associates for use in myocardial scanning, has become the tracer of choice for imaging the normal myocardium. It appears to be intermediate between potassium and cesium in its myocardial uptake and secondary redistribution.[64,2a]

Other tracer substances that have been investigated as potential intravenous myocardial scanning agents include radioiodinated free fatty acids.[36] This approach stems from the observation that the free fatty acids are major nutrients of the myocardium. Oleic acid has been studied most, probably because its doubly bonded carbon group is readily radioiodinated. For intravenous administration, the labeled fatty acid is solubilized by absorption onto serum albumin. Although this approach to myocardial scanning was first investigated in the early 1960s, it has not proved to be clinically useful. The results in Fig. 3-2 show that the tissue distribution of the labeled fatty acid is inferior to that of the alkaline metals. Also, it is difficult to prepare, and the radioiodine is readily lost from the labeled product. The radioiodine also appears to be rapidly lost from the myocardium, and the blood levels of the tracer remain high. Many of the problems associated with halogenated fatty acids have been overcome with the introduction of a terminal halogen. ^{77}Br or ^{123}I 17-haloheptadecanoic acid shows considerable promise as a myocardial imaging agent.[66a]

One of the most exciting areas of investigation with nuclear methods is the evaluation of regional myocardial metabolism using "physiologically labeled" tracers (e.g., labeled with isotopes of nitrogen, carbon, or oxygen), which can be imaged with a positron camera. ^{11}C-labeled palmitic acid has been synthetized and its biological properties in the myocardium investigated by Sobel and his colleagues (Chapter 9).

Diffusible agents

Unlike the other radiopharmaceuticals reviewed in this chapter, a number of diffusible agents have been in clinical use for years. Hence several reviews of their preparation, handling, and use in assessing regional blood flow are already available.[56,63,120] Flow is determined from the measured washout rate either after intratissue deposit of the tracer in a small volume of isotonic saline solution, after an interarterial injection, or after a period of inhalation in the case of the noble gases. Flow can also be determined from measurement of uptake rate.

^{133}Xe, either diluted with air for inhalation or dissolved in isotonic saline solution for injection, has become one of the most widely used of the diffusible tracers. The rapid biologic excretion of this tracer, its favorable gamma-ray energy, and its availability have contributed to its acceptance. ^{81m}Kr, a tracer with a 13 sec half-life, can be obtained from a sterile minigenerator. This tracer is administered through a small-bore catheter. Saline solution is passed through the minigenerator directly into the catheter for the injection.[55]

[^{123}I] iodoantipyrine has more recently been developed for use as a diffusible tracer.[91a,112a]

Choosing the indicator

Diagnostic nuclear medical procedures determine the spatial or time-spatial distribution of a tracer in a patient. The distribution depends on the following types of variables: (1) those determined by the tracer, (2) those determined by the measuring system, and (3) those determined by the patient's physiologic makeup and metabolism. It is crucial to minimize the first two types of variables so that the measurements more sensitively reflect what is occurring as a result of the patient's disease or lack of disease. In general, the simpler, more re-

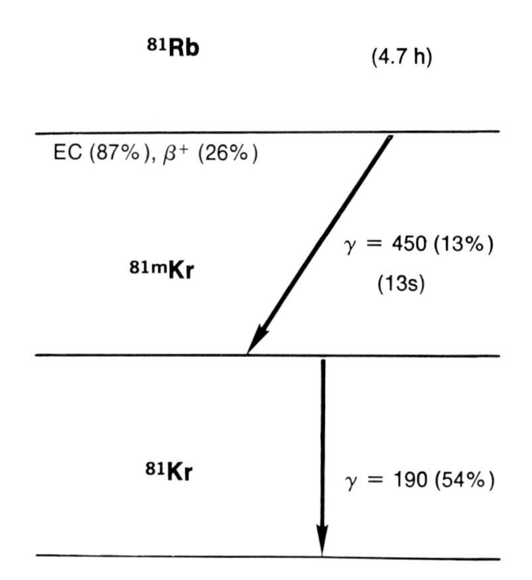

Fig. 3-3. Simplified decay scheme of ^{81}Rb (an alkaline metal) and its daughter ^{81m}Kr (a noble gas). ^{81}Rb parent is potentially useful as myocardial scanning agent. ^{81m}Kr is potentially useful as blood flow indicator. Simultaneous measurement of both radionuclides, that is, both gamma rays, may provide additional functional information about myocardial circulation.

producible the tracer is, the less likely it is to generate artifacts that degrade the sensitivity of diagnostic procedures. Thus tracers such as ^{201}Tl, which can be administered in its most simple chemical form and which traces the kinetics of a critically important biologic substance, tend to be the most ideal tracers. Unfortunately, the nuclear properties of ^{201}Tl are not perfectly matched to available detection systems, and its half-life is not matched to the biologic time course of the events to be measured, such as myocardial uptake. The major photons arising from the mercury x ray are too low to be precisely localized. The half-life of 73 hr is too long for the measurement of myocardial perfusion. These factors encourage exploration for analogs with better economic and nuclear properties. However, to date none have been found. One of the myocardial imaging tracers, ^{81}Rb, has the unique property of decaying to the short-lived diffusible tracer ^{81m}Kr. Thus it can be expected that ^{81}Rb, as indicated by its 450 keV gamma rays, will be initially distributed in proportion to cardiac output, whereas ^{81m}Kr, as indicated by its 190 keV gamma rays, will clear from muscle in proportion to subsequent blood flow. By measuring both gamma rays, it is possible to obtain additional blood flow parameters and increase the diagnostic power of the tracer test. Although this Rb-Kr relationship is true, it is difficult to measure unless high spectral resolution instrumentation is employed. To date, the Rb-Kr system has not been widely used clinically.

Many of the myocardial scanning agents under investigation are cyclotron-produced nuclides, which have the general advantage, when compared to reactor-produced tracers, of being carrier free. However, a disadvantage of these cyclotron-produced tracers is that they may not be entirely free from other radionuclides and trace chemicals. Also, they are usually expensive as a result of both production and distribution problems. $^{201}Tl^{+}$ is currently the myocardial scanning agent of widest choice, despite of its great expense.

BLOOD CLOT INDICATORS

Three general approaches to the use of radioactive tracers in the detection of peripheral blood clots have been proposed. First, blood pool tracers can be used for monitoring the rates and magnitude of venous volume changes in response to gravity or exercise.[28,34,95,110] Thrombosis is indicated by a slowing of the emptying of venous pools in the calf region of the leg when patients are tilted so that the pools are drained by gravity. Blood volume indicators ^{113m}In, ^{99m}Tc-labeled albumin, and ^{99m}Tc-labeled red blood cells, which are discussed in detail in earlier sections of this chapter, are suitable for the required venous hemodynamic measurements. Second, tracers that are involved in blood clot formation are used in the detection of thrombogenesis; ^{125}I-labeled fibrinogen is the notable example for use in this approach. Third, tracers that are involved in fibrinolysis are being explored for the detection and localization of clots with radionuclide imaging. The types of tracer studies used in the study and diagnosis of thrombosis are as follows: (1) lung scanning for the detection of pulmonary emboli, (2) measurement of the relative rates of peripheral venous emptying in testing for venous obstruction, (3) detection of sites of radiofibrinogen and antifibrin accumulation for demonstration of clot formation, and (4) scanning for the detection and localization of preformed blood clots. The substances that have been labeled in attempts to develop suitable radiopharmaceuticals are fibrinogen; plasmin; plasmin-streptokinase complex; antifibrin; kinases such as urokinase, streptokinase, and fibrinokinase (tissue activator); particles; platelets; and leukocytes.

Table 3-8. Normal values for fibrinogen distribution and metabolism*

Parameter	Mean value and standard deviation
Plasma fibrinogen concentration	284 ± 71 mg/100 ml
Total plasma fibrinogen pool	119 ± 40 mg/kg
Plasma fibrinogen as fraction of total body fibrinogen	0.72 ± 0.07
Fibrinogen half-life	4.14 ± 0.56 days
Fractional catabolic rate of plasma fibrinogen pool	0.24 ± 0.04 day^{-1}

*Data from Collen, D., Tytgat, E. N., Claeys, H., and Piessens, R.: Metabolism and distribution of fibrinogen. I. Fibrinogen turnover in physiological conditions in humans, Br. J. Haematol. **22**:681, 1972. The values were for thirty-five healthy subjects.

Radioiodinated fibrinogen

The high death rate from pulmonary embolism[17,74] and the inaccuracy of the clinical assessment of venous thrombosis (Lambie and associates, for example, found a 63% clinical false-negative diagnosis rate[61]) place a high priority on the development of an effective and simple screening test for thrombosis. This has become increasingly important because the results of the use of low-dose heparin for the prevention of thrombosis in high-risk groups have been disappointing.[32,40,54] It was found that a screening test with ^{125}I-labeled fibrinogen detected up to 90% of cases of established deep-vein thrombosis.[8] ^{125}I-labeled fibrinogen has been used as a radiopharmaceutical for several years in England and other countries,[3,30,44] but it was only recently approved for use in the United States because of concern about the transmission of hepatitis. The hepatitis virus can follow fibrinogen through the protein separation procedures.[50]

Fibrinogen is obtained either from the plasma of hepatitis-free donors or from the patient who is to be tested. The fibrinogen is then radioiodinated with care to avoid denaturation, alterations in rate of catabolism, loss of clotability, and aggregation. Krohn and associates presented data suggesting that, of the several methods available, the iodine monochloride method results in the least alteration in the chemical properties of the tagged fibrinogen.[59a] Hagen and coworkers have developed a rapid kit method of labeling a patient's own fibrinogen by this method.[38a] Although ^{125}I has been the preferred radioisotope for clinical studies in the past because of its longer half-life, lower self-radiation, and low radiation exposure, ^{131}I and ^{123}I are being explored to make the test more suitable to scanning procedures.[20,23] ^{123}I-soluble fibrin and highly iodinated fibrinogen show promise as scanning agents. The advantage of highly iodinated fibrinogen is that its blood clearance is increased yet its clotting behavior is preserved.[41a]

Radioiodinated fibrinogen is also useful for the study of fibrinogen kinetics.[88a] Table 3-8 lists recently reported control values.

Labeled clot-binding substances

To be useful for locating blood clots, a radiopharmaceutical should diffuse into the clot and bind irreversibly to it. It should not lyse the clot, at least not during the observation period. It should clear from the bloodstream fast enough that background radioactivity is reduced sufficiently to permit scanning within an hour or so after administration of the tracer, but it should not be cleared so fast that there is insufficient time for the labeled material to diffuse into the clot in appreciable amounts. The tracer should also be thrombus specific.[86] If active clot formation is taking place, the radioiodinated fibrinogen can be used. Since it does not clear rapidly from the blood, scanning is difficult in regions other than the extremities. Later, when the deposition of fibrin is no longer taking place, radiolabeled fibrinogen is not useful. Thus, at later stages of thrombosis, a substance that adheres to or invades the clot is required. The various pathways are being explored in current investigations. Although some of the early reports have been encouraging, an acceptable clinical procedure has not yet evolved.

Leukocytes apparently are involved in the early stages of thrombosis. A relative concentration of leukocytes in clot fifteen times that in the surrounding blood is achieved about 13 hr after the initiation of thrombosis, and the concentration remains at higher levels in clot than in blood for approximately a week.[43] Thus radiolabeled leukocytes are potentially useful in clot localization. The initial studies with ^{51}Cr of Kwaan and Grumet in twelve subjects with deep vein thrombosis are encouraging.[60]

The detection of venous thrombi by radioactive particle adhesion was introduced by Webber and associates in 1969.[115] This technique requires the injection of the radioactive particles downstream from the site of thrombosis. Apparently, as the particles pass the thrombi, some become entrapped because of an electrostatic attraction.[116] Both ^{99m}Tc-labeled microspheres and macroaggregates are being used for isotope venography, usually as adjuncts to lung scanning.

Labeled antibodies to human fibrinogen have been studied by Spar and associates,[100] who demonstrated that the antibodies can be used to detect thrombi by scanning.[101] This method has a number of inherent limitations. Fibrinogen is ubiquitous in the blood; therefore the antibodies will, most likely, all become bound to circulating proteins before they have a chance to react with fibrin of thrombi. Furthermore, fibrin deposits are not limited to thrombi. High levels of radiopharmaceutical in the blood, difficulty in radiopharmaceutical preparation, and the use of a foreign protein are also handicaps to this approach.

Radiolabeled plasmin has also been evaluated with limited success.[19a,32a] Either plasminogen can be labeled and then reacted with something like streptokinase to generate the labeled plasmin, or the streptokinase can be labeled and reacted with plasminogens to generate a labeled plasminogen-streptokinase complex.

It was first shown by Gross that labeled streptokinase labels thrombi.[35] Subsequently, others showed that by this approach experimental thrombi and thromboemboli in dogs could be detected by scanning.[24,28,98] As of the present, however, a suitable technique for routine clinical use in humans has not been established in the United States. Labeled urokinase also has been evaluated as a scanning agent.[116a] Its advantage over streptokinase is that it is a native human protein; however, it appears to have a lower clot-binding affinity than either streptokinase or tissue plasminogen activator.[108] This latter protein, if it could be prepared in radiopharmaceutical quality and quantities by tissue culture synthesis, would probably be the most thrombus-specific tracer of those discussed so far.

Choosing the indicator

The choice with ^{125}I-labeled fibrinogen is whether to use the tracer on all high-risk, postsurgical patients or to treat all high-risk surgical patients with low-dose heparin. Probably many institutions will take an intermediate position and use the ^{125}I-labeled fibrinogen in selected patients.

Often, the choice with isotope venography is when should it be used in combination with the lung scan. Certainly it provides additional information that can be useful. The question is, in which patients is the information worth the additional expense and time required to carry out the procedure?

Radiopharmaceuticals suitable for clot detection by scanning are still under development. Since thrombosis is a dynamic process with the thrombi undergoing a gradual metamorphosis from a platelet-rich gel to a fibrin network that either lyses or gradually becomes organized, epithelialized, and recanalized to finally become scar tissue, probably several types of tracers could be used in providing diagnostic information, depending on the stage of the disease.

INFARCT INDICATORS

The observation that ^{99m}Tc-labeled bone-scanning agents and other ^{99m}Tc complexes accumulate in infarcted muscle has led to the development of a new scanning procedure. These agents provide a direct means of infarct detection by scanning for "hot spots." Areas of ischemia or infarction can also be detected indirectly, using tracers of myocardium perfusion such as ^{201}Tl, which reveal the lesion as a "cold spot." The ^{99m}Tc, or "hot spot," scan delineates irreversibly damaged tissue, whereas the perfusion scan reveals as "cold spots" underperfused regions of the myocardium. Since the two types of images provide different information, both can be valuable in patients with ischemic heart disease.

When $^{99m}TcO_4^-$ is reduced with stannous ions in the presence of tetracycline, glucoheptonate, pyrophosphate, diphosphonate, or methylene diphosphonate, a radiopharmaceutical that will localize in infarcted muscle is produced. The mechanism of localization is not established. It may be related to increases in calcium in the necrotic tissues or to the change in the milieu of ions, which causes the complex to release the reduced technetium where, by ligand exchange, it either precipitates or binds to structural proteins or other nonmobile molecules.

Although a variety of tracers can be used to detect infarcted tissue, more experience has been reported with stannous pyrophosphate. Kits are commercially available for preparing this tracer. The radiopharmaceutical occasionally may be unstable due to oxidation, which causes a buildup of free pertechnetate with time. Also, it can contain insoluble technetium, which accumulates in the liver. Standard quality control tests for free and hydrolyzed technetium can be used to eliminate inferior preparations prior to patient administration.

REFERENCES

1. Adatepe, M. H., Welch, M., Archer, E., Studer, R., and Potchen, E. J.: The laboratory preparation for indium-labeled compounds, J. Nucl. Med. **9:**426, 1968.
2. Arnold, R. W., Subramanian, G., McAfee, J. G., et al.: Comparison of ^{99m}Tc complexes for renal imaging, J. Nucl. Med. **16:**357, 1975.

2a. Atkins, H. L., Budinger, T. F., Lebowitz, E., et al.: Thallium-201 for medical use. Part 3: human distribution and physical imaging properties, J. Nucl. Med. **18:**133, 1977.

3. Atkins, P., and Hawkins, L. A.: Detection of venous thrombosis in the legs, Lancet **3:**1217, 1965.
4. Beierwaltes, W. H., Ice, R. D., Shaw, M. J., et al.: Myocardial uptake of labeled oleic and linoleic acids, J. Nucl. Med. **16:**842, 1975.
5. Berlin, N. I.: Red-cell labeling agents. In Andrews, G. A., et al., editors: Radioactive pharmaceuticals, United States Atomic Energy Commission CONF 651111, 1966, pp. 337-354.
6. Bolles, T. F., Kubiatowicz, D. O., Evans, R. L., Grotenhuis, I. M., and Nora, J. C.: Tc-99m labeled albumin (human) microspheres (15.30 μM): their preparation, properties and uses, International Atomic Energy Agency and World Health Organization Symposium on New Developments in Radiopharmaceuticals and Labeled Compounds, Copenhagen, March 26-30, 1973.
7. Bonte, F. J., Parkey, K. W., Graham, K. D., et al.: A new method for radionuclide imaging of myocardial infarcts, Radiology **110:**473, 1974.
8. Browse, N. L., Clapham, W. F., Croft, D. N., Jones, D. J., Thomas, M. L., and Williams, J. O.: Diagnosis of established deep vein thrombosis with the ^{125}I-fibrinogen uptake test, Br. Med. J. **4:**325, 1971.
9. Buchanan, J. W., Rhodes, B. A., and Wagner, H. N., Jr.: Labeling albumin microspheres with indium-113m, J. Nucl. Med. **10:**487, 1969.
10. Buchanan, J. W., Rhodes, B. A., and Wagner, H. N., Jr.: Labeling iron-free albumin microspheres with indium-113m, J. Nucl. Med. **12:**616, 1971.
11. Burdine, J. A., Jr.: ^{113m}Indium radiopharmaceuticals for multipurpose imaging, Radiology **93:**605, 1969.

12. Callahan, R. J., McKusick, K. A., Lamson, M., III, et al.: Technetium-99m-human serum albumin: evaluation of a commercially produced kit, J. Nucl. Med. **17**:47, 1976.
13. Camin, L. L.: Personal communication, 1976.
14. Carr, E. A., Jr., Gleason, G., Shaw, J., and Krontz, B.: The direct diagnosis of myocardial infarction by photoscanning after administration of cesium-131, Am. Heart J. **68**:627, 1964.
15. Chien, S.: A theory for the quantification of transcapillary exchange in the presence of shunt flow, Circ. Res. **29**:173, 1971.
16. Collen, D., Tytgat, E. N., Claeys, H., and Piessens, R.: Metabolism and distribution of fibrinogen. I. Fibrinogen turnover in physiological conditions in humans, Br. J. Haematol. **22**:681, 1972.
17. Coon, W. W., and Willis, P. W., III: Deep venous thrombosis and pulmonary embolism: predictions, prevention, and treatment, Am. J. Cardiol. **4**:611, 1959.
18. Cooper, J. F., and Wagner, H. N., Jr.: Preparation and control of ^{113m}In radiopharmaceuticals. In Radiopharmaceuticals from generator-produced radionuclides, International Atomic Energy Agency, Vienna, 1971, p. 83.
19. Cooper, M.: Personal communication.
19a. Darte, L., Olsson, C. G., and Persson, R. B. R.: Rapid detection of deep vein thrombosis with ^{99m}Tc-plasmin using hand detector or scintillation camera, Personal communication, 1976.
20. DeNardo, G. L.: Personal communication.
21. Ducassau, D., Arnaud, D., Bardy A., et al.: A new stannous agent kit for labelling red blood cells with ^{99m}Tc and its clinical application, Br. J. Radiol. **49**:344, 1976.
22. Duffy, G. J., D'Auria, D., Brien, T. G., Osmond, D., and Mehigan, S. A.: New radioisotope test for detection of deep venous thrombosis in the legs, Br. Med. J. **1**:712, 1973.
23. Dugan, M. A., Kozar, J. J., III, Charkes, N. D., Maier, W., and Budzynski, A.: The use of iodinated fibrinogen for localization of deep venous thrombi by scintiscanning, Radiology **106**:445, 1973.
24. Dugan, M. A., Kozar, J. J., III, Ganse, G., and Charkes, N. D.: Localization of deep vein thrombi using radioactive streptokinase, J. Nucl. Med. **14**:233, 1973.
25. Eckelman, W. C., Richards, P., Hauser, W., and Atkins, H.: Technetium-labeled red blood cells, J. Nucl. Med. **12**:22, 1971.
26. Endo, M., Yamazaki, T., Konno, S., Hiratsuka, H., Akimoto T., Tanaka, T., and Sakakibara, S.: The direct diagnosis of human myocardial ischemia using ^{131}I-MAA via the selective coronary catheter, Am. Heart J. **80**:498, 1970.
27. Evans, J. R., Gunton, R. W., Baker, R. G., Beanlands, D. S., and Spears, J. C.: Use of radioiodinated fatty acid for photoscans of the heart, Circ. Res. **16**:1, 1965.
28. Fabricant, J. I., Anlyan, W. C., Baylin, G. J., and Isley, J. K.: Isotope studies for evaluation of venous diseases of the lower extermity, J. Nucl. Med. **3**:136, 1962.
29. Fischer, J., Wolf, R., and Leon, A.: Technetium-99m as a label for erythrocytes, J. Nucl. Med. **8**:229, 1967.
30. Flanc, C., Kakkar, V. V., and Clark, M. B.: Detection of venous thrombosis of the legs using ^{125}I-labeled fibrinogen, Br. J. Surg. **55**:742, 1968.
30a. Ford, L., Shroff, A., Benson, W., et al.: SNM drug problem reporting system, J. Nucl. Med. **19**:116, 1978.
31. Frisbie, J. H., Tow, D. E., Sasahara, A. A., et al.: Noninvasive detection of intracardiac thrombosis-131-I fibrinogen cardiac survey, Circulation **53**:988, 1976.
32. Gallus, A. S., Hirsh, J., Tuttle, R. J., Trebilcock, R., O'Brien, S. E., Carroll, J. J., Minden, J. H., and Hudecki, S. M.: Small subcutaneous doses of heparin in prevention of venous thrombosis, N. Engl. J. Med. **288**:545, 1973.
32a. Gomez, R. L., Wheeler, H. B., Belko, J. S., et al.: Observations on the uptake of a radioactive fibrinolytic enzyme by intravascular clots, Ann. Surg. **158**:905, 1963.
33. Gray, S. J., and Sterling, K.: The tagging of red cells and plasma proteins with radioactive chromium, J. Clin. Invest. **29**:1604, 1950.
34. Greyson, N. D., Rhodes, B. A., Williams, G. M., and Wagner, H. N., Jr.: Radiometric detection of venous function and disease, Surg. Gynecol. Obstet. **137**:220, 1973.
35. Gross, R.: Findings with labeled streptokinase in vitro and in vivo. In Bortrag, Proceedings 9th Congress European Society of Haematology, Lisbon, 1963.
36. Gunton, R. W., Evans, J. R., Baker, R. G., Spears, J. C., and Beanlands, D. S.: Demonstration of myocardial infarction by photoscans of the heart in man, Am. J. Cardiol. **16**:482, 1965.
37. Gutkowski, R. F., Dworkin, H. J., Porter, W. C., et al.: Preparation of ^{99m}Tc-labeled red blood cells, J. Nucl. Med. **17**:942, 1976.
38. Hagan, P. L., and Krejcarek, G. E.: Personal communication, 1976.
38a. Hagan, P. L., Krejcarek, G. E., Taylor, A., et al.: A rapid method for the labelling of albumin microspheres with In-113 and In-111: concise communication, J. Nucl. Med. **19**:1055, 1978.
39. Hagan, P. L., Loberg, M. D., Rhodes, B. A., et al.: Kit preparation of radioiodinated autologous fibrinogen using ^{131}I-monochloride, J. Nucl. Med. **15**:974, 1974.
39a. Hamilton, R. G., and Alderson, P. O.: A comparative evaluation of techniques for rapid and efficient in vivo labeling of red cells with [^{99m}Tc] pertechnetate, J. Nucl. Med. **18**:1010, 1977.

40. Handley, A. J.: Low dose heparin after myocardial infarction, Lancet **2**:623, 1972.
41. Harwig, J. F., Coleman, R. E., Harwig, S. S. L., et al.: Highly iodinated fibrinogen: a new thrombus-localizing agent, J. Nucl. Med. **16**:756, 1976.
41a. Harwig, J. F., Harwig, S. S. L., Eichling, J. O., et al.: ^{125}I-labeled soluble fibrin: preparation and comparison with other thrombus imaging agents, Int. J. Appl. Radiat. Isot. **28**:157, 1977.
42. Henkin, R. E., Yao, J. S., Westerman, B. R., Quinn, J. L., III, Bergon, J. J., and McKoveck, L.: Radionuclide venography (RNV) with albumin microspheres, Scientific Exhibit Society of Nuclear Medicine Meeding, June 12-15, 1973.
43. Henry, R. L.: Leukocyte and thrombosis, Thromb. Diath. Haemorrh. **13**:35, 1965.
44. Hobbs, J. T., and Davies, J. W. L.: Detection of venous thrombosis with ^{131}I-labeled fibrinogen in the rabbit, Lancet **2**:134, 1960.
45. Holman, B. L., Dewanjee, M. K., Idoine, J., Fliegel, C. P., Davis, M. A., Treves, S., and Eldh, P.: Detection and localization of experimental myocardial infarction with ^{99m}Tc-tetracycline, J. Nucl. Med. **14**:595, 1973.
46. Holman, B. L., Lesch, M., Zweiman, F. G., et al.: Detection and sizing of acute myocardial infarcts with ^{99m}Tc(Sn) tetracycline, N. Engl. J. Med. **29**(1):159, 1974.
47. Hosain, F., McIntyre, P. A., Poulose, K. P., Stern, H. S., and Wagner, H. N., Jr.: Binding of trace amounts of ionic indium-113m to plasma transferrin, Clin. Chim. Acta **24**:69, 1969.
48. Hosain, P., Hosain, F., Iqbal, Q. M., Carulli, N., and Wagner, H. N., Jr.: Measurements of plasma volume using ^{99m}Tc and ^{113m}In labeled proteins, Br. J. Radiol. **42**:627, 1969.
49. Huddlestun, J. E., Mishkin, F. S., Carter, J. E., DuBois, P. D., and Reese, I. C.: Planental localization by scanning with 113m-indium, Radiology **92**:587, 1969.
50. Hume, M., Sevitt, S., and Thomas, D. P.: Venous thrombosis and pulmonary embolism, Cambridge, Mass., 1970, Harvard University Press.
51. Hurley, P. J., Cooper, M., Reba, R. C., Poggenburg, K., and Wagner, H. N., Jr.: ^{43}KCl: a new radiopharmaceutical for imaging the heart, J. Nucl. Med. **12**:516, 1971.
52. Jansen, C. Quoted by Wagner, H. N., Jr., and Rhodes, B. A.: Radioactive tracers in diagnosis of cardiovascular disease, Progr. Cardiovasc. Dis. **15**:1, 1972.
53. Kaihara, S., Natarajan, T. K., Maynard, C. D., and Wagner, H. N., Jr.: Construction of functional image from spatially localized rate constants obtained from serial camera and rectilinear scanner date, Radiology **93**:1345, 1969.
54. Kakkar, V. V., Corrigan, T., Spindler, J., Fossard, D. P., Flute, P. T., and Crellin, R. Q.: Efficacy of low doses oh heparin in prevention of deep vein thrombosis after major surgery: a double-blind, randomized trial, Lancet **2**:101, 1972.
55. Kaplan, E., and Mayron, L. E.: Evaluation of perfusion with the ^{81}Rb-^{81m}Kr generator, Semin. Nucl. Med. **6**:163, 1976.
56. Kety, S. S.: Theory of blood tissue exchange and its application to measurement of blood flow. In Methods in medical research, vol. 8, Chicago, 1960, Year Book Medical Publishers, Inc., pp. 223-227.
57. Khentigan, A., Garrett, M., Lun, D., et al.: Effects of prior administration of Sn(II) complexes on *in vivo* distribution of ^{99m}Tc-pertechnetate, J. Nucl. Med. **17**:380, 1976.
58. Kitani, K., and Taplin, G. V.: Rapid hepatic turnover of radioactive human serum albumin in sensitized dog, J. Nucl. Med. **15**:938, 1974.
59. Korubin, V., Maisey, M. N., and McIntyre, P. A.: Evaluation of technetium-labeled red cells for determination of red cell volume in man, J. Nucl. Med. **13**:760, 1972.
59a. Krohn, K., Sherman, L., and Welch, M.: Studies of radioiodinated fibrinogen. I. Physiochemical properties of the ICl, chloramine-T and electrolytic reaction products, Biochim. Biophys. Acta **285**:404, 1972.
60. Kwaan, H. C., and Grumet, G.: Clinical use of ^{51}Cr-leukocytes in detection of deep vein thrombosis, Circualtion **46**(suppl. 2):52, 1972.
61. Lambie, J. M., Mahoffy, R. G., Barber, D. C., Karmody, A. M., Scott, M. M., and Matheson, N. A.: Diagnostic accuracy in venous thrombosis, Br. Med. J. **2**:142, 1970.
62. Lamson, M., III, Callahan, R. J., Castronovo, F. P., Jr., et al.: A rapid index of free activity in preparations of ^{99m}Tc-albumin, J. Nucl. Med. **15**:1061, 1974.
63. Lassen, N. A.: Studies of peripheral circulation. In Belcher, E. H., and Vetter, H., editors: Radioisotopes in medical diagnosis, London, 1971, Butterworth & Co., Ltd., pp. 474-499.
64. Lebowitz, E., Green, M. W., Bradley-Moore, P., Atkins, H., Ansari, A., Richards, P., and Belgrave, E.: ^{201}Tl for medical use, J. Nucl. Med. **14**:421, 1973.
65. Lin, M. S., Kruse, S. L., Goodwin, D. A., et al.: Albumin-loading effect: a pitfall in saline paper analysis of ^{99m}Tc-albumin, J. Nucl. Med. **15**:1018, 1974.
66. Love, W. D., Romney, R. B., and Burch, G. E.: Comparison of the distribution of potassium and exchangeable rubidium in the organs of the dog using rubidium-86, Circ. Res. **2**:112, 1954.
66a. Machulla, H.-J., Stöcklin, G., Kupfernagel, C., et al.: Comparative evaluation of fatty acids labeled with C-11, Cl-34m, Br-77, and I-123 for metabolic studies of the myocardium: concise communication, J. Nucl. Med. **19**:298, 1978.
67. Malek, P., Ratusky, J., and Vavrejn, B.: Ischaemia detecting radioactive substances for scanning

cardiac and skeletal muscle, Nature **214:**1130, 1967.
68. Mayron, L. W., Friedman, A. M., Kaplan, E., et al.: A sterile, multicomponent ^{81}Rb-^{81m}Kr minigenerator and delivery system, Int. J. Nucl. Med. Biol. **2:**141, 1975.
69. McAfee, J. G., Stern, H. S., Fueger, G. F., Boggish, M. S., Holzman, G. B., and Zolle, I.: ^{99m}Tc-labeled serum albumin for scintillation scanning of the placenta, J. Nucl. Med. **5:**936, 1964.
70. McLaughlin, P., Coates, G., Wood, D., et al.: Detection of acute myocardial infarction by technetium-99m polyphosphate, Am. J. Cardiol. **35:**390, 1975.
71. McLean, J. R., and Welsh, W. J.: Measurement of unbound ^{99m}Tc in ^{99m}Tc-labeled human serum albumin, J. Nucl. Med. **17:**758, 1976.
72. McRae, J., Sugar, R. M., Shipley, B., et al: Alterations in tissue distribution of ^{99m}Tc-pertechnetate in rats given stannous tin, J. Nucl. Med. **15:**151, 1974.
73. Meinken, G., Stivastava, S. C., Smith, T., et al.: Is there a "good" Tc-99m-albumin? abstracted, J. Nucl. Med. **17:**537, 1976.
74. Morrell, M. T., Truelane, S. C., and Barr, A.: Pulmonary embolism, Br. Med. J. **2:**830, 1963.
75. Müller, T., and Andersen, K. W.: Quality control problems with particle suspensions for lung scintigraphy, International Atomic Energy Agency and World Health Organization Symposium on New Developments in Radiopharmaceuticals and Labeled Compounds, Copenhagen, March 1973.
76. Noonan, T. R., Fenn, W. O., and Haege, L.: The distribution of injected radioactive potassium in rats, Am. J. Physiol. **132:**474, 1941.
77. Parkey, R. W., Bonte, F. J., Meyer, S. L., et al.: A new method for radionuclide imaging of acute myocardial infarction in humans, Circulation **50:**540, 1974.
78. Parkey, R. W., Bonte, F. J., Stokely, E. M., et al.: Acute myocardial infarction imaged with ^{99m}Tc-stannous pyrophosphate and ^{201}Tl: a clinical evaluation, J. Nucl. Med. **17:**771, 1976.
78a. Pavrel, D. G., Zimmer, A. M., and Patterson, V. N.: In-vivo labeling of red blood cells with ^{99m}Tc: a new approach to blood poor visualization, J. Nucl. Med. **18:**305, 1977.
79. Poe, N. D.: Comparative myocardial uptake and clearance characteristics of potassium and cesium, J. Nucl. Med. **13:**557, 1972.
80. Poe, N. D., Robinson, G. D., Jr., Graham, L. S., et al.: Experimental basis for myocardial imaging with ^{123}I-labeled hexadecenoic acid, J. Nucl. Med. **17:**1077, 1976.
81. Porter, W. C., Dworkin, H. J., and Gutkowski, R. F.: The effect of carrier technetium in the preparation of ^{99m}Tc-human serum albumin, J. Nucl. Med. **17:**704, 1976.
82. Prokop, E. K., Strauss, H. W., Shaw, J., et al.: Comparison of regional myocardial perfusion determined by ionic potassium-43 to that determined by microspheres, Circulation **50:**978, 1974.
83. Raban, P., Gregora, V., Sindelor, J., and Alvarez-Cervera, J.: Two alternate techniques of labeling iron-free albumin microspheres with ^{99m}Tc and ^{113m}In, J. Nucl. Med. **14:**344, 1973.
84. Rhodes, B. A.: Blood flow through arteriovenous anastomoses. In Horst, W., editor: Frontiers of nuclear medicine, Berlin, 1971, Springer-Verlag, p. 262.
85. Rhodes, B. A.: Considerations in the radiolabeling of albumin, Semin. Nucl. Med. **4:**281, 1974.
86. Rhodes, B. A., Bell, W. R., Malmud, L. S., Siegel, M. E., and Wagner, H. N., Jr.: Labeling and testing of urokinase and streptokinase: new tracers for detection of thromboemboli, International Atomic Energy Agency and World Health Organization Symposium on New Developments in Radiopharmaceuticals and Labeled Compounds, Copenhagen, March 26-30, 1973.
87. Rhodes, B. A., Bell. W. R., and Som, P.: Tracer methods for the detection of blood clots. In Subramanian, G., Rhodes, B. A., Cooper, J. F., and Sodd, V., editors: Radiopharmaceuticals, New York, 1975, Society of Nuclear Medicine.
88. Rhodes, B. A., Rutherford, R. B., Lopez-Majano, V., Greyson, N. D., and Wagner, H. N., Jr.: Arteriovenous shunt measurements in extremities, J. Nucl. Med. **13:**357, 1972.
88a. Rhodes, B. A., and Smith, J.: Radiotracer techniques to study fibrin deposition and lysis. In Collombetti, L. G., editor: Radiopharmacology, Cleveland, 1979, CRC Press.
89. Rhodes, B. A., Turahi, K. S., Bell, W. R., and Wagner, H. N., Jr.: Radioactive urokinase for blood clot scanning, J. Nucl. Med. **13:**646, 1972.
90. Rhodes, Z. A., Zolle, I., Buchanan, J. W., and Wagner, H. N., Jr.: Radioactive microspheres for the study of the circulation, Radiology **92:**1453, 1969.
91. Robinson, G. D., Jr., and Lee, A. W.: Radioiodinated fatty acids for heart imaging: iodine monochloride addition compared with iodide replacement labeling, J. Nucl. Med. **16:**17, 1975.
91a. Robinson, G. D., Jr., and Lee, A. W.: Kit method for preparation of 4-iodoantipyrine (I-123) from Na ^{123}I: concise communication, J. Nucl. Med. **17:**1093, 1976.
92. Rollinson, C. L.: Problems of chromium reactions. In Andrews, G. A., et al., editors: Radioactive pharmaceuticals, United States Atomic Energy Commission CONF 651111, 1966, pp. 429-446.
93. Rosenthal, L., and Greyson, N. D.: Observations on the use of ^{99m}Tc-albumin macroaggregates for detection of thrombophlebitis, Radiology **94:**413, 1970.
94. Rossman, D. J., Strauss, H. W., Siegel, M. E., et al.: Accumulation of ^{99m}Tc-glucoheptonate in acutely infarcted myocardium, J. Nucl. Med. **16:**875, 1975.

95. Rutherford, R. B., Reddy, C. M. K., Walker, A. G., and Wagner, H. N., Jr.: A new quantitative method of assessing the function status of the leg veins, Am. J. Surg. **122:**594, 1971.
96. Sapirstein, L. A.: Fractionation of the cardiac output of rats with isotopic potassium, Circ. Res. **4:**689, 1956.
97. Siegel, M. E., Giargiana, F. A., Jr., Rhodes, B. A., White, R. I., Jr., and Wagner, H. N., Jr.: Effect of reactive hyperemia on the distribution of radioactive microspheres in patients with peripheral vascular disease, Am. J. Roentgenol. Radium Ther. Nucl. Med. **118:**814, 1973.
98. Siegel, M. E., Malmud, L. S., Rhodes, B. A., Bell, W. S., and Wagner, H. N., Jr.: Scanning of thromboemboli with ^{131}I-streptokinase, Radiology **103:**695, 1972.
99. Smith, T. D., and Richards, P.: A simple kit for the preparation of ^{99m}Tc-labeled red blood cells, J. Nucl. Med. **17:**126, 1976.
100. Spar, I. L., Bole, W. F., and Marrach, D.: ^{131}I-labeled antibodies to human fibrinogen: diagnostic studies and therapeutic trials, Cancer **20:**865, 1967.
101. Spar, I. L., Perry, J. M., Benz, L. L., et al.: Detection of left atrial thrombi: scintillation scanning after administration of ^{131}I rabbit antibodies to human fibrinogen, Am. Heart J. **78:**731, 1969.
102. Spence, R. J., Rhodes, B. A., and Wagner, H. N., Jr.: Regulation of arteriovenous anastomotic and capillary blood flow in the dog leg, Am. J. Physiol. **222:**326, 1972.
103. Steigman, J., Williams, H. P., and Soloman, N. A.: The importance of the protein sulfhydryl group in HSA labeling with technetium-99m, abstracted, J. Nucl. Med. **16:**573, 1975.
104. Stern, H. S., Goodwin, D. A., Scheffel, U., and Wagner, H. N., Jr.: In^{113m} for blood pool and brain scanning, Nucleonics **25:**62, 1967.
105. Stern, H. S., Goodwin, D. A., Wagner, H. N., Jr., and Kramer, H.: In^{113m}—a short-lived isotope for lung scanning, Nucleotides **24:**57, 1966.
106. Strauss, H. W., Harrison, K., Langan, J. K., et al.: Thallium-201 for myocardial imaging: relation of thallium-201 to regional myocardial perfusion, Circulation **51:**641, 1975.
107. Strauss, H. W., and Pitt, B.: Common procedures for the noninvasive determination of regional myocardial perfusion, evaluation of regional wall motion and detection of acute infarction, Am. J. Cardiol. **38:**731, 1976.
108. Thorsen, S., Glas-Greenwalt, P., and Astrup, T.: Differences in the binding of urokinase and tissue plasminogen activator, Thromb. Diath. Haemorrh. **28:**65, 1972.
108a. Thrall, J. H., Freitas, J. E., Swanson, D., et al.: Clinical comparison of cardiac blood pool visualization with technetium-99m labelled red blood cells labelled in vivo and with technetium-99m human serum albumin, J. Nucl. Med. **19:**796, 1978.
109. Tomashelfski, J. F., and Zorn, S.: Pulmonary embolism: a new look at an old problem, Geriatrics **28:**76, 1973.
110. Tow, D. E., Wagner, H. N., Jr., and North, W. A.: Detection of venous obstruction in legs with ^{99m}Tc-albumin, J. Nucl. Med. **8:**277, 1967.
111. Treves, S., and Collins-Nakai, R. L.: Radioactive tracers in congenital heart disease, Am. J. Cardiol. **38:**711, 1976.
112. Urbanek, J., and Graf, M.: Placenta scanning using ^{99m}Tc-albumin and ionic ^{113m}In, J. Nucl. Med. **12:**825, 1971.
112a. Uszler, J. M., Bennett, L. R., Mena, I., et al.: Human CNS perfusion scanning with ^{123}I-iodoantipyrine, Radiology **115:**197, 1975.
113. Wagner, H. N., Jr., editor: Principles of nuclear medicine, Philadelphia, 1968, W. B. Saunders Co., p. 838.
114. Wagner, H. N., Jr., Stern, H. S., Rhodes, B. A., Reba, R. C., Hosain, F., and Zolle, I.: The design and development of new radiopharmaceuticals, International Atomic Energy Agency Symposium on Medical Radioisotope Scintigraphy, Salzburg, Austria, August 6-15, 1968.
115. Webber, M. M., Bennett, L. R., Crogin, M., and Webbs, R., Jr.: Thrombophlebitis—demonstration by scintiscanning, Radiology **92:**620, 1969.
116. Webber, M. M., and Vitery, W.: MAA-studies of electrophoretic mobility and charge: relationship to thrombosis affinity, J. Nucl. Med. **14:**463, 1973.
116a. Weir, G. J., Jr., Roberts, R. C., Wenzel, F. J., and Saulter, R. D.: Visualization of thrombi with technetium-99m urokinase: a negative report, Lancet **2:**341, 1976.
117. Welch, M. J., and Krohn, K. A.: A critical review of radiolabeled fibrinogen: its preparation and use. In Subramanian, G., Rhodes, B. A., Cooper, J. F., Sodd, V., editors: Radiopharmaceuticals, New York, 1975, Society of Nuclear Medicine.
118. Weller, D. A., Adolph, R. J., Wellman, H. N., Carroll, R. G., and Kim, O.: Myocardial perfusion scintigraphy after intracoronary injection of ^{99m}Tc-labeled human albumin microspheres, Circulation **45:**963, 1972.
119. Williams, M. J., and Deegan, T.: ^{99m}Tc-labeled serum albumin in cardiac output and blood volume studies, Thorax **26:**460, 1971.
120. Winberg, M. M.: Use of radioisotopic tracers in the study of nutritional circulation. In Cohen, Y., editor: Radionuclides in pharmacology. Vol. I, International encyclopedia of pharmacology and therapeutics, New York, 1971, Pergamon Press, Inc., pp. 399-488.
121. Wiseman, J., Strauss, H. W., Pitt, B., Rigo, P., Larson, S. M., and Wagner, H. N., Jr.: Gallium-67 citrate for heart scanning in bacterial endocarditis, J. Nucl. Med. **14:**694, 1973.

122. Wochner, R. D., Adatepe, M., van Ambeerg, O., and Potchen, E. J.: New method for estimation of plasma volume with the use of the distribution space of transferrin-113m-indium, J. Lab. Clin. Med. **75:**711, 1970.
123. Wright, F. W.: Placental localization by isotope scanning with ^{113m}In: results in 200 patients, Br. Med. J. **2:**436, 1970.
124. Zaret, B. L., Stenson, R. E., Martin, N. D., et al.: Potassium-43 myocardial perfusion scanning for the noninvasive evaluation of patients with false-positive exercise tests, Circulation **48:**1234, 1973.
125. Zolle, I., Rhodes, B. A., and Wagner, H. N., Jr.: Preparation of metabolizable radioactive human serum albumin microspheres for studies of the circulation, Int. J. Appl. Radiat. Isot. **21:**155, 1970.

4 □ Image display and analysis

Computer methods in nuclear cardiology

D. E. Lieberman

For many years, the justification of nuclear medical computer systems for image enhancement has been debated. But all clinicians definitely agree on one point—computer systems are mandatory for performing nuclear medical cardiac studies. It is clear that the demands of cardiac data acquisition and processing require a minicomputer- or microcomputer-controlled environment. This chapter examines the following topics:

1. General configuration of nuclear medical computer systems
2. Acquisition modes, including multiple-gated acquisition
3. Clinical applications

NUCLEAR MEDICINE COMPUTER SYSTEMS

A complete system for the acquisition and processing of nuclear medical computer studies must provide these basic functions:

1. Acquisition of data from a gamma camera
2. Storage of acquired or processed data on a bulk storage medium
3. Manipulation and processing of data
4. Display of images and results

Fig. 4-1 illustrates a block diagram for a basic nuclear medical data processing system. In general, two types of computer systems exist in the market: "hard-wired" and "software-based" systems. A hard-wired system is preprogramed at the factory, and its programing can only be altered by the manufacturer. The operator communicates with the system using push buttons or switches on the front panel. A "software-based" system uses a minicomputer or a programable microcomputer as the controlling element. The operation of the system can be altered by the operator with a new series of programing instructions placed in the memory of the computer. The operator communicates with the computer primarily by keyboard, although push buttons or switches are sometimes included.

In Fig. 4-1, the organization of a software-based computer system is illustrated. The computer is placed in the center of the diagram, representing control over the flow of information to and/or from other devices contained in the system. The programing instructions that control the computer are placed in its memory using a disk cartridge (pp. 78 to 80), or sometimes by means of a punched paper tape. The discussions in the following sections are primarily oriented toward software-based nuclear medical computer systems.

Acquisition of data

The data from a gamma camera are received at very high rates by the computer. Typical data rates may reach from 10,000 to 200,000 events/sec, depending on the camera in use and the study being performed. The computer must be capable of receiving each individual event, storing the data in memory, and transferring blocks of data to a bulk storage medium such as disk cartridge. If the computer is not capable of processing events as they are received, events will be lost. This loss of data is often called "computer dead time."

The data being received from the gamma

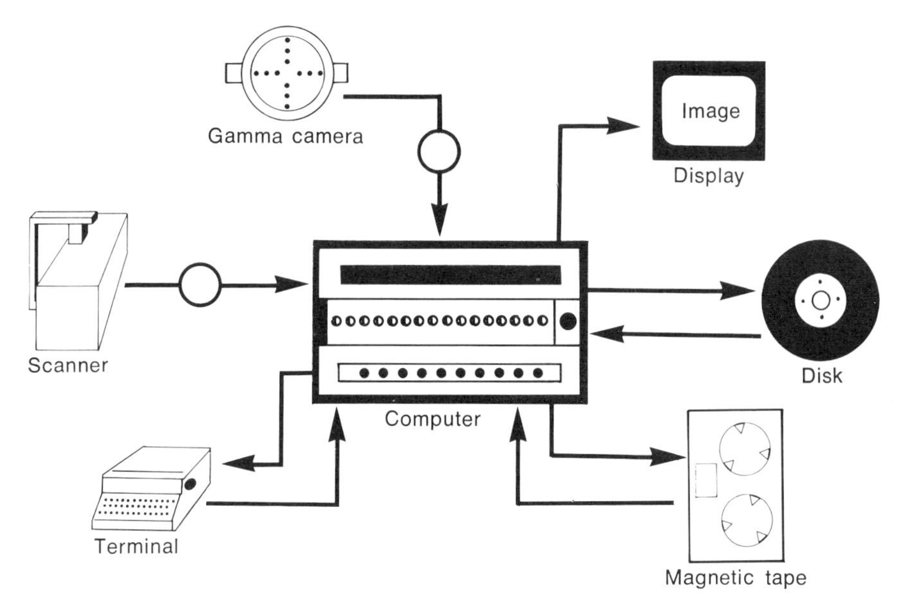

Fig. 4-1. General configuration of nuclear medical computer system.

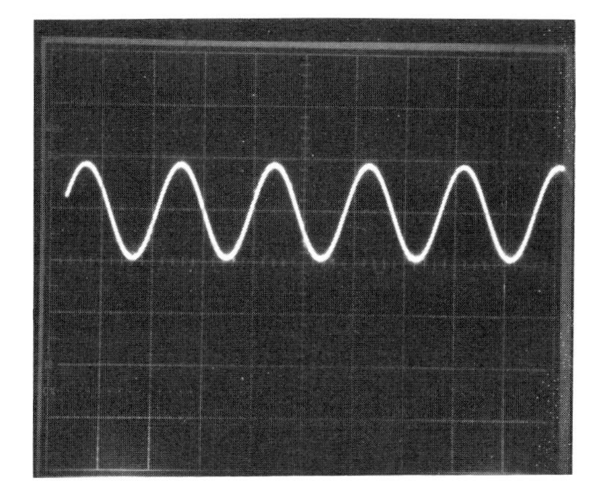

Fig. 4-2. Oscilloscope trace of "sine" wave.

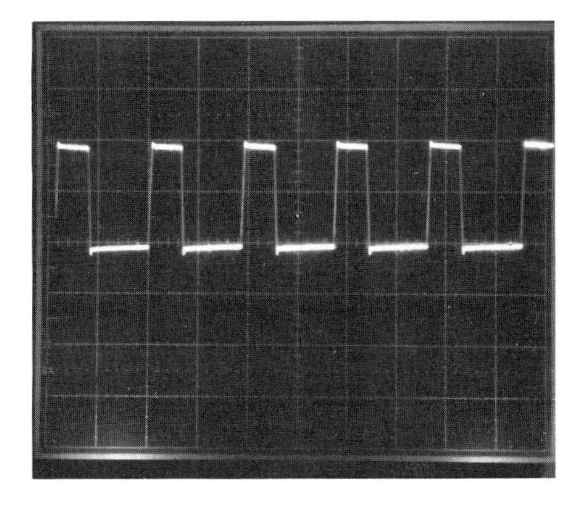

Fig. 4-3. Oscilloscope trace of digital signal.

camera are *not* in a computer-compatible form. These camera signals are referred to as "analog" signals. The amplitudes of analog signals are completely random in nature and may have no predictable characteristics. The computer, on the other hand, accepts only "digital" signals. Digital signals are highly formatted and predictable waveforms and must meet several stringent requirements before they are classified as digital. Fig. 4-2 illustrates an analog signal waveform called a "sine" wave. Note that the amplitude of the signal varies as a function of time. Fig. 4-3 illustrates a digital signal waveform. Note that the amplitude of the signal has one of two possible values. Furthermore, the switching time between one amplitude to the next is extremely small. Digital signals always remain at one of two defined voltage levels and switch quickly between these two levels. If the number "1" is assigned to one voltage level and "0" is assigned to the other, digital signals can be used to represent binary numbers. For this reason, only digital signals are successfully interpreted by the computer.

For the computer to interpret camera signals, a device called the "analog to digital converter" (ADC) is placed between the camera and the computer system. The ADC accepts the analog waveforms from the camera, decodes the signal, and feeds a series of digital equivalents to the computer. The conversion time of the ADC must be fast enough for events not to be lost from the camera.

Generally, the events received from the gamma camera are evenly spaced. Occasionally, however, a rapid burst of activity will result in several events very close together. Most computer systems include a small memory buffer between the camera and the computer to accommodate these rapid bursts of activity. The memory buffer saves the information until the computer can "catch up" with the camera. This process is often called "derandomization."

Storage of data

As data are acquired from the gamma camera, the ADC converts the signals to digital form, and the digital information is then stored in the computer's memory. In the case of dynamic flow studies or list-mode acquisition, the computer's memory is incapable of holding the large volume of data being received from the camera. While acquisition is in progress, the computer must transfer the contents of its memory to a bulk storage device, clear the memory, and resume acquisition of events.

The fastest bulk storage device available in today's market (that is suitable for this application) is the disk. Even so, the transfer process from computer memory to disk is still considered slow when compared to the rate of data coming from the camera. Events from the camera typically arrive approximately every 10 μsec, whereas transfer to a disk may require up to 40 msec. The difference in speed is approximately three orders of magnitude, demonstrating that the disk is much slower than the computer.

The problem of spooling data to disks while acquisition is in progress is solved by dividing the computer's memory in half. While data is acquired into the first half, the second half of memory transfers data to the disk. While data is acquired in the second half, the first half of memory is, in turn, transferred to the disk. Most minicomputer designs enable these two operations to occur simultaneously. This technique is often called "ping-ponging" buffers because the acquisition of data alternates between the halves of memory. Even using this technique, the framing rate of dynamic studies is severely restricted by the slow speed of disks (with respect to the camera's data rates).

Several types of disk systems are available for storage of nuclear medical studies. Fig. 4-4 illustrates a single-platter disk cartridge. Fig. 4-4, *A*, shows the cartridge with the platter removed. The platter is coated with a magnetic recording material, much like cassette tapes. It is on the surface of the platter that data are recorded or retrieved. Fig. 4-4, *B*, illustrates a disk inserted into the drive. The disk drive spins the platter with a rotational velocity in the range of 1500 rpm. A recording head moves radially along the top and bottom surface of the platter, reading or writing information from and/or to the disk. The head floats on a cushion of air above the platter but never touches the magnetic material. If the head *does* come in contact with the platter, the disk fails to operate, and the heads (along with the platter) are usually destroyed. This occurrence is called a "head crash."

Fig. 4-5 illustrates a multiplatter disk configuration. Instead of a single platter, several platters are stacked on a single spindle. When the cartridge is placed in the drive, all platters spin together. Each platter, however, requires its own pair of recording heads to read and write information onto the surfaces of that platter. Usually all heads are also tied to a single spindle. Thus when one recording head is moved into position, all heads move to the same location on their respective platters. A multiplatter disk system has a vastly in-

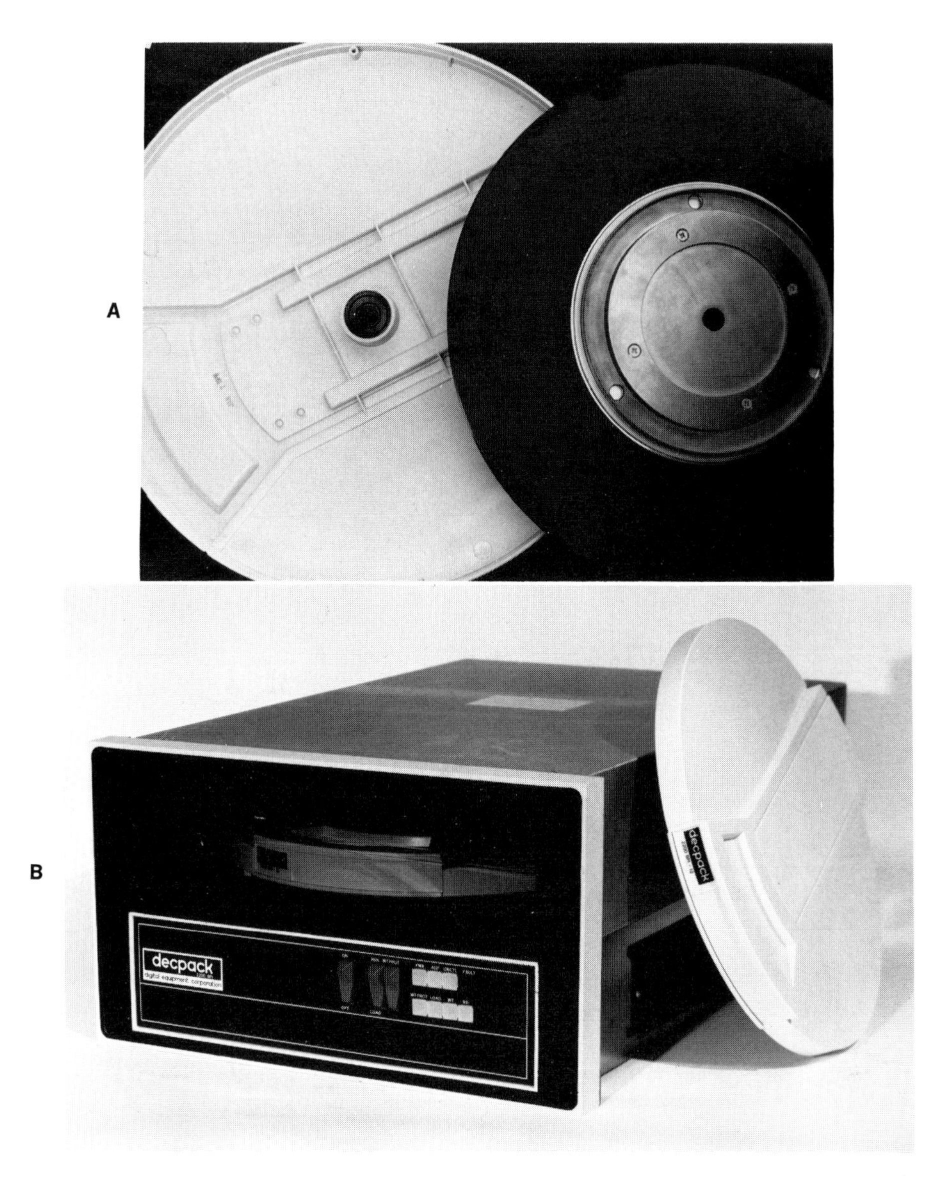

Fig. 4-4. Single-platter disk system. **A,** Platter removed from cartridge holder. **B,** Disk cartridge placed in drive. (Courtesy Digital Equipment Corp.)

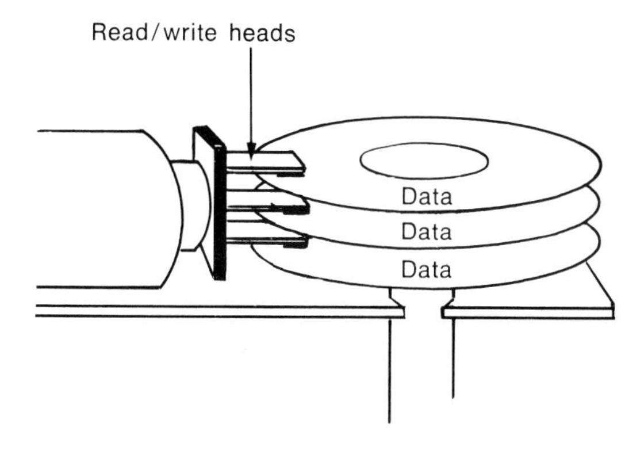

Fig. 4-5. Structure of multiple-platter disk systems.

Fig. 4-6. Floppy disk and disk drive.

creased storage capacity over single platter disk drives and is usually able to store data at higher data rates.

A third form of disk storage is the "floppy" disk. The floppy derives its name from the fact that the platter is actually flexible, or "floppy," in nature. As illustrated in Fig. 4-6, the floppy is physically much smaller than a single-platter disk cartridge—its size closely resembles a 45 rpm music record. The floppy's storage capacity is about 10% of a single-platter disk. Instead of floating above the platter, the recording heads of a floppy disk drive actually come in contact with the surface of the platter. This tends to greatly decrease the data rate and limit the lifetime of the floppy disk cartridge. The floppy disk has proved to be a useful and convenient form of bulk storage for small volumes of data.

Display of data

An important function of nuclear medical computer systems is the presentation of image data on the display screen. Because of the many thousands of count levels that exist in a typical nuclear medical image, the image is not viewed exactly as it is acquired. Instead, a transformation is performed on the image data that groups various ranges of activity in the image and reduces the number of intensity levels used to display the result. This transformation performed on the images is usually invisible to the operator; however, certain parameters of the transformation may be controlled under program operation, and thus digital image enhancement techniques may be applied.

Images are viewed on a display screen or recorded in hard-copy form using a high-resolution printer. The most popular mode of display being used in today's market is the color video display system. Fig. 4-7 illustrates a block diagram of typical color video interface connections. The nuclear medical image is usually transformed into an image containing a fixed number of intensity levels, ranging from 16 to 256. The color video interface scans the image, as-

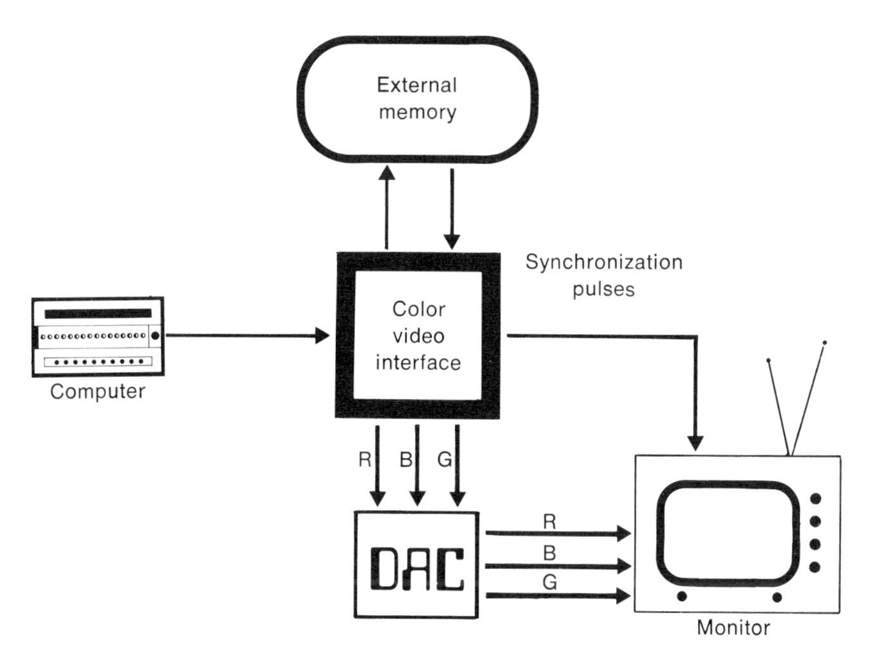

Fig. 4-7. Typical color video interface connections.

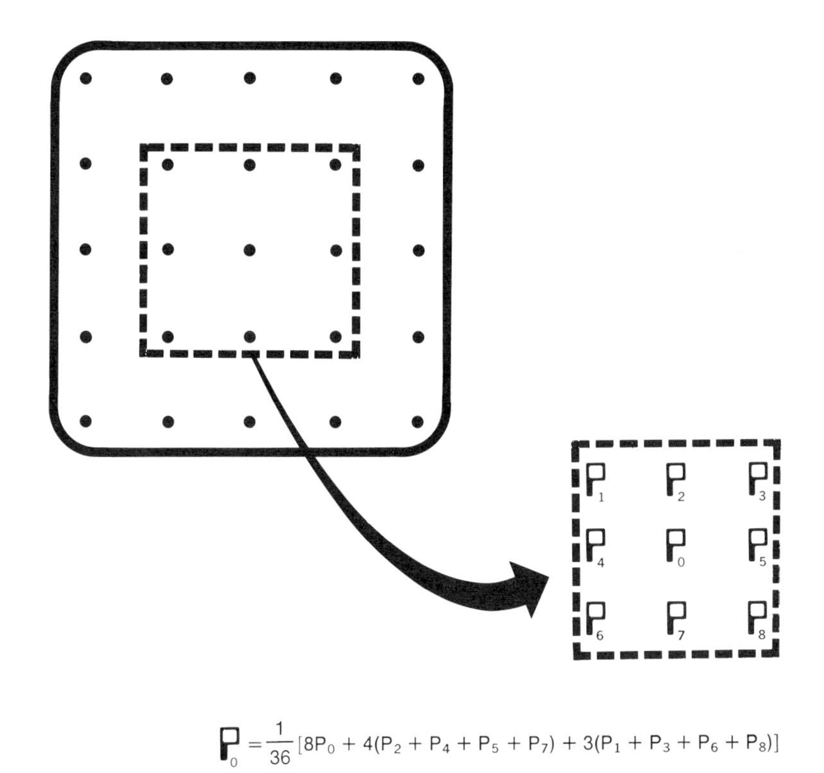

$$P_0 = \frac{1}{36}[8P_0 + 4(P_2 + P_4 + P_5 + P_7) + 3(P_1 + P_3 + P_6 + P_8)]$$

Fig. 4-8. Nine point–weighted smoothing algorithm.

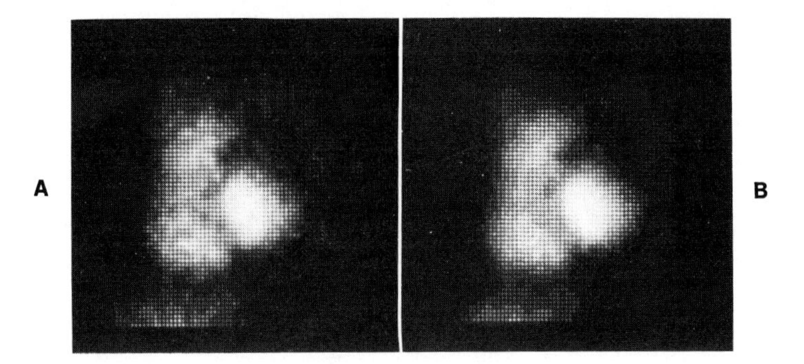

Fig. 4-9. Effect of smoothing on cardiac images. **A,** Before smoothing. **B,** After smoothing.

signs a specific color to each corresponding intensity level, and displays the result as a video picture. Each color is some mixture of varying intensities of three primary colors—red, blue, and green.

The primary advantage of video display is the ability to present large image matrices on the screen with no flicker effect. Additionally, the "digital" appearance of computer images, apparent in early display systems, is minimized with video display. Dynamic studies, images, and final results can be displayed on the video screen, recorded on videotape cassettes using standard videotape recording systems, and filed for later reference.

Many digital enhancement techniques are applied to images to improve their appearance. A simple background subtraction procedure, for example, can greatly enhance the readability of a clinical image. A digital enhancement technique often applied to cardiac images is smoothing, as illustrated in Fig. 4-8. Each point in the digital matrix is modified according to the "weighting value" of its immediate neighbors. If nine points are involved in the smoothing process, then the algorithm is called a "nine point–weighted smooth." Other techniques use different algorithms. The result of smoothing is the removal of statistical fluctuation patterns from the image display. Images appear less patchy and in general "smoother." Fig. 4-9 shows two images—one before and one after a nine point–weighted smooth. Although smoothing adds little *quantitative* value to the data, it greatly enhances image appearance when visualizing dynamic sequences, such as left ventricular wall motion, for example.

ACQUISITION MODES

The data from the gamma camera arrive as three separate signals. The "x" signal indicates the x position of the event that has most recently interacted with the camera crystal. Similarly, the "y" signal indicates the y position of the most recent event. A third signal, called the "z" or "strobe" pulse, indicates when the event recognized by the camera falls within the energy window specified by the operator. In other words, the z signal tells the computer which events to accept and which events to ignore.

The computer is able to process these signals in several different ways. The three modes of acquisition currently employed in nuclear medicine are frame-mode, list-mode, and multiple-gated acquisition.

Frame-mode acquisition

The philosophy of frame mode is to accept the x-y pairs from the ADC and build an image in the computer's memory as the pairs are generated. In effect, a digital image of the data is created as the counts interact with the gamma camera crystal. When acquisition is complete, images are immediately available for processing or display.

Fig. 4-10 illustrates the concept of a

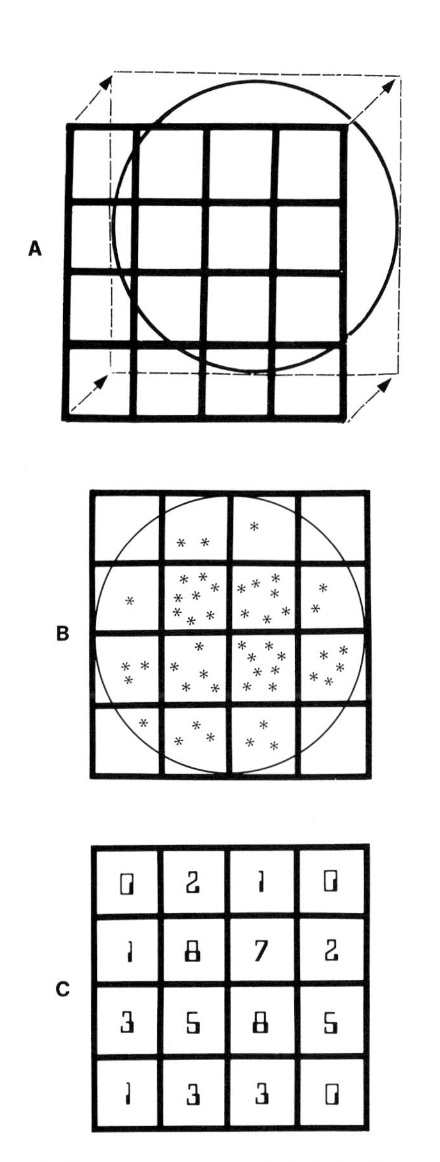

Fig. 4-10. Digitizing the camera field. **A,** Partitioning the camera crystal. **B,** Acquiring events into each partition. **C,** Digital map of the events.

digital image. Imagine that a wire mesh or screen is placed over the face of the camera crystal. In Fig. 4-10, *A*, a wire mesh divides the crystal into sixteen squares. As counts interact with the crystal (Fig. 4-10, *B*), the individual events fall into one of the sixteen areas partitioned by the wire screen.

By totaling the number of events in each square of the wire grid, a digital map of the image is generated. The totals for Fig. 4-10, *B*, are arranged in four rows and four columns in Fig. 4-10, *C*. Each square of Fig. 4-10, *B*, has a corresponding total in the digital matrix of Fig. 4-10, *C*. By observing the digital values in the map, the viewer can determine the distribution of activity in the camera field.

The process of digitization divides the camera field into areas exactly like those in Fig. 4-10. In practice, however, the camera field is usually divided into much finer partitions by using larger matrix sizes.

The advantage of the frame mode is the availability of images immediately following acquisition of the data. Additionally, certain matrix modes offer practically unrestricted count rate limitations for static images. Due to the limited transfer of images to disk, however, temporal resolution of frame mode is relatively poor. Maximum framing rates are usually limited to 25 frames/sec in frame mode.

List-mode acquisition

List-mode acquisition is technically less sophisticated than frame mode but yields no readily available images when acquisition is complete. In list mode, x-y pairs from the camera are transferred directly to the computer's memory. No modification or processing of the x-y pairs takes place at the time of acquisition. Table 4-1 illustrates list-mode acquisition. The six x-y pairs from the camera are simply arranged sequentially in the computer's memory. No matrices or images result as a product of list-mode acquisition.

In addition to x-y pairs, other types of data may be inserted in the list data stream. Table 4-2 shows the six x-y pairs of Table 4-1 with time markers and physiologic triggers included in the data list. The time markers indicate passage of time intervals. Typically, time markers are inserted every 1 or 10 msec. The physiologic trigger marks the occurrence of random events—not necessarily synchronized with time. In cardiac studies, the occurrence of the R wave is used as the physiologic trigger. The R wave can be detected using an ECG monitor.

Table 4-1. List-mode acquisition

Memory location	Data
1	(1,5)
2	(2,3)
3	(2,2)
4	(8,4)
5	(1,1)
6	(9,2)

Table 4-2. List-mode acquisition with time markers and physiologic triggers

Memory location	Data
1	(1,5)
2	Physiologic trigger
3	(2,3)
4	(2,2)
5	Time marker
6	(8,4)
7	Physiologic trigger
8	(1,1)
9	Time marker
10	(9,2)

Although no images are immediately produced, list-mode acquisition is extremely useful. Since the x-y pairs from the camera are recorded on disks, the list-mode data can be formatted into digital image matrices after acquisition is complete. The formatting process offers extremely flexible control over imaging parameters (Chapter 1, pp. 33 to 43). Unfortunately, it may require several minutes to format list-mode data, whereas frame-mode images are available immediately after acquisition. Additionally, only list-mode acquisition is able to store time markers and physiologic triggers as part of the data stream.

Apart from the need to format images after acquisition, list-mode acquisition introduces another serious drawback. Since every x-y pair from the camera is stored on disks, a study of 10 to 15 sec could conceivably fill a disk cartridge to capacity. The introduction of multiplatter disk systems has somewhat relieved the strain of storing list-mode data; however, this mode of acquisition is known to consume much disk storage for long studies.

Multiple-gated acquisition

A revolutionary approach to acquisition combines the convenience of frame mode with the ability to recognize time markers and physiologic triggers. This approach is called multiple-gated acquisition and has appeared only recently in the commercial market. In this mode, data from the camera are channeled to a series of image buffers located in the computer's memory (Fig. 4-11). A trigger signal (usually a physiologic trigger such as an R wave in the cardiac cycle) initiates the distribution of activity among the buffers in memory. Immediately after the trigger, data from the camera are placed in matrix 1 for a fixed duration of time. When this time interval has elapsed (typically 40 msec), data from the camera are then channeled to matrix 2. The process continues until all matrices in the memory have been used. The computer then waits for the next occurrence of the trigger signal and repeats the process.

Multiple-gated acquisition is used to achieve very high framing rates for *repetitive* dynamic processes. In fact, it is the repetitive property of the cardiac cycle that permits the application of the multiple-gating technique. To increase temporal resolution, the number of matrices used in each cycle can be increased and the time intervals can be made larger or smaller as the application dictates. A common application of multiple-gated acquisition is the visualization of cardiac wall motion. An ECG monitor detects the R wave, which is used to trigger each gated cycle. After the R wave is detected, data are distributed over as many as 127 frames throughout one cardiac cycle. The study may be continued over many cardiac cycles, yielding a series of gated images with excellent counting statistics.

The limitations on framing and image resolution for this acquisition mode are strictly functions of the memory available in the computer. Because of the high framing rates involved, the disk cannot be successfully utilized as an intermediate storage area for the gated images. Every data frame used

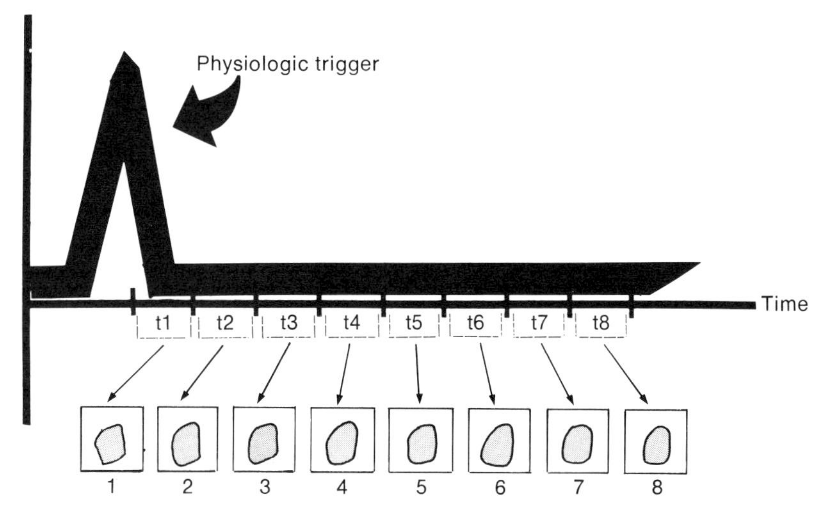

Fig. 4-11. Theory of multiple-gated acquisition.

to hold data in the gated acquisition cycle must remain in memory until acquisition is complete.

Clinical applications

The attractive application of nuclear cardiology is the ability to quantitate certain parameters of cardiac function noninvasively. The full potential of nuclear cardiology has not yet been reached, but current analysis techniques yield valuable quantitative data in the following areas: (1) calculation of ventricular ejection fraction and (2) observation of ventricular wall motion.

Two general approaches are used to obtain data in these areas. The first approach requires the observation of a bolus of activity as it passes through the ventricle. The second approach uses physiologic triggers to apply gate acquisition techniques after the isotope has reached equilibrium in the bloodstream. The following sections discuss these two approaches, stressing the computer's role in quantitation of the data.

Single first-pass cardiac studies

The first-pass technique involves the observation of the activity of a bolus as it passes through the ventricle. Immediately after bolus injection, the gamma camera records data as the bolus passes through the ventricle of interest. A typical study might record 20 sec of data at a framing rate of 25 frames/sec. The study can be recorded in frame mode if the disk transfer rates will support 25 frames/sec, or the study can be recorded in list mode and formatted into images after the study has terminated.

After acquisition is complete, a region of interest is flagged over the ventricle, and a time-activity curve for that region is generated by the computer. An example of such a curve is illustrated in Fig. 4-12. The peaks of the curve correspond to the end of diastole, whereas the dips in the curve correspond to the end of systole. By using the formula indicated in Fig. 4-12, the ejection fraction can be computed. This formula simply expresses the difference in activity at end of systole and end of diastole as a percentage.

The first-pass method has several advantages that make it a favorable alternative to multiple-gated studies (pp. 86 to 89). They are as follows:

1. A physiologic triggering signal is not required. This eliminates both the need to interface detection electrodes to the patient and the variability of triggering artifacts. End of systole and

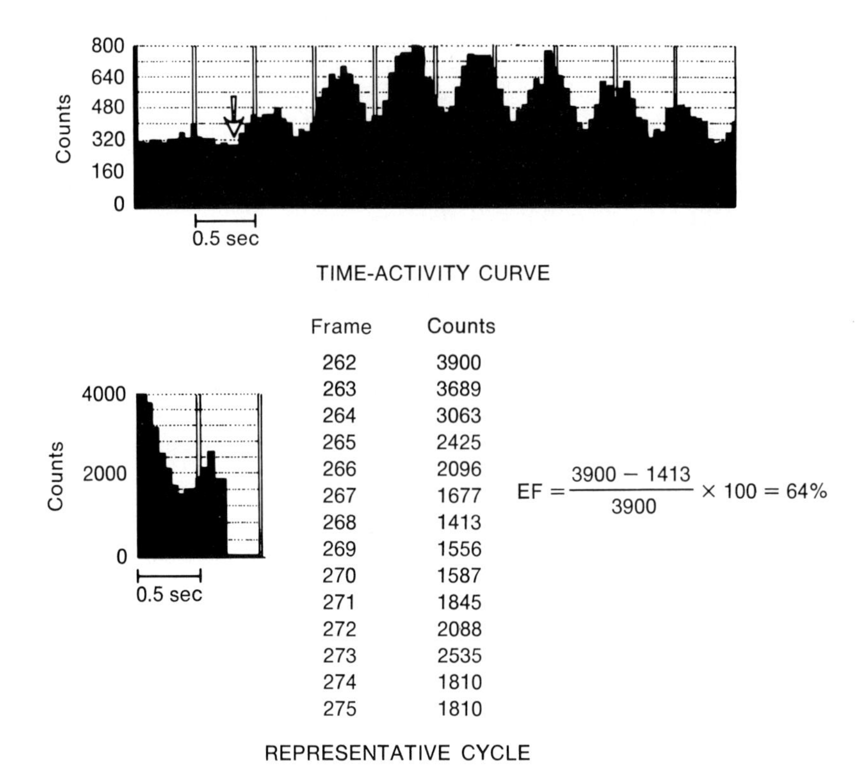

Fig. 4-12. Time-activity curve for first-pass cardiac study.

end of diastole are determined by direct observation of the time-activity curve.

2. Because each cardiac cycle is recorded separately, the results of the first-pass technique provide more instantaneous observations than do multiple-gated images.
3. Background activity is low.
4. Because the bolus passes through the full field of view of the camera, the count rates are reasonably high.

Following are some disadvantages to first-pass cardiac techniques:

1. Individual frames in the study are subject to statistical variations.
2. Only one view can be obtained per injection.
3. Rest/stress testing is not possible without additional injections.

Multiple-gated cardiac studies

Multiple-gated cardiac studies also provide the ability to compute ejection fraction as well as several other useful cardiac parameters. The theory of multiple-gated acquisition is discussed on pp. 84 and 85. After the multiple-gated images have been acquired, the computer searches through each image in the gated series and traces an outline around the ventricle of interest. Fig. 4-13 illustrates twelve images of a multiple-gated series. The frames corresponding to end of systole and end of diastole are labeled by the computer. Fig. 4-14, *A*, illustrates computer-determined edge selection for the end-systolic and end-diastolic images in the gated series. As the reader might suspect, the algorithm for edge detection of the ventricle is the crucial step.

There are several algorithms that have been applied to edge detection. Fig. 4-15 illustrates a simplified algorithm for the determination of edges using second derivatives. A series of profiles are selected across the area of the ventricle, and a family of digital curves is produced. Each profile curve is then examined for the presence of

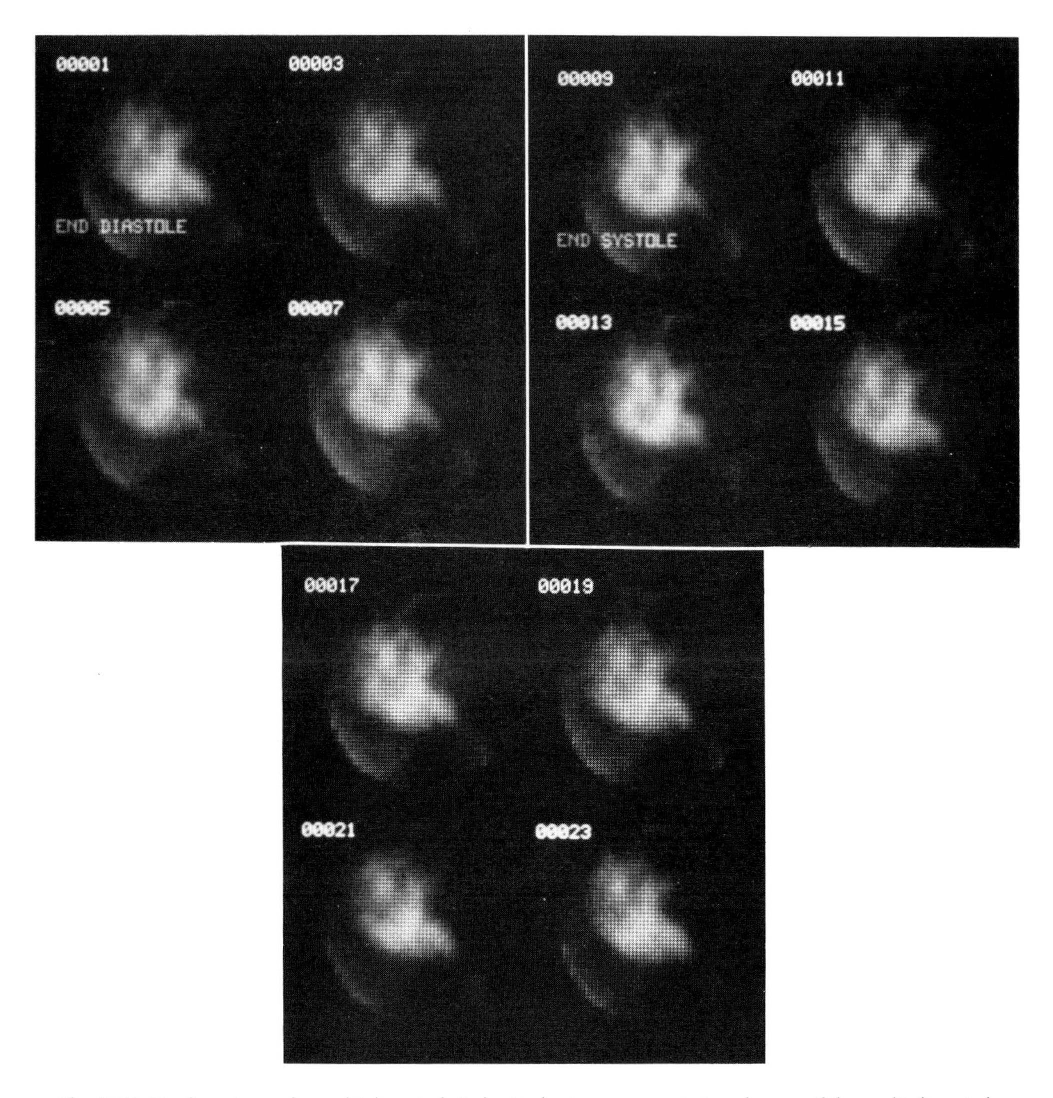

Fig. 4-13. Twelve views of a multiple-gated study. Each view represents two frames of the multiple-gated series. End of systole and end of diastole are labeled by the computer.

second derivatives. Mathematically speaking, the second derivative indicates inflection points in the curve, that is, points on the curve where the slope changes direction. In theory, these inflection points determine the edge of the ventricle in a consistent manner. In practice, however, statistical fluctuations in the data introduce artifacts and limit the effectiveness of this approach. Other criteria for edge detection must be combined with the second-derivative edge detection technique to fully define the perimeter of the ventricle.

When the perimeter of the ventricle has been determined for all images in the gated series, the computer sums the activity in each ventricle and plots the results as a function of time. The resulting curve is called a ventricular volume curve and is illustrated in Fig. 4-14, *B*. Using the equation from Fig. 4-12, the ejection fraction can be computed. In Fig. 4-14, *B*, the computer has carried the analysis several steps forward. The first derivative of the volume curve is computed and plotted below the volume curve near the axis. The first derivative of the volume curve represents ejection velocity and is also a parameter utilized in the

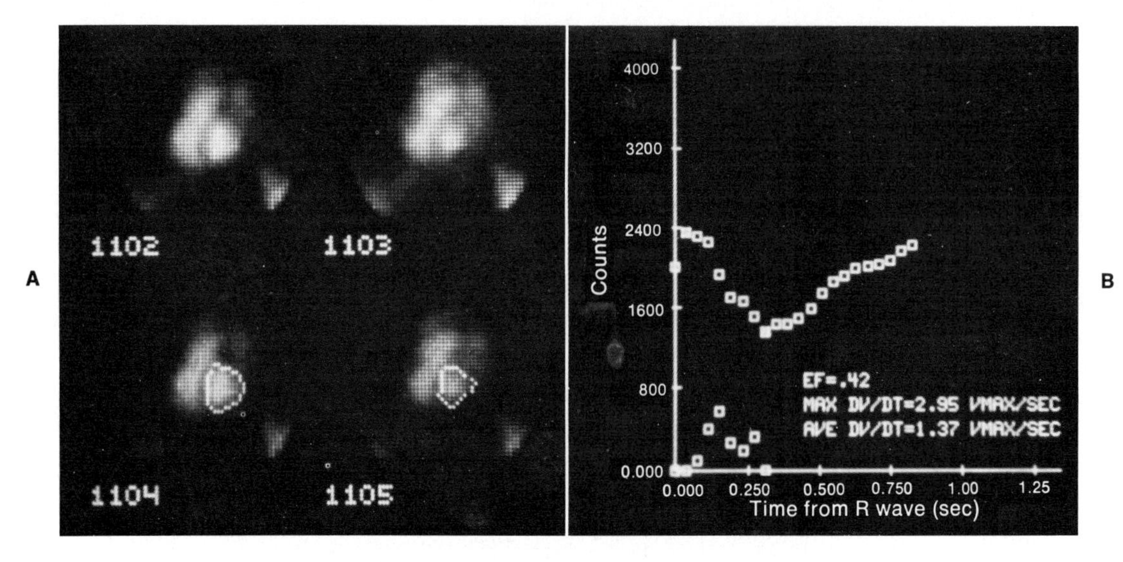

Fig. 4-14. Computer-determined edge selection. **A,** End of systole and end of diastole views shown together with computer-detected edges of ventricle in each view. **B,** Ventricular volume curve generated from multiple-gated series.

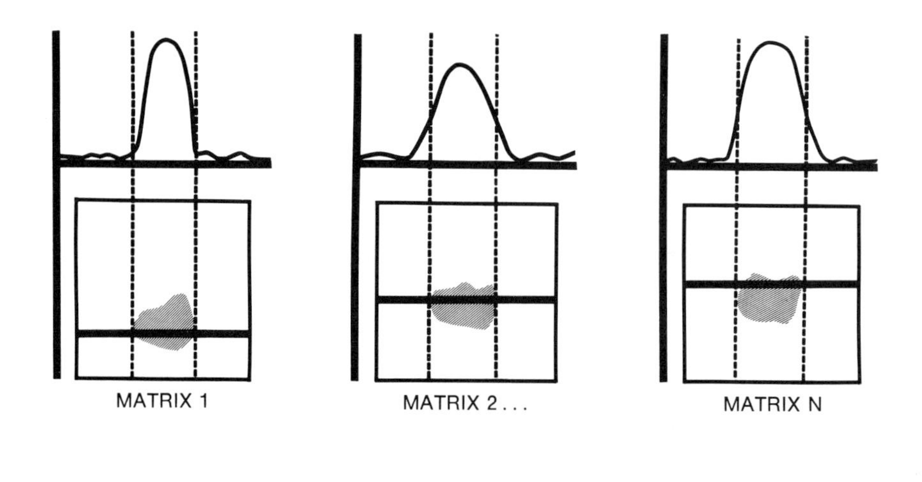

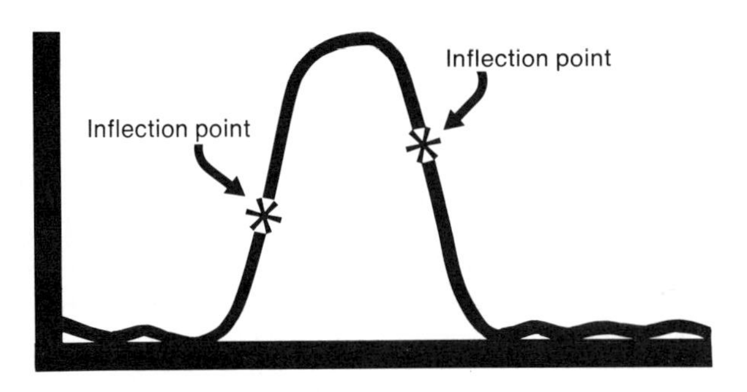

Fig. 4-15. Edge-detection algorithm using second derivative approach.

evaluation and comparison of ventricular performance.

The advantages of multiple-gated acquisition over other approaches are as follows:

1. Since each image in the multiple-gated series is averaged over many cardiac cycles, images with valid statistics can be produced for subsequent analysis. The long acquisition time of the study, however, (on the order of several minutes) increases the sensitivity of the analysis to background correction. A background area must be selected near the ventricle and the background activity subtracted from the images before cardiac parameters are computed. The high statistics images in this approach yield consistent and reproducible average measurements of ventricular ejection fraction.
2. The gated images may be played back in a "movie" mode in *real* time, that is, at the rate that it was originally acquired. In fact, several researchers at the National Institutes of Health suggest that only real time display resolves the ventricular, atrial, and other anatomic motions present in the movie-mode display.[1]
3. Multiple-gated studies are performed after the radioisotope has equilibrated in the bloodstream. If necessary, the study may be repeated without need for additional injections. This property of multiple-gated studies makes it ideally suited for rest/stress testing or drug intervention studies.

Following are some disadvantages to multiple-gated studies:

1. The patient must be interfaced to an ECG monitor for detection of the R wave. Movement of the electrode against the skin or other electrical interference can falsely trigger the gate and cause erroneous data to be acquired.
2. An arrhythmia of the heart can cause false triggering of the gate and, once again, causes erroneous data to be acquired.
3. Multiple-gated images are composite views of hundreds of heart cycles and represent an average cardiac cycle over a specific time interval.

Summary of computer methods

Computer systems have greatly improved the diagnostic potential of nuclear medical imaging techniques. Computers play an integral role in the acquisition, processing, and display of nuclear cardiology procedures.

1. Nuclear medical computer systems acquire data from the gamma camera in analog format, convert the signals to digital form, and store the resulting data in the computer's memory.
2. Bulk storage of data is handled by multiplatter disks, single-platter disks, or floppy disks.
3. Images are typically displayed for observation using color video display systems. The assignment of colors to images can be controlled by the operator. Images can also be recorded on videotape cassettes or hard-copy printers.
4. Three acquisition modes used to collect nuclear medical studies are frame-mode, list-mode, and multiple-gated acquisition.
5. First-pass cardiac analyses do not require physiologic triggering signals and yield consistent results in the computation of ejection fraction. Only one view can be obtained per injection.
6. Multiple-gated techniques require input of a physiologic signal and provide statistically valid images of the composite cardiac cycle. The determination of ventricular volume is subject to computer algorithms used to locate the boundaries of the ventricle.

Description of an image display and analysis system

T. K. Natarajan

Since general-purpose computers provide the flexibility in terms of programing new algorithms or changing existing algorithms as the requirements change, an image display and analysis (IDA) system was developed at the Johns Hopkins Medical Institutions, designed around a small general-purpose digital computer. The IDA system performs operations on data from scanners and cameras and has three basic functions: image enhancement, regional quantification, and data reduction. Similar systems are now available commercially.

The display in the system is in a form suitable for visual perception. The characteristics of the display such as brightness, contrast, and size can be varied while the image is being observed. The variability of these characteristics aids in the subjective interpretation of the information in the image. This is qualitative analysis of the image.

The image can also be analyzed quantitatively. This can be done in single and serial images. Single images provide the count and area information. Serial images, in addition to supplying counts and area, provide temporal information. Such quantitative analysis aids in the objective interpretation of the information in the image.

SYSTEM DESCRIPTION

The IDA system setup is shown in Fig. 4-16, *A*. The major components of the system are interconnected as shown in Fig. 4-16, *B*. In the heart of the system is a small general-purpose computer. The use of a general-purpose computer allows easy development of new programs or modifications of existing programs to suit a particular need as it arises. Other components of the system are as follows:

1. Teletypewriter for initiating various programs and hard-copy output of pertinent answers
2. High-speed paper-tape punch and reader
3. Magnetic disk recorders for programs and data manipulation
4. Magnetic tape recorder for storing data
5. Floating point processor for high-speed arithmetic manipulations
6. Display system consisting of a video monitor and other hardware to produce the image. Controls are provided for variation of the image contrast and size continuously
7. Light pen for observer selection of areas of interest from the image for further analysis and quantification

All the high-speed components of the system are connected to the computer through a multiplexer with priority assignments. Fig. 4-16, *B*, shows only the hardware components of the system. Equally important are the "software" components, which are programs of instruction for various operations on the data.

Fig. 4-17 shows an operator delineating a region of interest by means of the light pen in a cardiac blood pool image. This operation is called flagging. Areas of any size and shape can be flagged. The area delineated is simultaneously intensified in the image as a guide for the operator. If the area flagged is not right, the operator can easily erase the flags and start over again. This process of flagging does not in any way alter the original data that formed the image.

In essence, the IDA system consists of hardwares and softwares to derive from the images the three parameters: counts, area, and time. These three parameters are the fundamental quantities of measurement in nuclear medicine. All the measurements in nuclear medicine, including imaging, involve these three parameters.

The IDA system is interfaced both with moving-detector scanners and with stationary-detector cameras (Fig. 4-18). The camera is better suited for fast dynamic studies in general because of its ability to observe an entire area of interest. The scanner has its place in slow dynamic studies and static studies.

CAMERA INTERFACE

The selected x and y coordinate signals are simultaneously digitized by two analog-to-digital converters in 4 μsec. Immediately

A

B

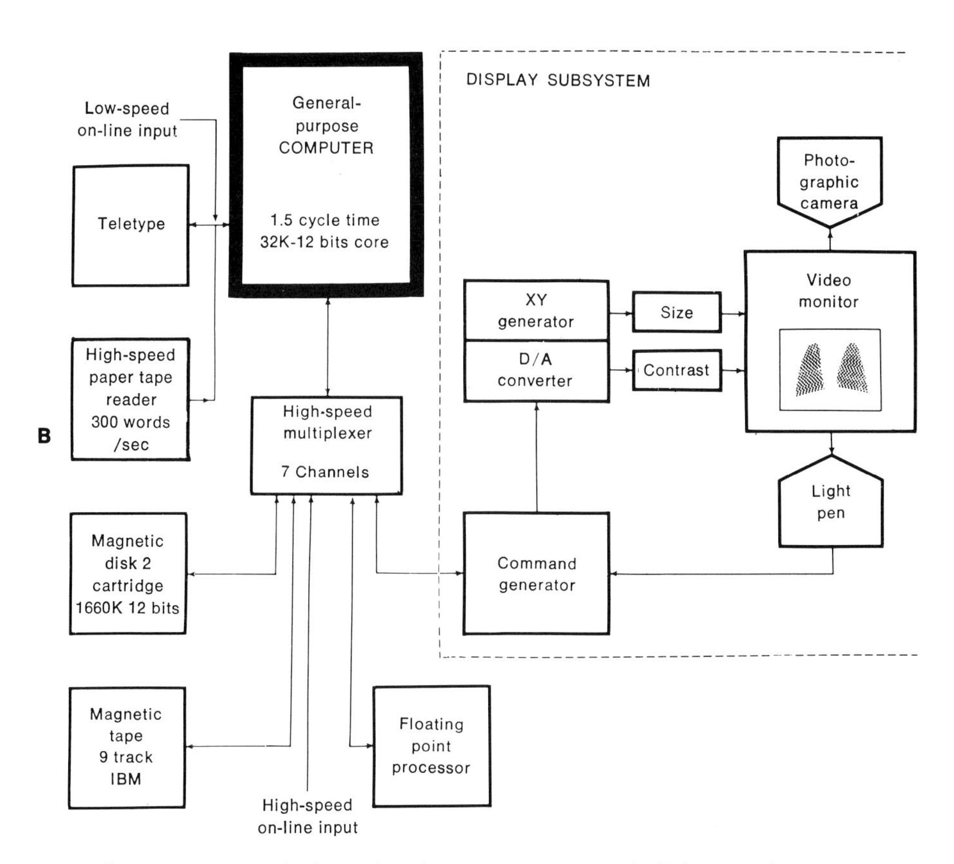

Fig. 4-16. A, Image display and analysis (IDA) system. **B,** Block diagram of IDA system.

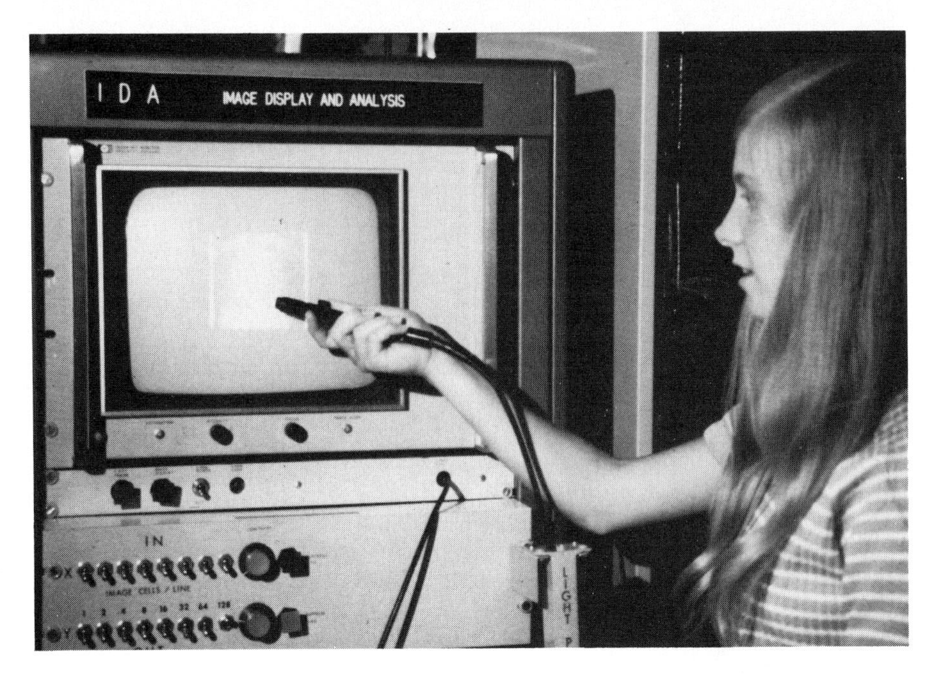

Fig. 4-17. Display subsystem. Operator is outlining region of interest in cardiac blood pool image.

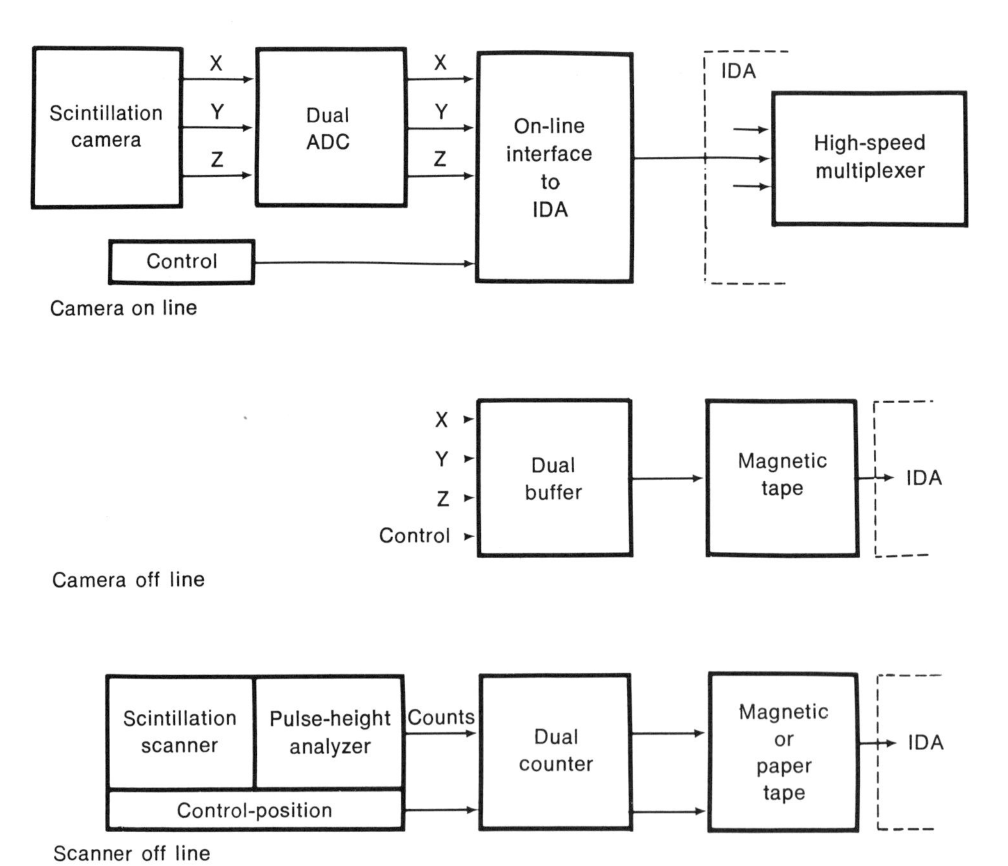

Fig. 4-18. Digitization of scintillation camera image data.

after the conversion, synchronized with the read-write cycles of the core memory, the location in memory addressed by the x and y digitized signals is incremented by one count to build a histogram of the activity distribution seen by the camera. This incremental operation takes 1.5 μsec. During this 1.5 μsec, another pair of x and y signals could be digitized. These operations are illustrated diagrammatically in Fig. 4-19 for two x-y pairs. The circular area of the camera crystal is shown fitted inside a 64 by 64 cell square matrix formed by the memory.

If the camera is putting out signals at the rate of 50,000 events/sec, the computer time needed to process these signals is $(50 \times 10^3 \times 1.5 \times 10^{-6})$, or 0.075, sec in every second. So in every second, (1 − 0.075), or 0.925, sec of computer time, during which the count-incrementing operation is not performed, is available for other tasks. One task that is performed by the system is live time display of the current memory content as an analog image. This is achieved by having the computer execute a program of instructions to transfer sequentially and cyclically the contents of the memory to the display subsystem and letting the event signals from the camera momentarily interrupt the display operation to increment a count at the proper location as the counts occur. Thus, if the counts in each location of the memory are begun at zero and the display operation started, the buildup of the image can be seen. At the conclusion of the

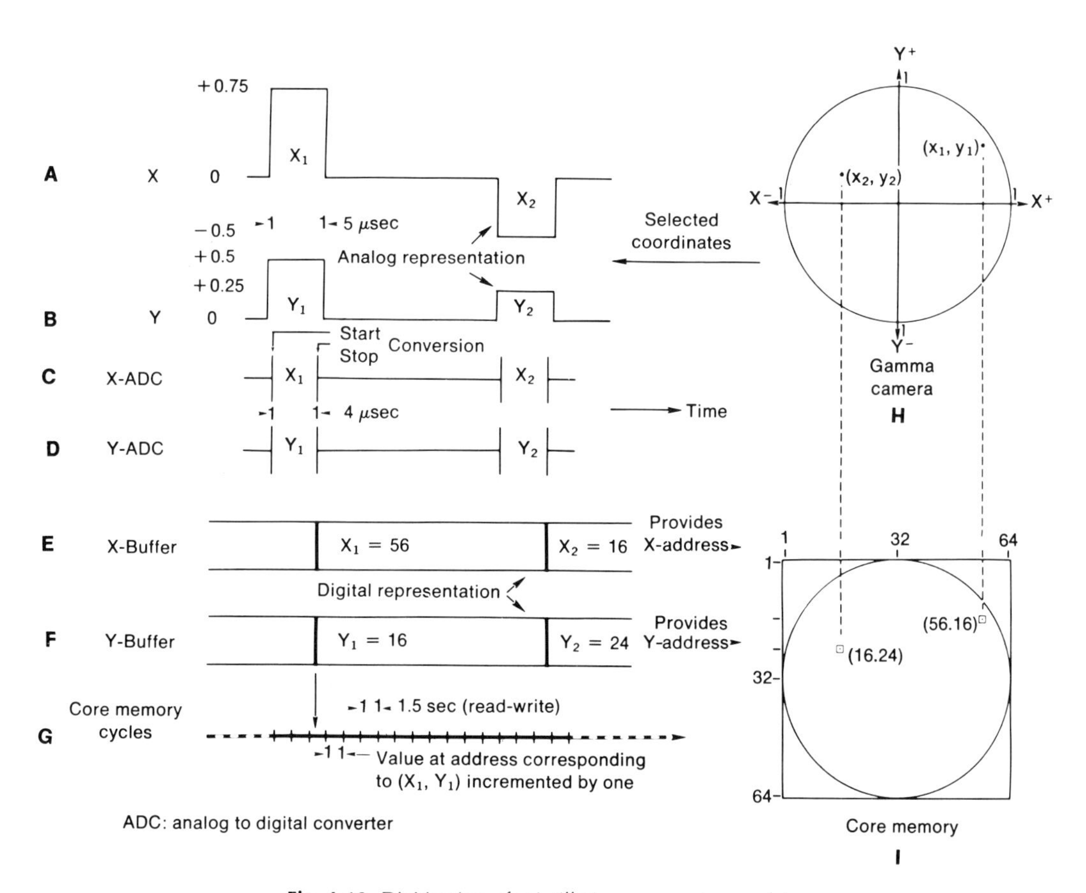

Fig. 4-19. Digitization of scintillation camera image data.

preset time or preset count, the memory data are recorded on a storage device such as magnetic tape or disk.

SYSTEM PROGRAMS

Following are some typical programs that are useful. They are written in modular units that can be combined for a particular clinical application to obtain pertinent data from single or serial images.

Load. This program is used for loading the selected image-frame data into the computer memory from the selected storage device.

Display. This program cycles the image data between specified limits of the memory and the display subsystem to produce on the video monitor a flicker-free tonal image of the radionuclide distribution.

Flag (light pen). Areas of interest in the image can be selected with the light pen while the image is observed. Flagged cells are intensified in the image. Use of the light pen allows regional discrimination in the image irrespective of count rate.

Flag above limit. This program flags all image cells with counts above a desired limit.

Remove flag above limit. This program removes the flag of all flagged cells with counts above a desired limit.

Range. This is a combination of the two previously given programs and is used for displaying isocount zones. This allows count rate discrimination irrespective of its regional distribution.

Remove flag. This removes the flag of all image cells.

Transfer flagged cells. This program enables the observer to select any region in an image by flagging and transfer the data in the flagged region to a separate field to be viewed or analyzed separately.

Transfer unflagged cells. This program is similar to the previously given program except that it operates on the unflagged image cells.

Store flagged addresses. This program is used for storing the addresses of the image cells flagged either with the light pen or by programs.

Flag stored addresses. This program is used for flagging the image cells at the addresses stored.

Sum. All flagged cells are summed to give the total counts in the flagged area and the size of the area.

Add image. Multiple images are added cell by cell by this program, and the resultant data are stored as a composite image.

Subtract constant. A constant number is subtracted from all the image cells. This program is used for subtracting background activity.

Subtract image. This program subtracts the data in one image from that in another image cell by cell. The result can be displayed as an image to portray the difference in the distribution of activity between any two image frames.

Maximum and minimum. The maximum and minimum counts per image cell are obtained by this program over either a flagged or an unflagged region.

Negative images. This program is used for selecting either black on white or white on black images. This selection can be made either on a whole image or on the flagged part of an image.

Flagged cell counts. This program gives the counts in each flagged cell for statistical analysis.

Average. This program is used for obtaining the average counts per cell in a flagged region of an image.

Time activity. This program plays back all the image frames of a dynamic study one by one; the counts in selected regions are typed out frame by frame. In an optional mode, these counts can be typed out normalized to a fixed area size, such as 100 image cells, for all the regions and with or without background subtraction.

Smooth. This program is used for statistical smoothing of the image data.

Ratio. The cell-by-cell ratio between two image frames is obtained by this program. This ratio can be displayed as an image to portray the fractional change in the regional distribution of activity between two image frames.

Normalize. This program normalizes the value of the image cells in proportion to the maximum value in the image by setting the maximum value equal to a given number. By use of this program, different images with different maximum counts per cell can be viewed over the same range of contrast of the video monitor.

Draw horizontal line. This program intensifies the image cells along a specified horizontal line.

Remove flag above horizontal line. This program removes the flag of image cells above the horizontal line.

Remove flag below horizontal line. This program removes the flag of image cells below the horizontal line.

Draw vertical line. This program intensifies the image cells along a specified vertical line.

Remove flag left of vertical line. This program removes the flag of image cells left of the vertical line.

Remove flag right of vertical line. This program removes the flag of image cells right of the vertical line.

Profile. This program types the counts per image cell along a specified horizontal or vertical line averaged over the number of lines specified.

Mirror flag. Corresponding to an area flagged, another area equal in size and shape but positioned symmetrically from a specified horizontal or vertical line is flagged automatically.

SYSTEM APPLICATION

Static studies

Fig. 4-20 is a static image of the myocardium obtained with a scanner and displayed through the IDA system. The regional intensities are proportional to ^{43}K distribution, which in turn is a function of blood flow. There is an area of decreased activity in this image (indicated by the *arrow*), but it is difficult to see at this level of image contrast. There is also uptake in the liver. The information is in the form of count densities or counts and areas.

One of the simplest operations that can be done using the IDA is image enhancement by alteration of the contrast. This is shown in Fig. 4-21. The area of decreased activity (indicated by the *arrow*) is more obvious.

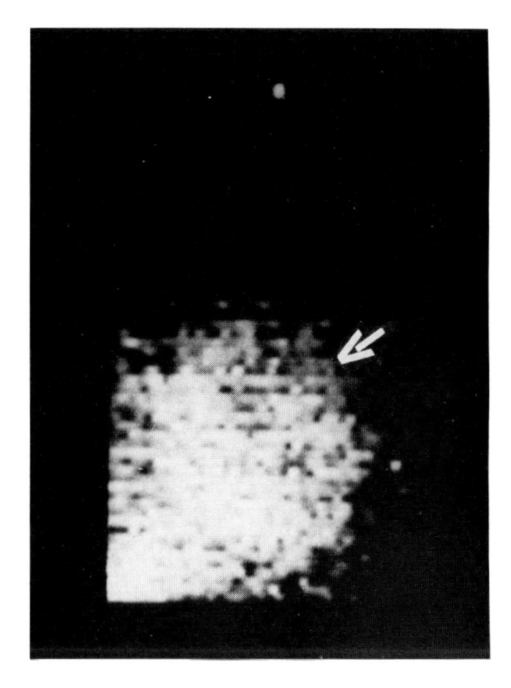

Fig. 4-20. Anterior view of myocardial image obtained with ^{43}K and scanner and displayed through IDA system.

The next interest in a static image like this is to quantify the extent of myocardial involvement. This can be done simply by use of the light pen. The normal and abnormal regions can be flagged separately, and then an index can be derived from the areas and counts in these two regions for myocardial involvement. This method, although quantitative, needs subjective judgment about the location and extent of the involved area before it can be flagged. The image enhancement feature is helpful in this procedure.

To make it more objective, operations on this data can be performed on the basis of count rate. By means of programs, the computer can be instructed to pick out and flag areas in the image in certain count rate bands. From this, areas that are statistically different can be computed and an index formed. However, the computer at present is not that good in distinguishing areas in the myocardium from areas in the liver that

Fig. 4-21. Anterior view of myocardial image obtained with ^{43}K and scanner (same as Fig. 4-20) and displayed through IDA system with contrast enhancement.

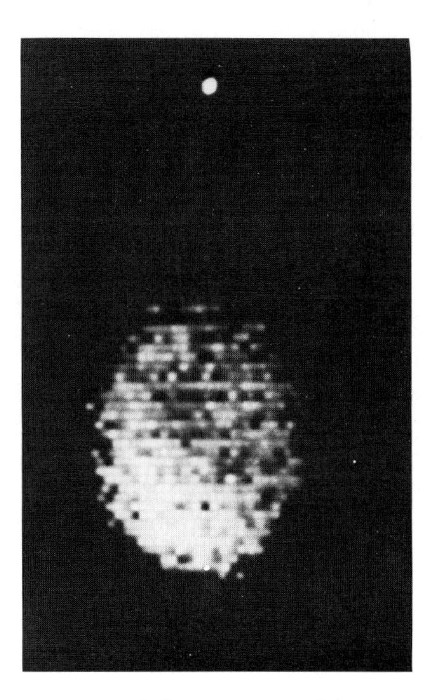

Fig. 4-22. Myocardial region stripped from Fig. 4-20.

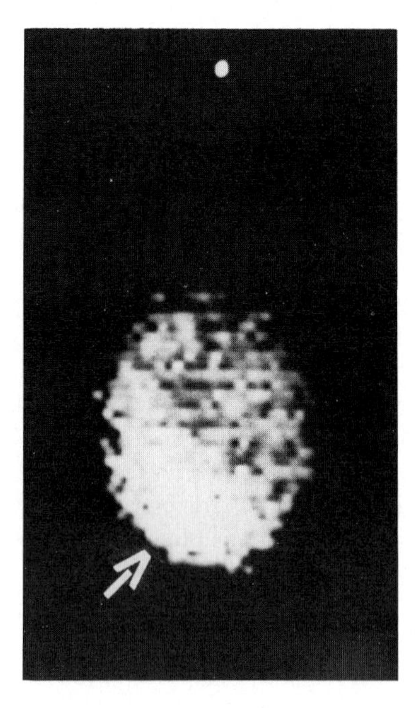

Fig. 4-23. Myocardial region with a sample of normal area flagged as indicated by *arrow*.

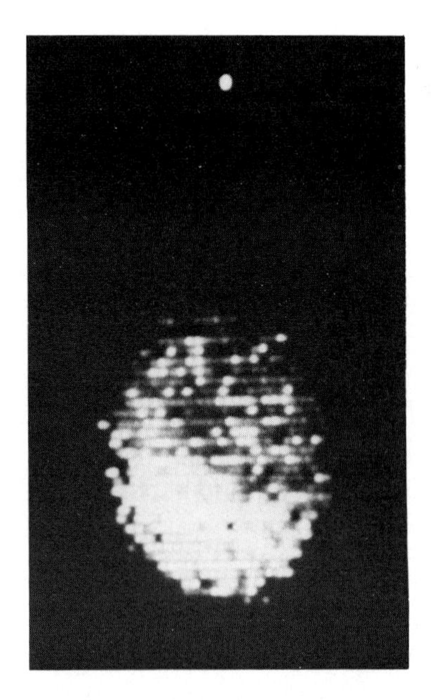

Fig. 4-24. Stripped myocardial image from Fig. 4-22, with regions within two standard deviations of normal (from Fig. 4-23) flagged by computer.

have similar count densities. The only parameter it knows well and looks for is counts. The operator's knowledge of the anatomy provides the ability to distinguish the region of the liver from the myocardium.

The region of the myocardium is flagged with the light pen, and then by means of programs, this flagged region is transferred to another block of memory for display as shown in Fig. 4-22. The data in this stripped image are only from the myocardium. Then, if a region that is normal as shown in Fig. 4-23 is flagged with the light pen, the computer can pick out regions significantly different from this normal region on the basis of count densities. Fig. 4-24 shows areas flagged by computer that have counts within two standard deviations of the normal region. From this, an index can be derived for the ischemic region.

Dynamic studies

Fig. 4-25 shows diagrammatically how a dynamic study is done using IDA and the scintillation camera. Data are integrated for a preset time duration and recorded on tape as digitized image frames. In the bottom of Fig. 4-25, the selected images are pulled out, showing different phases during the passage of a bolus of radionuclide through the heart. Each frame by itself has the counts and area information. All these frames in sequence provide the temporal information.

These frames can also be recorded on the disk and played back in a movie format with provisions for fast and slow speeds, reverse motion, and stop action. Areas of interest can be flagged in any one of these frames or combination of frames that is optimal for a particular region. These flagged areas can then be automatically duplicated in all the frames to study the time course of the tracer.

Fig. 4-26 shows the time course of a bolus of ^{99m}Tc-labeled albumin given intravenously to a patient through the regions of right atrium (RA), right ventricle (RV), lung (LU), pulmonary artery (PA), and left ven-

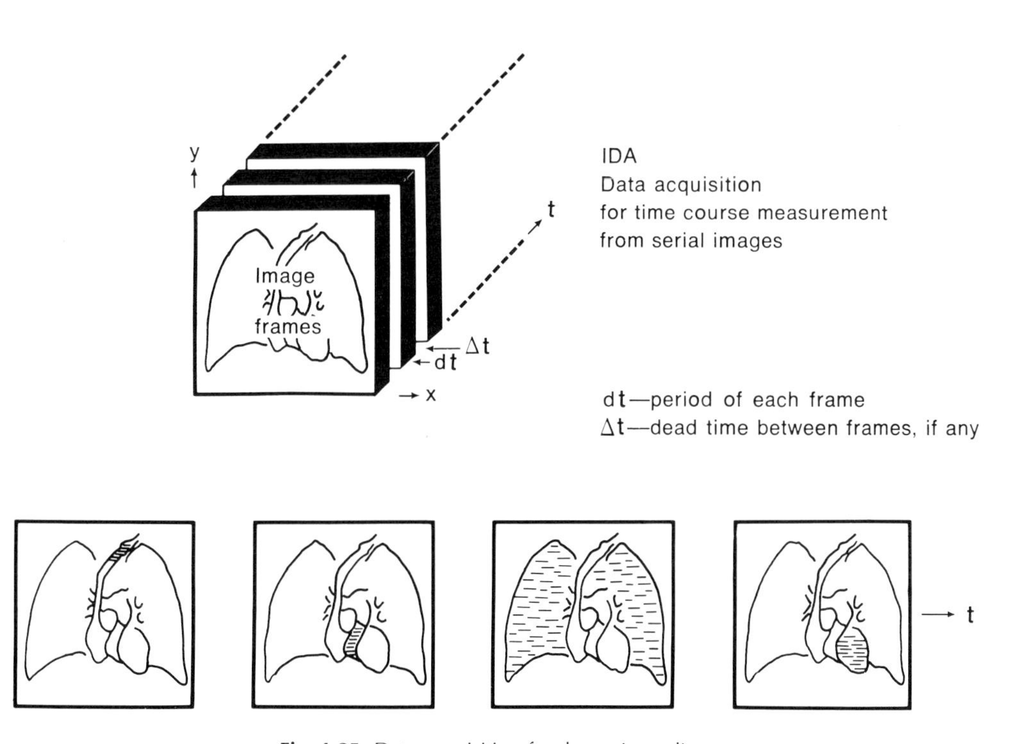

Fig. 4-25. Data acquisition for dynamic studies.

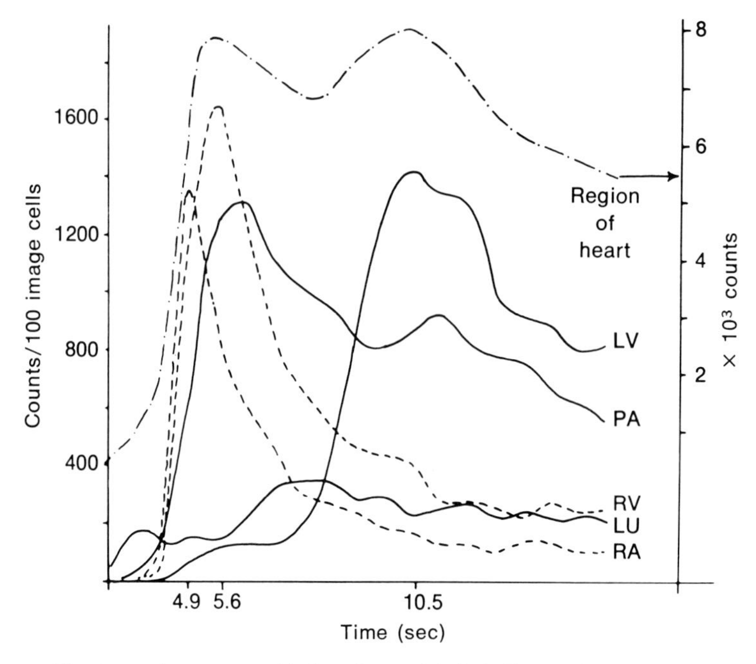

Fig. 4-26. Time course of bolus of ^{99m}Tc-labeled albumin through heart.

tricle (LV). Before cameras and computerized systems were used, simple collimated detectors were used to measure the time course of a tracer through the heart; this situation is simulated by flagging of the whole region of the heart, as indicated by the top curve in Fig. 4-26. The positions of the peaks corresponding to the right and left sides of the heart can still be seen in the composite curve. This composite curve could be stripped for analysis into its component parts with a model. However, with the present camera and computerized systems, much better resolution is achieved, as seen in the component curves. Through observation of the relative positions of the peaks and their washout characteristics, a pathologic state can be diagnosed.

LEFT VENTRICULAR VOLUME CURVE

In 1970 a technique for imaging the heart at selected portions of the cardiac cycle was introduced at the Johns Hopkins Medical Institutions. The mechanics of it was to turn on the image-producing cathode ray tube (CRT) of the scintillation camera only during selected intervals of the cardiac cycle. By gating during the intervals as shown in Fig. 4-27 and after a sufficient number of cardiac cycles, two images are obtained, one at the end of systole and one at the end of diastole. One of the quantitative measurements worked out with this technique is the ejection fraction. This was derived from the areas of the left ventricle in the two images. This method is somewhat tedious and requires subjective determination of the periphery of the left ventricle.

With the computerized system, the technique can be modified to look at the count rate change instead of the area change as the left ventricular volume changes. The CRT and photographic camera have simply been replaced with the computer and magnetic tape, respectively (Fig. 4-28). The output of this system is digitized images instead of the analog images of the previous technique. If the left ventricle and background are selected in the end-systolic and end-diastolic images, the ejection fraction can be calculated from the net left ventricular counts in the two images. Data from twenty patients correlate well with contrast angiography (r = 0.80).

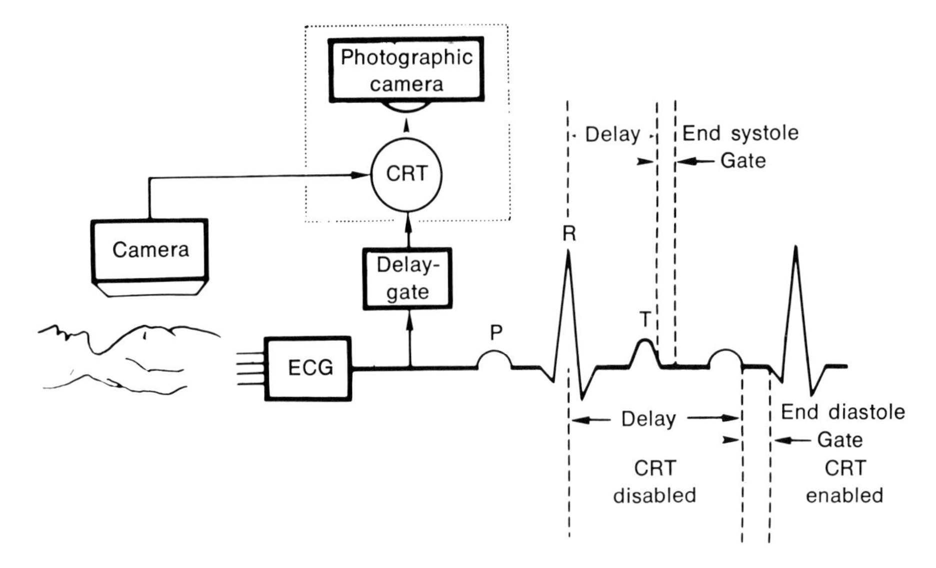

Fig. 4-27. Electrocardiographic gating of scintillation camera (analog technique).

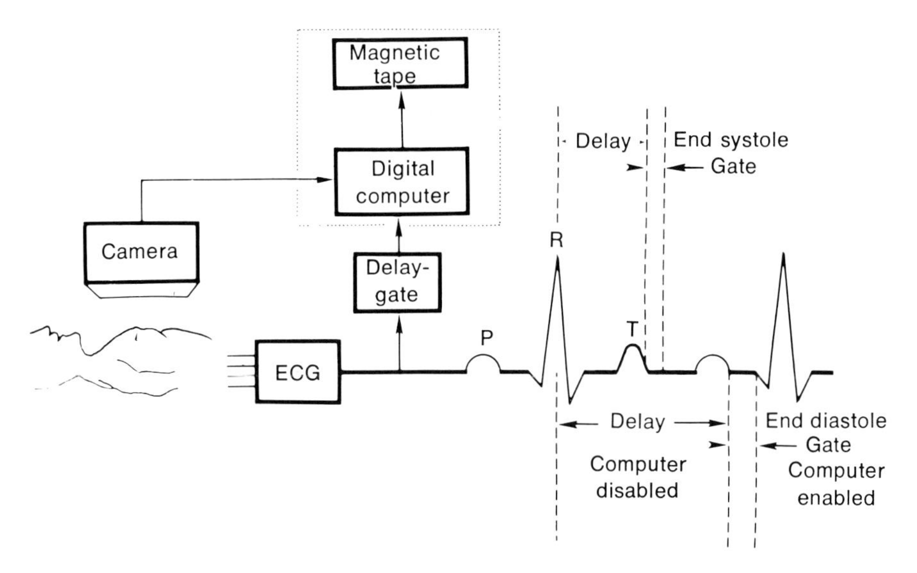

Fig. 4-28. Electrocardiographic gating of scintillation camera (digital technique).

When only the end-systolic and end-diastolic intervals are studied, considerable data are thrown away. To utilize all the data available and to avoid the slight risk in preselecting and fixing the two intervals, the system diagramed in Fig. 4-29 is being investigated. When the computer is used in the system, a series of counters numbered 1 through 100 can be set up in the computer memory. These chains of counters correspond to discrete intervals during the cardiac cycle, starting with R peak, and accumulate counts from selected regions. So that enough counts are accumulated, the operation with the counters is repeated whenever an R peak occurs. At the end of a sufficient number of cardiac cycles, the counters have counts proportional to the volume change of the regions they represent. The left ventricular curve from a

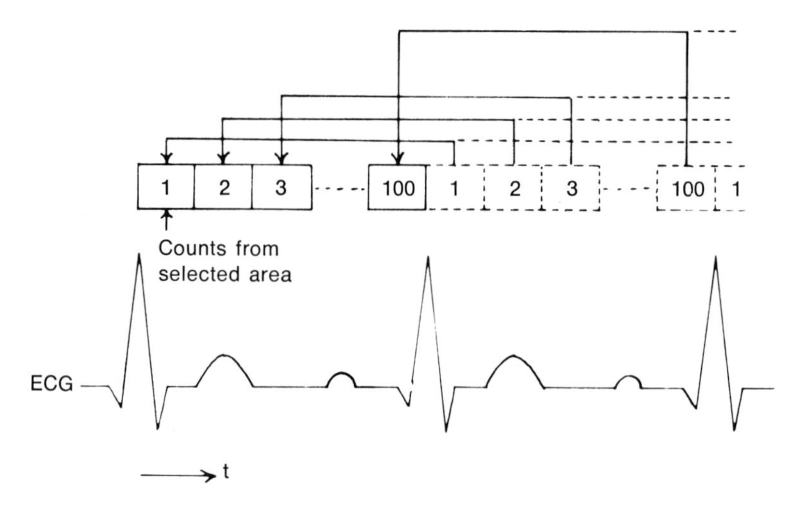

Fig. 4-29. Electrocardiographic gating with multiscaler operation for generating volume curves.

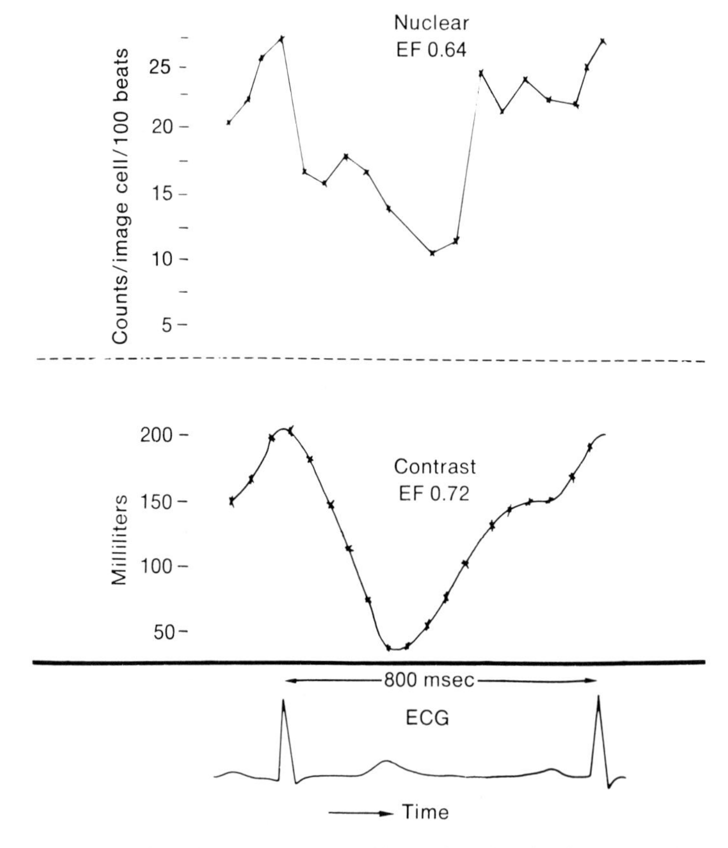

Fig. 4-30. *Top,* Left ventricular volume curve obtained by nuclear digital technique; *middle,* volume curve calculated from contrast angiography; and *bottom,* ECG.

patient is shown in Fig. 4-30 along with his ECG and the volume curve calculated from contrast angiography.

FUNCTIONAL IMAGE

With improvements in technology in terms of speed and spatial resolution, volume curves can be hypothetically constructed for all the cells in the matrix. Then, by means of the computer, values for the volume rate of change (dv/dt) are calculated for each cell, and these values are used as the parameters for an image. This image can be in black and white with the gray scale in some proportion to the rate values or in color with a color scale in some proportion to the rate values. This functional image portrays the regional contractility of the myocardium. This is a special form of data reduction.

Two examples of functional images are shown in Fig. 4-31. The two images were obtained from a series of 0.5 sec digitized image frames. Fig. 4-31, *A*, shows the functional image of differential regional clearance of ^{133}Xe injected into the coronary artery of a normal dog. Fig. 4-31, *B*, shows the same functional image from the same dog when two proximal branches of the left anterior descending coronary artery were partially occluded. The slower washout of the tracer is revealed by decreased intensity in the occluded region (indicated by *arrow*). These two images each have in them all the information contained in their respective series of images, in terms of the tracer clearance, and are easier to interpret.

REFERENCE

1. Bacharach, S. L., Green, M. V., Borer, J. S., et al.: A real-time system for multi-image gated cardiac studies, J. Nucl. Med. **18**:1, 1977.

BIBLIOGRAPHY

Ashburn, W. L., Kostuk, W. J., Karliner, J. S., et al.: Left ventricular volume and ejection fraction determination by radionuclide angiography, Semin. Nucl. Med. **3**(2):165, 1973.

Bacharach, S. L., Green, M. V., Borer, J. S., et al.: A real-time system for multi-image gated cardiac studies, J. Nucl. Med. **18**:1, 1977.

Berger, H. J., Matthay, R. A., Marshall, R. C., et al.: Noninvasive radionuclide technique for right ventricular ejection fraction in man, Circulation **53**(54): 11, 1976.

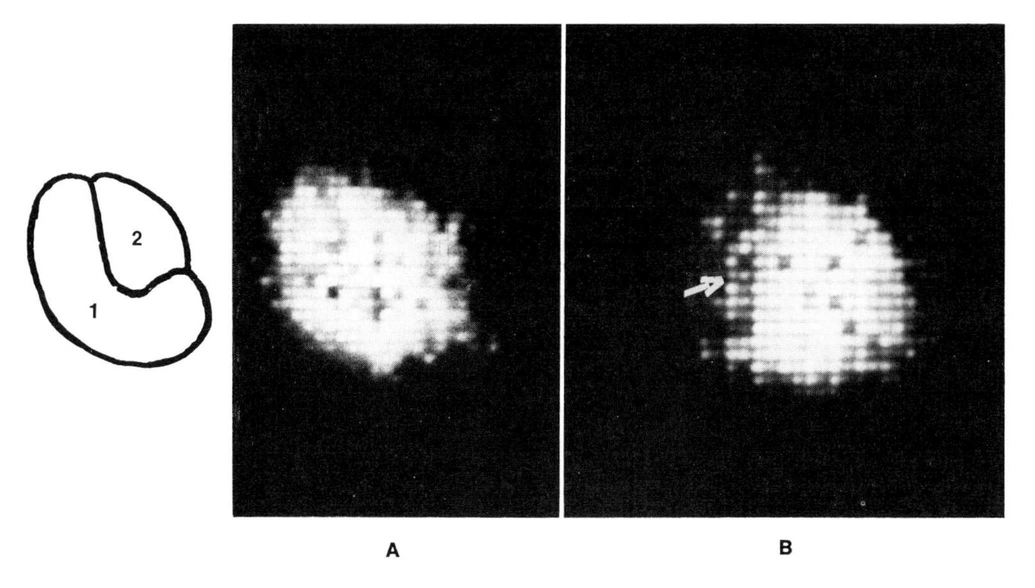

Fig. 4-31. A, Functional image of differential regional clearance of ^{133}Xe injected into the coronary artery of a normal dog. *Left,* regions of, *1,* left anterior descending branch and, *2,* circumflex branch for view used. **B,** Functional image of same dog in same view as **A** but with two proximal branches in left anterior descending branch partially occluded.

Borer, J. S., Bacharach, S. L., Green, M. V., et al.: Realtime radionuclide cineangiography in the noninvasive evaluation of global and regional left ventricular function at rest and during exercise in patients with coronary artery disease, N. Engl. J. Med. **296:**839, 1977.

Burow, R. D., Strauss, H. W., Singleton, R., et al.: Analysis of left ventricular function from multiple gated acquisition (MUGA) cardiac blood pool imaging: comparison to contrast angiography, Circulation **56:**1024, 1977.

Folland, E. D., Hamilton, G. W., Larson, S. M., et al.: Radionuclide ejection fraction: a comparison of three radionuclide techniques with contrast angiography, J. Nucl. Med. **18:**1159, 1977.

Green, M. V., Brody, W. R., Douglas, M. A., et al.: Count rate measurement of left ventricular ejection fraction from gated scintigraphic images, J. Nucl. Med. **17:**557, 1976.

Grenier, R. P., Bender, M. A., and Jones, R. H.: A computerized multicrystal scintillation gamma camera. In Hine, G., editor: Instrumentation in nuclear medicine, New York, 1968, Academic Press, Inc.

Lieberman, D. E.: Computer methods: the fundamentals of digital nuclear medicine, St. Louis, 1977, The C. V. Mosby Co.

Marshall, R. C., Berger, H. J., and Zaret, B. L.: Quantitative radionuclide angiocardiography for assessment of left and right ventricular performance. In Lieberman, D. E., editor: Computer methods: the fundamentals of digital nuclear medicine, St. Louis, 1977, The C. V. Mosby Co.

Natarajan, T. K., and Wagner, H. N., Jr.: A new image display and analysis system (IDA) for radionuclide imaging, Radiology **93:**823, 1969.

Natarajan, T. K., and Wagner, H. N., Jr.: Functional images of the lungs, IAEA-SM-185/15, 1975, International Atomic Energy Agency.

Parker, J. A., Secker-Walker, R., Hill, R., Siegel, B. A., and Potchen, E. J.: A new technique for the calculation of left ventricular ejection fraction, J. Nucl. Med. **13:**649, 1972.

Strauss, H. W., Natarajan, T. K., Sziklas, J. J., Poulose, K. P., Fukushima, T., and Wagner, H. N., Jr.: Computer assistance in the interpretation and quantification of lung scans, Radiology **97:**277, 1970.

Strauss, H. W., and Pitt, B.: Common procedures for the noninvasive determination of regional myocardial perfusion, evaluation of regional wall motion and detection of acute infarction, Am. J. Cardiol. **38**(6):731, 1976.

Strauss, H. W., Zaret, B. L., Hurley, P. J., Natarajan, T. K., and Pitt, B.: A scintiphotographic method for measuring left ventricular ejection fraction in man without cardiac catheterization, Am. J. Cardiol. **28:** 575, 1971.

Wagner, H. N., Jr., and Natarajan, T. K.: The computer in nuclear medicine, CONF 710425, Proceedings Symposium on Sharing of Computer Programs and Technology in Nuclear Medicine, Oak Ridge, Tenn., 1971.

Wagner, H. N., Jr., and Natarajan, T. K.: Computer in nuclear medicine, Hosp. Practice **7:**121, 1972.

Wagner, H. N., Jr., Wake, R., Nickoloff, E., and Natarajan, T. K.: The nuclear stethoscope—a simple device for generation of left ventricular volume curves, Am. J. Cardiol. **38:**747, 1976.

SECTION TWO

CLINICAL APPLICATIONS IN VIVO

5 □ Evaluation of central circulatory dynamics with the radionuclide angiocardiogram

Peter Steele
Dennis Kirch
Robert Vogel

In 1962 Folse and Braunwald reported a method of using radionuclide indicator dilution with a single precordial detector and introduction of the radionuclide directly into the left ventricle that permitted estimation of the fraction of the left ventricular end-diastolic volume (EDV) ejected per beat.[7] The values for ejection fraction (EF) in their patients were smaller than those generally observed by angiocardiographic means. The cross talk and streaming responsible for this discrepancy are discussed later. Fig. 5-1 is an example of a normal precordial left ventricular radionuclide dilution curve obtained by Folse and Braunwald from a patient who had no aortic valve disease or mitral regurgitation. This can be compared to the schematic illustration in Fig. 5-2 of the variations of activity in a ventricle into which a sudden single injection of radioactive tracer has been made.

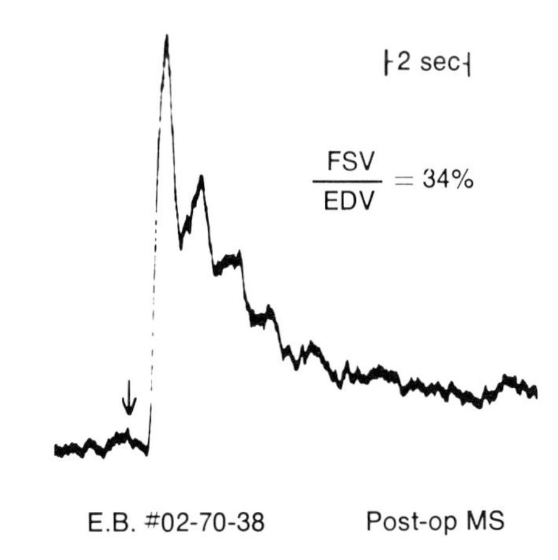

Fig. 5-1. Precordial radionuclide dilution curve obtained from patient studied six months after mitral commissurotomy. Patient had no aortic valve disease or mitral regurgitation. FSV = forward stroke volume; EDV = end-diastolic volume. (From Folse, R., and Braunwald, E.: Circulation **25:**674, 1962.)

Although Folse and Braunwald found the technique suitable for determination of fraction of left ventricular volume ejected per beat and of ventricular EDV and residual volume, it was not widely applied clinically because of the necessity of performing the relatively difficult procedure of introducing the radionuclide directly into the left ventricle.

There have been many attempts to make the procedure noninvasive by injecting the tracer into the vena cava and detecting the tracer by precordial counting.[5,6] A typical example of the (low-frequency) tracing obtained after vena cava injection is shown in Fig. 5-3. An example of one of the many analog models of the cardiopulmonary circulation used in analysis of such a tracing is shown in Fig. 5-4. Because of its great practical appeal, the analysis of these precordial

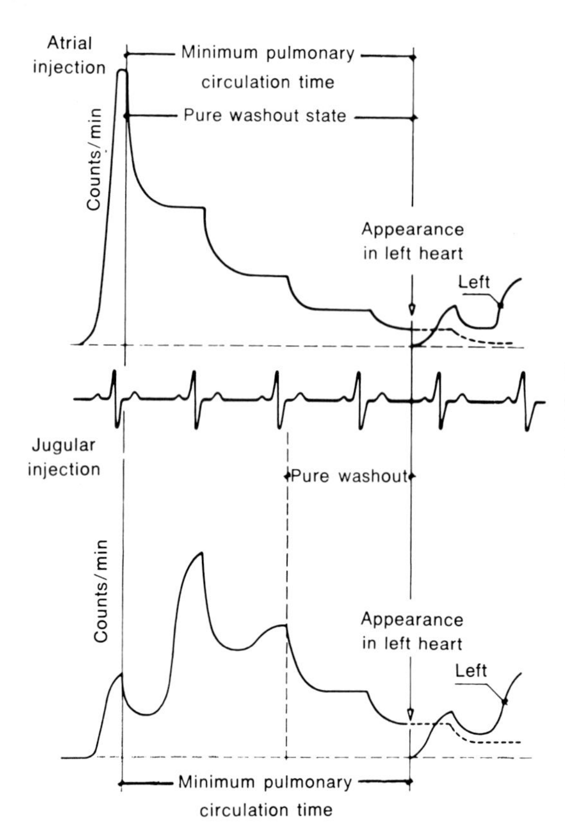

Fig. 5-2. Schematic representation of precordial activity after injection of radioiodinated human serum albumin into right atrium *(upper tracing)* and into external jugular vein *(lower tracing)*. (From Donato, L.: Prog. Cardiovasc. Dis. **5**:1, 1962. By permission.)

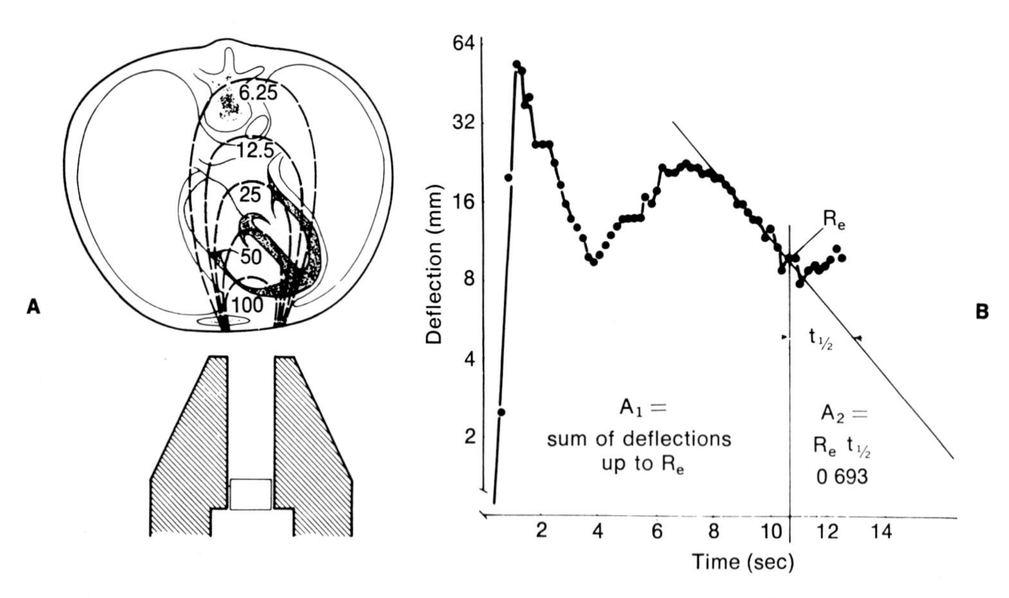

Fig. 5-3. A, Geometric relationship between counter and chest in radiocardiography. **B,** Semilogarithmic plot of radiocardiogram and extrapolation of downslope of left curve. (From Donato, L.: Semin. Nucl. Med. **3**:111, 1973. By permission.)

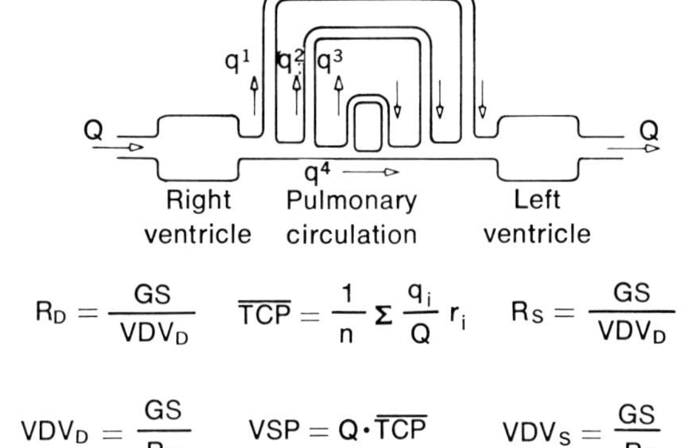

$$R_D = \frac{GS}{VDV_D} \qquad \overline{TCP} = \frac{1}{n} \Sigma \frac{q_i}{Q} r_i \qquad R_S = \frac{GS}{VDV_D}$$

$$VDV_D = \frac{GS}{R_D} \qquad VSP = Q \cdot \overline{TCP} \qquad VDV_S = \frac{GS}{R_S}$$

Fig. 5-4. Analog model of cardiopulmonary circulation. (From Donato, L.: Prog. Cardiovasc. Dis. **5:**1, 1962. By permission.)

curves has intrigued many of the ablest investigators for the last ten years but has eluded a fully satisfactory solution. The compartment of interest must be discretely labeled and monitored, or else adequate information must be obtained from each of the compartments involved to obtain a full solution from *compartment-system analysis*.

With the scintillation camera supplemented with data storage and computer systems it is possible to discretely monitor each of the components of the central circulation on the basis of visual identification of anatomic sites.[11,23,31,32] An example of the data obtained with such a system is shown in Fig. 5-5. Such a system provides a great deal of information for input into a comprehensive analysis of the chamber radionuclide dilution curves. However, analysis of "low-frequency" data from the combined curves does not contribute materially to

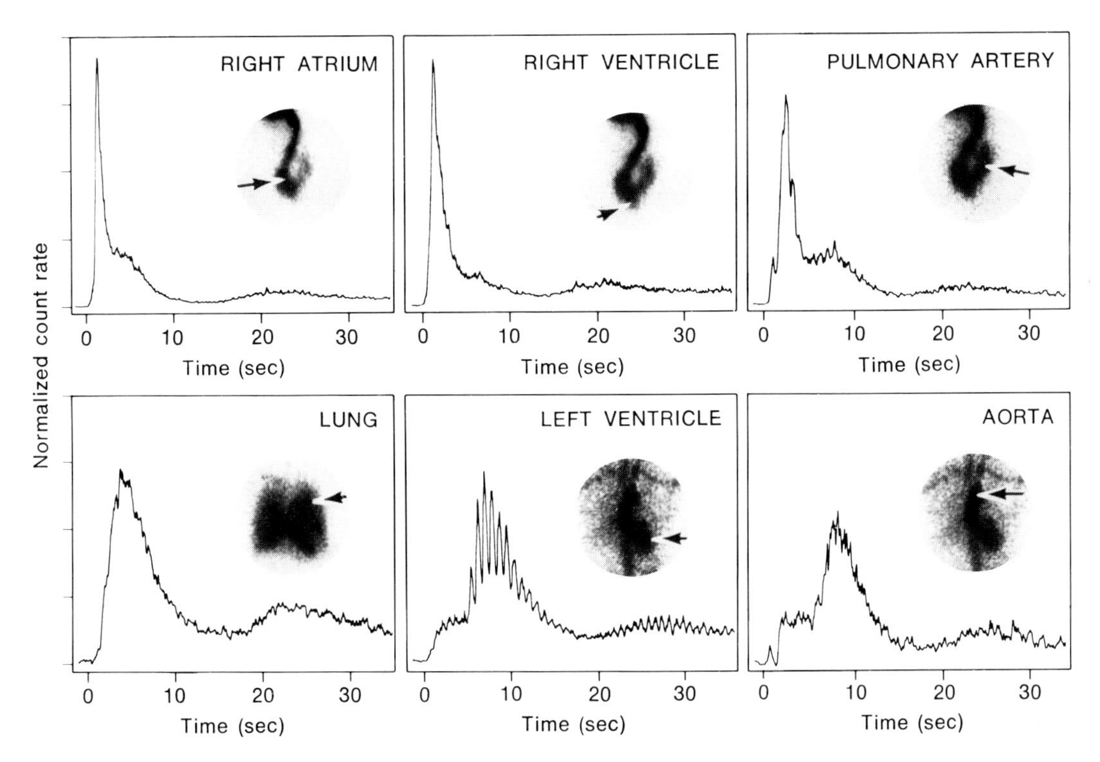

Fig. 5-5. Strip-chart recordings of passage of radionuclide through six areas of interest in central circulation. Arrows indicate anatomic structure selected for each recording. (From Van Dyke, D. C., Anger, H. O., Sullivan, R. W., Vetter, W. R., Yano, Y., and Parker, H. G.: J. Nucl. Med. **13:**585, 1972.)

evaluation of left ventricular performance. It has been established that even with low-frequency radionuclide dilution curves from the selected areas of interest obtained with the camera, there is still a very sizable interference between counts from the various chambers of the central circulation. Adequate correction of these curves for their mutual interference is still an unsolved problem. The relative counting efficiencies needed for cross-talk background correction might be obtained with an injection into each major region; however, this would mean that it is necessary to have access to all heart chambers with a catheter, and in addition, this leads to some very complicated analysis.

Left ventricular ejection fraction has been accurately measured from static radionuclide images[30] or from electrocardiographically gated images[29] obtained following peripheral venous injection of a radionuclide. The gated technique in particular has been widely applied to the clinical problems of patients with heart disease. With this technique images of the left ventricle are generated at end of diastole and end of systole and volumes computed from the image areas using methods similar to analysis of contrast ventriculography.[4,13] Left ventricular ejection fraction has also been successfully measured from radionuclide time-activity curves obtained during the first transit of the tracer through the left ventricle. Several methods of first-transit (or dynamic) radionuclide angiocardiography that differ as to recording device and site of tracer injection have been developed.

DYNAMIC RADIONUCLIDE ANGIOCARDIOGRAPHY USING THE ANGER SCINTILLATION CAMERA WITH COMPUTER ANALYSIS

An inherent advantage of cardiac chamber volume measurement with radionuclides is that count rate is a function of volume without depending on chamber geometry (Chapter 1). Contrast radiographic measurement of volume assumes that the left ventricle is an ellipsoid of revolution.[2] An inherent disadvantage of radionuclide methods is the necessity of correcting the chamber count rate for radiation scattered from adjacent tissues. Van Dyke and associates[31] recognized the contribution of scattered radiation to the left ventricular count rate and developed an empirical correction so that left ventricular ejection fraction (LVEF) calculated from time-radioactivity curves was comparable to EF obtained from contrast left cineventriculography. Using a scintillation camera with a television camera recording the passage of the radionuclide through the heart and lungs, Van Dyke and co-workers determined the amount of radioactivity contained in the left ventricle as a function of time by measuring the light output from the television monitor during videotape playback (Fig. 5-5). Correction for radiation scatter was made by recording a time-activity curve from a ring-shaped area surrounding the left ventricle and subtracting this curve from that obtained from the left ventricular area. Using this method, an accurate measurement of LVEF was obtained in respect to contrast cineventriculography ($r = 0.85$; $N = 16$). This method used [^{99m}Tc] albumin as the radionuclide injected into either the superior vena cava or a peripheral vein. With LVEF, EDV can be calculated from cardiac output (CO) and heart rate (HR) as follows:

$$\mathrm{LVEDV} = \frac{\mathrm{CO/HR}}{\mathrm{LVEF}} \quad (1)$$

LVEDV calculated from LVEF is only accurate in patients without mitral or aortic regurgitation. Weber and associates[32] have successfully applied the method of Van Dyke and have developed computer techniques to simplify the calculations.

Steele and colleagues[26] have modified the technique of Van Dyke by using a wedged pulmonary artery catheter for delivery of the radionuclide and a scintillation camera–computer system for data acquisition and analysis.[2,6] With the patient in the 40-degree right anterior oblique projection, 10 to 15 mCi of ^{99m}Tc as pertechnetate was injected, and the study was recorded at 20 frames/sec. with an Anger camera interfaced to a digital computer. For generation of time-activity curves, an area of interest coincident with the left ventricle at the end of diastole was constructed with the computer (Fig. 5-6). A semiannular ring surrounding the left ventricle was also constructed as a background area (Fig. 5-6). Time-activity curves were obtained from each area; following subtraction of the background curve from the left ventricular curve, a corrected left ventricular time-activity curve was generated (Fig. 5-7).

The LVEF was calculated from the beat-

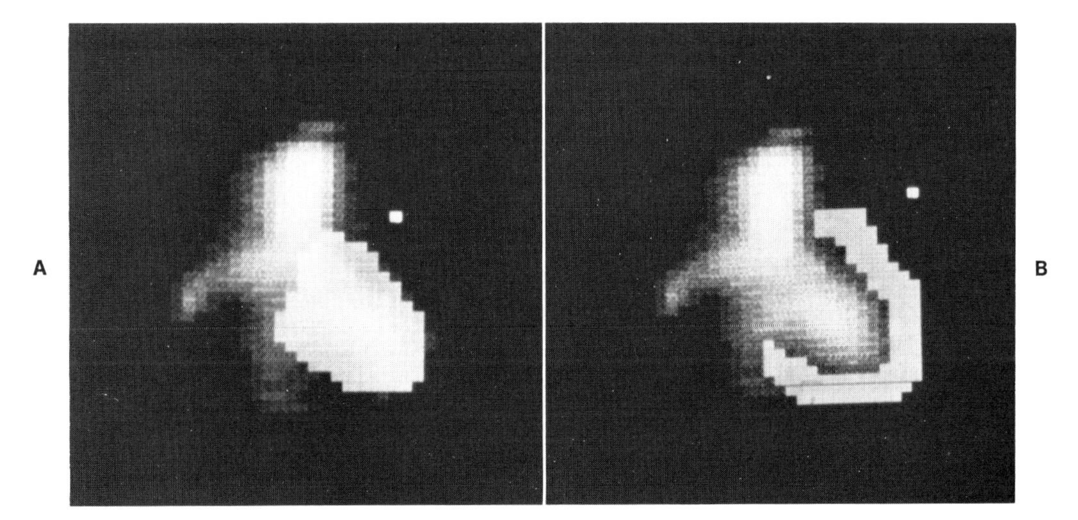

Fig. 5-6. Areas of interest defined for left heart study. **A,** Left ventricular area. **B,** Semiannular ring area for background correction. (From Steele, P., Kirch, D., Matthews, M., and Davies, H.: Am. J. Cardiol. **34:**179, 1974.)

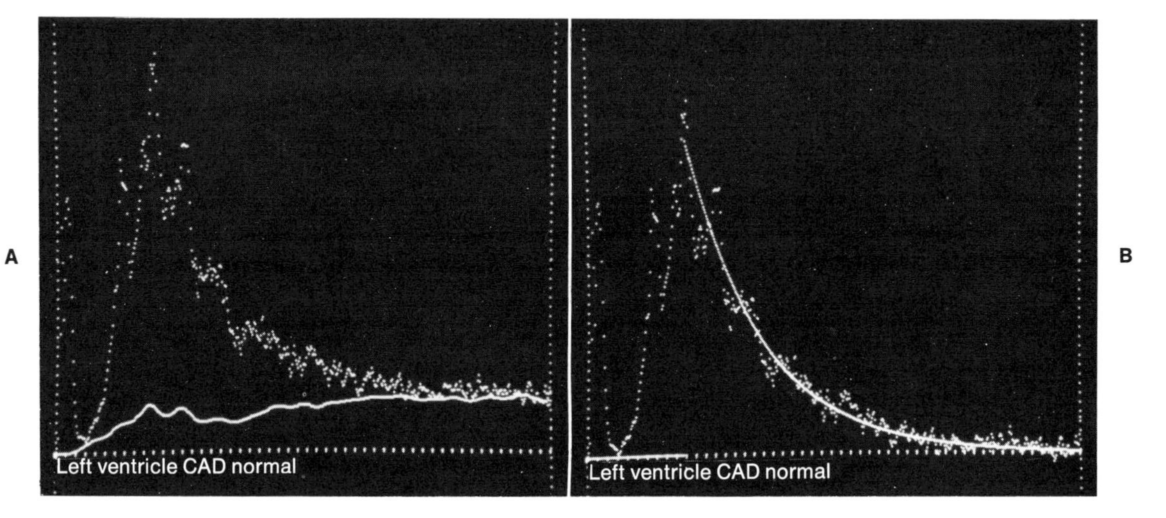

Fig. 5-7. Time-activity curves extracted from areas of interest coincident with left ventricle and background. Spike at far left in each photograph represents radioactivity in catheter as it traverses right heart. Prompt entry of activity into left ventricle is apparent. **A,** Curves from left ventricle and its background *(solid line)*. Ejection fraction is 0.53, $\frac{d - s}{d}$. **B,** Curve corrected for background and fitted with single exponent. (From Steele, P., Kirch, D., Matthews, M., and Davies, H.: Am. J. Cardiol. **34:**179, 1974.)

to-beat fractional fall in count rate (end of diastole to end of systole) divided by the background-corrected end-diastolic count rate (Fig. 5-7). In twenty-nine patients, the LVEF determined in this way correlated ($r = 0.84$) with that obtained by contrast cineventriculographic studies.

A second inherent disadvantage of dynamic studies is the low count rate obtained from the left ventricular area of interest. In gated studies there is no effective limit to counts obtained from the left ventricle because the tracer remains in the vascular system and data acquisition can continue to a predetermined number of counts. An advantage of the wedge injection dynamic method is the relatively high count rates obtained from the left ventricle because the tracer does not traverse the pulmonary vasculature. Between 500 and 1000 counts/50 msec (10 to 20 kcounts/sec) from the left ventricular area of interest are obtained, which exceeds the count rate usually achieved from the left ventricle following right-sided injection.

The high count rate obtained from the left ventricle following injection of tracer into a wedged catheter allows formation of a time-activity curve from an area of interest coincident with the left ventricular minor axis. This curve can be used to compute left ventricular mean circumferential fiber-shortening velocity (VCF) and systolic ejection rate (SER).[27] For calculation of VCF an area of interest was constructed coincident with the minor axis by bisection of the left ventricular major axis (apex to aortic valve) (Fig. 5-8) and a time-activity curve generated (Fig. 5-9). Mean VCF was calculated in circumferences per second, as follows:

$$\mathrm{VCF} = \frac{\sqrt{\mathrm{CED}} - \sqrt{\mathrm{CES}}}{\mathrm{ET}} \cdot \frac{1}{\sqrt{\mathrm{CED}}} \qquad (2)$$

where

$\sqrt{\mathrm{CED}}$ = Square root of count rate at end of diastole

$\sqrt{\mathrm{CES}}$ = Square root of count rate at end of systole

ET = Ejection time

Ejection time was calculated from the time-activity curve by identifying the frame at the onset of ejection and at end of systole and then counting the intervening frames. The square roots of the end-diastolic and end-systolic count rates were used to convert an area function (counts from minor axis) to a linear function to correspond with the contrast ventriculographic method for measuring VCF.[12]

Left ventricular mean SER can be calculated from the left ventricular time-activity curve in end-diastolic volumes per second by a modification of the method of Peterson[20] as follows:

$$\mathrm{SER} = \frac{\mathrm{LVC_{ED}} - \mathrm{LVC_{ES}}}{\mathrm{ET}} \cdot \frac{1}{\mathrm{LVC_{ED}}} \qquad (3)$$

where

$\mathrm{LVC_{ED}}$ = End-diastolic count rate
$\mathrm{LVC_{ES}}$ = End-systolic count rate

In computing VCF and SER, the time-activity curves were not corrected for scattered radiation, and consequently the indices derived from the radionuclide data were not numerically identical with indices obtained from contrast ventriculography, but the correlation was quite good ($r = 0.90$ and $r = 0.98$, respectively; $N = 31$).[27] Thus a dynamic radionuclide method for measuring LVEF, VCF, and NSR from time-activity curves is available, and these measurements are accurate, compared with contrast studies. Since left heart catheterization is not required, the method is relatively atraumatic, and the short half-life of ^{99m}Tc (6 hr) would allow serial studies of left ventricular performance at frequent intervals.

Injection of the radionuclide into a wedged catheter results in a left ventricular area of interest count rate that approaches the maximum count rate (50 kcounts/sec) that an Anger scintillation camera can handle without image distortion. Although this count rate can be achieved with wedge injection, it is still not adequate to obtain time-activity curves for left ventricular axes other than the major and minor axes. Consequently, it is impossible to dynamically

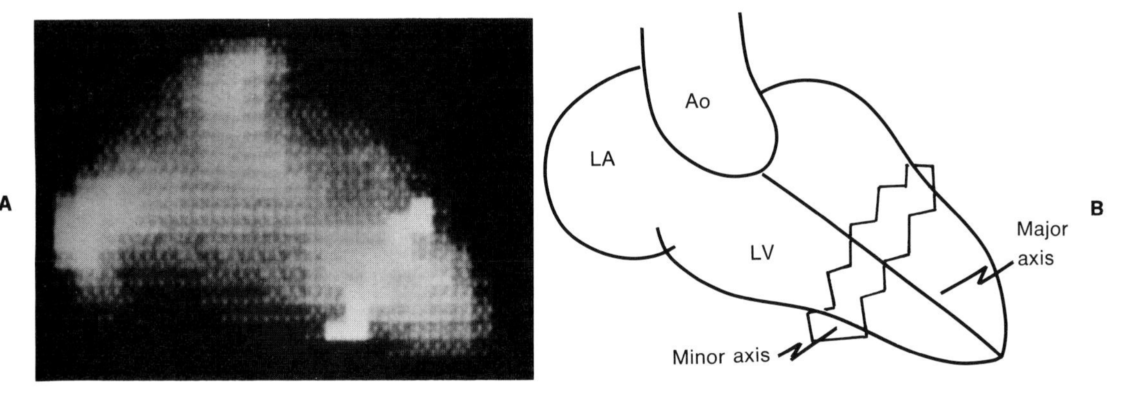

Fig. 5-8. RAO summed image, **A,** and drawing, **B,** illustrate area of interest coincident with left ventricular *(LV)* minor axis, which bisects left ventricular major axis–aortic valve to left ventricular apex. Left atrium *(LA)* and ascending aorta *(Ao)* are identified. (From Steele, P., LeFree, M., and Kirch, D.: Am. J. Cardiol. **37**:388, 1976.)

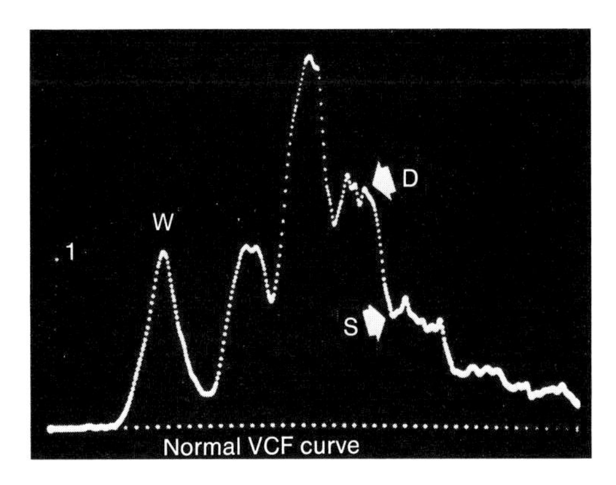

Fig. 5-9. Time-activity curve obtained from minor axis area of interest in normal patient. End of diastole *(D)* (onset of ventricular ejection) and end of systole *(S)* are indicated. Counts are on vertical axis with time on horizontal axis. Fractional reduction in count rate for beat indicated is 0.54 and for following beat, 0.51. Peak at left *(W)* represents activity in catheter as it traverses left ventricular area of interest. (From Steele, P., LeFree, M., and Kirch, D.: Am. J. Cardiol. **37**:388, 1976.)

acquire data for analysis of left ventricular segmental contraction with the conventional Anger scintillation camera.

The reason for the count rate limitation of the Anger camera is the serial nature of the detection scheme; only one scintillation event can be dealt with and any coincidental events must either be ignored or, if detected, will produce erroneous data (Chapter 1). Recognition of this aspect of the Anger scintillation camera performance suggested the possibility of developing an imaging system that operates by a parallel detection scheme. In preliminary studies with an image-intensifier camera, left ventricular segmental contraction can be studied with dynamically acquired data.[16] Segmental contraction abnormalities have been adequately studied using the gated approach, in which images are obtained without time constraint.[33]

An additional advantage of wedged pulmonary artery injection of a radionuclide is that bolus entry of the tracer into the left heart results, which allows the left ventricular time-activity curve to be fitted with a single exponent for measurement of end-diastolic volume. The time constant (a) of the curve is computed and end-diastolic volume (EDV) calculated from the following equation:

$$EDV = \frac{SV}{[1 - e^{-at}]} \quad (4)$$

where

SV = Stroke volume calculated from cardiac output and heart rate
t = 60/HR in seconds

The total ejection fraction (TEF) for the ventricle is computed from the fractional drop in the corrected curve relative to the baseline curve between successive points of end of diastole and end of systole.[7,31] This calculation is performed using as many beats as can be detected and then averaged. The end-diastolic volume can then be calculated from the ejection fraction and stroke volume by the following equation, which is implicit in the definition of ejection fraction:

$$EDV = \frac{SV}{TEF} \tag{5}$$

Equation 4 is derived from the work of Newman and associates,[19] who showed that for an idealized reciprocating pumping chamber with competent valves the time constant (a) of the exponent fitted to the indicator washout curve is described as follows:

$$-a = (\log A_n - \log A_o)/t_n \tag{6}$$

where

A_o = Amount of indicator injected
A_n = Amount of indicator after n beats
t_n = Time for n beats

Newman and co-workers also showed that an equivalent relation for a is as follows:

$$-a = (HR) \log \left[\frac{ESV}{(ESV + SV)}\right] \tag{7}$$

where ESV is the end-systolic volume of the pumping chamber. Rearrangement yields the following:

$$-a\,(HR) \log \left[\frac{(EDV - SV)}{EDV}\right] \tag{8}$$

and

$$e^{-(60)\left(\frac{a}{HR}\right)} = 1 - \left(\frac{Sv}{EDV}\right) \tag{9}$$

and

$$1 - e^{-at} = \frac{SV}{EDV} \tag{10}$$

Solving equation 7 for EDV, equation 1 is obtained as follows:

$$EDV = \frac{SV}{[1 - e^{-at}]} \tag{11}$$

The equivalence of equations 4 and 5 for determining EDV is valid only under the presumption of valvular competence. If one assumes that the outflow valve only is incompetent, this work can be extended readily so that quantity $(1 - e^{-at})$ is seen to be equal to the forward ejection fraction (FEF). Thus equation 4 provides the correct value of EDV even under the conditions of insufficiency of the outflow valve of the chamber under study. By this rationale, a discrepancy of greater than 10% between the EDV computed from equations 4 and 5 is the fundamental quantitative indication of valvular insufficiency. (In the work presented here, equation 1 has been found to provide accurate values for the EDV of the left ventricle in the presence of aortic insufficiency.)

With wedge injection of radionuclide, bolus entry into the left atrium also occurs, and it can therefore be assumed that this relation also applies to the left atrium under conditions of mitral insufficiency. With a pair of time-activity curves obtained following wedge injection (bolus entry into left heart) from the left atrium and left ventricle, quantitation of combined mitral and aortic regurgitation has been accomplished.[15]

Kenny and associates[14] have developed an interesting method for achieving bolus entry of radionuclide into the left heart in which ^{15}O as carbon dioxide was inhaled. An accelerator was required to make ^{15}O, which might limit its application to patients, but preliminary results are interesting.

Measurement of LVEF from first-transit (dynamic) radionuclide data requires that the LV time-activity curve be corrected for the effects of scattered radiation (background). This was recognized by Van Dyke and associates[31] and an empirical correction devised so that an accurate EF, as compared with contrast cineventriculography, was obtained. The correction consisted of generation of a time-activity curve from tissues surrounding the LV and subtraction of this curve from the LV curve to obtain the correct time-activity curve. Schelbert and co-workers[22] explored the effect that variation in the background area of interest had on the calculation of LVEF. In six patients LVEF from cineventriculography averaged 0.59. LVEF from an uncorrected LV time-activity curve underestimated LVEF, average 0.45. With a semiannular ring background area, in which the ring avoided the aorta, LVEF was in close agreement with contrast, average 0.58. A background area, which formed a ring around the LV, including the ascending aorta, resulted in overestimation of LVEF, average 0.90. Schelbert and associates[22] also studied the effect of imprecise definition of the LV area of interest on LVEF and observed that LVEF was overestimated when the LV area failed to include the entire LV, and when the ascending aorta was included in the LV area, LVEF was underestimated. LVEF was only slightly underestimated when the LV area was only minimally larger than the LV and avoided the aorta. Additional useful data from this study were that LVEF obtained from time-activity curves generated following peripheral venous or central venous injection was not different.

Kurtz and associates[18] applied first-transit radionuclide methodology to children and obtained an accurate measurement of LVEF. These authors injected ^{99m}Tc through a peripheral vein and undertook experiments to develop precise definition of the left ventricular and background areas (Chapter 7). What is clear is that accurate measurement of LVEF from first-transit dynamics requires background subtraction and careful definition of the LV area.

The radionuclide methods that have been developed to measure LVEF can be applied to measurement of right ventricular ejection

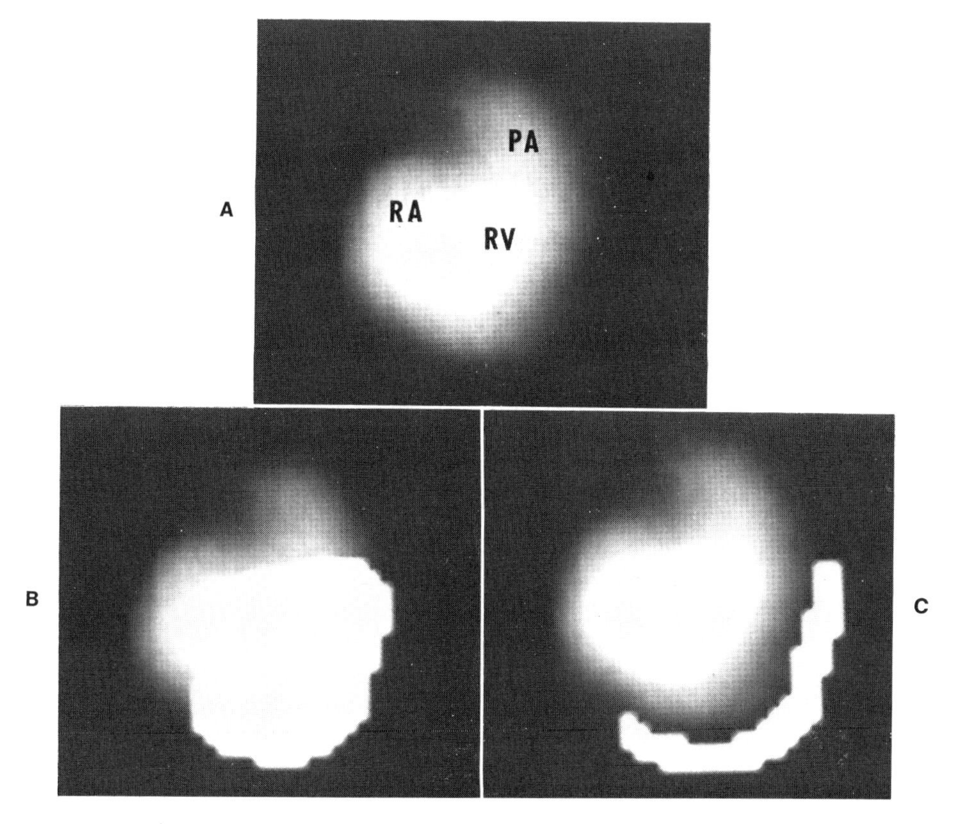

Fig. 5-10. Summed image of right heart in RAO projection, **A,** with right atrium *(RA),* right ventricle *(RV),* and pulmonary artery *(PA)* indicated. Right ventricular area of interest, **B,** and semiannular background, **C,** are shown.

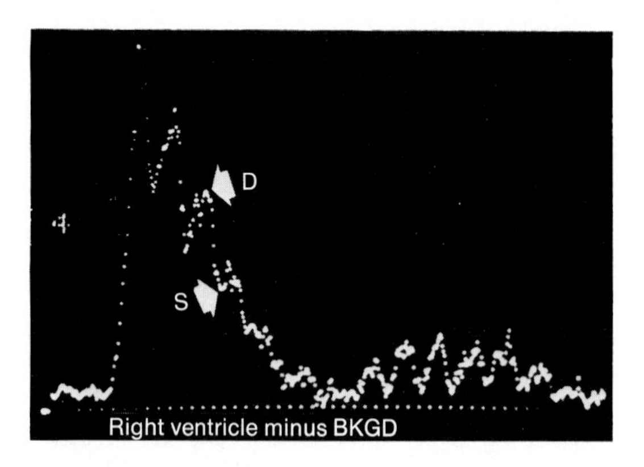

Fig. 5-11. Background-corrected time-activity curve from right ventricular area of interest. Ejection fraction is calculated from fractional fall in counts from end of diastole *(D)* to end of systole *(S)* divided by end-diastolic count rate. Ejection fraction is 0.47 for indicated beat, 0.49 for preceding beat, and 0.49 for following beat (mean, 0.48). (From Steele, P., Kirch, D., LeFree, M., and Battock, D.: Chest **70:**51, 1976.)

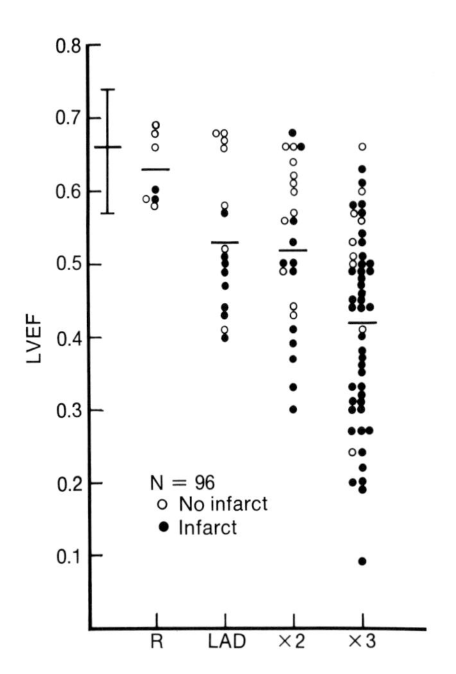

Fig. 5-12. Relationship between LVEF and extent and location of coronary disease. Solid circles identify patients with electrocardiographic evidence of myocardial infarction. Horizontal lines represent average values. Normal average and range are indicated at left. R is isolated obstruction of right coronary artery; LAD, of left anterior descending; ×2, of two vessels; ×3, of three vessels. (From Steele, P., Kirch, D., LeFree, M., and Battock, D.: Chest **70:**51, 1976.)

fraction (RVEF). Steele and colleagues[25] applied the wedge injection radionuclide method to the right heart with injection of ^{99m}Tc into the superior vena cava. Time-activity curves were obtained from a right ventricular area of interest and a semiannular background area with an Anger camera interfaced to a digital computer in a manner analogous to that for the left ventricle (Figs. 5-10 and 5-11). Background correction for the right and left ventricles should be identical, since the injection sites, central vein and pulmonary wedge, were similar. This radionuclide technique was compared to the biplane contrast cineventriculography, area-measurement method of Arcilla and associates[1] in forty-three men with good agreement ($r = 0.80$).

An example of the clinical application of dynamic radionuclide angiocardiography using an Anger scintillation camera and a digital computer is the measurement of RVEF and LVEF in patients with coronary artery disease.[25] In this study the wedge injection of ^{99m}Tc was used for measurement of LVEF and superior vena caval injection for RVEF. In ninety-six men with coronary disease, a relationship between LVEF and the number of diseased coronary arteries as defined by coronary arteriography was observed (Fig. 5-12). All seven men with involvement of only the right coronary artery had normal LVEF. LVEF was decreased in nine of fifteen men (60%) with involvement of the left anterior descending artery, and in fourteen of twenty-three men (61%) with double-vessel and in forty-three of fifty-one men (84%) with triple-vessel obstruction (Fig. 5-12). A history of myocardial infarction was associated with impairment of LVEF in these men (Fig. 5-12).

RVEF was preserved relative to LVEF in these men. RVEF was normal in the men with isolated right coronary disease, and was normal in twelve men (80%) with left anterior descending obstruction. RVEF was normal in 83% of those with double-vessel and in 57% of those with triple-vessel obstruction (Fig. 5-13).

RVEF was preserved in comparison to

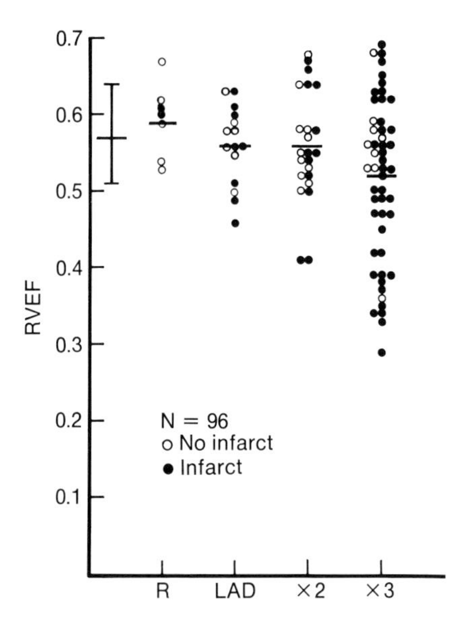

Fig. 5-13. Relationship between RVEF and extent and location of coronary disease. Solid circles identify patients with electrocardiographic evidence of myocardial infarction. Average values are indicated by horizontal lines. Normal average and range are indicated on left. R is isolated obstruction of right coronary; LAD, of left anterior descending; ×2, of two vessels; ×3, of three vessels. (From Steele, P., Kirch, D., LeFree, M., and Battock, D.: Chest **70:**51, 1976.)

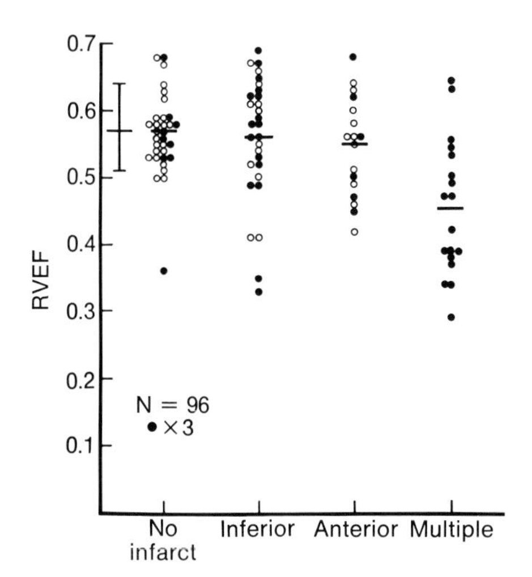

Fig. 5-14. Relationship between RVEF and occurrence, location, and extent of myocardial infarction. Solid circles indicate patients with triple-vessel disease. (From Steele, P., Kirch, D., LeFree, M., and Battock, D.: Chest **70:**51, 1976.)

LVEF in men with a history of infarction. Only three men (9%) without a history of infarction had abnormal RVEF, and one of these had triple-vessel involvement (Fig. 5-14). Twenty-six of sixty-four men (41%) with a history of infarction had depressed RVEF, and twenty (77%) of these had triple-vessel involvement (Fig. 5-14).

Measurements of ejection fraction with a radionuclide has the advantage that the tracer does not disturb cardiac performance. Radionuclide methods can be used to sequentially measure LVEF and RVEF (as in the study just discussed), or EF can be serially measured at frequent intervals.

PORTABLE SCINTILLATION PROBE METHOD FOR EVALUATION OF CENTRAL CIRCULATORY DYNAMICS AT THE BEDSIDE

The studies discussed in this chapter have shown that LVEF can be measured accurately with a scintillation camera for recording of high-frequency time-activity curves from an area of interest coincident with the LV. Since location of the scintillation camera and its data storage system at the patient's bedside is not easily achieved, the use of a portable scintillation probe that is collimated to record activity from the LV has been investigated.

The initial studies in which the scintillation camera with data storage and area-of-interest capability was used showed that EF, read from recording of the left ventricular area of interest without modification of the baseline, was approximately half the true value.[31] Dynamic studies of a single-chamber model demonstrated that this was not a fault of the recording system but was due to activity in adjacent structures.

Since an accurate measure of EF is a prime object of the single-probe bedside method, success of the method again depends on the establishment of the baseline from which the EF is to be measured. The true baseline must result in a constant EF throughout the left ventricular phase. This requirement is met by any one of a family of baseline curves (Fig. 5-15). From compari-

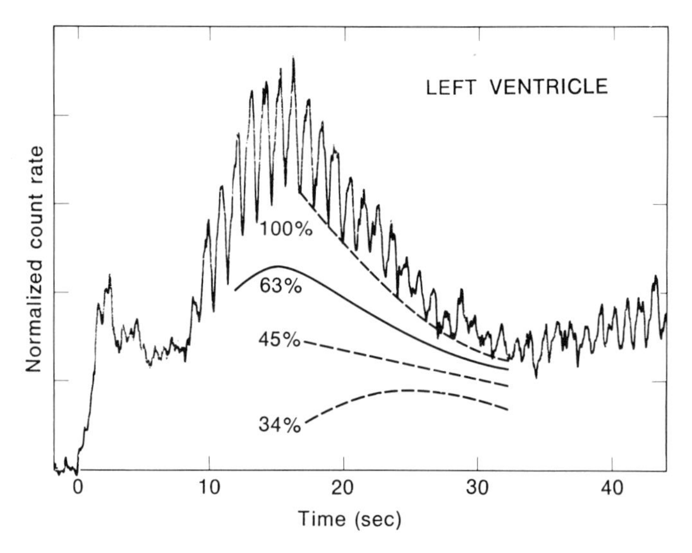

Fig. 5-15. Recording from LV. Some of family of curves that result in constant EF through left ventricular phase. Solid curve (63%) represents correct baseline obtained by simultaneous quantitative angiocardiography. (From Van Dyke, D. C., Anger, H. O., Sullivan, R. W., Vetter, W. R., Yano, Y., and Parker, H. G.: J. Nucl. Med. **13**:585, 1972.)

son with the results of contrast angiography, it is apparent that the correct baseline intercepts the left ventricular curve at its low point after passage of the bolus and before recirculation. All curves above a certain value meet this requirement; this indicates that the true value must be established from each patient's record. Ideally, one must look in front of, behind, and around the ventricle to measure the cross talk and background interference. In practice, this can be approximated by the recording of the transit of tracer through a ring-shaped area of interest around the LV (Fig. 5-16). The optimal width of the ring is determined by trial and error. The recording of the activity in the ring immediately surrounding the ventricle shows fluctuations due to cross talk when the ventricle is full. If only the low points of that curve (when the ventricle itself is contracted and out of sight) are taken and if the recording is done at a sensitivity that brings the low point to the same value as the low point of the left ventricular curve, a satisfactory baseline curve is established (Fig. 5-19).

Equipment and methods

Some of the many advantages of a simple precordial scintillation counter are as follows: (1) portability, allowing utilization of the technique at the bedside of desperately sick patients; (2) 10 to 100 times better statistics, which makes possible reduction of administered dose; (3) minimum data manipulation and analysis time, making routine clinical application feasible; (4) usableness in serial studies that do not exceed allowable dosage to the patient or linear range of the instruments; (5) minimal cost, making wide application possible; and (6) usableness in conjunction with radionuclides such as ^{113m}In, which is not suitable for conventional scintillation cameras.

LVEF can be measured with the use of [^{99m}Tc] pertechnetate; however, for measurement of cardiac output (CO) and for calculation of left ventricular end-diastolic volume and pulmonary blood volume (PBV), a tracer that remains within the circulation is required. In preference to ^{99m}Tc-labeled albumin, we tend to use ^{113m}In. Indium binds with transferrin and can therefore be used

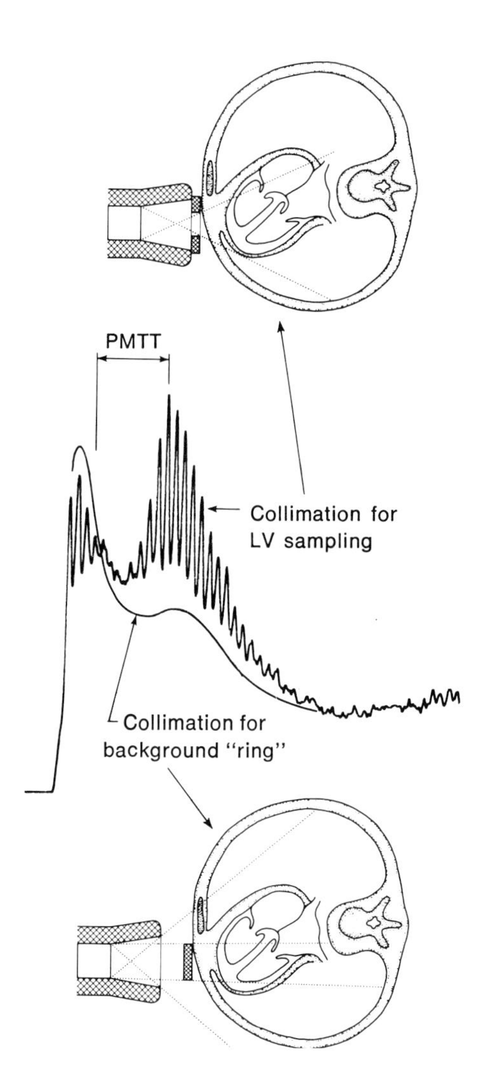

Fig. 5-16. Schematic representation of single-probe collimator configuration for recording ventricular radionuclide concentration curve and background ring. Pattern in center is characteristic of those obtained in normal subject.

to measure blood volume (BV) and CO.[10]* ^{113m}In is easily prepared from a commercially available long-lived generator and is short lived (half-life is 1.7 hr), allowing serial studies within a short period of time without exceeding allowable radiation limits. Alternatively, in the absence of indium, CO can be measured at the bedside by thermal dilution with the Swan-Ganz balloon-tipped, flow-directed catheter[8]; this allows pertechnetate to be used for measurement of LVEDV and PBV.

The portable scintillation probe used is a standard, commercially available unit with a 5 cm by 5 cm NaI crystal. A portable stand carries the high-voltage power supply, count rate meter with a capacity of 10 million counts/min and time constant of less than 0.1 sec, and high-speed strip-chart recorder. Collimation for obtaining the left ventricular radionuclide concentration curve is shown at the top of Fig. 5-16.

The scintillation probe technique requires determination of the midpoint of the LV for positioning of the probe. This is accomplished by the process of taping a radiopaque marker (paper clip) over the estimated midpoint of the LV and obtaining a chest film with the patient supine in mid-inspiration. With the use of the roentgenogram, the midpoint of the LV relative to the marker is determined. The correct precordial position is then marked with a felt-tip pen to facilitate accurate repositioning for serial studies.

A central venous (superior or inferior vena cava) catheter is required for delivery of radionuclide. Studies are performed with the patient supine and the probe perpendicular to the precordium. Radionuclide in the amount of 1 mCi is injected into the catheter, the catheter is flushed with saline solu-

---

*Indium is very insoluble except in acid pH. Thus care must be taken not to dilute the injected material with, or introduce it into a line containing, solutions that will even transiently raise the pH, since the indium will rapidly precipitate. The best way to avoid this problem is to withdraw the patient's blood into the catheter just prior to introduction of the indium.

tion or the patient's own blood, and the precordial radiocardiogram is recorded. At 5 min another recording is made for determination of the equilibrium count rate, and a blood sample is obtained for measurement of BV.

Collimation of the probe is then modified to the configuration shown in the lower part of Fig. 5-16 for recording activity immediately surrounding the LV. For this purpose, a second 500 μCi bolus of ^{99m}Tc or ^{113m}In is given. This record is matched in amplitude at the low point of the curve to the record from within the left ventricular area of interest to provide the correct baseline from which EF is calculated. LVEF is calculated as the ratio of peak-to-valley and peak-to-background deflection for the first four beats after the peak of the LV curve.

Quantitation of the bedside strip-chart recording, which is essentially the same as that presented for the scintillation camera method, is as follows: Cardiac output is determined, as with other indicator dilution methods, from the planimetric area A under the first pass with the LV washout replotted semilogarithmically to correct the recirculation. The count rate at equilibrium per unit time (C_{eq}) is then determined, and the CO is calculated from the following equation:

$$CO = \frac{C_{eq}}{A} \times BV \qquad (12)$$

where $A = \int C(t)dt$.

Pulmonary blood volume is obtained from the product of the CO and the pulmonary transit time (PTT). The PTT is estimated as the interval from ejection of label from the RV (75% of peak activity) to the peak activity in the LV.[28]

Stroke volume (CO divided by heart rate) is related to EF and EDV of the LV as follows:

$$LVEDV = \frac{SV}{LVEF} \qquad (13)$$

This equation is not valid when aortic or mitral regurgitation is present.

At the outset of our studies, we validated the measurements of BV, CO, and LVEF. BV was determined in nine patients with ^{113m}In and with radioiodinated (^{125}I-labeled) serum albumin with good correlation (r = 0.79) (Fig. 5-17). CO measured with the scintillation probe and ^{113m}In correlated (r = 0.78, N = 35) with CO measured with indocyanine green dye (by pulmonary artery injection, sampling from ascending aorta or brachial artery, and a modification of the Stewart-Hamilton method) (Fig. 5-18). LVEF was measured with the scintillation probe and ^{113m}In in thirty-three patients who also underwent LV cineangiography. The ventriculograms were filmed in the right anterior oblique projection, and the EF was calculated by the area-length method of Dodge and associates.[4] Correlation between the two techniques was good (r = 0.92) (Fig. 5-19).

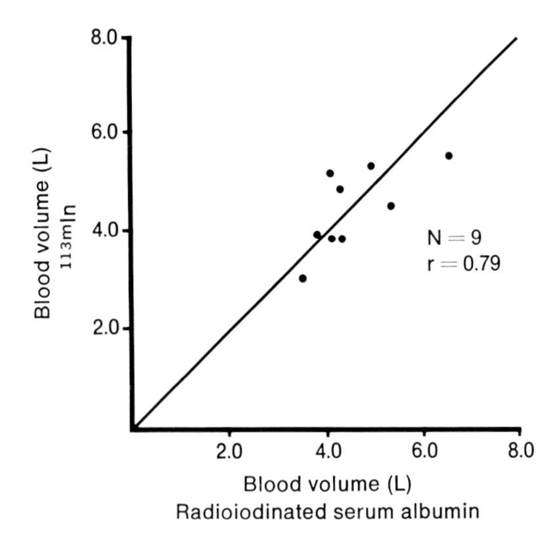

Fig. 5-17. Correlation of ^{113m}In BV with BV determined with ^{125}I-labeled serum albumin.

Confirmation of the external probe radiocardiographic measurement of CO with the Fick method and with dye dilution with arterial sampling has been established.[17,21] In addition, Donato recently presented data from Maseri and co-workers comparing consecutive measurements of CO performed by radiocardiography in the same subject by change of collimator type or placement and by peripheral or right atrial injection of tracer.[6]

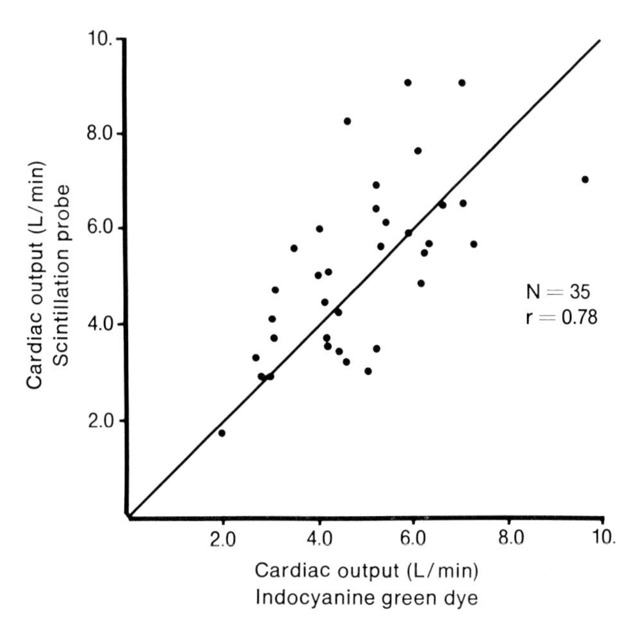

Fig. 5-18. Correlation of CO measured with scintillation probe and ^{113m}In with CO measured with indocyanine green dye.

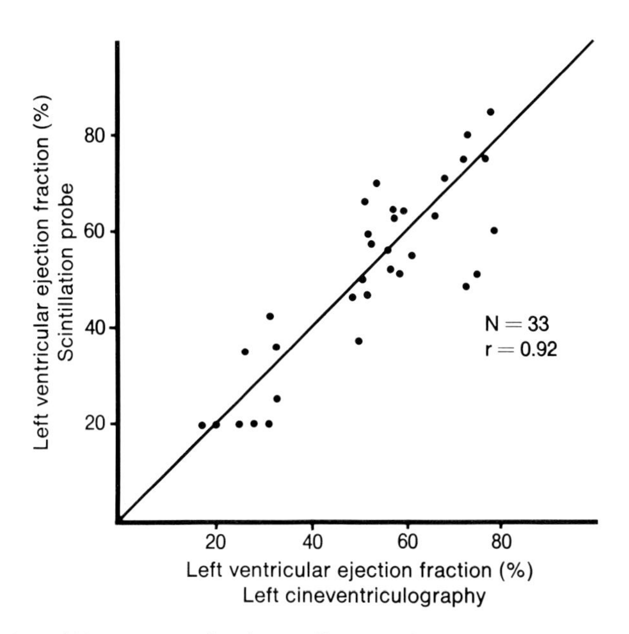

Fig. 5-19. Correlation of LVEF measured with scintillation probe and ^{113m}In with EF measured from LV cineangiography.

Further direct comparisons of the EF measured by the single-probe and collimator system used in these studies with the EF determined by contrast angiography and Fick CO have been reported by Berndt and associates.[3] We concluded that these results validate the usefulness of the technique of measuring the EF concomitantly with the CO at the bedside.

Features of radiocardiogram obtained with probe

A precordial radiocardiogram obtained with a typical portable probe and ^{113m}In from a normal subject is shown in Fig. 5-20. The essential features of such a strip-chart recording, which is immediately available at the patient's bedside, are as follows:

1. Overall transit time through the central circulation (time from injection to peak of LV): prolonged in circulatory failure and in the presence of increased PBV
2. Rate of appearance of label in RV (flow from SVC to RV): prolonged in severe circulatory failure
3. Downslope of RV washout curve: slow in circulatory failure and in the presence of dilated right atrium and right ventricle
4. PTT: decreased with decreased PBV and increased in circulatory failure (Fig. 5-22)
5. Loss of pulmonary valley between RV and LV: obliterated with markedly decreased PBV or markedly dilated right atrium and right ventricle (Figs. 5-22 and 5-23)
6. Area under the first pass of the indicator: inversely related to CO when related to the concentration after mixing in the blood (feature 11)
7. Counts from surrounding tissues after eclipsing the LV: provides the baseline against which EF is measured (Fig. 5-16)
8. EF: related to feature 7 (Figs. 5-15, 5-16 and 5-20)
9. Washout slope of LV: slow in failure and in left-to-right shunt (Fig. 5-24)
10. First recirculation: shortened in AV fistula, blunted in circulatory failure, and obliterated in left-to-right shunt
11. Concentration after complete mixing: small as compared to area under first pass (feature 6) in low CO or exceptionally large BV

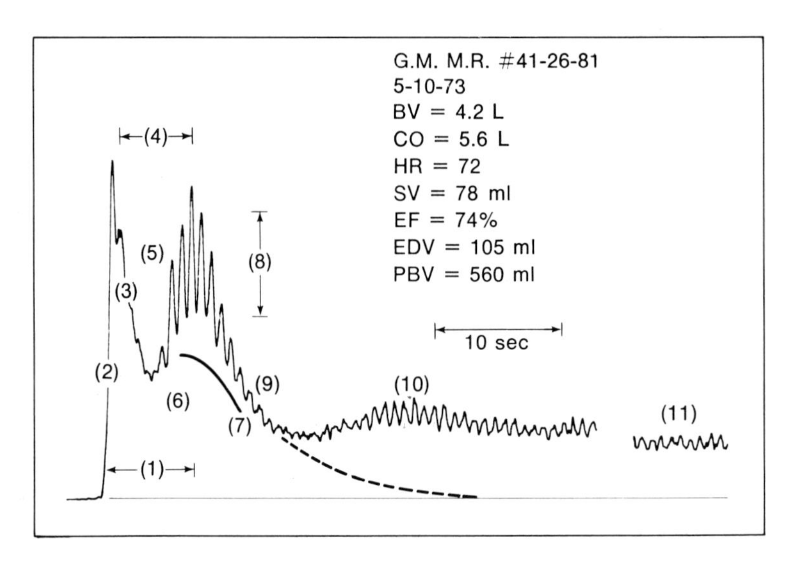

Fig. 5-20. Strip-chart recording of a precordial radiocardiogam in normal individual with the essential features enumerated. (See text for description.)

Clinical studies

We have performed more than 500 scintillation probe studies in over 200 patients. Most of the patients were seriously ill and were studied at the bedside with the scintillation probe recording device and ^{113m}In. The following examples illustrate the clinical usefulness of the scintillation probe technique.

This technique has been utilized in approximately seventy-five patients with acute respiratory failure (ARF) superimposed on chronic airway obstruction (CAO).[24] Patients with CAO frequently suffer episodes of respiratory infection (bronchitis, pneumonia), increased bronchospasm (asthma), pulmonary thromboembolism, and congestive heart failure. It can be exceedingly difficult to delineate on clinical grounds the pathophysiologic disturbances present in a given case of ARF. In particular, the presence or absence of LV dysfunction in these patients can be very difficult to determine. The definitions of the specific abnormalities of central circulatory function are of value because they make it possible to assess the effects of the differing modalities of therapy.

Fig. 5-21 is the precordial radiocardiogram of a man who was admitted with severe ARF caused by acute bronchitis with marked hypoxemia (Pa_{O_2}: 43 mm Hg) superimposed on serious CAO. Except for the rapid heart rate, the pattern was normal in all respects with a normal LVEF. The clinical value of this technique is that pattern recognition alone often provides the clinician with the information necessary for specific treatment of the pathophysiologic disturbances present. In this patient it was apparent from the pattern alone that neither major decrease or increase of PBV nor LV dysfunction was present. It was also demonstrated in this patient that decreased LVEF is not necessarily present in severe ARF with marked hypoxemia.

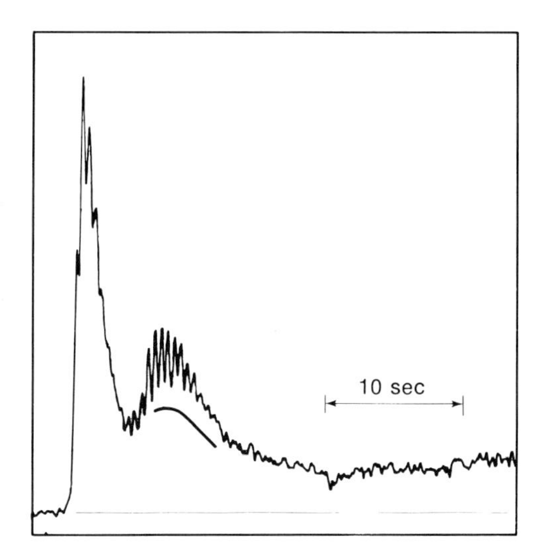

Fig. 5-21. Radiocardiogram obtained shortly after patient was admitted to hospital for severe ARF. Radiocardiogram is normal in all respects.

About 45% of our patients with ARF have a normal radiocardiographic pattern. There may be minor reduction of LVEF in some patients, but the average LVEF is within the normal range. The normal pattern allows the clinician to avoid drugs designed to increase cardiac performance, and therapy can be directed toward improving ventilatory function.

Fig. 5-22 is the precordial radiocardiogram obtained from a 75-year-old man who collapsed two days after cystectomy and ileal loop diversion for bladder carcinoma. Arterial blood gas measurement showed a Pa_{O_2} of 73 mm Hg and Pa_{CO_2} of 26 mm Hg. Ventilation-perfusion lung scintigraphy revealed multiple perfusion defects in regions with normal ventilation. The bedside radiocardiogram was grossly abnormal with greatly diminished pulmonary valley and rapid transit of the label from RV to LV (decreased PTT). Washout from the RV was not unduly prolonged; this indicated that loss of the pulmonary valley was primarily due to loss of PTT. This pattern is characteristic of marked reduction in pulmonary vasculature (PBV: 250 ml), characteristically due to massive embolus. Note the respiratory fluctuations in the LV portion of the record and the moderately reduced LVEF, which indicate some degree of LV dysfunction. The diagnosis of pulmonary embolic disease was further substan-

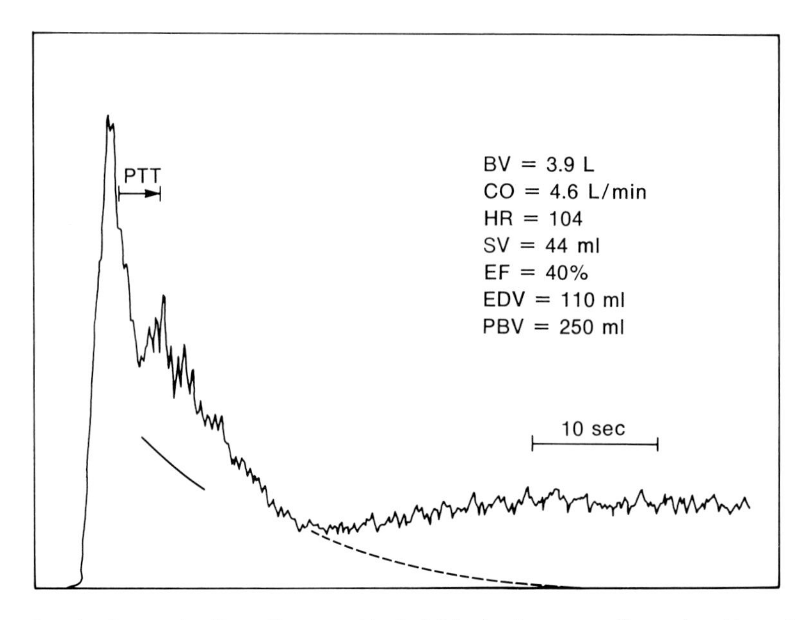

Fig. 5-22. Grossly abnormal radiocardiogram with diminished pulmonary valley and rapid transit of label from RV to LV (decreased PTT). Pattern is characteristic of marked reduction in pulmonary vasculature (PBV of 250 ml), characteristically due to massive embolus. Note respiratory fluctuations in LV portion of record and moderately reduced LVEF, which indicate some degree of LV dysfunction.

tiated with pulmonary angiography. This pattern (short PTT) is seen in major pulmonary embolism. Pulmonary embolism has been confirmed by perfusion lung scanning, pulmonary arteriography, or autopsy in all nine patients in whom short PTT has been observed. Serial studies reveal normalization of PTT in those who survive.

The scintillation probe has been useful in the clinical problem of cardiovascular collapse in which major pulmonary embolism and major LV dysfunction such as acute myocardial infarction can be difficult to differentiate. Pattern recognition alone will differentiate these two, since PTT is maintained in major LV dysfunction and LVEF is reduced.

Preliminary results of serial studies of patients with ARF suggest that PBV (CO × PTT) is quite stable and that a relatively small reduction in PBV (20% to 30%) may reflect episodes of less-than-massive pulmonary embolism. Appreciation of changes in PBV of this magnitude requires calculation of PBV, since these changes cannot be recognized from the patterns alone.

Fig. 5-23 is the bedside radiocardiogram

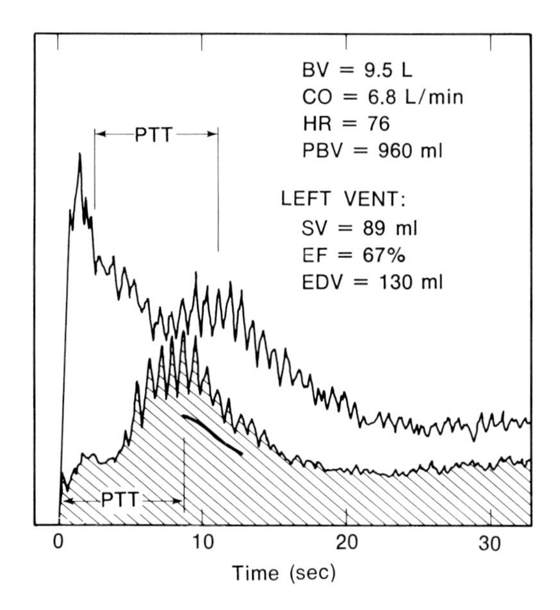

Fig. 5-23. Bedside radiocardiogram of patient in which the normal pulmonary valley is reduced *(unshaded portion)*. Injection of label directly into pulmonary artery *(shaded portion)* confirmed maintenance of PTT. Reduction or obliterations of pulmonary valley with maintenance of PTT is characteristic record of marked dilation of chambers of right side of heart.

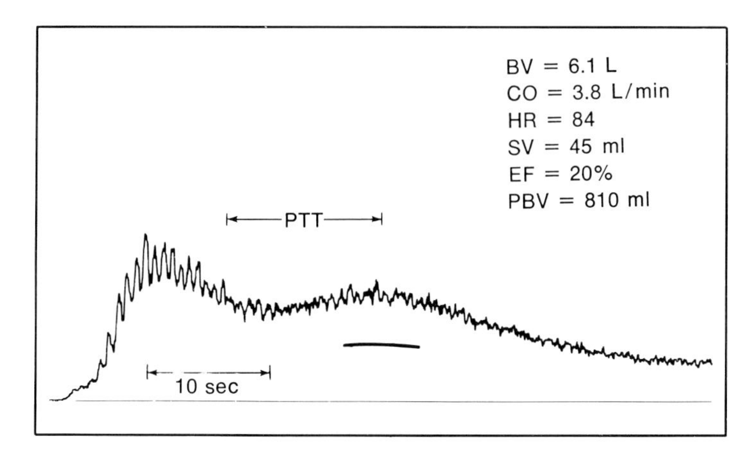

Fig. 5-24. Radiocardiogram from patient with ARF. Marked reduction in LVEF and slowed central circulation.

of a patient in which the normal pulmonary valley is absent (top unshaded portion of the radiocardiogram). Clinically, the patient had marked hypoxemia with physical findings of tricuspid regurgitation, hepatic enlargement, massive peripheral edema, and documented decline in his vital capacity and 1 sec forced expiratory volume from 3.5 to 2.9 L and 1.4 to 0.9 L, respectively, over the period of a year. Injection of the label directly into the pulmonary artery (bottom shaded portion of Fig. 5-23) indicated maintenance of the PTT, with a calculated PBV of approximately 10% of the BV. This was found by Giuntini and associates[9] to be a consistent value for PBV in patients with chronic lung disease as well as in normal subjects. Obliteration of the pulmonary valley with maintenance of PTT and PBV is the characteristic record of marked dilatation of the right atrium and right ventricle. About 35% of our patients with ARF have a pattern of absent pulmonary valley. To define the PTT in these patients, it may be necessary to inject radionuclide into a catheter placed in the pulmonary artery. With short PTT, very rapid appearance of the label in the LV will occur, whereas if loss of the pulmonary valley is due to slowed RV washout, a normal PTT will be recorded after injection of radionuclide into the main pulmonary artery.

We have observed the pattern of marked RV dilatation in patients with severe pulmonary hypertension such as Eisenmenger's syndrome, primary pulmonary hypertension, and mitral stenosis. Interestingly, in most patients there is preservation of a normal LVEF even in the presence of marked RV dilatation.

In about 20% of our patients with ARF, the radiocardiogram is similar to that in Fig. 5-24. This man had marked hypoxemia and did not have clinical evidence of coronary artery disease (angina, history of myocardial infarction, abnormal initial force deformity in ECG). Furthermore, the chest x-ray film did not suggest LV failure, which can be very difficult to appreciate in the chest x-ray films of patients with CAO. Bedside catheterization of the right side of the heart with the Swan-Ganz catheter revealed marked elevation of the pulmonary artery wedge pressure (LV filling pressure) and only modest elevation of the RA pressure (8 mm Hg). Serial radiocardiograms did not show improvement, and the patient died. At autopsy, marked coronary artery disease with extensive old and recent myocardial infarction was demonstrated.

In our patients with ARF, reduction of LVEF to less than 25% is associated with an unfavorable prognosis, and at autopsy severe coronary artery disease is usually demonstrable. In these patients, clinical evidence of either coronary disease or LV

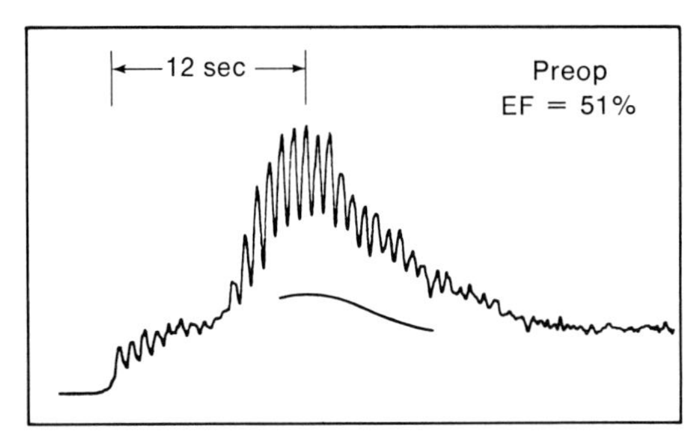

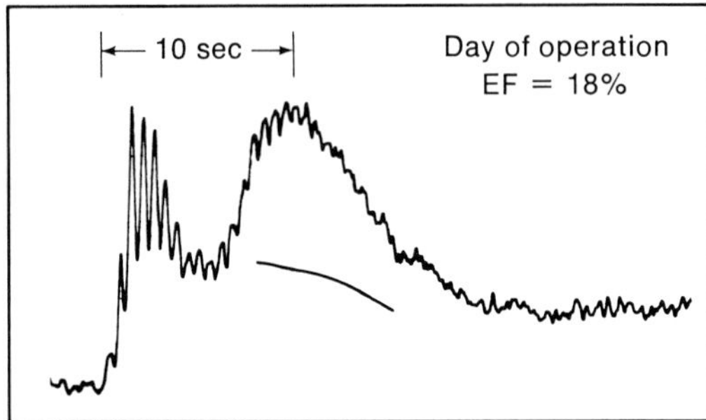

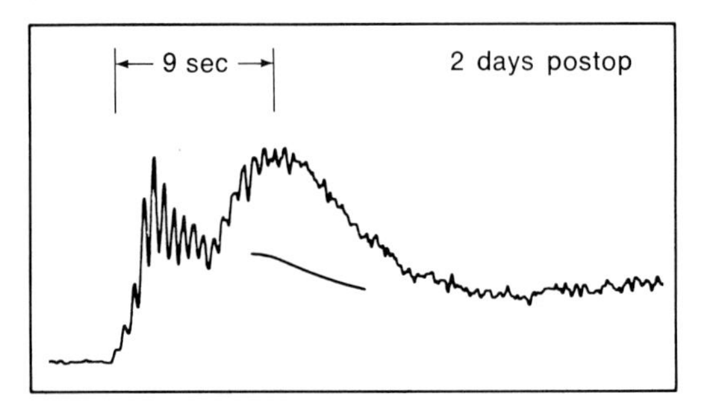

Fig. 5-25. Effect of aortic valve replacement on EF in immediate postoperative period.

dysfunction is frequently absent. We have seen improvement of LVEF in ARF in two patients, both of whom recovered. Whether or not this represents myocardial hypoxia requires further substantiation.

The ability to quantitate LV function repeatedly at the bedside is an important aspect of the probe technique.[28] In the viewing of sequential recordings, it is possible to recognize improvement or deterioration visually from the shape of the pattern alone. Fig. 5-25 consists of three studies illustrating the effect of aortic valve replacement on EF in the immediate postoperative period.

Summary

We believe that the bedside scintillation probe technique can provide clinicians with valuable information concerning the central circulation in seriously ill patients with respiratory failure. In addition, this technique can distinguish major pulmonary embolism from marked left ventricular dysfunction and should prove useful as an aid for this difficult clinical problem. Not only is this a clinically useful technique, but also it should prove to be a powerful investigative tool for the serial study of seriously ill patients.

ACKNOWLEDGMENT

We are grateful to Mrs. Carol Vandello for her excellent assistance.

REFERENCES

1. Arcilla, R. A., Tsai, P., Thilenius, O., and Ranniger, K.: Angiographic method for volume estimation of right and left ventricles, Chest **60:**466, 1971.
2. Arvidsson, H.: Angiocardiographic determination of left ventricular volume, Acta Radiol. (Stockh.) **56:**321, 1961.
3. Berndt, T., Alderman, E. L., Wasnich, R., Hsieh, S., Van Dyke, D., and Harrison, D. C.: Evaluation of portable radionuclide method for measurement of left ventricular ejection fraction and cardiac output, J. Nucl. Med. **16:**289, 1975.
4. Dodge, H. T., Sandler, H., Ballew, D. W., and Lord, J. D.: The use of biplane angiocardiography for measurement of left ventricular volume in man, Am. Heart J. **60:**762, 1960.
5. Donato, L.: Selective quantitative radiocardiography, Prog. Cardiovasc. Dis. **5:**1, 1962.
6. Donato, L.: Basic concepts of radiocardiography, Semin. Nucl. Med. **3:**111, 1973.
7. Folse, R., and Braunwald, E.: Determination of fraction of left ventricular volume ejected per beat and of ventricular end-diastolic and residual volumes, Circulation **25:**674, 1962.
8. Ganz, W., and Swan, H. J. C.: Measurement of blood flow by thermodilution, Am. J. Cardiol. **29:** 147, 1970.
9. Giuntini, C., Lewis, M. L., Sales Luis, A., and Harvey, R. M.: A study of the pulmonary blood volume in man by quantitative radiocardiography, J. Clin. Invest. **42:**1589, 1963.
10. Hosain, P., Hosain, R., Iqbal, Q. M., Carulli, N., and Wagner, H. N., Jr.: Measurement of plasma volume using ^{99m}Tc and ^{113m}In labelled proteins, Br. J. Radiol. **42:**627, 1969.
11. Ishii, Y., and MacIntyre, W. J.: Measurement of heart chamber volumes by analysis of dilution curves simultaneoulsy recorded by scintillation camera, Circulation **44:**37, 1971.
12. Karliner, J. S., Gault, J. H., Eckberg, D., Mullins, C. B., and Ross, J. Jr.: Mean velocity of fiber

shortening. A simplified measure of left ventricular myocardial contractility, Circulation **44:**323, 1971.

13. Kennedy, J. W., Trenholme, S. E., and Kasser, I. S.: Left ventricular volume and mass from single-plane cineangiocardiogram. A comparison of anteroposterior and right oblique methods, Am. Heart J. **80:**343, 1970.
14. Kenny, P. J., Watson, D. D., Janowitz, W. R., Finn, R. D., and Gilson, A. S.: Left heart imaging following inhalation of ^{15}O-carbon dioxide: concise communication, J. Nucl. Med. **17:**965, 1976.
15. Kirch, D. L., Metz, C. E., and Steele, P. P.: Quantitation of valvular insufficiency by computerized radionuclide angiocardiography, Am. J. Med. **34:**711, 1974.
16. Kirch, D. L., Steele, P. P., LeFree, M. T., Evans, P. L., Stern, D. M., and Vogel, R. A.: Application of a computerized image-intensifier radionuclide imaging system to the study of regional left ventricular dysfunction, IEEE Trans. Nucl. Sci. **NS-23:** 507, 1976.
17. Kloster, R. E., Bristow, J. D., Starr, A., McCord, C. W., and Griswold, H. E.: Serial cardiac output and blood volume studies following cardiac valve replacement, Circulation **33:**528, 1966.
18. Kurtz, D., Ahnberg, S., Freed, M., LaFarge, C. G., and Treves, S.: Quantitative radionuclide angiocardiography. Determination of left ventricular ejection fraction in children, Br. Heart J. **38:**966, 1976.
19. Newman, E. V., Morrell, M., Genecin, A., Monge, C., Milnor, W. R., and McKeever, W. P.: The dye dilution method for describing the central circulation, Circulation **4:**735, 1951.
20. Peterson, K. L., Skloven, D., Ludbrook, P., Uther, J. B., and Ross, J., Jr.: Comparison of isovolumic and ejection phase indices of myocardial performance in man, Circulation **49:**1088, 1974.
21. Pritchard, W. H., MacIntyre, W. J., and Moir, T. W.: The determination of cardiac output by the dilution method without arterial sampling. II, Validation of precordial recording, Circulation **18:** 1147, 1958.
22. Schelbert, H. R., Verba, J. W., Johnson, A. D., Brock, G. W., Alazraki, N. P., Rose, F. J., and Ashburn, W. L.: Nontraumatic determination of left ventricular ejection fraction by radionuclide angiocardiography, Circulation **51:**902, 1975.
23. Shames, D. M., and Weber, P. M.: A general logical structure for quantitative analysis of radiocardiographic data, Clin. Res. **22:**210, 1972.
24. Steele, P., Ellis, J. H., Jr., Van Dyke, D., Sutton, F., Creagh, E., and Davies, H.: Left ventricular ejection fraction in severe chronic obstructive airways disease, Am. J. Med. **59:**21, 1975.
25. Steele, P., Kirch, D., LeFree, M., and Battock, D.: Measurement of right and left ventricular ejection fractions by radionuclide angiocardiography in coronary artery disease, Chest **70:**51, 1976.
26. Steele, P., Kirch, D., Matthews, M., and Davies, H.: Measurement of left heart ejection fraction and end-diastolic volume by a computerized, scintigraphic technique using a wedged pulmonary artery catheter, Am. J. Cardiol. **34:**179, 1974.
27. Steele, P., LeFree, M., and Kirch, D.: Measurement of left ventricular mean circumferential fiber shortening velocity and systolic ejection rate by computerized radionuclide angiocardiography, Am. J. Cardiol. **37:**388, 1976.
28. Steele, P., Van Dyke, D., Trow, R. S., Anger, H. O., and Davies, H.: A simple and safe bedside method for serial measurement of left ventricular ejection fraction, cardiac output, and pulmonary blood volume, Br. Heart J. **38:**122, 1974.
29. Strauss, H. W., Zaret, B. L., Hurley, P. J., Natarjan, T. K., and Pitt, B.: A scintiphotographic method for measuring left ventricular ejection fraction in man without cardiac catherization, Am. J. Cardiol. **28:**575, 1971.
30. Sullivan, R. W., Bergeron, D. A., Vetter, W. R., Hyatt, K. H., Haughton, V., and Vogel, J. M.: Peripheral venous scintillation angiocardiography in determination of left ventricular volume in man, Am. J. Cardiol. **28:**563, 1971.
31. Van Dyke, D. C., Anger, H. O., Sullivan, R. W., Vetter, W. R., Yano, Y., and Parker, H. G.: Cardiac evaluation from radioisotope dynamics, J. Nucl. Med. **13:**585, 1972.
32. Weber, P. M., dos Remedios, L. V., and Jasko, I. A.: Quantitative radioisotopic angiocardiography, J. Nucl. Med. **13:**815, 1972.
33. Zaret, B. L., Strauss, H. W., Hurley, P. J., Natarajan, T. K., and Pitt, B.: A non-invasive scintiphotographic method for detecting regional ventricular dysfunction in man, N. Engl. J. Med. **284:**1165, 1971.

6 □ Gated blood pool imaging

Techniques

H. William Strauss and Bertram Pitt

The gated blood pool scan is a means of imaging the blood pool of the heart during the cardiac cycle by synchronizing the recording of scintillation data with an indicator of cardiac contraction such as the electrocardiogram,[29] pulse-wave tracing or heart sounds.[4] The principle of using a physiologic signal that occurs at a fixed time in relation to the mechanical activity of the heart to trigger (gate) the recording of scintillation events permits repetitive sampling of a specific phase of the cardiac cycle from each of many cycles until an image of reasonable count density is recorded (Chapter 5). The gated blood pool technique requires several criteria to be met before it can work effectively:

1. The cardiac function should be constant during the interval of data recording.
2. During data recording, there should be no patient motion and diaphragmatic motion should be minimized.
3. The triggering signal and cardiac function should have a fixed relationship during the interval of imaging.
4. The tracer employed for imaging should remain confined in the same space during the interval of data recording.
5. The number of points selected for recording during the cardiac cycle should be adequate to characterize cardiac function.
6. The interval of data recording for each frame should be sufficiently short to effectively stop cardiac motion during data recording.
7. The count density of the image must be sufficient to provide resolution of the cardiac structures.
8. The mode of data recording and display should provide spatial resolution adequate to depict the smallest significant wall motion abnormality.
9. A sufficient number of views must be recorded to ensure that each chamber is adequately visualized.

If each of these criteria is met, then the gated blood pool data will permit an evaluation of both global and regional ventricular function. Function is determined by measuring the changes in either the activity or the volume of the left ventricular chamber from one portion of the cardiac cycle to another. In addition to visualizing the cardiac chambers, the myocardium, pericardium, and fat pad are seen as a zone of decreased tracer concentration in the images (due to their low blood volume compared to activity in the chambers and lungs), which makes it possible to determine changes in the thickness of the myocardium during the cardiac cycle. To perform the gated scan, a knowledge of the radiopharmaceutical, the imaging device, and the data recording and analysis instrumentation is important.

PERFORMANCE OF THE GATED SCAN

Radiopharmaceuticals

A high-quality blood pool label is critical to the acquisition of the best resolved data in the shortest interval of time. The relationship of activity in the cardiac blood pool to that in the pulmonary blood pool background will determine the certainty with which the borders of the chambers are

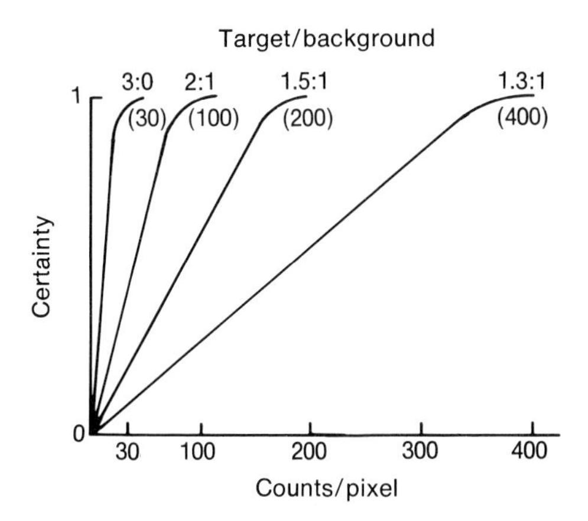

Fig. 6-1. Comparison of number of counts per pixel required for statistical certainty that perceived contrast difference is real.

defined for a given count density of data recorded.

In Fig. 6-1, the relationship of count density per picture element of computer matrix required to define with certainty that a difference between two adjacent areas exists is plotted as a function of the perceived ratio at the detector. The true ratio of activity between heart and lung that might be expected with a perfect imaging system and a perfect blood pool label is about 5:1, which would permit the recording of data with a density of 50 counts/picture element to provide excellent definition of borders. Since a high photon flux is required to perform these examinations, a radiopharmaceutical with a photon energy that will result in the highest intrinsic photopeak efficiency with the imaging device while providing adequate spatial resolution is desired. Thus, although ^{113m}In, when administered as the chloride, will react in vivo with transferrin to form a blood pool label, ^{113m}In has a photon energy of 393 keV, and that photon has abundance of only 64%, which results in a detected net photopeak photon flux that is only 24% of that found with a comparable dose of ^{99m}Tc. Therefore, of the radiopharmaceuticals available at this time, ^{99m}Tc appears to be the most appropriate for use with the scintillation camera. (The ¼-inch crystal in use in many mobile cameras makes this particularly true.) The pharmaceuticals that have been suggested as blood pool labels include both albumin[29] and red blood cells (Chapter 2).[14] Albumin has a relatively large distribution space compared to the red blood cell volume, particularly in the liver and to a lesser extent in the lungs; therefore images recorded with the radiopharmaceutical will have a lower heart-to-lung ratio than those recorded with labeled red blood cells.[30] However, even with the ^{99m}Tc-labeled red blood cell radiopharmaceuticals, a ratio of only 3:1 can be achieved in the image. Although both in vivo and in vitro methods have been proposed to achieve cell labeling, recent evidence suggests that there is little difference between the two methods in the results obtained clinically.[14] Administration of a stannous pyrophosphate solution intravenously, followed in 20 min by a bolus of [^{99m}Tc] pertechnetate, as described by Pavel,[20] appears to provide a high-quality label. The ^{99m}Tc activity appears to bind almost instantaneously to the globin portion of the hemoglobin with a labeling efficiency of well over 90% and does not clear from the cells for several hours. As described by Alderson,[1] the unlabeled portion of the injected dose appears to clear rapidly from the body in the urine; thus a high target-to-background activity level is maintained over the heart.

Scintillation camera

Both the Anger-type scintillation camera and the multicrystal-type instrument are capable of recording equilibrium-gated blood pool data. The selection of an instrument should be predicated on its resolution, field size, and count rate capacity. Data of importance to interpretation may be obtained from structures other than the heart itself, such as the size or shape of the great vessels or atrial size and function. To image these structures in most patients, a 35 cm diameter field-of-view instrument may be helpful. This is particularly true when consideration is given to multiangle collimators or tomographic imaging. However, the standard field camera providing a 25 cm diameter field of view is probably the most common instrument employed for cardiac investigation. The recent development of the gating system for the multicrystal instrument suggests that this too will be employed for the performance of equilibrium imaging in the future.

The uniformity of the instrument is critical for accurate quantification of ventricular function predicated on the number of counts recorded in the left ventricle after background correction. Instruments equipped with provisions for uniformity correction to less than 5% variation across the field should be used whenever possible.

Collimators for gated blood pool scanning

Parallel-hole collimator. The collimator most frequently employed for blood pool measurements is the parallel-hole all-purpose collimator because it has no significant spatial distortion with depth, although resolution will decrease at increasing depth (Fig. 6-2). The resolution of most parallel-hole collimators at the surface is similar, since hole size and hole number are similar. However, changes in collimator sensitivity are accomplished by varying the length of a collimator hole; thus when a collimator of higher sensitivity is made, its

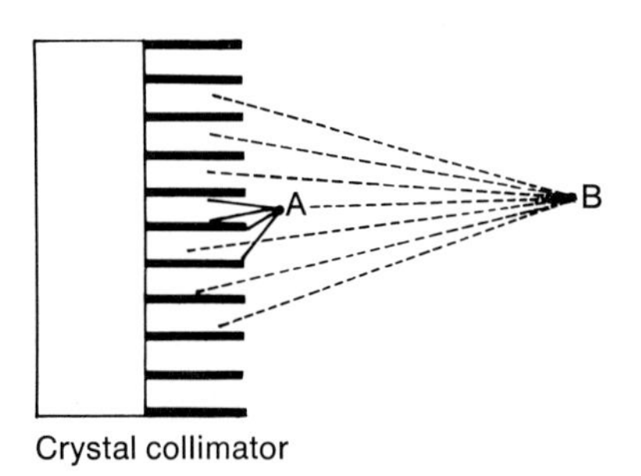

Fig. 6-2. Effect of depth on collimator resolution. Photons emanating from *point A* have only one hole in collimator that they can successfully pass through and strike crystal. Photons coming out in other directions will strike collimator septum and be totally absorbed. However, photons emanating from *point B,* at some distance from face of collimator, can enter any one of several holes and strike crystal. Thus localization of photons emanating from *point B* is significantly less than localization of those emanating from *point A.*

holes will be shorter than those of a collimator of greater resolution. When these collimators are employed clinically, the resolution at depth varies inversely with hole length. Thus higher sensitivity collimators provide for a maximal photon collection (high count rate) but significant loss of resolution at depth. For example, for a 2 mm hole size, 0.2 mm septal with a 10 mm length, the resolution at the surface is 2 mm; at 5 cm from the collimator surface the resolution is 10 mm, whereas, if the collimator were 20 mm long, the surface resolution would be 2 mm, but at 5 cm the resolution would be 5 mm. This increase in resolution is obtained at a loss of ~ 50% in sensitivity. The selection of the best trade-off between collimator sensitivity and resolution should be based on anticipated count rate and imaging time. To achieve resolution, some counts are required, but the relationship of count density to resolution is nonlinear. If no events are recorded, then no resolution is possible. If too few events are recorded, poor resolution will occur. Beyond some point, depending on the size and shape of the heart, no increase in resolution will occur with a further increase in recorded counts. Thus it is important to define the minimal amount of spatial resolution required in terms of the clinical problem. In the left ventricle, although it is desirable to have a spatial resolution of fractions of a millimeter, lesions of less than 1 cm are probably inconsequential. However, before selecting a collimator with a 1 cm resolution as the limiting factor for system resolution at a depth of 7 cm (the likely depth of the heart from the collimator in many moderately obese subjects), the additional image degrading effects of Compton scatter, scintillation camera positioning signal noise, and computer matrix resolution must be considered. When these other factors are taken into account, the collimator resolution at depth should be no more than 8 mm to provide a system resolution of at least 1 cm.

Converging collimator. The converging collimator improves spatial resolution by magnifying the object placed between the focal point and the surface of the collimator. The amount of magnification will vary with the distance from the collimator face; thus objects close to the focal point (usually the focal point is 70 cm from the collimator surface) are magnified to a greater extent and detected with a relatively greater sensitivity than objects placed close to the surface. However, the attenuation of photons, which must pass some distance through the object, makes it possible to view primarily only that point closest to the detector. This collimator produces a type of distortion for a volume source characterized by an increase in sensitivity and magnification for a distant source, which, however, is seen with poor resolution in comparison to a closer source, which is better resolved, but seen with less magnification and lower sensitivity. The result of this distortion is both a loss of image contrast and a nonlinear relationship between perceived counts and activity present in the volume of the object. These distortions make this type of collimator less useful than a parallel-hole collimator for imaging the adult heart. However, in a smaller heart, such as that of an infant or young patient, this distortion decreases and the collimator becomes more desirable.

Bifocal collimator. To define the motion of the heart, multiple views are usually recorded. Since each view requires imaging for some time, the greater number of views, the longer the imaging time required. In instances of short-term interventions like exercise, it is impossible to collect multiple views sequentially to define wall motion. To optimize the likelihood of detecting these changes, data must be recorded from multiple vantage points simultaneously. This can be accomplished by using multiple detectors or by using a collimator that has holes placed at several angles. Multiple detectors are impractical at present due to their size and expense, whereas a "biplane" collimator can be attached to any scintillation camera. This type of collimator will limit the size of each view recorded. Since it is important to record data from the entire heart, a two-view collimator with a larger field of view was designed. This "bifocal" collimator permits the collection of data from larger segments of the field but with a slight loss of spatial resolution (Fig. 6-3). The diverging collimator causes a distortion opposite that of the converging collimator. Although this type of distortion is more acceptable for cardiac blood pool imaging than that resulting from a converging collimator, it is still more desirable to employ parallel-hole collimation if at all possible. The two-view collimators also create a problem of data digitization. The data collected over the heart in each image are only placed in a small portion of the computer matrix. This can be overcome by recording the data in a larger matrix. Preliminary results with the bifocal collimator appear promising[28] and suggest that this approach will be particularly useful for exercise studies.

Tomography. An alternative approach to multiview imaging is to employ tomography. The tomograms can be recorded as a series of planar images or transverse images using either "positron" or "singles"

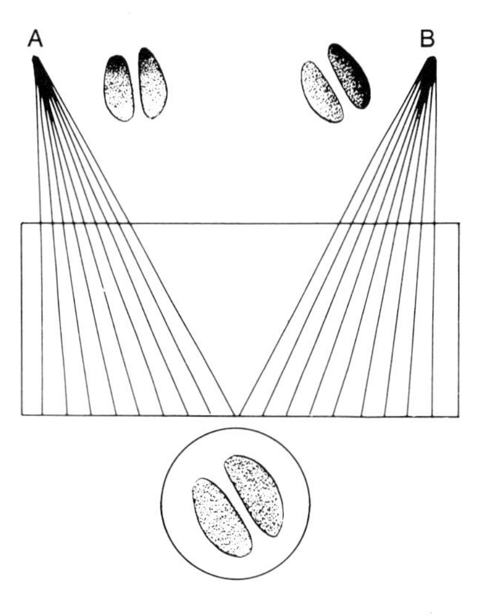

Fig. 6-3. Diagrammatic representation of bifocal-diverging collimator. Shading within chambers of object imaged represents uniform distribution of activity in structure. Smaller reproductions of these zones on crystal side of collimator (with selected shading) illustrate how this object would be perceived by collimator. Areas of more intense shading indicate zones detected with greater sensitivity and resolution.

events. For a more complete discussion of transverse section position tomography, see Chapters 2 and 9 for singles tomography. "Singles" *planar* tomography can be done with either a moving collimator[11] or a multiaperture device.[31] The moving collimator devices result in tomograms that have characteristic effects determined by the type of motion employed to blur the off-plane image data. To date, little work has been done with these devices in the heart. The multiaperture pinhole approach employs a *stationary* detector, which views the object from several vantage points. Reconstruction is accomplished by moving and combining these multiple stationary images in the computer. The planar tomograms obtained from the multiaperture pinhole device result in resolution of about 5 mm in a plane about 7 cm from the central pinhole aperture and have a thickness of 10 mm. Reconstruction can be effectively performed in any plane 0 to 10 cm from the collimator surface. The application of this technique to multiple-gated blood pool imaging will make it possible to observe the motion of any portion of the heart during the cardiac cycle, while recording data from one position.

Trigger

The most common triggering signal employed to record gated scan data is the R wave of the electrocardiogram. The signal has several features that make it desirable: (1) it is easy to obtain in most patients; (2) a signal the size and shape of the R wave only occurs once in the cardiac cycle of most patients; and (3) it bears a fixed relationship to the mechanical events of cardiac contraction in most individuals. Any signal that fulfills these criteria, however, could be used. It is most important to realize that in some patients there may be a second signal that appears to the gate identical to the R wave—such as a pacemaker spike in a patient with a conduction abnormality or a tall peaked T wave in a patient with a low amplitude R wave and electrolyte abnormalities. In these individuals, the gate may trigger twice in the course of a single cardiac cycle.

In addition, if the electrodes are not firmly connected to the patient, the gate may trigger on noise impulses, which can occur at surprisingly regular intervals but are unrelated to the patient's cardiac cycle. The result of these gating problems is that the gated images reveal diminished cardiac function. To be certain that the gate is functioning properly, a trace or film recording of the patient's electrocardiogram and the superimposed triggering signal should be recorded with each examination.

DATA RECORDING

The information from the gated blood pool scan can be recorded directly on film, into the memory of a computer, or simultaneously on both. The disadvantage of recording data onto film is that there are only limited changes that can be made in the display of gray shades and in target to background. In addition, showing the data as a "continuous-loop movie," a very important step for data interpretation, is difficult. Also, if the data are only recorded on film, quantification of changes in activity in the left ventricle is impossible. During data recording a region of interest may be set over the left ventricle, which will permit assessment of the changes in activity versus time using a small analog or digital computer in the scintillation camera. However, this method is limited, since the actual data from which the calculations are performed are not preserved and subsequent recalculation of the data is impossible. The optimum method of performing and recording multiple-gated image data is a digital computer recording all the image data. Recent evidence of Hamilton and associates[13] suggests that at least twenty frames are required to define the function parameters in each cardiac cycle. At that temporal resolution, a spatial resolution of 8 mm/picture element appears to be satisfactory, since the heart has some motion during image acquisition (there is incomplete stopping of cardiac motion with a 20 msec or greater exposure time), and the repetitive nature of sampling the cardiac cycle while the patient is breathing quietly contributes to some additional blurring of the images, which makes high spatial-resolution data matrices less useful.[27] However, if cardiac motion is stopped more effectively by using a frame interval of less than 15 msec/data point (at average resting heart rate), then higher spatial-resolution recording is possible.

Clinically interpretable gated blood pool data can be recorded in as short an interval as 30 sec or over several minutes. The selection of a short or long collection interval should be predicated on the quality of data desired and the rapidity of the occurrences to be observed. If the measurements are made with the patient in a "steady state" condition such as at rest, then thc measurements can be made for any interval necessary to record a sufficient information density in the region of the ventricle. However, if the circumstances require a measurement following administration of a short-lived drug such as nitroglycerin or an intervention such as exercise to induce ischemia, then the measurements must be completed in an interval dictated by the expected duration of the changes in ventricular function. To make these short records most interpretable, the data matrix selected for recording must be coarse. Recording data in a matrix of 8 mm^2 elements will provide a data density per element four times greater than that obtained with a matrix of 4 mm^2 elements. The net effect of this difference, aside from the loss of spatial resolution, is a marked decrease of apparent noise in the data matrix, which renders the image far more interpretable. To make the coarser data matrices suitable for viewing, interpolation is frequently used. No significant loss of lesion detection with coarse data matrices has been observed in a comparison of coarse and fine data matrix recording of ten cardiac studies.[27]

Analysis

The analyses of data from the multiple-gated blood pool examination are performed in two parts: (1) inspection of the data while they are played as a continuous-loop cinematic display of the average cardiac cycle and (2) quantification of volumes and function.

Tests of adequacy of data recording

Before proceeding to a detailed analysis of the information depicted by the changes in chamber volume per unit time, it is important to establish the technical adequacy of the data presented for interpretation. The technical adequacy can be established by evaluating the sharpness of the borders of the structures depicted in the image and the adequacy of the triggering signal. Patient motion, inadequate complexation of the ^{99m}Tc to the red blood cells, marked obesity, and improper setting of the scintillation camera pulse-height analyzer will all contribute to degrading the quality of the structure borders in the image. If the edges are well demarcated, the image should then be evaluated for the adequacy of gating.

There are two types of problems that occur with gating—regular and irregular. Regular occurrences (those that occur with a fixed relationship to the true R wave) will produce a "step" in the data. This will appear as a sudden coarsening in the image quality as the initial portion of the cycle with higher counts is replaced by the portion with fewer events per frame. In computers that do not normalize each display frame for brightness, this type of misgating is also apparent as a change in the display brightness. If this type of gating problem is suspected, the total number of events recorded in each data frame should be evaluated, and a variation of less than 10% should be found when data from the entire field of view are recorded. (Although the blood volume of the heart changes during the cardiac cycle, the atria fill out of phase with the ventricles and the blood ejected by the heart enters the lungs and aorta, which are also in the field of view, thus no change in total counts is observed. However, if the image is made up mostly of the left ventricle, as occurs when "zoom-mode" recording is employed, then the counts per frame may change substantially, depending on the phase of the cardiac cycle.)

Irregular occurrences in gating are associated with atrial fibrillation and multifocal premature contractions. Many of the gating devices have filters to remove aberrant beats, which exceed some threshold of an R-R interval recorded during an initial sampling period. In some circumstances, the rate may vary widely; therefore only one beat in ten is recorded, resulting in marked prolongation of imaging time. In patients with this problem, a decision must be made either to record the data with the filter disabled or not to perform the gated scan until the rate is made more regular by treatment. The latter, of course, is preferable, but occasionally some information about the patient's average ventricular function is helpful. These problems are difficult to detect from a review of the data and should be noted during data recording. The result of this problem is an underestimation of the patient's cardiac function.

Occasionally, an electrode may fall off and the gate will trigger on noise or some other random pulse that bears no relationship to the cardiac cycle—these images will pass the counts test, but the heart will appear to be akinetic. Although some cardiomyopathies may have such dilated chambers and little motion that they are difficult to perceive, the condition is rare, and patients with these findings are in extremis. Thus the finding of an akinetic heart, in which none of the chambers appear to function well, should raise the prospect of misgating.

In these instances of regular gating problems, the data must be "normalized" before they are analyzed to measure ejection fraction. This is extremely important, since ejection fraction will be determined from the number of counts recorded in the left ventricle—if the total number of events in each data frame is not the same, then the calculated ejection fraction will be erroneously high and therefore misleading.

Inspection

The rate of image replay should be adjustable from stop motion to faster than real time, since some wall motion abnormalities may be more visible at one rate than another.

The optimum replay speed appears to be 1 cardiac cycle/0.75 sec. If 300,000 counts are recorded in each frame and thirty frames are recorded in the cardiac cycle, then the data rate appearing on the display is in excess of 11 million events/sec. The images should be displayed in a size that makes the data in the region of interest appear on the retina in a zone of 3 degrees of arc. At a viewing distance of 1 m, the images should be 5 cm in height and width. A minimum of sixty-four gray shades is particularly important when high count-density images are viewed. In gated cardiac studies at rest, the peak events recorded in one picture element may be as high as 1000, whereas the minimum may be 0. Thus the gray studies would each represent 1000/64 or 15 counts/level. However, some systems use only sixteen gray shades; therefore each gray shade represents 1000/16 or 63 counts/level, and the images will appear to have "isocount" lines—which make interpretation difficult. The image of the average cardiac cycle should be viewed initially at full gray scale to define the relative activity in the heart and background and to identify each chamber, then the image should be displayed with increasing contrast enhancement to assist in defining the relationship between each of the chambers and structures in the image. (For additional information on data recording see Chapter 1.)

APPEARANCE OF NORMAL SCAN

Immediately before atrial systole, the blood volume of the normal adult heart is approximately 400 ml.

Right atrium

The right atrium has an end-diastolic volume of 60 ml and a wall thickness of 2 mm. A fat pad is commonly adjacent to the right atrium, separating the atrium from the right lung. Thus, although atrial myocardium is not seen, the fat pad is usually well defined on gated equilibrium images. The plane of the tricuspid valve is about 30 degrees to the right of anterior. Since the right atrium is very anterior in location and triangular in shape, the tricuspid valve plane is usually well defined in the anterior view. Due to the shape and position of the right atrium, enlargement of this structure will result in a bulging of the chamber in the lateral and superoinferior dimensions.

The right atrium is generally best defined on the right anterior oblique (RAO) view, although it is also well seen in the anterior view and can usually be defined in the 45-degree LAO view at the end of ventricular systole as a bulge superimposed on the anterior aspect of the right ventricle.

The right atrium contracts from its lateral aspect toward the valve plane. The atrial contraction on scan appears short, occupying less than 100 msec, and appears as a movement of the lateral margin of the chamber toward the valve plane. The inferior and superior aspects of the chamber do not move significantly during contraction. The chamber appears to have a two-phase motion: following atrial systole, there is continuous filling during which the tricuspid valve moves medially; maximal atrial volume occurs at the end of ventricular systole. At the opening of the tricuspid valve, the atrium decreases in size as the tricuspid valve returns to its rest position. Thereafter the chamber remains relatively unchanging in size until atrial systole, when the lateral wall of the chamber contracts. The atrium will not contract in atrial fibrillation or in the presence of pressure or volume overload.

Right ventricle

The musculature of the ventricle is greater in the inflow portion at the tricuspid valve than at the outflow portion near the septal surface and pulmonic valve. The end-diastolic volume is about 165 ml in a 70 kg adult. Thickness of the right ventricular free-wall myocardium is 5 mm, which cannot be visualized on the gated scan. During ventricular systole, the tricuspid valve plane moves toward the left ventricle, whereas the apex of the right ventricle moves slightly or not at all. The outflow tract portion of the ventricle close to the tricuspid valve contracts slightly in some patients. Qualitative evaluation of the right ventricle is usually performed in the 30-degree RAO, anterior, and 45-degree LAO views. Quantification of right ventricular function requires performance of a first-transit study (Chapters 5 and 16).

The common response of the chamber to acute or subacute increase in load is dilatation. Due to the pancake shape of the chamber, it is likely that any increase in volume would result in an increase in anteroposterior (AP) diameter, followed only at a later time by an increase in lateral dimension, primarily toward the left, since the tricuspid valve is a fibrous border that is difficult to displace. Under circumstances of chronic, gradually increasing load, the chamber will hypertrophy, and volume changes will be minimal. The outflow tract is about 2 to 3 cm in diameter and does not ordinarily contract to a significant degree unless either the function of the remainder of the chamber is reduced or the pressure required of the right ventricle is increased. The region of the pulmonic valve is usually seen as a thin zone of decreased tracer concentration high in the outflow tract best defined in the anterior view. When a portion of the right ventricular myocardium is severely diseased, it will cease contracting. However, akinesis of a portion of the right ventricle is not commonly seen, and the most frequent appearance on the gated scan of right ventricular dysfunction is ventricular dilation with a decrease in ejection fraction.

Synchrony. The right ventricle contracts in synchrony with the left, except in conditions of bundle-branch block or volume or pressure overload. Under the former condition, the ventricle activated first will begin contracting before the other. This is commonly seen in patients with demand pacemakers who are studied on two occasions, one when the pacer is active during the data recording and the second when the pacer is inactive during data recording. Patients with volume or pressure overload, on the other hand, have electrical activation of the chambers at the same time but frequently have a delay in the completion of ventricular systole in the overloaded chamber compared to the chamber with a normal load.

Septum. In the LAO position, the right ventricle appears crescentic, frequently with the right atrium forming the anterior border of the cardiac silhouette during ventricular systole. The septal surface of the chamber commonly appears curved and moves toward the left ventricle during systole. Septal motion is characterized by both thickening of this portion of muscle and by total movement.

Left atrium

The left atrium is slightly smaller than the right and has walls that are slightly thicker, measuring 3 mm in thickness. The atrium is supplied by four pulmonary veins and is located posterior to the right ventricle and pulmonary artery. The normal left atrium is not seen in the anterior view; in the LAO view, the atrial appendage is usually seen to a variable degree. Usually, the pulmonary veins are not defined in any view. When the left atrium enlarges, it extends superiorly above the base of the left ventricle and laterally. In cases of extreme enlargement, the left atrium may form the apparent right border of the heart.

Contraction of the left atrium is very difficult to define well on the equilibrium-gated blood pool scan because of superimposed structures. The view that defines the left atrium best is the left posterior oblique (LPO). In this position, the right atrium lies beneath the left atrium, the mitral and tricuspid valve planes are aligned, and the left ventricle is separated from the atrium. Even in this position it is difficult to define discrete contractile motion of the left atrium.

Left ventricle

The left ventricle comprises the major weight of the heart and, like the right, is also composed of both an inflow and an outflow portion. The walls of the ventricle are 15 mm thick, and the end-diastolic volume is approximately 150 ml.

The thickness of the muscle of the left ventricle is sufficiently great to produce a "halo" around the chamber. The thickness of this halo can be used to estimate the mass of the left ventricle. The chamber is usually ovoid in shape in the anterior or LAO position, becoming more spherical as dilation occurs. Contraction occurs along the entire perimeter of the chamber with less motion along the long axis from the base of the heart to the apex than in the minor axis. The anterolateral wall moves slightly more than the inferior wall. The rate and amount of emptying of the ventricle will vary with the heart rate and with the physical conditioning of the patient. The normal ejection is between 50% and 65% by the gated method, and the end-diastolic volume is 150 ml in a 70 kg man. In some patients, the posterior papillary muscle causes a well-defined indentation on the blood pool image (best seen in the LAO view), and the anterior leaflet of the mitral valve and its associated papillary muscles and chordae are seen as a zone of decreased tracer concentration in the superior portion of the ventricle in the LAO position. The left ventricle is rigidly attached to the rest of the heart at the base where the valves are located and the septum. Therefore, when the chamber enlarges, it will increase in size in the superoinferior dimension, laterally, and in its long axis.

The left ventricle is best defined in the anterior, 45-degree LAO and 60-degree LPO positions (Fig. 6-4). The RAO position, although viewing the chamber in its longest axis, does not permit the recording of a high spatial-resolution image with a

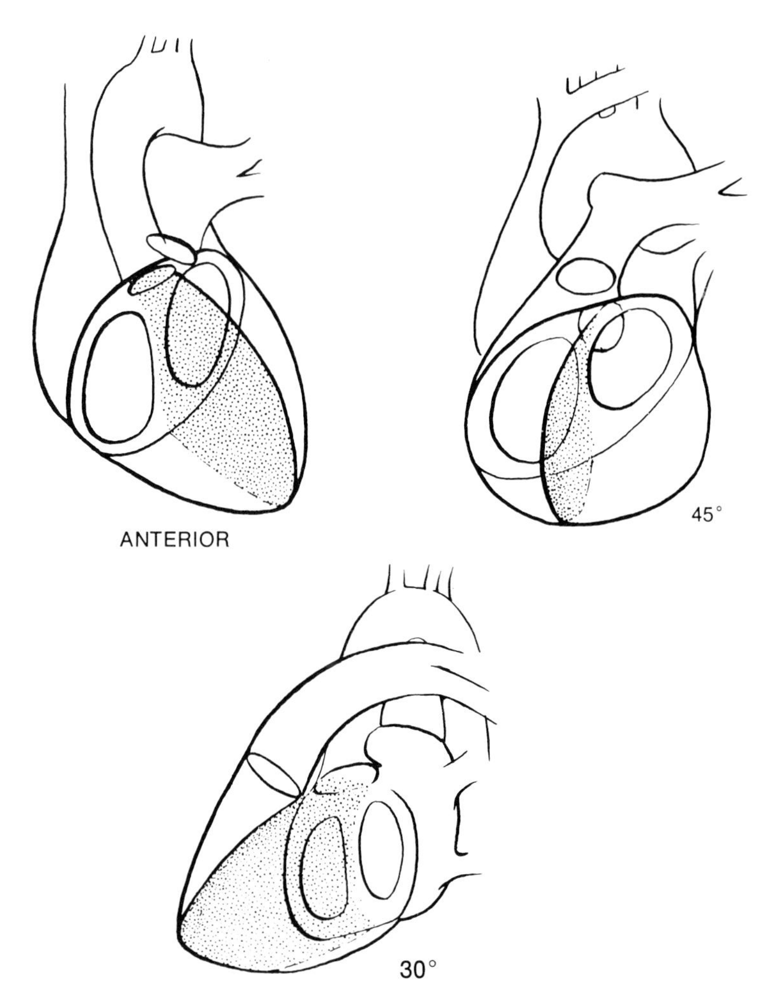

Fig. 6-4. Diagram of left ventricle in anterior, LAO (45-degree), and LPO (30-degree) positions, depicting relationship of septum and cardiac chambers to one another.

parallel-hole collimator due to the distance from the collimator to the left ventricle. For this reason, the 30-degree LPO position, which brings the chamber closer to the detector and minimizes the chance for obscuring borders due to overlying right ventricle, is preferred. Regional disease of the muscle manifests itself as a focal abnormality of wall motion, usually with a compensatory increase in motion of the remaining normal portions of the chamber (Fig. 6-5). These focal abnormalities are most commonly associated with coronary artery disease but may occur with focal cardiomyopathies such as sarcoidosis. Thus in patients with suspected focal disease, two findings should be sought; the first is the location and extent of the abnormal wall motion. This will vary, depending on the extent of the ventricular wall involved. Myocardial infarction, for example, may not involve the total thickness of the wall. Frequently a portion of the thickness of the muscle is spared; thus some motion of the diseased area is possible. In addition, even in the case of a total-thickness scar, the adjacent muscle may contract and passively move the diseased area. Since all viable muscle must thicken, it is frequently helpful to look for local thickening of the myocardial halo

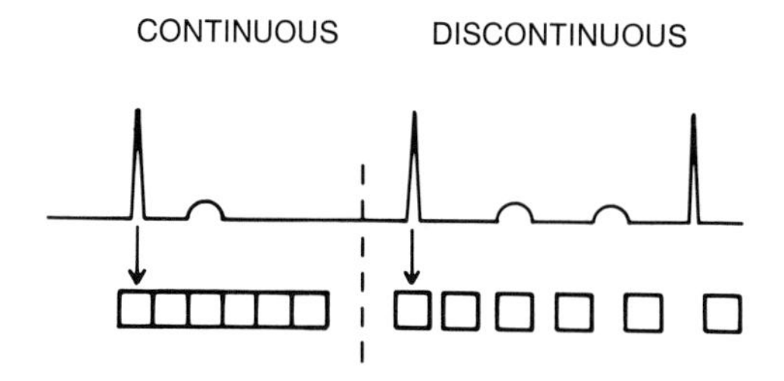

Fig. 6-5. R wave triggers computer to file image data in memory one. After a preset imaging interval, data are filed in memory two and thereafter in memory three, etc., and information is recorded continuously throughout cardiac cycle. When new R wave arrives, data in first interval are again filed in memory one, and process is repeated. Many nuclear medical computers have limited memory capacity, and a limited number of frames can be recorded each cycle. To optimize data recording under these circumstances, data may be recorded discontinuously, with information recorded during specified portions of cardiac cycle.

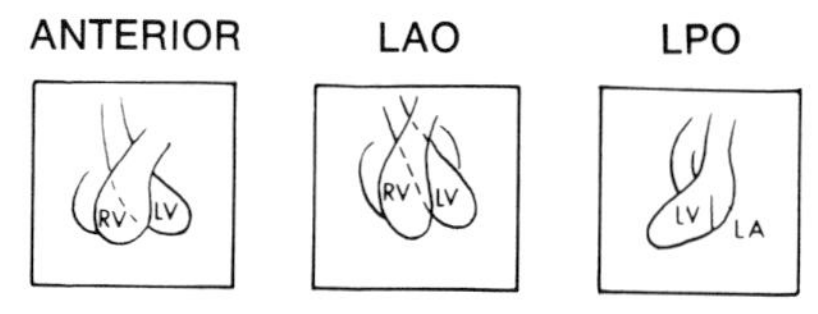

1. Identify left ventricle on end-diastolic image
2. Calculate end-diastole volume using formula:

$$V = \frac{\pi \cdot L \cdot M_1 \cdot M_2}{6} \qquad M = \frac{4A}{\pi \cdot L}$$

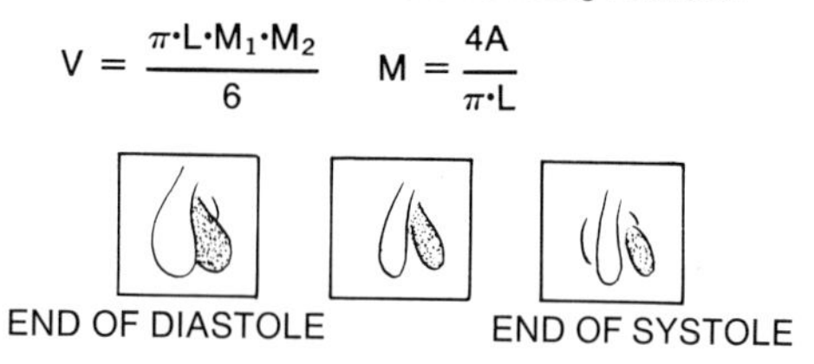

3. From LAO multiple-gated scan calculate time-activity curve
4. Combine data

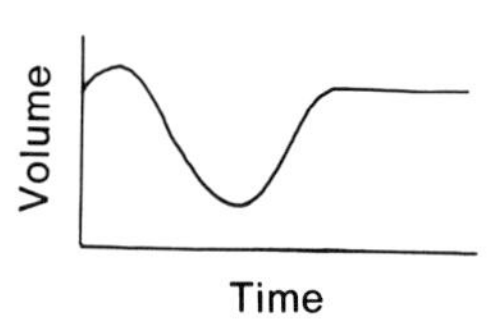

Fig. 6-6. Method of computing volumes and ejection fraction from gated blood pool scan data.

to be certain that all walls that are moving are doing so because the myocardium in that area is contracting and not because of passive motion. The second point to be sought is the motion of the remainder of the chamber—in cases of abnormal wall motion, the adjacent myocardium should move excessively to "make up" for the lost motion of the damaged zone. If this increase in motion is not observed, then disease of the "normally moving" zone may be inferred. After subjective evaluation of the data, the examination should be quantified to measure end-diastolic volumes, ejection fraction, ejection and filling rates, and regional ventricular function.

QUANTIFICATION OF VENTRICULAR FUNCTION

The values that can be calculated from the blood pool image include the end-diastolic volumes of the left ventricle, right ventricle, and right atrium. The left atrial volume could be calculated but is considerably less reliable, since it is difficult to obtain good visualization of the chamber in two views. The other chambers can be readily defined in two positions during the course of the gated examination.

Calculation of the chamber volumes can be done either from the spatial relationship of picture element size to real space when a parallel-hole collimator is employed to record the examination or the relationship of perceived activity at the detector to the volume in the chamber (Fig. 6-6). The latter method requires corrections for both attenuation within the patient and for background activity in structures surrounding the heart. These corrections can be readily performed with positron-imaging devices but are far more difficult with conventional gamma-imaging techniques. Since the geometric method of calculating chamber volumes has been validated by comparison of angiographic measurements to casts of the ventricular chambers, it is the preferred method. Although the diseased ventricle does not have the shape of a prolate ellipsoid or a sphere at end of systole, the shape at end of diastole is usually spheroidal, rendering the geometric assumptions for that portion of the calculation valid. Calculation

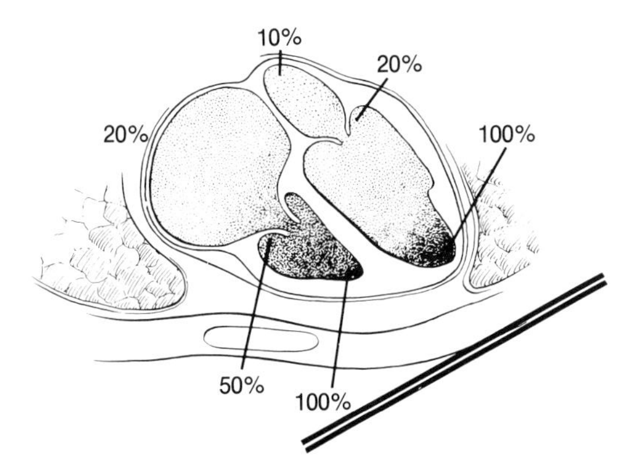

Fig. 6-7. Measurement of RV and LV function by counts method relies on changes in activity perceived by detector to reflect changes in volume of chamber. However, in heart, structures overlap, and activity in ventricles has an underlying atrial contribution. After allowance for attenuation in chest wall, average efficiency for detecting photons from specific portions of ventricle in an average-sized heart at middle of systole is indicated. Note large atrial contribution to right ventricular counts. For this reason, right-sided ejection fraction cannot be readily determined from LAO gated scan at equilibrium.

of ejection fraction is relatively independent of geometry because the changes in ventricular activity are employed as the means of measuring the changes in volume at each point in the cardiac cycle. On theoretic grounds, the measurement of activity in the ventricle also includes activity in overlapping structures, such as chest wall, atrium, and lung (Fig. 6-7). The contribution of structures far from the detector is mitigated by gradual loss of collimator sensitivity with distance and attenuation of photons as they traverse tissue. This latter point is important; in water-equivalent tissue, the photon attenuation is $e^{-0.15d}$, where d is distance in centimeters and 0.15 is the attenuation coefficient for a 140 keV photon for ^{99m}Tc. Thus this value is the attenuation photons originating behind the ventricle in the atrium, which must traverse the ventricle to reach the detector in the LAO position. Since the average normal ventricle is 9 cm at its longest length, photons originating in the atrium will be attenuated, and fewer than 20% of the photons arising from the atrium will reach the detector. The contribution of these photons to lowering the ejection fraction will be minimal. This is important, since the atrial contribution of activity will always be to a lower ejection fraction because the atrium fills and ejects blood out of phase with the ventricle. In patients with mitral stenosis, in which the ventricle is small and the atrium enlarged, however, this may present an important problem.

The pulmonary contribution, on the other hand, is quite different. The contribution from deep in the lungs is far less attenuated than that from the heart. The pulmonary background represents tissues that are outside the heart and that really do not exist in the region of the cardiac silhouette. However, from a pragmatic point of view, the use of a lung background permits an accurate calculation of the ejection fraction and of global ventricular function curves.

The calculation of the ventricular time-activity curve requires identification of both a background and a ventricular region of interest. There are two methods of defining the ventricular region of interest—manual and automatic.[8] With the automatic methods, the operator usually must identify the region of the image that contains the left ventricular blood pool. Since the heart-to-background-activity relationship averages 2.5 to 3.1, the ability to identify the region of the left ventricle is enhanced if the background activity is subtracted from the images prior to defining the left ventricle. Identification of the ventricle is also enhanced if the statistical noise (high frequency) is minimized by smoothing the data. On the smoothed background-subtracted data, the ventricle is identified by placing either a box around it or a series of points at the perimeter. The programs then search within the boundaries of the chamber to identify the picture elements in the ventricle with the highest activity. Once these elements and their location have been found, the programs then search outward on a line-by-line basis to identify the perimeter of the chamber on that line. The virtue of the derivative approach is that it is automatic, whereas that of the threshold is more controlled by the user. The correlations obtained with both of these methods to contrast angiography are high and suggest that it is really more a matter of personal preference than of a real methodological difference regarding which is used in a specific laboratory.

Following are the most common problems in quantifying ventricular function with the counts approach:

1. The background area should be defined in a zone of lung tissue with no overlying or underlying great vessels and should be far enough away from the cardiac chambers to be certain that they are not erroneously included. To determine the adequacy of the background zone selected, a curve of activity in the background versus time should be generated. If only the pulmonary parenchyma is included in the background, the curve will be flat. Although the activity in the lung increases slightly with each cardiac cycle, the activity is so spread out that the change in one small zone of the lung is immeasurable.
2. The definition of the left ventricular region of interest should change from image to image as the ventricular volume changes. This is more important for normally contracting ventricles than for chambers that are diseased and have decreased motion. If a single region of interest is used to define the left ventricle at end of diastole, it may include a significant portion of atrial appendage and possibly right ventricle by the time the heart reaches the systolic portion of the cycle, which will result in erroneous overestimation of ventricular activity at end of systole and a low ejection fraction. On the other hand, if the heart moves vigorously, it may actually move out from under the region of interest during systole, resulting in an overestimation of ejection fraction. For these reasons, it is important to apply a method of quantification that provides for careful characterization of the background and permits a frame-by-frame identification of the left ventricular region of interest (which the operator can alter if required). Thus, the quantified data will prove a correct and reproducible measurement of left ventricular function. Recent reports cite a 5% variation in the measurement of ejection fraction as the level of confidence that can be achieved on repetitive measurements from the same patient.

Quantification of right ventricular function

Right ventricular size can be estimated from the end-diastolic images using geometric methods, but quantification of right ventricular function from the changes in activity is difficult. The right ventricle lies close to the atrium in the LAO position and has the left atrium and ventricle beneath it in the anterior and RAO positions (Fig. 6-8). To separate the right ventricle successfully from its surrounding structures requires recording a first-pass study in the RAO position during passage of the bolus through the right heart and lungs. In this manner, the right atrium is clearly defined and separated from the right ventricle by the tricuspid valve plane, and the right ventricle is not obscured by the left heart because the

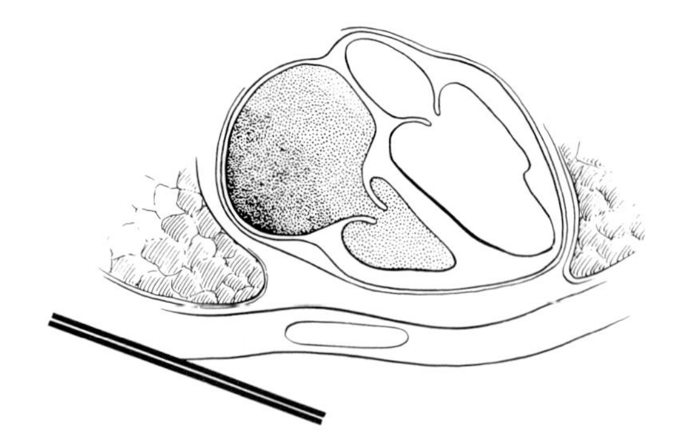

Fig. 6-8. Gated first pass. Relationship of chambers and activity in initial transit RAO view. There is no chamber overlap, and long axes of both RA and RV are well defined. Changes in activity with each cardiac cycle are related to filling and emptying of heart. If data are recorded and gated by R wave during initial transit, RVEF can be calculated.

bolus is restricted to the right heart and lungs. To record a sufficient number of events in the right heart, the data are recorded with electrocardiographic synchronization for five to seven beats after injection of the bolus, then recording is terminated. These data can then be analyzed both subjectively and quantitatively in similar fashion to that of the left ventricle.[19]

Wall motion

Characterization of wall motion subjectively divides ventricular function into normal, hypokinetic, akinetic, and dyskinetic, based on a comparison of end-diastolic to end-systolic wall outline. The tracer techniques do not employ realignment of the chamber outline at end of systole to that found at end of diastole. Similarly, the quantification methods employed to evaluate regional wall motion do not have a realignment component. This lack of realignment can be considered a flaw of the tracer procedure. However, by treating all data in a similar manner, there is less variation from center to center in the interpretation of regional wall motion by the tracer methods than that found in contrast ventriculography. Rather then adopt a realignment strategy for the tracer techniques, it is likely that methodology will be defined that will both employ the change in the cavitary blood pool outline and require thickening of the ventricular wall to define normal wall motion.

Quantification of regional ventricular function can be achieved using regional radial shortening quantification as suggested by Warner and co-workers[32] for contrast angiograms or by calculation of regional ejection fraction.[16,18] Both methods have been attempted, but greater success has been found with the regional ejection fraction approach, since it is less subject to the statistical fluctuations that can occur while following the motion of one small point on the wall. The presentation of this quantified data can be in the form of a numerical result for each zone of the chamber, an image of the changes in regional volume from end of diastole to end of systole (a stroke volume image),[16] or a stroke volume image divided by an end-diastolic image (an ejection fraction image). In addition, "functional images" of wall motion can be generated using the slope of ejection at each point in the chamber to define a shade of gray in the image or using the time from the onset of systole to maximal ejection at that point to define the gray shades of the image. Although it is unclear at this time which of these methods will be the most useful, it is apparent that many approaches to depicting and quantifying regional ventricular function will be studied until one or several are found to be most useful and survive the test of time.

EVALUATION OF INTERVENTION ON VENTRICULAR FUNCTION

Data can be recorded with short interventions provided ventricular function is stable during the interval of data recording and there is minimal patient motion. These measurements can be made with drugs, pacing, and exercise. To be certain that the opportunity to define an alteration in ventricular function is maximized, the interval of data recording should be as long as possible, in keeping with the intervention under investigation. Since most interest is in defining the overall response of ventricular function to the intervention, it is important to record several points during the intervention if at all possible, since just starting and ending point data may not adequately describe the response.

CONCLUSION

When careful attention is paid to the technical details of data recording and analysis, the gated scan provides reproducible data of both right and left heart function. Because of the simplicity of the procedure, it is likely that it will be used increasingly in the future to define ventricular function in anticipation of cardiac catheterization in patients with coronary, congenital, valvular, and cardiomyopathic disease.

Clinical applications

Bertram Pitt, James H. Thrall, and
H. William Strauss

Since its introduction in 1971, the noninvasive determination of ventricular function by gated cardiac blood pool imaging has found wide application.[29,33] Although controversy continues to exist concerning the relative merits of "first-pass" and equilibrium-gated blood pool techniques, it is becoming increasingly clear that left ventricular ejection fraction calculated from either technique correlates well with that obtained by standard contrast left ventriculography.[2,29] Correlation coefficients of over 0.9 have been reported from several laboratories throughout the world. The choice as to which technique to employ depends in part on available instrumentation and experience. Fast count-rate instruments such as the multicrystal scintillation camera lend themselves to the first-pass technique, whereas with the Anger single-crystal scintillation camera image quality and shorter processing time for analysis favor equilibrium-gated blood pool techniques.

The time-activity curves generated from the semiautomatic edge detection programs used with either first-pass or equilibrium-gated techniques are identical in shape to the left ventricular volume curve obtained from standard contrast left ventriculography.[8,21] There is an excellent correlation between left ventricular ejection fraction obtained from the time-activity curve by subtracting minimal activity during the cardiac cycle from maximal activity and dividing by maximal activity and that obtained by standard contrast angiography. In addition, there is a good correlation between the rates of ventricular emptying and filling as well as positive and negative $\frac{dV}{dt}$ obtained with the two techniques (Fig. 6-9). Although contrast angiography is the standard against which gated blood pool imaging is judged, there is reason to believe that the left ventricular ejection fraction obtained from the multiple-gated blood pool images is more accurate and reproducible

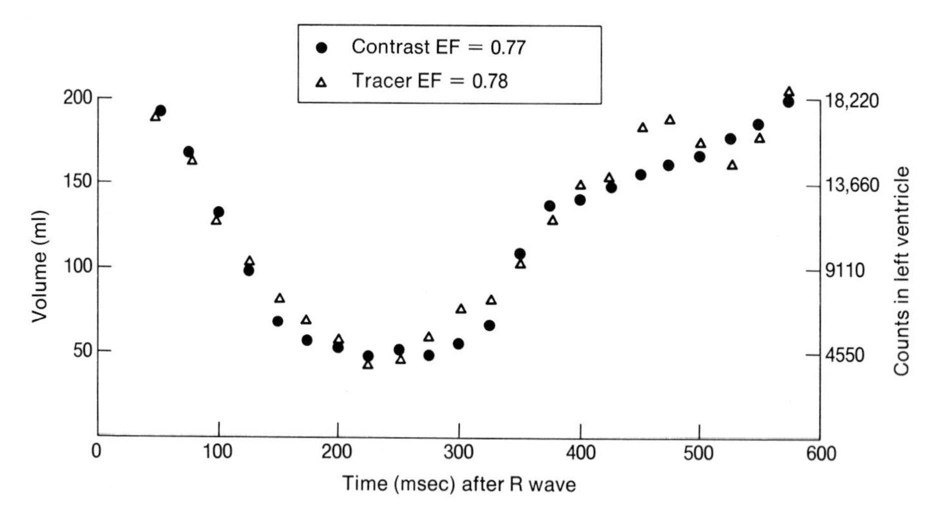

Fig. 6-9. *Closed circles,* Typical correlation between contrast ventriculogram analysis of left ventricular volume versus time curve obtained from analysis of biplane contrast angiograms. *Open circles,* Activity versus time curves obtained by semiautomatic edge-detection methods from left ventricle in multiple-gated blood pool scan. The ordinate on left is calibrated from contrast data in terms of milliliters, whereas that on right is calibrated in terms of counts in left ventricle obtained from tracer study. Since there is a linear relationship between activity in ventricle and absolute volume over wide range of values (Chapter 1), it is possible to calibrate activity curve at end of diastole and to calculate dV/dt for both filling and emptying portions of curve directly from tracer procedure.

than that obtained by contrast angiography. Calculation of left ventricular ejection fraction from a time-activity curve is independent of any assumption as to ventricular geometry, whereas calculation from a contrast angiogram assumes an ellipsoidal shape, unless the more time-consuming Simpson's rule is applied. Since in patients with ischemic heart disease abnormalities of regional myocardial wall motion are frequent, the calculation of ventricular volume using standard angiographic formula, assuming an ellipsoidal shape must of necessity be in error. Interobserver variation in calculation of ejection fraction is also less with the semiautomatic edge-detection programs used to analyze the gated blood pool image than with contrast angiography and careful tracing of the ventricular outline.

Regional myocardial wall motion can also be determined from the gated cardiac blood pool image.[33] The standard angiographic approach to analysis of regional myocardial wall motion with calculation of percent change in major and minor hemiaxes from end of diastole to end of systole has been applied to the gated blood pool images and shown to correlate well with standard contrast left ventriculography. Semiautomatic regional myocardial wall-motion programs have also been developed. After determining the edge of the left ventricle at end of diastole from the previously described semiautomatic edge-detection program, the center of activity within the left ventricle is calculated. Radii are then constructed from the center of activity to the edge of the ventricle, and the change in activity over time is calculated in several quadrants defined by radii set by the operator to define the major areas of coronary artery distribution. In the normally contracting ventricle, regional ejection fractions calculated from the regional time-activity curves are approximately equal in magnitude, and the time to minimal activity is almost identical in all areas. Patients with ischemic heart disease and abnormalities of regional myocardial wall motion show areas with a reduced regional ejection fraction or occasionally areas with a normal regional ejection fraction but a delayed time to minimal activity (tardokinesis) in comparison to other myocardial areas. Areas of ventricular dyskinesis can also be detected, often more easily than with contrast angiography, by finding an area with increasing activity at the time of overall minimal ventricular activity. Another approach to detecting regional myocardial wall motion abnormalities involves the use of subtraction images, that is, the subtraction of the end-systolic from the end-diastolic image. In the normally contracting ventricle, the subtraction image has uniform activity over the entire circumference of the left ventricular outline (Fig. 6-10). In patients with areas of akinesis or severe hypokinesis, ventricular circumference defects can be detected corresponding to areas of reduced myocardial wall motion. Although sensitive to severe hypokinesis and akinesis, the approach does not allow the detection of more subtle abnormalities of myocardial wall motion such as tardokinesis.

ACUTE INFARCTION

Gated cardiac blood pool imaging to evaluate ventricular function has been found to be of value in a variety of clinical situations. The gated cardiac blood pool scan can detect the presence of myocardial infarction by detecting an abnormality of regional myocardial wall motion. In an early study of thirty-eight patients with acute myocardial infarction documented by serial ECGs and serum enzyme measurements, abnormal regional myocardial wall motion was detected in all thirty-eight.[24] A definite area of akinesis was found in thirty-six, whereas in two patients hypokinesis was detected. Although not as sensitive to old myocardial infarction, the gated cardiac blood pool scan appears to be more sensitive than the electrocardiogram or resting ^{201}Tl myocardial image. Determination of left ventricular volume and left ventricular ejection fraction early in the course of acute myocardial infarction has also been of value and more sensitive in detecting left ventricular dys-

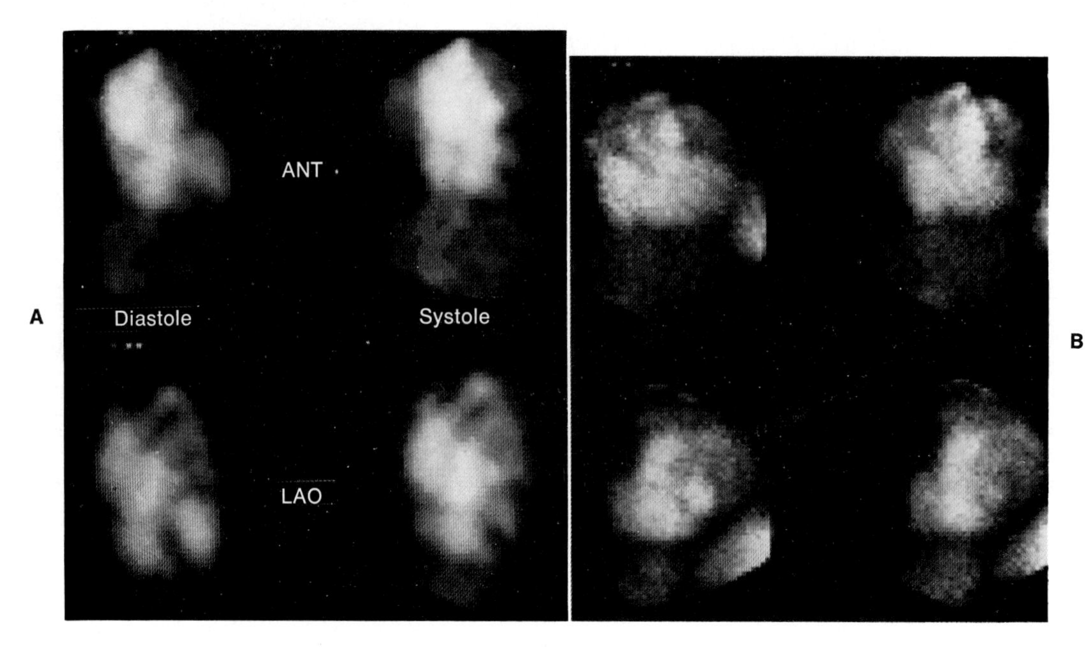

Fig. 6-10. A, Anterior and LAO views at end of diastole and end of systole of patient with hypertension, admitted with congestive cardiac failure. Usually left ventricle is enlarged with poor ejection fraction under these circumstances. In this patient, however, there is excellent ventricular function with normal ejection fraction. Left ventricular myocardium is hypertrophied, as determined by wall thickness at end of diastole (primarily of septum). Aorta is dilated, a finding that increases at end of systole. This constellation of findings suggests aortic aneurysm and hypertrophic myopathy. Patient died several days later and was found to have hypertrophic myopathy. **B,** *Top,* Anterior and *bottom,* LAO views of gated blood pool scan at end of diastole *(left)* and end of systole *(right)* in patient with severe right ventricular dysfunction. Right atrium is enlarged and noncontractile. Right ventricle is enlarged and has some contraction in outflow portion at end of systole. Left ventricle is small and contracts vigorously. No focal wall motion abnormalities are observed. This type of scan is observed in patients with mitral valve disease, right ventricular infarction, pulmonary hypertension, or severe acute pulmonary diseases such as adult respiratory distress syndrome.

function than routine hemodynamic monitoring with measurement of left ventricular filling pressure, cardiac index, and stroke work.[24] In uncomplicated cases of patients with acute myocardial infarction who had a normal left ventricular filling pressure and cardiac index, left ventricular end-systolic volume was significantly increased in comparison to normal. In patients with pulmonary congestion complicating their acute infarction, both left ventricular end-diastolic and end-systolic volumes were significantly increased, whereas ejection fraction was decreased. In patients with cardiogenic shock, there was a further significant increase in end-diastolic and end-systolic volumes and a fall in ejection fraction. Increases in left ventricular end-diastolic volume were usually accompanied by a concomitant rise in left ventricular filling pressure. Occasionally, however, left ventricular filling pressure may be markedly increased with only a small increase in end-diastolic volume, reflecting a change in ventricular compliance early during the course of infarction. Serial changes in ejection fraction are also of value in providing objective evidence of the patient's progress and prognosis. Although release of creatine phosphokinase (CPK) has been shown to be a good guide to prognosis in patients with acute myocardial infarction (Chapter 9), there is suggestive evidence that left ventricular ejection fraction and the extent of regional wall motion abnormality may be even better indications.[25] The left ventricular ejection fraction and extent of ventricular akinesis reflect overall ventricular damage, whereas release of CPK reflects only acute damage. In a patient with a previous infarction, prognostic indices based on CPK release alone may be misleading, since it is the total rather than the acute damage alone that determines ultimate prognosis.

Cardiogenic shock can be caused by either left or right ventricular dysfunction (Fig. 6-10). When cardiogenic shock is associated with right ventricular failure, enlargement of the right ventricle is common. The gated cardiac blood pool scan in the LAO position has been used to assess the relative size of the right and left ventricles to determine whether the procedure could be used to differentiate cardiogenic shock due to right ventricular involvement from that due primarily to the left.[23] In six patients with cardiogenic shock, three had electrocardiographic evidence of anterior infarction and three had inferior infarction. Of the three with anterior infarction, there was marked left ventricular enlargement: ratio of right to left ventricular area to 0.62. In those with inferior infarction, the ratio was increased to 2.05 (normal, 1.1), suggesting primarily right ventricular dilation and dysfunction. One of the patients with increased right ventricular size died and at autopsy was found to have massive right ventricular enlargement. The other two patients with enlarged right ventricles were treated with volume loading and improved clinically. All three patients with enlarged LV and shock died. The study supports the contention that patients with cardiogenic shock due to right ventricular infarction have a better prognosis than those with primarily left ventricular involvement. Since the therapy for these two conditions is quite different, it is important to differentiate between them early in the course of the patient's illness. The gated scan can do this, even in acutely ill patients.

The gated cardiac blood pool scan may also be of value in the evaluation of patients with cardiogenic shock due to rupture of a papillary muscle or of the intraventricular septum. Cardiogenic shock in these patients can be due to either a small strategically placed infarct or to massive myocardial damage. The finding of a relatively good left ventricular ejection fraction in a patient with severe pulmonary congestion or cardiogenic shock due to papillary muscle or intraventricular septal rupture suggests a good prognosis if the defect can be surgically repaired. In such a patient we would institute vigorous medical therapy, including insertion of an intra-aortic balloon pump and vasodilators. If the patient does not improve or deteriorates, we move rapidly to surgery. On the other hand, if the left ventricular ejection fraction is low, we attempt to delay surgical intervention, since prognosis is poor with or without surgical intervention. In these patients with a low ejection fraction, it is the extent of myocardial damage rather than the rupture alone that determines outcome.

CONGESTIVE HEART FAILURE

In patients with congestive heart failure following myocardial infarction, the gated blood pool scan has been used to differentiate between those with localized left ventricular aneurysm and those with diffuse left ventricular hypokinesis. Since patients with diffuse left ventricular hypokinesis have a higher mortality rate if a surgical intervention such as coronary revascularization or aneurysmectomy is permitted, these patients are not considered optimum surgical candidates. Patients with localized disease are better candidates for surgery, since they have a lower mortality rate and frequently improve following operation. The gated blood pool scan has been used to screen out those with diffuse hypokinesis and to spare them the expense and possible further illness from catheterization, whereas those patients found to have localized akinesis or aneurysm on the gated blood pool scan are referred for cardiac catheterization to obtain detailed information about left ventricular and coronary anatomy in anticipation of possible surgery. In an evaluation of twenty-two patients with congestive failure

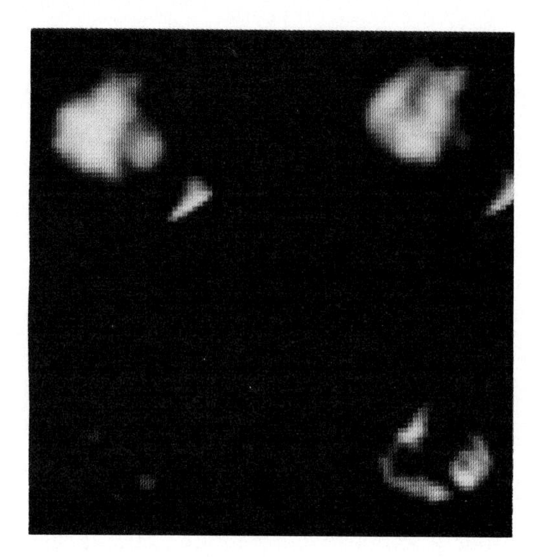

Fig. 6-11. *Top,* End-diastolic and end-systolic images in LAO position and *bottom,* subtraction image (end of diastole minus end of systole). This subtraction image represents areas of changes during ejection of blood. This image is also called stroke volume image.

following myocardial infarction, the gated scan successfully separated those with diffuse hypokinesis from those with a localized aneurysm.[22] The gated scan was more sensitive than the presence of an abnormal cardiac impulse, S-3 gallop, persistent ST-segment elevation, or plain chest radiograph in this evaluation, since all these findings were distributed almost equally between those patients with aneurysm and those with diffuse hypokinesis.

One of the most valuable applications of gated cardiac blood pool imaging in patients with congestive heart failure is the ability to obtain an objective assessment of the effects of therapeutic interventions to improve ventricular function. With the introduction of integrated portable camera-computer systems, serial measurements of ejection fraction and regional myocardial wall motion can be made at the bedside of acutely ill patients in the CCU, surgical recovery room, or emergency room. The need for and efficacy of vasodilators and inotropic agents can be quickly and reliably assessed. The study may be repeated during patient follow-up examinations to determine the need for continued therapy.

VENTRICULAR FUNCTION RESERVE

Multiple-gated cardiac blood pool imaging has recently been applied to exercise. Borer and associates have shown the utility of exercise multiple-gated cardiac blood pool imaging in patients with suspected ischemic heart disease.[5] With this technique the patient lies beneath the detector of the scintillation camera, and resting RAO and LAO multiple-gated blood pool images are obtained. The detector is then placed in the LAO position and the patient exercised to his maximal level with a supine bicycle ergometer. Although motion during exercise is a potential limitation of this technique, proper positioning of the patient and the use of a supine bicycle ergometer appear to minimize the difficulty.

In a recent study, left ventricular ejection fraction obtained by exercise multiple-gated blood pool imaging was found to correlate well with that obtained by exercise contrast left ventriculography.[17] Normal patients tend to increase their left ventricular ejection fraction with exercise from a mean of approximately 65% to 75%. Patients with ischemic heart disease either do not show the normal increase or have an actual decrease in ejection fraction.[5] The decreases in ejection fraction is, in general, associated with the development of myocardial ischemia, whereas the failure to increase ejection fraction suggests previous myocardial damage without ischemia.

Patients with asymptomatic significant coronary artery disease also show an abnormal exercise response with this technique. Patients with chest pain of noncardiac origin show a normal increase in ejection fraction during exercise. This differentiation is of clinical importance, since patients with atypical chest pain pose a difficult clinical dilemma, which results in the need for invasive coronary arteriography with its concomitant risks and expense. The sensitivity for detection of ischemic heart disease by exercise multiple-gated cardiac blood pool imaging has been reported to be greater than 90%.[5] In comparison, the sensitivity for detection of coronary artery disease by exercise electrocardiography is on the order of 65% and with conventional nontomographic exercise ^{201}Tl myocardial imaging, 80% to 85% at best.[3]

Although the results of exercise multiple-gated cardiac blood pool imaging for the detection of ischemic heart disease as reported by Borer and co-workers[5] have been most encouraging, these excellent results have not been uniformly duplicated in other centers. In part, we believe failure to detect an abnormal exercise response in a patient with significant coronary artery disease may be due to an inadequate technique. Failure to achieve maximal exercise levels may result in falsely negative results. Recent experience at the University of Michigan has pointed out the utility of graded exercise multiple-gated cardiac blood pool imaging. Normal individuals tend to increase their ejection fractions to a plateau value in the range of 75% to 85%. Patients with ischemic heart disease may, however, show a number of patterns. In some, the left ventricular ejection fraction falls progressively with exercise as left ventricular function deteriorates with the development of exercise-induced ischemia. In others, however, left ventricular ejection fraction initially increases and falls only at maximal exercise levels. Failure to achieve maximal exercise levels either because of inadequate technique, development of leg pain, or breathlessness may result in the recording of an increase in ejection fraction and therefore a reduced sensitivity to coronary artery disease.

Exercise multiple-gated cardiac blood pool imaging has also been applied to a number of other clinical situations. Changes in ejection fraction during exercise are being used to screen patients prior to hospital discharge for possible coronary angiography as well as to plan and evaluate cardiac rehabilitation programs. One of the most sensitive ways to detect cardiac dysfunction following infarction and to evaluate the effect of therapeutic agents such as vasodilators is the response of ejection fraction during exercise.[6] Exercise multiple-gated cardiac blood pool imaging is also finding increasing use in the initial evaluation and follow-up study of patients undergoing coronary artery bypass grafting.[15] Prior to surgery, patients with angina pectoris tend to decrease their ejection fraction during exercise. The finding of a return to a normal exercise response following surgery suggests graft patency and relief of ischemia. The finding of a decrease in resting ejection fraction and the presence of a new wall motion abnormality postoperatively suggests relief of the patient's symptoms by infarction of the previously ischemic area. The finding of a postoperative resting ejection fraction unchanged from preoperative value with persistence of an abnormal decrease in ejection fraction during exercise suggests persistence of an ischemic region.

Exercise multiple-gated cardiac blood pool imaging is also proving valid in assessing ventricular function and planning for surgery in patients with aortic regurgitation.[7] Patients with symptomatic aortic regurgitation have an abnormal exercise response in comparison to normal individuals. Patients with asymptomatic aortic regurgitation may have a normal or abnormal exercise response. It has been suggested that those with an abnormal exercise response may be candidates for valve replacement despite the fact that they are asymptomatic. Although as yet unproved, this hypothesis is worthy of further study, since current clinical criteria for valve replacement in patients with aortic regurgitation often result in delay and irreparable myocardial damage, which is not corrected by valve replacement.

The potential of exercise multiple-gated blood pool imaging can be appreciated by a recent study in patients with a history of viral myocarditis.[10] The patients in this study had all successfully recovered from an episode of viral myocarditis and had returned to full activity. They had normal exercise electrocardiograms, chest x-ray results, and echocardiograms. Exercise multiple-gated cardiac blood pool imaging revealed an abnormal response in the majority of these patients, reflecting prior myocardial damage. This experience combined with that of patients with coronary artery disease suggests that exercise multiple-gated blood pool imaging will be of value in screening asymptomatic individuals in high-risk occupations such as pilots or astronauts.

CONCLUSION

The application of rest and exercise multiple-gated blood pool imaging to patients with acute infarction, congestive heart failure, and valvular and congenital heart disease is in its infancy. The ability to detect ventricular dysfunction at rest and exercise, to objectively assess the effect of therapeutic intervention at the bedside in acutely ill patients, and to assess prognosis have added a new dimension to clinical cardiology, which is not as yet fully exploited.

REFERENCES

1. Alderson, P. O., and Hamilton, R. G.: A comparative evaluation of techniques for rapid and efficient in vivo labelling of red cells with [^{99m}Tc] pertechnetate, J. Nucl. Med. **18:**1008, 1977.
2. Ashburn, W. L., Schelbert, H. R., and Verba, J. W.: Left ventricular ejection fraction—a review of several radionuclide angiographic approaches using the scintillation camera, Prog. Cardiovasc. Dis. **20:**267, 1978.
3. Bailey, I. K., Griffith, L. S. C., Rouleau, J., Strauss, H. W., and Pitt, B.: Thallium-201 myocardial perfusion imaging at rest and exercise: comparative sensitivity to electrocardiography in coronary artery disease, Circulation **55:**79, 1977.
4. Berman, D. S., Salel, A. F., Denardo, G. L., et al.: Clinical assessment of left ventricular regional contraction patterns and ejection fraction by high resolution gated scintigraphy, J. Nucl. Med. **16:**695, 1975.
5. Borer, J. S., Bacharach, S. L., Green, M. V., et al.: Real time radionuclide cineangiography in the non-invasive evaluation of global and regional left ventricular function at rest and during exercise in patients with CAD, N. Engl. J. Med. **296:**839, 1977.
6. Borer, J. S., Bacharach, S. L., Green, M. V., Kent, K. M., Johnston, G. S., and Epstein, S. E.: Effect of nitroglycerin on exercise induced abnormalities of left ventricular regional function and ejection fraction in coronary artery disease. Assessment by radionuclide cineangiography in symptomatic and asymptomatic patients, Circulation **57:**314, 1978.
7. Borer, J. S., Bacharach, S. L., Green, M. V., Kent, K. M., Rosing, D. R., Seides, S. F., McIntosh, C. L., Conkle, D. M., Morrow, A. G., and Epstein, S. E.: Left ventricular function during exercise before and after aortic valve replacement, abstracted, Circulation **55,56:**98, 1977.
8. Burow, R. D., Strauss, H. W., Singleton, R., Pond, M. P., Rehn, T., Bailey, I. K., Griffith, L. S. C., Nickoloff, E., and Pitt, B.: Analysis of left ventricular function from multiple gated acquisition (MUGA) cardiac blood pool imaging: comparison to contrast angiography, Circulation **56:**1024, 1977.
9. Caldwell, J. H., Williams, D. L., Kennedy, J. W., et al.: Quantitative semi-automated technique for determination of regional left ventricular function during rest and exercise, J. Nucl. Med. **19:**710, 1978.
10. Das, S. K., Brady, T. J., Thrall, J. H., and Pitt, B.: Evidence of cardiac dysfunction in asymptomatic patients with prior myocarditis, abstracted, Circulation **57 & 58** (suppl. II):II24, 1978.
11. Freedman, G. S., editor: Tomographic imaging in nuclear medicine, New York, 1973, Society of Nuclear Medicine.
12. Green, M. V., Ostrow, H. G., Douglas, M. A., et al.: High temporal resolution ECG-gated scintigraphic angiocardiography, J. Nucl. Med. **16:**95, 1975.
13. Hamilton, G. W., Williams, D. L., and Gould, L.: Selection of appropriate frame rates for radionuclide angiography, J. Nucl. Med. **17:**556, 1976.
14. Hegge, F. N., Hamilton, G. W., Larson, S. M., et al.: Cardiac chamber imaging: a comparison of red blood cells labeled with Tc99m in vitro and in vivo, J. Nucl. Med. **19:**129, 1978.
15. Kent, K. M., Borer, J. S., Green, M. V., Bacharach, S. L., McIntosh, C. L., Conkle, D. M., and Epstein, S. E.: Effects of coronary revascularization on left ventricular function during exercise, N. Engl. J. Med. **298:**1434, 1978.
16. Kirschenbaum, H.: Unpublished observations.
17. Lo, K., Brady, T., Thrall, J., and Pitt, B.: Correlation of exercise gated blood pool imaging with exercise contrast angiography (submitted for publication).
18. Maddox, D. E., Holman, B. L., Wynne, J., et al.: Ejection fraction image: a noninvasive index of regional left ventricular wall motion, Am. J. Cardiol. **41:**1230, 1978.
19. McKusick, K. A., Bingham, J., Pohost, G., et al.: The gated first pass radionuclide angiogram: a method for measurement of right ventricular ejection fraction, abstracted, Circulation **57 & 58** (suppl. II):II130, 1978.
20. Pavel, D. G., Zimmer, A. M., and Patterson, V. N.: In vivo labeling of red blood cells with ^{99m}TC: a new approach to blood pool visualization, J. Nucl. Med. **18:**305, 1977.

21. Pitt, B., and Strauss, H. W.: Current concepts: evaluation of ventricular function by radioisotopic technics, N. Engl. J. Med. **296:**1097, 1977.
22. Rigo, P., Murray, M., Strauss, H. W., and Pitt, B.: Scintiphotographic evaluation of patients with suspected left ventricular aneurysm, Circulation **50:**985, 1974.
23. Rigo, P., Murray, M., Taylor, D. R., Weisfeldt, M. L., Kelly, D. T., Strauss, H. W., and Pitt, B.: Right ventricular dysfunction detected by gated scintiphotography in patients with acute inferior myocardial infarction, Circulation **52:**268, 1975.
24. Rigo, P., Strauss, H. W., Taylor, D. R., Kelly, D. T., Weisfeldt, M. L., Strauss, H. W., and Pitt, B.: Left ventricular function in acute myocardial infarction evaluated by gated scintigraphy, Circulation **50:**678, 1974.
25. Schulze, R. A., Strauss, H. W., and Pitt, B.: Sudden death in the year following myocardial infarction: relation to late hospital phase VPC's and left ventricular ejection fraction, Am. J. Med. **62:**192, 1977.
26. Steinberg, S., Thrall, J. H., and Pitt, B.: Unpublished observations.
27. Strauss, H. W., McKusick, K. A., Alpert, N., et al.: Multiple gated blood pool scans—effect of pixel and frame time on image resolution (submitted for publication).
28. Strauss, H. W., Boucher, C., and Okada, R.: The bifocal diverging collimator: a method for simultaneous biplane imaging of the heart, abstracted, J. Nucl. Med. (in press).
29. Strauss, H. W., Zaret, B. L., Hurley, P. J., Natarajan, T. K., and Pitt, B.: A scintiphotographic method for measuring left ventricular ejection fraction in man without cardiac catheterization, Am. J. Cardiol. **28:**575, 1971.
30. Thrall, J. H., Freitas, J. E., Swanson, D., Rogers, W. L., Clare, J. M., Brown, M. L., and Pitt, B.: Clinical comparison of cardiac blood pool visualization with technetium-99m red blood cells labeled in vivo and with technetium-99m human serum albumin, J. Nucl. Med. **19:**796, 1978.
31. Vogel, R. A., Kirsh, D., LeFree, M., et al.: A new method of multiplanar emission tomography using a seven pinhole collimator and an Anger scintillation camera, J. Nucl. Med. **19:**648, 1978.
32. Warner, H., et al.: Axis approach to measure regional wall motion, personal communication.
33. Zaret, B. L., Strauss, H. W., Hurley, P. J., Natarajan, T. K., and Pitt, B.: A noninvasive scintiphotographic method for detecting regional ventricular dysfunction in man, N. Engl. J. Med. **284:**1165, 1971.

7 □ Detection and quantification of intracardiac shunts

Salvador Treves
J. A. Parker

Radionuclide angiocardiography is a useful method for the evaluation of patients with congenital heart disease because it provides information on a significant number of hemodynamic parameters. The advent of the gamma scintillation camera,[2,17] computer systems, and short-lived radionuclides in recent years has rapidly accelerated and expanded the use of this technique in children.[12,24,25,27] There are radionuclide techniques for measurement of transit time, cardiac output, right and left ventricular ejection fraction, stroke volume, end-diastolic volume, and myocardial blood flow and mass; detection, localization, and quantitation of intracardiac shunts; and evaluation of myocardial wall motion. Newer mobile gamma camera-computer systems permit the study of severely ill patients, including premature and newborn infants in neonatal or intensive care units.[22] Specially designed magnifying collimators and the development of ultrashort-lived radionuclides[23] should result in an improvement in the diagnostic capabilites of this technique.

Radionuclide angiocardiography has the advantage of being relatively nontraumatic, requiring a small intravenous injection. It carries no risk, produces no hemodynamic disturbance, and the radiation dose is low. Therefore it can be used repeatedly if necessary. However, this technique also has limitations. For example, its anatomic resolution is limited, and it does not provide information about intracardiac pressures or oxygen saturation.

METHOD

To accurately record, display, and analyze these studies, it is necessary to use a gamma camera with good intrinsic resolution equipped with adequate collimators (high sensitivity, high resolution) and an on-line digital computer system.

We will describe the technique used at The Children's Hospital Medical Center in Boston with some detail. The patient is premedicated with oral potassium perchlorate at 6 mg/kg of body weight to block the uptake of ^{99m}Tc as sodium pertechnetate by the thyroid and other sites and enhance its urinary excretion.

RADIOPHARMACEUTICAL

^{99m}Tc as sodium pertechnetate in a dosage of 200 μCi/kg of body weight is used. A minimal total dose of 2 mCi is necessary for adequate counting statistics. Other technetium radiopharmaceuticals can also be used with the same dosage schedule. [^{99m}Tc] stannous diethylenetriamine pentacetic acid (DTPA) does not require premedication with perchlorate and exhibits a more rapid blood disappearance than radiotechnetium. This results in lower background radiation, which is valuable if a second injection is desired. ^{99m}Tc-labeled human serum albumin (HSA) is used for determination of cardiac output, blood pool studies, and ECG-gated angiocardiography (ventriculography). The whole-body dosage with this agent is similar to that of [Tc] DTPA.

INJECTION TECHNIQUE

An intravenous scalp needle is inserted in a vein close to the heart such as an external jugular vein. This is important because for adequate temporal resolution and accurate quantitation a small, single, and short ($\cong$ 3 sec) bolus of radionuclide solution should be injected. After the needle has been inserted in the external jugular vein, the patient is placed under the gamma scintillation camera. The chest is viewed on the anterior or LAO projection. We developed a disposable injector with a one-way valve (Fig. 7-1), which is useful in delivering a small bolus. The radionuclide solution (0.1 to 0.5 ml) is rapidly injected into a stream of flowing saline solution (0.5 to 10 ml).[21] The adequacy of the bolus is checked on a time-activity curve generate over the superior vena cava (Fig. 7-2). If the bolus is inadequate, a second injection is given through the same or from another vein. Injections in the right external jugular vein result in satisfactory studies in at least 95% of the cases. Injections into the antecubital veins result in a 15% to 20% rate of unsatisfactory studies (prolonged or fragmented bolus). In premature or small infants, it is possible to obtain a high rate of adequate studies by injection into antecubital veins.

DATA RECORDING

The angiocardiogram is recorded on a digital computer using either matrix-mode (64 × 64, 20 frames/sec) or list-mode ac-

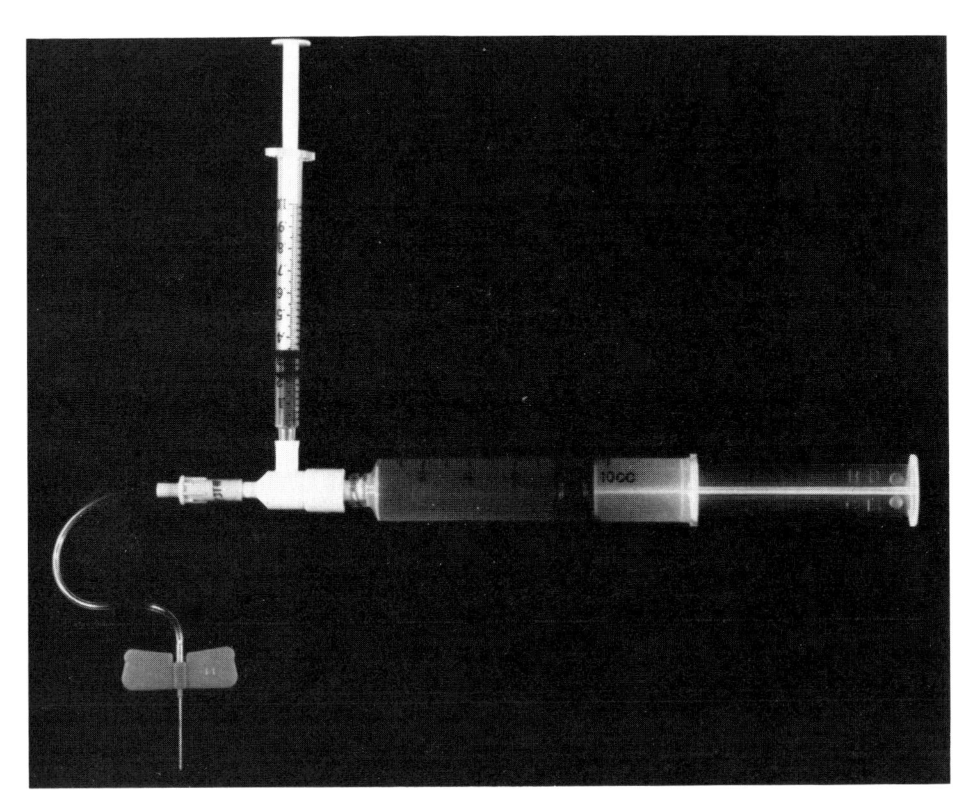

Fig. 7-1. Radionuclide injector. Injection is performed by initiating a flush of saline solution with one hand while holding syringe of radiopharmaceutical with other hand. Radiopharmaceutical is then rapidly injected without interrupting saline flush. (From Treves, S., and Collins-Nakai, R. L.: Am. J. Cardiol. **38:**711, 1976.)

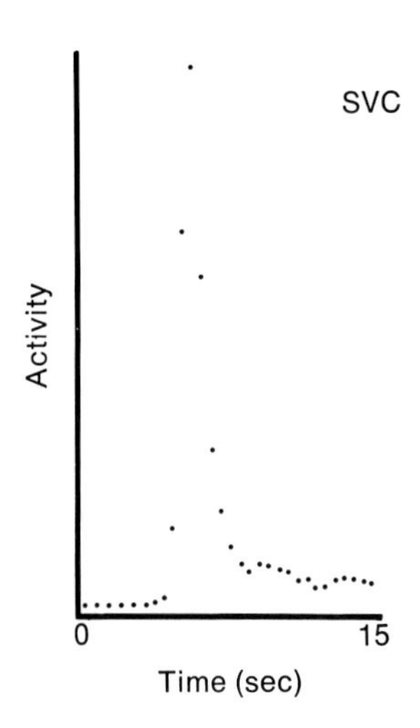

Fig. 7-2. Time-activity curve over superior vena cava. Curve was taken at 2 frames/sec and represents an adequate bolus of radioindicator. Fragmented or prolonged bolus should be discarded because it results in inadequate temporal resolution of various phases of radionuclide angiogram and produces inaccurate quantitative information (i.e., overestimation of left-to-right shunting or calculation of left-to-right shunt curve when there is no shunt).

quisition (Chapter 4). For shunt detection, localization, and quantitation and first-pass ejection fraction measurements, 25 sec of recording are sufficient. A recording period of 5 min at 1 frame/sec is added for estimation of equilibrium for calculation of cardiac output.

DATA DISPLAY AND ANALYSIS

For anatomic evaluation of the angiocardiogram series, images of 0.5 or 1.0 sec are displayed on the computer oscilloscope. The images can be interpolated into a fine matrix (i.e., 128 × 128) with or without contrast enhancement to facillitate anatomic evaluation. Also, it is possible to add and subtract images corresponding to various phases of the angiogram to facillitate the selection of regions of interest without overlap of adjacent cardiovascular structures. As an example, the images corresponding to the right and left sides of the heart and the superior vena cava, pulmonary artery, and aorta are subtracted from the lung phase in selecting pulmonary regions of interest for quantitation of left-to-right shunts. In addition, some computer systems allow a cinelike display, which can be replayed at various speeds.

DETECTION, LOCALIZATION, AND QUANTITATION OF LEFT-TO-RIGHT SHUNTS

With an intravenous bolus of a suitable radiopharmaceutical, a series of images of the anterior (or LAO) chest are taken with the gamma camera and recorded by the computer. Either 2 or 4 frames/sec are generally adequate for shunt quantitation. More rapid acquisition rate (i.e., 100 frames/sec) with subsequent reformatting at various frame rates can be helpful. Following a check for adequacy of the bolus on a curve generated over the superior vena cava, a visual analysis of the angiogram is carried out to assess gross radionuclide flow through the cardiopulmonary system and to define regions of interest for the generation of time-activity curves. The bolus peak should be unique and last about 3 sec (Fig. 7-2). The appearance of a normal radionuclide angiogram has been described previously, and its serial images reveal circulation of the radioactivity within the superior (or inferior vena cava), right atrium, right ventricle, pulmonary artery, lungs (Fig. 7-3), left atrium (usually not well delineated on the anterior projection), left ventricle, and aorta (Fig. 7-4). Patients with small left-to-right shunts (for example, with a pulmonary-to-systemic flow ratio less than 1.5) may reveal a pattern that would be difficult to distinguish from normal.

Patients with left-to-right shunting characteristically show a clearly defined right side of the heart and pulmonary artery (almost identical to normal), although in large left-to-right shunts there may be an apparent reduction of the activity as the bolus is diluted with nonradioactive blood passing through the shunt. Later frames in the study reveal persistent radioactivity within the lungs (Fig. 7-3) and the right side of the heart (at levels depending on the level of the shunt) and also poor visualization of the left side of the heart and aorta (Fig. 7-5). In addition, in left-to-right shunting, there is a more rapid transpulmonary transit time

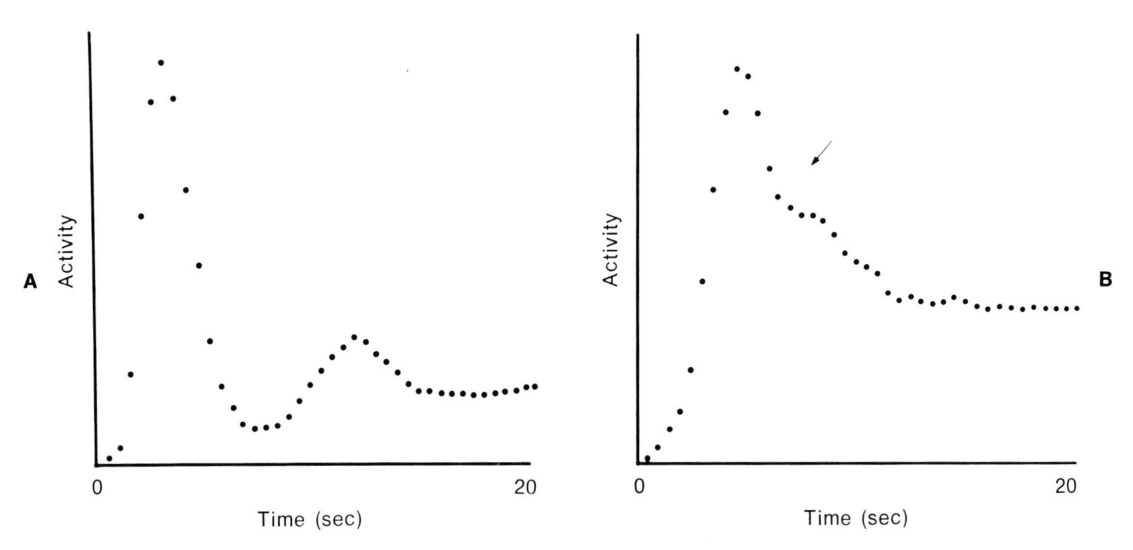

Fig. 7-3. Pulmonary time-activity curves. Curves were taken at 2 frames/sec. **A,** Normal curve. **B,** Pulmonary curve in left-to-right shunt. Note early pulmonary recirculation *(arrow)* due to radiolabeled blood returning prematurely to lungs.

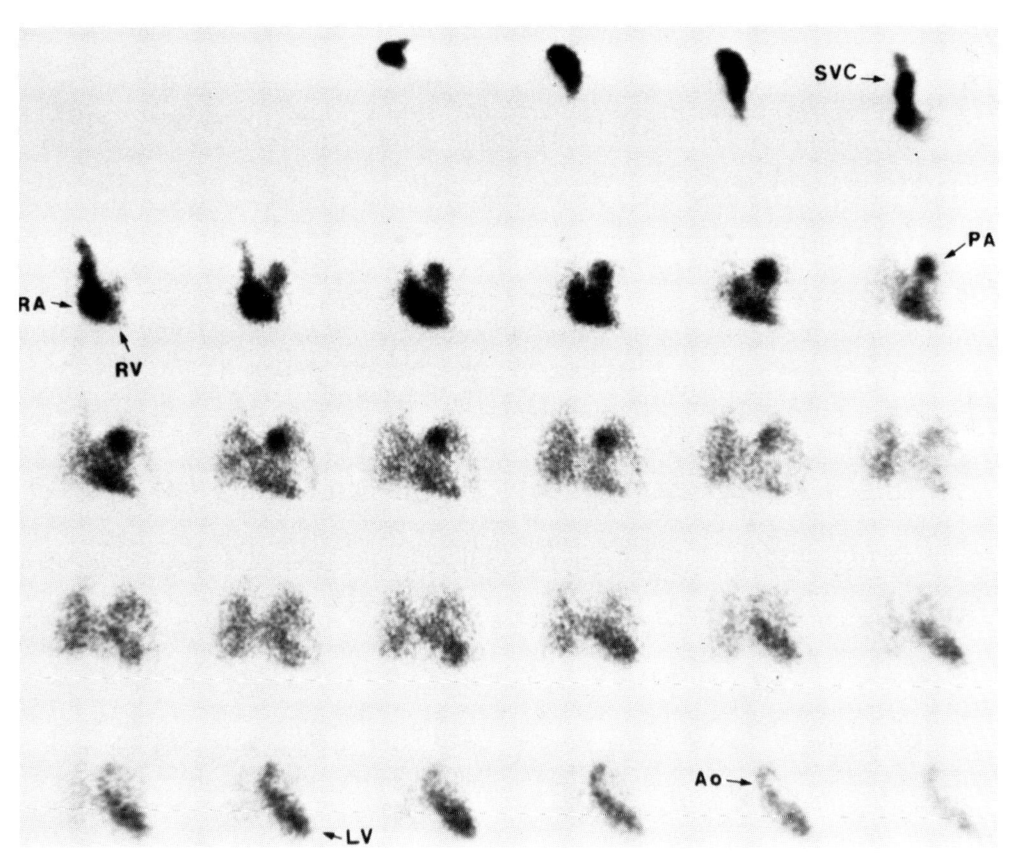

Fig. 7-4. Radionuclide angiogram in patient without left-to-right shunt. Images were taken at 0.3 sec intervals. Patient was injected in right external jugular vein. Radioindicator is seen as it circulates within superior vena cava *(SVC),* right atrium *(RA),* right ventricle *(RV),* pulmonary artery *(PA),* and lungs. Then radioactivity returns to left atrium, left ventricle *(LV),* and aorta *(Ao).* As just noted, patient had no left-to-right shunting; however, poststenotic dilation of pulmonary artery can be seen. Clinically, differential diganosis was atrial septal defect versus pulmonary artery stenosis.

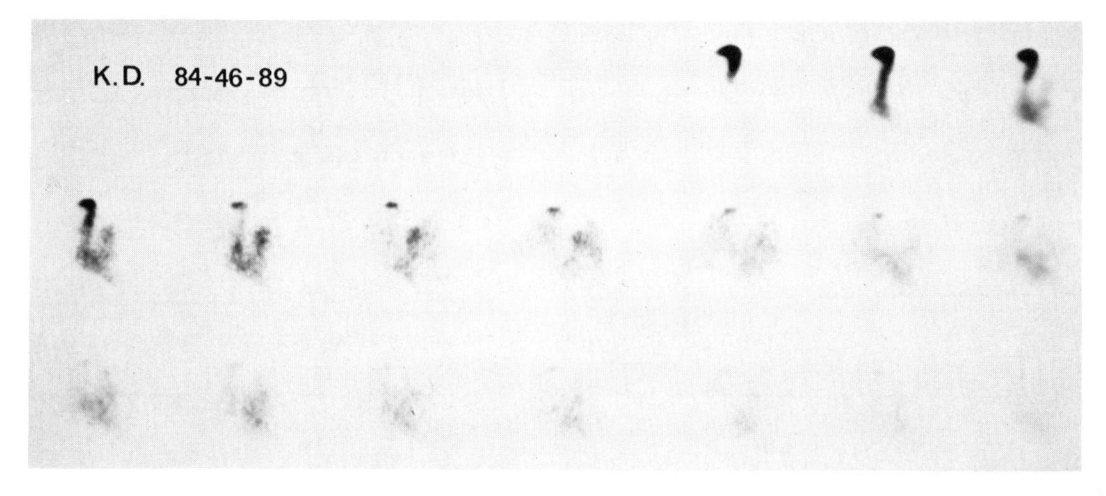

Fig. 7-5. Left-to-right shunt. Radionuclide angiogram was obtained on patient with left-to-right shunting through PAPVR. Study was recorded at 0.5 sec intervals. *Top row:* Radioindicator in superior vena cava, RA, RV, and PA. Note also apparent reduction of radioactivity within superior vena cava, presumably due to dilution of bolus with nonradioactive blood through large left-to-right shunt. Remainder of images reveal persistence of radioactivity within lungs and poor visualization of left side of heart and aorta.

than normal because of high pulmonary blood flow, although this characteristic is difficult to appreciate visually. These scintigraphic findings are diagnostic of left-to-right shunting.

With careful selection of small regions of interest over the right atrium and right ventricle, it is often possible to determine the level at which the left-to-right shunt occurs,[15] provided there is no overlap of the left and right sides of the heart on the projection used. For example, in partial anomalous pulmonary venous return (PAPVR) to the superior vena cava, it is possible sometimes to detect a small second peak of early recirculation due to the left-to-right shunt. Similarly, there would be a recirculation peak on a curve obtained from a region of interest over the right atrium in atrial septal defect. Recirculation peaks will also be seen over the right ventricle and the lungs. This approach can only detect the most proximal level of shunting. In ventricular septal defects, the right atrial curve will generally not show a recirculation peak, but this will be seen at the venticular level and over the lungs. Finally, in patent ductus arteriosus, early recirculation peaks will be observed over the lungs but not over the right atrium and right ventricle. Following are two other characteristics of patent ductus arteriosus: (1) frequently there appears to be greater pulmonary circulation to the left lung and (2) despite left-to-right shunting of a certain magnitude with early pulmonary recirculation, the left side of the heart is reasonably well visualized. The left pulmonary time-activity curve in patent ductus arteriosus reveals a slower downslope than that of the right pulmonary curve (Fig. 7-6). Likewise, in palliative left-to-right shunts (Potts, Waterston, Blalock), the downslope of the curve of the lung receiving the shunt is slower compared to the contralateral lung's curve, if the shunt is patent.

Analysis of pulmonary time-activity curves can also serve to quantify the amount of shunting. One method developed in our laboratory uses a gamma-variate model to separate the activity caused by the first pulmonary transit from that of early recirculation due to the shunt.[3,14] Thompson and co-workers[20] and Stramer and Clark[18] found that the slope of a gamma-variate curve closely resembles the shape of an indicator-dilution curve. A "least squares" fit can be used to match the

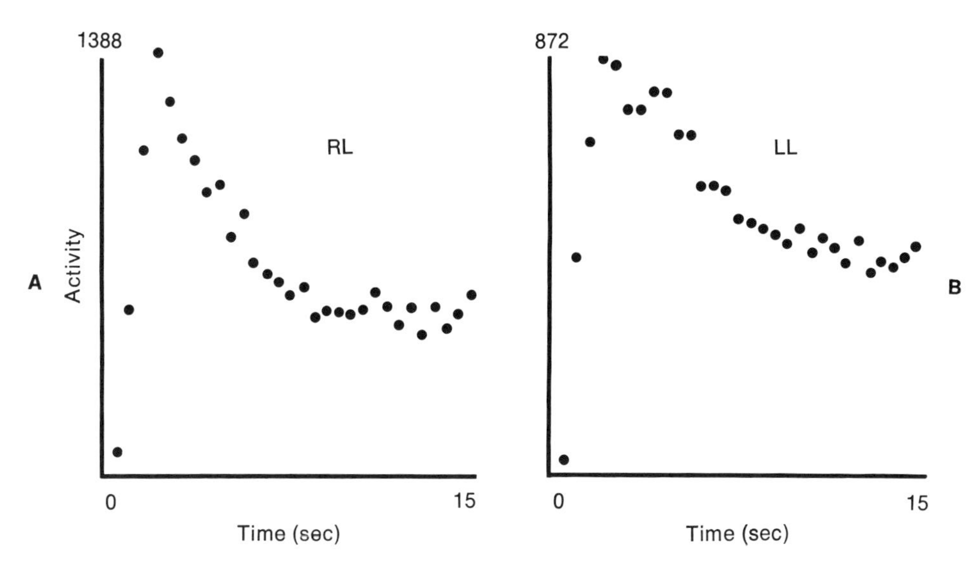

Fig. 7-6. Patent ductus arteriosus (PDA) in 2-month-old child. Pulmonary time-activity curves from the right lung (RL), **A,** and left lung (LL), **B.** With certain frequency in patent ductus arteriosus, downslope on left-lung curve is slower than on right-lung curve. Calculation of pulmonary-to-systemic flow ratio will reveal a higher value using left-lung curve. It is believed that an average of pulmonary-to-systemic flow ratios of RL and LL would provide net left-to-right shunting.

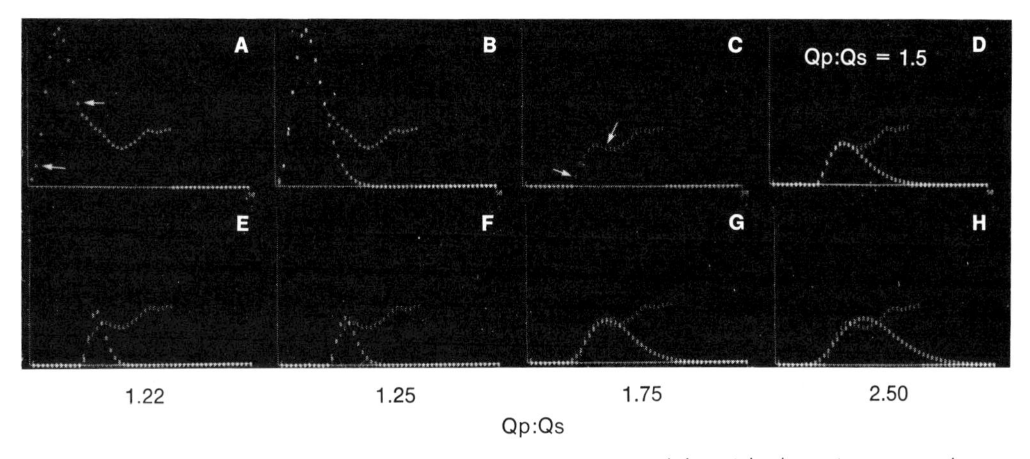

Fig. 7-7. Calculation of pulmonary-to-systemic flow ratio (Qp:Qs) in left-to-right shunt. A gamma variate function is fitted to initial portion of pulmonary curve from a point 5% to 10% of the peak to a point on downslope where decay is no longer monotonous (*arrows* in **A**) to define first pulmonary transit area, A_1, **B.** Area A_1 is subtracted from remainder of curve to obtain another curve, which represents shunt and subsequent systemic and shunt transits, **C.** Initial portion of curve is defined (*arrows* in **C**), and another gamma variate function is interpolated to define first shunt flow area (area A_2 in **D**). $A_1/(A_1\text{-}A_2)$ = Qp:Qs. Segment defined includes point at about 10% of peak on upslope to another point at or just past peak, if there is small downslope. This part of analysis is critical. For good results it is necessary to obtain close fit, **D.** Second gamma variate fits, **E** to **H,** will result in underestimation or overestimation of shunt size. (From Treves, S., and Collins-Nakia, R. L.: Am. J. Cardiol. **38:**711, 1976.)

parameters of the experimental data to a gamma variate. A significant advantage of the gamma-variate approach is that all the data up to the point at which recirculation interrupts the shape of the curve can be used to predict the final portion of the curve. If one takes a pulmonary time-activity curve in a left-to-right shunt and fits a gamma variate to its first peak (Fig. 7-7, *A*), an area (A_1) under the curve is defined. This area, A_1, is assumed to represent the first circulation of the radiotracer through the pulmonary vascular circuit (Qp = pulmonary flow) without recirculation. If this area, A_1, is subtracted from the original curve, another area (A_2) is defined. This area, A_2, is believed to represent early reappearing radioactivity in the lung due to the left-to-right shunt. If one subtracts the area corresponding to the shunted radioactivity (area A_2) from the first pulmonary transit activity (area A_1), what remains is an area that is proportional to the flow in the systemic circuit (Qs = systemic flow). A ratio of areas $A_1/A_1 - A_2$ should then represent the pulmonary-to-systemic flow ratio (Qp:Qs). This method has been applied to a number of patients and has been found accurate for calculation of Qp:Qs between 1.2:1 to 3.0:1 (r = 0.94) (Fig. 7-8). The region of interest from which the pulmonary curve is generated views the chest wall and will also reflect normal bronchial arterial circulation, which may be, in part, responsible for the reduced accuracy in the zone between Qp:Qs of 1.0:1 to 1.2:1. Thus it is easy to understand how collateral or intercostal flow could affect the pulmonary curves, especially in the postoperative course of tetralogy of Fallot. Shunts with a Qp:Qs greater than 3.1 are not accurately quantitated with this method because the pulmonary curve of such short magnitude cannot be accurately fit with the gamma variate to represent the first radioindicator transit.

A C_2/C_1 ratio was proposed by Folse and Braunwald.[8] C_1 is defined as the peak activity of the pulmonary curve, and C_2 is the activity at a point after the peak equal in time to the interval between the first appearance of activity and the peak. Normal limits for values generated with this technique have varied from institution to institution, probably relating to injection techniques, equipment, probe placement, and criteria for identifying the appearance of activity on

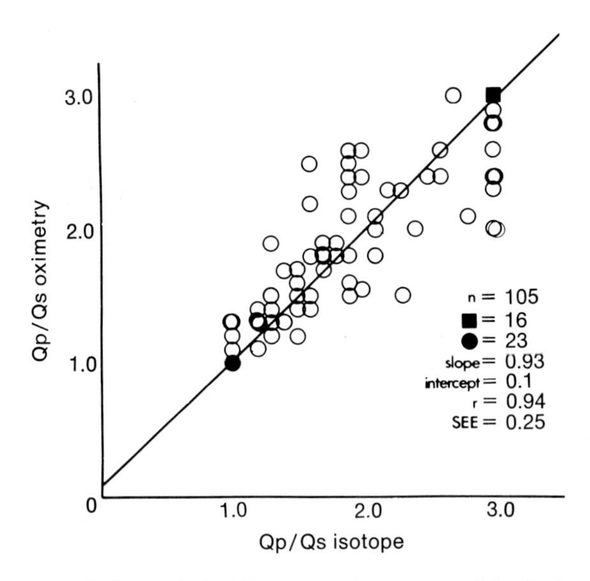

Fig. 7-8. Pulmonary-to-systemic flow ratio in 105 patients. Comparison of Qp:Qs in group of patients at catheterization and with radionuclide angiogaphy reveals good correlation. (From Askenazi, J., Ahnberg, D. S., Korngold, E., LaFarge, C. G., Maltz, D. L., and Treves, S.: Am. J. Cardiol. **37**:382, 1976.)

the curve. The C_2/C_1 ratio is greater in patients with left-to-right shunts than in normal subjects, but there is an overlap between the values in patients with and without left-to-right shunts, especially if there are associated defects (i.e., valvular insufficiency, stenosis, etc.).

Another method developed by Alderson and associates[1] also analyzes the pulmonary curve after intravenous injection of $^{99m}TcO_4$. They define an area, Y, which is delineated by the area resulting from semilogarithmic extrapolation of the initial portion of the downslope to a point at 1% of the peak and a line from the baseline to the peak. Another area, X, is defined as the area between the experimental and extrapolated curves to a vertical line traced at the 1% point. They calculate the percentage of left-to-right shunting using a ratio X/Y. Using a quadratic regression, they found a good correlation with shunt flow ($r = 0.95$).

DETECTION, QUANTITATION, AND LOCALIZATION OF RIGHT-TO-LEFT SHUNTS

A right-to-left shunt is identified by the early appearance of systemic activity.[10,11] The right and left sides of the heart are not well visualized due to early mixing of activity within the heart and relative pulmonary underperfusion. It is possible, however, to define gross cardiac anatomy by careful inspection of the angiogram. Multiple projections, whenever possible, aid in this morphologic evaluation. Intravenous injection of an inert gas can also permit detection of right-to-left shunts. Normally, inert gases are almost completely diffused into the alveolar air in their first pass through the lungs. In right-to-left shunts there is systemic appearance of activity.[4]

The magnitude of right-to-left shunts can be calculated using radioactive particles, inert gases, or nondiffusible indicators. In the case of radioactive particles, it is assumed that they are completely trapped in the pulmonary capillary bed after intravenous injection. In right-to-left shunting, systemic microembolization of intravenously injected particles occurs. Whole-body imaging with the gamma camera and counting over the lungs and the rest of the body using regions of interest provide an estimate of pulmonary-to-systemic flow ratio. Good agreement with catheterization data has been reported.[9] It is possible to calculate right-to-left shunting using left ventricular time-activity curves. A first peak of activity from blood shunted right-to-left is seen, followed by a second peak of activity due to blood that has circulated through the pulmonary circuit. Using an exponential extrapolation of the downslope following the two peaks, a pulmonary-to-systemic flow ratio can be calculated by the ratio of the area under the first peak to the whole area under both peaks.[26] Another suggested method is the so-called forward triangles from indicator dilution theory, again using curves obtained over the left ventricle.[19]

CARDIAC OUTPUT

Cardiac output can be calculated by analysis of a time-activity curve generated from a region of interest over the heart. A bolus of ^{99m}Tc-labeled human serum albumin ([^{99m}Tc] HSA) is given intravenously, and recording proceeds for 5 min (vide supra). Two initial peaks are seen on the precordial curve; the first peak corresponds to the radioactivity circulating within the right side of the heart and the second peak to the activity as it passes within the left side of the heart. The curve then levels off rather promptly as equilibrium of the indicator on the blood is reached (Fig. 7-9). However, ^{99m}Tc-labeled HSA preparations tend to break down slightly in vivo, and there is indeed usually a very slow downslope in the equilibrium portion of the precordial curve. A semilogarithmic extrapolation of the downslope of the second peak determines an area (A) (Fig. 7-9); an equilibrium value (e) is determined on the stable portion of the curve, and the total blood volume (TBV) is determined independently using ^{125}I-labeled HSA. With these parameters cardiac output

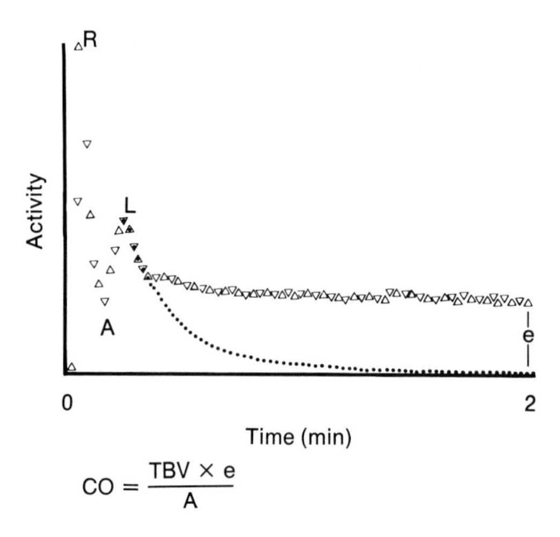

Fig. 7-9. Cardiac output (CO). Time-activity curve is obtained over heart. First spike corresponds to radioactivity as it circulates within right heart; second spike, less prominent, is activity through left heart. "Valley" between two spikes corresponds to maximum pulmonary activity. Following second peak, curve tends to reach uniform level when radioindicator is in equilibrium in blood pool *(e)*. Semilogarithmic extrapolation of downslope of second peak delimits area under radioindicator passing through right and left sides of heart. Total blood volume (TBV) is determined independently.

(CO) is calculated in milliliters per minute, as follows:

$$CO = \frac{TBV \times e}{A}$$

At equilibrium a portion of the activity arises outside of the heart chambers (e.g., chest wall). Empirically it has been discovered that the value obtained by this equation should be multiplied by 0.83 to correct for this effect.[21] For an excellent review of the basic concepts of radiocardiography, including cardiac output, refer to Donato.[6] We have studied seventeen pediatric patients[16] in the catheterization laboratory with simultaneous determinators of cardiac output by radionuclide angiocardiography and oxygen consumption (Fick) and found a good agreement between the two techniques (r = 0.88) (Fig. 7-10).

DETERMINATION OF VENTRICULAR EJECTION FRACTION AND RADIONUCLIDE VENTRICULOGRAPHY

There are two major approaches to the measurement of left ventricular ejection

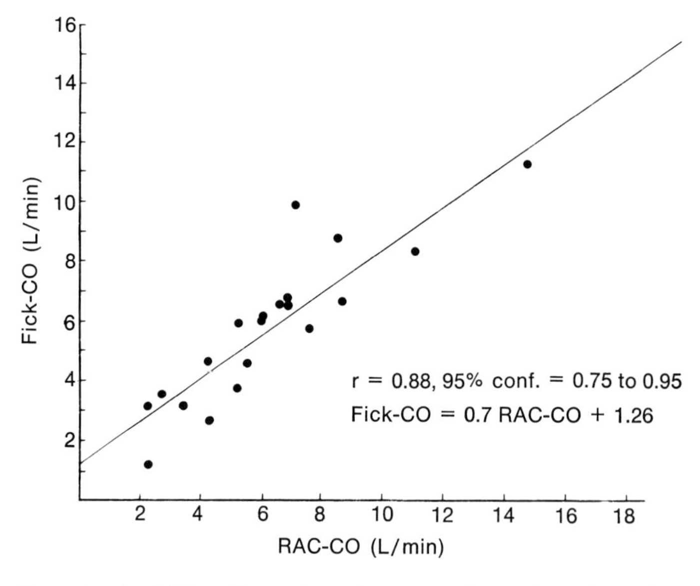

Fig. 7-10. Cardiac output in children. Comparison of twenty studies performed in catheterization laboratory using Fick method and radionuclide angiography simultaneously. Good correlation was obtained (r = 0.88).

fraction. One method uses ^{99m}Tc-labeled HSA or ^{99m}Tc-labeled red blood cells with simultaneous recording of the electrocardiogram and collection of a large number of heart beats (500 to 1000). These data are then reconstructed in such a way that a cardiac cycle can be divided into several frames between end of diastole and end of systole. These studies, which can be performed on several projections, provide good spatial resolution and are helpful in evaluating intracardiac chamber size, ventricular performance, and areas of abnormal ventricular contraction (e.g., aneurysms). This technique can also provide ventricular curves, which reflect intraventricular changes of volume as a function of time. Consequently, it is possible to calculate ejection fraction, ejection rate, etc.

A second method obtains data during the first pass of intravenous radionuclide through the heart. With rapid recording rates (20 to 40 frames/sec) it is possible to define end of systole and end of diastole without electrocardiographic control. End-diastolic, end-systolic, and intermediate images can be constructed by adding points (or frames) at the appropriate times (Fig. 7-11). The ejection fraction can be measured for single heart beats or it can be calculated by counting the radioactivity over the ventricular area at end of systole and over another area at end of diastole. Variations of these two approaches with improvements are being reported in the literature. We have had the opportunity to apply the first-transit approach for the measurement of left ventricular ejection fraction to a group of normal and abnormal pediatric patients. There was a good correlation between biplane angiography and the radionuclide method in a small group of patients (n = 12) studied in the catheterization laboratory (4 = 0.91, range = 0.18 to 0.78) (Fig. 7-12). In addition, we were able to determine left ventricular ejection fraction on thirty-four normal children.[13] The average left ventricular ejection fraction was 0.66 ±

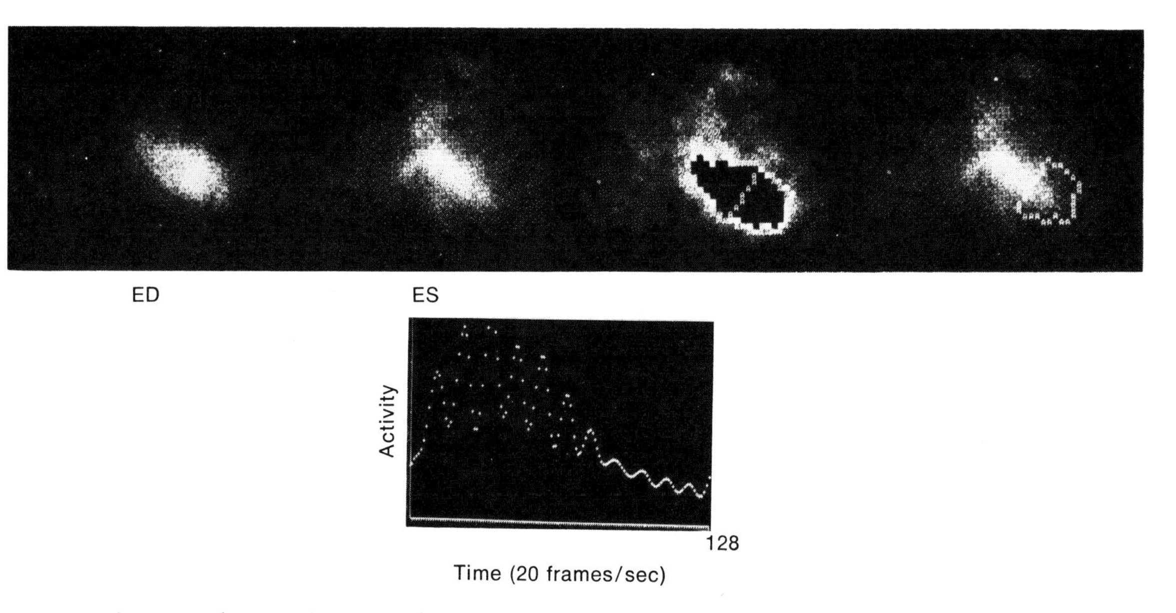

Fig. 7-11. Left-ventricular ejection fraction. End-diastolic *(ED)* and end-systolic *(ES)* images of left side of heart (left) are summed images obtained during first pass of radioindicator through left side of heart. Time-activity curve below has been obtained for time reference and was recorded at 20 frames/sec and filtered with a fast fourier transform Isocontour map (55% of maximum counts in left ventricle) is used to outline ventricle in end of diastole. Another region is similarly outlined for end-systolic frame. Ratio of end-diastolic counts to end-systolic counts provides an estimate of LVEF. (From Kurtz, D., et al.: Br. Heart J. **38:**966, 1976.)

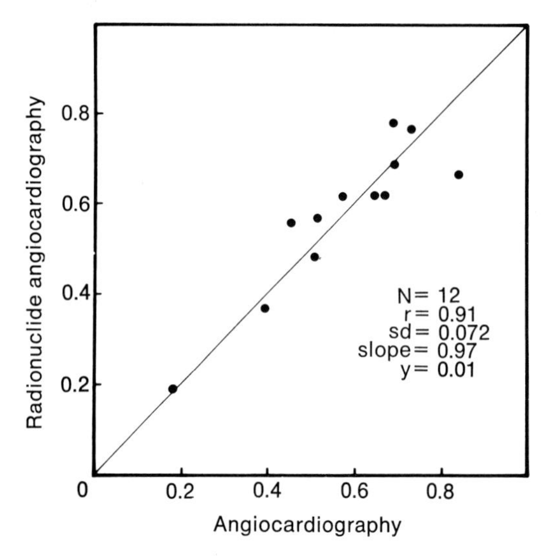

Fig. 7-12. Left-ventricular ejection fraction in children. Comparison of LVEF in twelve patients estimated by radionuclide angiography and angiography during catheterization. Note good agreement (r = 0.91) in this small number of observations. (From Kurtz, D., et al.: Br. Heart J. **38:**966, 1976.)

0.065 (SD) in the thirty children over age 2 years; the average for the four patients under age 2 years was 0.70 ± 0.05 (SD). Refer to Chapters 5 and 6 for a much more detailed review of the methods of ejection fraction and a complete reference list.

CLINICAL APPLICATIONS

The newborn infant

Radionuclide angiography can be helpful in differentiating cardiac from pulmonary disorders as the cause of cyanosis in the newborn. In pulmonary disorders, the radionuclide study is usually normal except in cases in which there is left-to-right shunting through a patent foramen ovale and/or a patent ductus arteriosus. Changes in the pulmonary vascular resistance have an important effect on the appearance of radionuclide angiograms and on the calculation of left-to-right shunting. For example, during the first days of life, pulmonary vascular resistance may be high and the net amount of left-to-right shunting through a ventricular septal defect may be imperceptible on the radionuclide angiogram or the pulmonary curves. Later, when the pulmonary artery resistance decreases, the left-to-right shunt flow through the defect could be appreciated (Fig. 7-13).

In the neonate, patent ductus arteriosus frequently coexists with prematurity and respiratory distress syndrome. In this clinical setting, it is often difficult to differentiate cardiovascular from pulmonary disease. At present, there are no quantitative methods to assess the presence and magnitude of left-to-right shunting in these infants, and there are no uniform guidelines as to optimal time for surgical interruption of a patent ductus arteriosus, when present. New small, mobile gamma scintillation cameras and computer systems allow study of these infants while in the neonatal care units without disruption of their monitoring and special care. Initial and serial quantitative estimation of pulmonary-to-systemic flow ratio have proved useful in the preoperative and postoperative management of these patients.[22]

For example, in newborn infants suffering from cyanosis due to cardiac disease with transposition of the great arteries or from a large right-to-left shunt with severe pulmonary stenosis, radionuclide angiocardiography will demonstrate rapid appearance of the radionuclide in the aorta at the same time or before it appears in the lungs. Precise anatomic diagnosis of these complex congenital abnormalities requires catheterization and angiography.

The patient with cardiac murmur

In some instances, it may be difficult to determine the origin of a cardiac murmur clinically. A systolic ejection murmur with wide splitting of the second heart sound may be encountered on both pulmonic stenosis and atrial septal defect. In addition, in both instances, there may be enlargement of the pulmonary artery segment in the chest x-ray film and signs of right ventricular overload in the electrocardiogram. At this point, a radionuclide angiocardiogram can be very helpful because it can firmly establish the diagnosis and a cardiac catheterization may be obviated.[21]

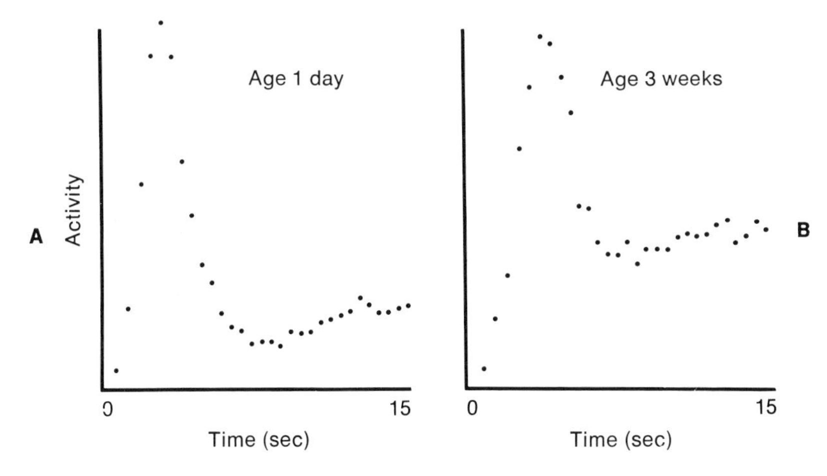

Fig. 7-13. Ventricular septal defect in newborn. Effect of pulmonary artery resistance in estimating pulmonary-to-systemic flow ratio. **A,** Pulmonary curve at age 1 day reveals no evidence of left-to-right shunt. **B,** Three weeks later, character of murmur has changed, pulmonary artery presumably decreased, and pulmonary curve reveals left-to-right shunt.

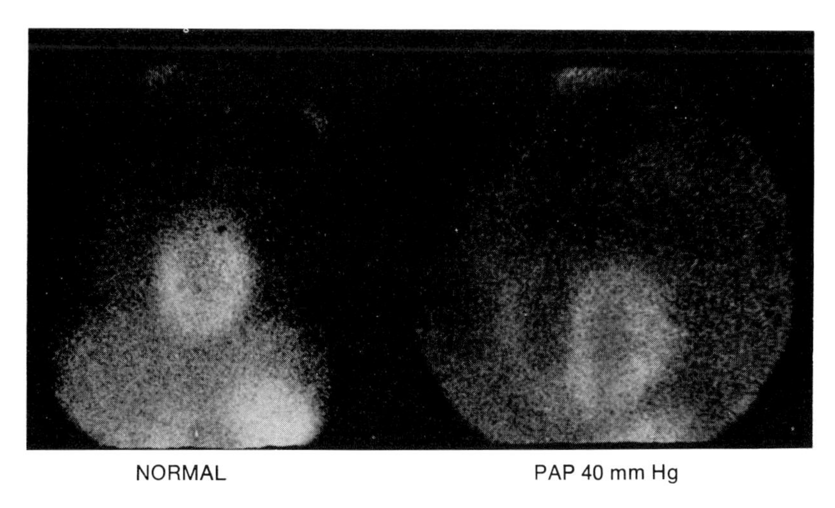

Fig. 7-14. Increased ^{201}Tl uptake in right myocardium secondary to increase in pulmonary artery pressure. *Left:* Normal patients, LAO projection. *Right:* Patient with simple pulmonary artery stenosis. Note increased uptake of ^{201}Tl in right ventricle. Left myocardial outline can be seen in both cases.

A decision for surgical closure may then be made, or the patient may be either followed clinically if the shunt is small or discharged with the diagnosis of pulmonic stenosis. Although not common, we have observed in one patient the natural closure of a small atrial septal defect previously documented by cardiac catheterization.

In ventricular septal defect, the character of murmur, the electrocardiogram, and the chest x-ray results may permit an approximation of the size of the shunt. Radionuclide angiocardiography provides precise estimation of the shunt flow, which can contribute significantly in the clinical management of patients. In a small left-to-right shunt through a ventricular septal defect without increase of pulmonary arterial pressure, radionuclide angiocardiography can objectively document decreasing shunt flow and complete closure. In patients with large ventricular or atrial septal defects, preoperative cardiac catheterization is necessary to diagnose the location and number of the defects and to assess pulmonary arterial pressures. Increase of the pulmonary artery pressure produces an increase in the right myocardial mass and flow. This results in increased uptake of ^{201}Tl in the right ventricle (Fig. 7-14). Noninvasive, external measurement of this increase in the right ventricular uptake of ^{201}Tl may prove to be an indicator of pulmonary artery pressure.[5,7]

Radionuclide angiocardiography can be used to calculate the net left-to-right shunting in patent ductus arteriosus. The distribution of flow in patent ductus arteriosus usually favors the left lung, thus a higher pulmonary-to-systemic flow rate is calculated using the left lung curve. The net amount of left-to-right shunting on partial anomalous venous return can also be calculated using the pulmonary time-activity curves.

The postoperative patient

The appearance of new murmurs or persistent murmurs during the early postoperative period may suggest patch detachment or valvular abnormalities. During this time cardiac catheterization is too hazardous, but radionuclide angiocardiography can rapidly, safely, and accurately detect and quantify the presence of residual or new shunts. Reoperation can then be carried out without catheterization if a shunt of large size is demonstrated. Late complications of pediatric cardiovascular surgery, such as residual or new shunts or obstruction of the superior vena cava (for example, in the Mustard operation), can also be demonstrated safely and noninvasively with this technique (Fig. 7-15).

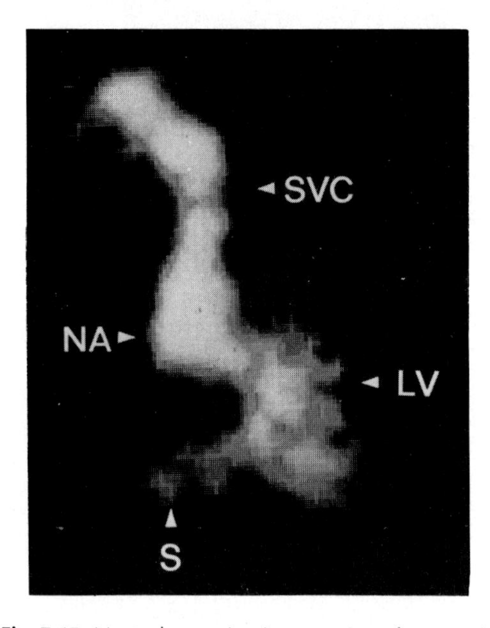

Fig. 7-15. Mustard operation in correction of transposition of great arteries. Residual shunt at level of neoatrium can be identified. Intravenous injection in right jugular vein. Superior vena cava *(SVC)*, neoatrium *(NA)*, and left ventricle *(LV)* can be seen. In addition, shunt *(S)* can be visualized quite clearly.

Evaluation of palliative shunts (Blalock-Taussig, Potts, Waterston) is possible by visual evaluation of pulmonary time-activity curves. If the shunt is patent, the curve of the lung on its side reveals a greater proportion of early pulmonary recirculation compared to the contralateral side.

ACKNOWLEDGMENTS

We wish to thank Miss Linda Gilman for her invaluable help during the preparation of this manuscript, Miss Yvette Lanoie and Mr. Ronald Grant for providing meticulous technical assistance, and Mr. Donald Sucher for the excellent illustrations.

REFERENCES

1. Alderson, P. O., Jost, R. G., Strauss, A. W., et al.: Radionuclide angiocardiography. Improved diagnosis and quantification of left-to-right shunts using area ratio techniques in children, Circulation **51:**1136, 1975.
2. Anger, H. O., Van Dyke, D. C., Gottschalk, A., et al.: The scintillation camera in diagnosis and research, Nucleonics **23:**51, 1965.
3. Askenazi, J., Ahnberg, D. S., Korngold, E., LaFarge, C. G., Maltz, D. L., and Treves, S.: Quantitative radionuclide angiocardiography. Detection and quantitation of left to right shunts, Am. J. Cardiol. **37:**382, 1976.
4. Bonsjakovic, B., Bennet, L., Vincent, W., et al.: Diagnosis of intracardiac shunts without cardiac catheterization, Circulation **44:**144, 1971.
5. Cohen, H. A., Baird, M. G., Roleau, J. R., Fuhrmann, C. F., Bailey, I. K., Summer, W. R., Strauss, H. W., and Pitt, B.: Thallium-201 myocardial imaging in patients with pulmonary hypertension, Circulation **54:**790, 1976.
6. Donato, L.: Basic concepts of radiocardiography, Semin. Nucl. Med. **3:**111, 1973.

7. Fischer, K. and Treves, S.: Unpublished data, 1977.
8. Folse, R., and Braunwald, E.: Pulmonary vascular dilution curves recorded by external detection in the diagnosis of left-to-right shunts, Br. Heart J. **24:**166, 1962.
9. Gates, G. I., Orme, H. W., and Dore, E. K.: Measurement of cardiac shunting with technetium-labelled albumin aggregates, J. Nucl. Med. **12:**746, 1971.
10. Greenfield, L. D., and Bennett, L. R.: Detection of intracardiac shunts with radionuclide imaging, Semin. Nucl. Med. **3:**139, 1973.
11. Greenfield, L. D., Vincent, W. R., Graham, L. S., et al.: Evaluation of intracardiac shunts, CRC Crit. Rev. Clin. Radiol. Nucl. Med. **6:**217, 1975.
12. Kriss, J. P., Yeh, S. H., Farrer, P. A., et al.: Radioisotope angiocardiography, abstracted, J. Nucl. Med. **7:**367, 1966.
13. Kurtz, D., Ahnberg, D. S., Freed, M., LaFarge, C. G., and Treves, S.: Quantitative radionuclide angiocardiography: determination of left ventricular ejection fraction in children, Br. Heart J. **38:** 966, 1976.
14. Maltz, D. L., and Treves, S.: Quantitative radionuclide angiocardiography: determination of Qp: Qs in children, Circulation **47:**1049, 1973.
15. McIlmoyle, G., Ahnberg, D., LaFarge, G., and Treves, S.: Localization of left-to-right shunts by radionuclide angiocardiography. In dynamic studies with radioisotopes in medicine, IAEA (Vienna) **2:**251, 1975.
16. Rabinovitch, M., Rosenthal, A., Ahnberg, D. S., Nadas, A., and Treves, S.: Cardiac output determination by radionuclide angiocardiography in patients with congenital heart disease, Am. J. Cardiol. **39:**309, 1977.
17. Rosenthal, L.: Applications of the gamma-ray scintillation camera to dynamic studies in man, Radiology **86:**634-639, 1966.
18. Stramer, F., and Clark, D. D.: Computer computations of cardiac output using the gamma function, J. Appl. Physiol. **28:**219, 1970.
19. Swan, H. J. C., Zapata Diaz, J., and Wood, E. H.: Dye dilution curves in cyanotic congenital heart disease, Circulation **8:**70, 1953.
20. Thompson, H. K., Stramer, F., Whalen, R., and McIntosh, H. D.: Indicator transit time considered as a gamma variate, Circ. Res. **14:**502, 1964.
21. Treves, S., and Collins-Nakai, R. L.: Radioactive tracers in congenital heart disease, Am. J. Cardiol. **38:**711, 1976.
22. Treves, S., Collins-Nakai, R., Ahnberg, D., and Lang, P.: Quantitative radionuclide angiocardiography (RAC) in premature infants with patent ductus arteriosus (PDA) and respiratory distress syndrome (RDS), J. Nucl. Med. **17:**554, 1976.
23. Treves, S., Kulprathipanja, S., and Hnatowich, D. J.: Angiocardiography with Iridium-191m. An ultrashort lived radionuclide (T−½=4.9 sec), Circulation **54:**275, 1976.
24. Treves, S., Lange, R. C., and Freeman, G. S.: Study of cardiopulmonary hemodynamics using a gamma camera and a computer, abstracted, J. Nucl. Med. **11:**369, 1970.
25. Treves, S., Maltz, D. L., and Adelstein, S. J.: The detection localization and quantitation of intracardiac shunts by radionuclide angiocardiography. In James, A. E., Wagner, H. N., Jr., and Cooke, R. E., editors: Pediatric nuclear medicine, Philadelphia, 1975, W. B. Saunders Co., pp. 231-246.
26. Weber, P. M., des Remedios, L. V., and Jasko, I. A.: Quantitative radioisotopic angiocardiography, J. Nucl. Med. **13:**815, 1972.
27. Wesselhoeft, H., Hurley, P. J., Wagner, H. N., Jr., et al.: Nuclear angiocardiography in the diagnosis of congenital heart disease in infants, Circulation **45:**77, 1972.

8 □ Historical perspectives and future needs in the measurement of coronary blood flow

Richard S. Ross

To gain a perspective on the methods used for measuring coronary blood flow, it is worthwhile to look back over fifteen years of experience in this field to see what general principles have held true and stood the test of time. Many have dropped by the wayside, but a few are still true. The following special problems in coronary heart disease or ischemic heart disease must be kept in mind so that there is understanding of what is really going on in the heart:

1. Coronary heart disease is a diffuse disease, although many of us tend to forget this. The heart of a patient with ischemic heart disease contains areas of good-looking muscle and areas of scar. There are areas of fatty replacement, and there are areas that are probably ischemic. Therefore it is a diffuse disease, and there are areas of perfectly normal tissue mixed in with severely diseased tissue.

2. The heart is surrounded by blood-containing tissue; therefore, to be studied, the heart must be isolated from its surroundings. A number of techniques have been used over the last fifteen years to effect this isolation. For example, selective sampling can be performed; this was one of the earliest ways of measuring coronary blood flow in humans. In this method, a catheter is placed in the coronary sinus, and the drainage from the left ventricle into the coronary sinus is sampled. Another possibility is selective injection, in which the indicator is introduced selectively into the input of the coronary circulation, namely, into the coronary arteries. A third possibility is the use of an indicator that will be selectively taken up by the myocardium and concentrated by the myocardium in excess of the concentration elsewhere. Systems of collimation also permit focusing on the heart and exclusion of the surrounding tissues. Also there can be a combination of all of these methods to increase the signal and get a better, sharper look at what is going on in the heart.

3. An important general principle of the diseased heart is that the flow may be normal at rest and that it takes some form of stress to uncover the deficit. The use of stress to uncover the disproportion between blood flow and demand was used by Zaret and associates in their exciting work.[9] This confirms what we know: a man with angina may be perfectly normal and have no symptoms at rest, but when he runs up three flights of stairs he develops chest pain. Thus coronary heart disease is a disease of stress.

4. Vascular lesions and flow deficit may not correspond. A lesion may be seen arteriographically in the anterior descending coronary artery, and the flow to the distribution of the anterior descending coronary artery may be perfectly normal because flow occurs through collateral vessels from the right circumflex branch. Lesions may look severe in one plane, but it is necessary

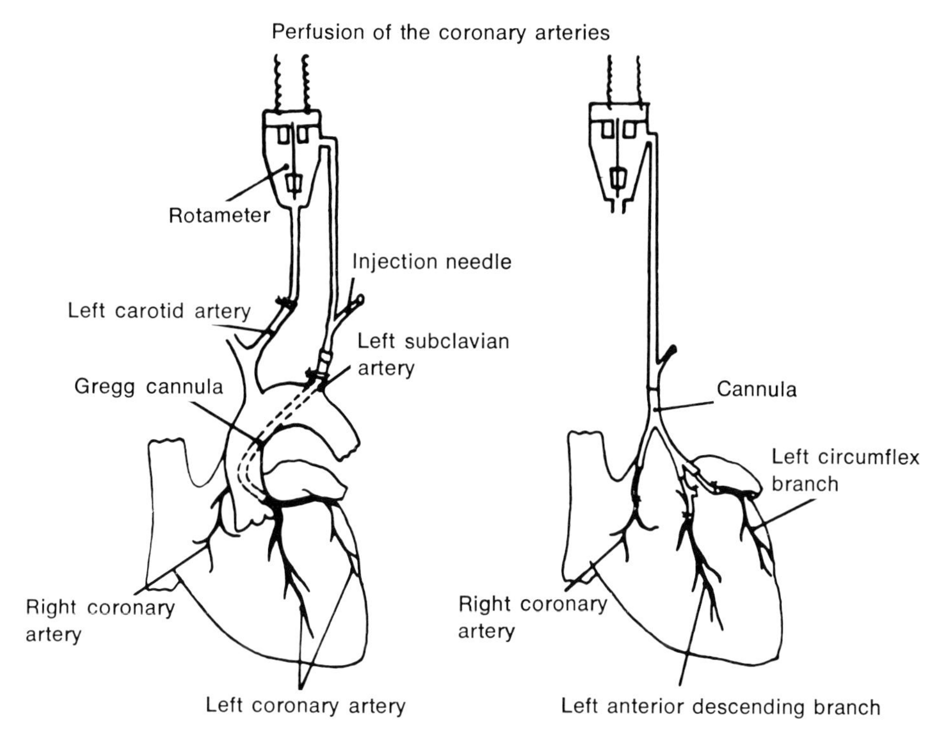

Fig. 8-1. Early experiment in measuring coronary blood flow with xenon.

to have a multidimensional look at the circulation of the heart to get all the information.

Blood flow measured by nitrous oxide is calculated in millimeters per hundred grams per minute. This concept induces considerable skepticism when first seen but can be understood in light of the experiments performed in 1962 in dogs.[8] Fig. 8-1 is an example of one of those experiments that illustrates these principles.

There are two heart preparations in a dog in which a device known as a rotameter was used to measure blood flow. Blood from the left carotid artery passed through this instrument and a metal cannula into the aorta and then into the left coronary artery. The rotameter, although a good instrument, has long since been replaced by the electromagnetic flowmeter. In the other preparation, the entire blood flow of the heart passed through this cannula. In both cases, there was a port in the input line into which an injection of a radioactive gas, xenon or krypton, could be made. This material would flow with the blood into the heart and diffuse. Since this was a dog heart, it was possible at the end of the experiment to cut it out and weigh it. In this particular instance, it was also possible to inject a dye through the coronary artery, then cut around the area that was dyed, and thus weigh the mass of myocardium that was perfused by this particular vessel. In this case, of course, it was the entire heart.

The slope of the washout curve is a function of the ratio of flow to volume. When the volume or mass is unknown, it is conventional to put in an arbitrary figure like per hundred grams; however, if it is possible to take out the entire heart or a section of the heart and weigh it, thus substituting an absolute value in grams for the volume, it is possible to measure flow in absolute terms. If this can be done, it should be possible to correlate flow as measured by xenon washout with flow measured by the rotameter, since the absolute value for volume is then known.

The data in Fig. 8-2 illustrate the blood

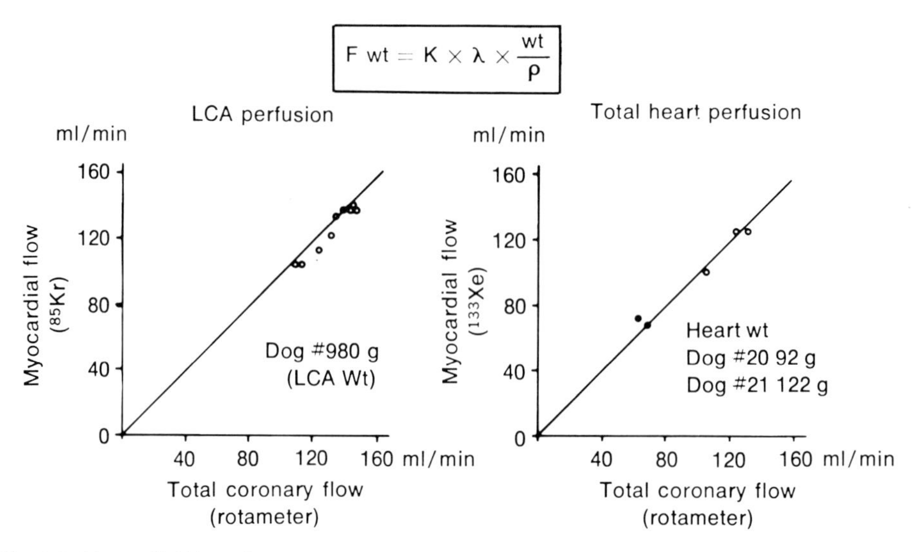

Fig. 8-2. Myocardial blood flow measured by rotameter compared to that measured by inert gas washout.

flow determined by inert gas washout and by rotameter. In this dog, the left coronary artery alone was perfused, and myocardial blood flow was measured with krypton. This krypton flow was plotted against myocardial blood flow, which was measured with the rotameter; this shows that there is a good correlation when the weight is measurable. Also, the same figure shows another situation in which the weight of the total heart was measured in two different dogs and the rotameter flow was plotted against myocardial blood flow. This figure illustrates the fundamental principle that the slope of washout curves is a function of the ratio of flow to volume and only if the volume is known can flow be expressed in absolute terms.

The heart sits in the chest in the middle of the lung and other vascular beds, and it is necessary to pick out the components coming from the heart. There are several methods of isolating the coronary circulation. At the beginning of interest in myocardial blood flow, researchers began looking at the possibility of splitting the decay curve and dissociating the coronary component temporally from the components of the surrounding tissues. This has proved to be totally unsatisfactory because there is so much overlap in the transit times between all the tissues within the chest that it is just not possible. The circulatory methods that can be used are coronary sinus sampling; arterial injection; and intramyocardial injection, which, although used, obviously has been practical only in the operating room when the heart is exposed. Another method is selective uptake. The anatomic methods of isolation are collimation and coincidence counting.

Isolation of the coronary circulation accomplished by sampling of the coronary sinus drainage with selective injection into the aorta is illustrated in Fig. 8-3.[7] Radioactive material is shown throughout the chambers of the heart.

Fig. 8-4 shows the method of selective injection, which has an advantage over the coronary sinus sampling method in that the right and left coronary arteries can be studied separately.

Fig. 8-5 shows the technique that is used for detecting anatomic abnormalities in which coronary artery catheterization and the selective arteriographic method of Sones are put together with the xenon method to measure the disappearance of xenon from the heart. This method was promising; however, Fig. 8-6 illustrates the failure of the method to yield the information sought.

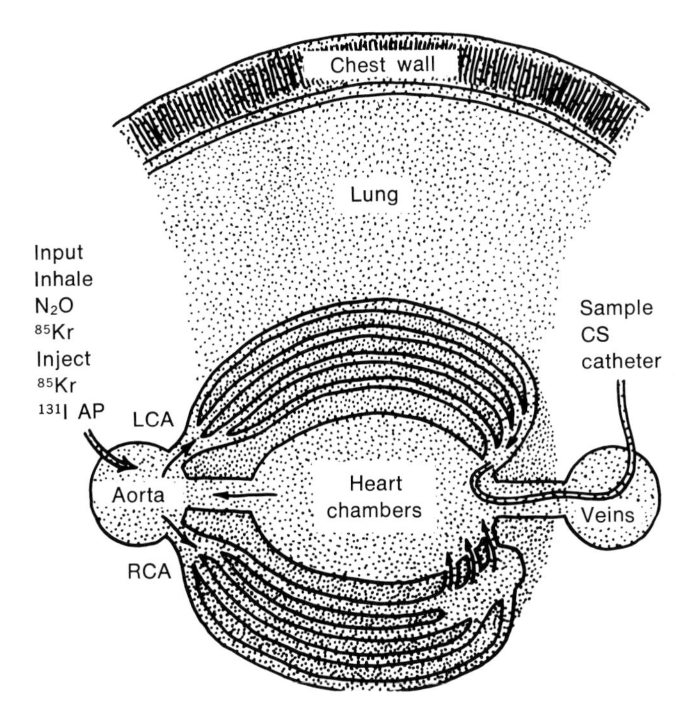

Fig. 8-3. Selective injection method of isolating heart from remainder of circulation with coronary sinus sampling.

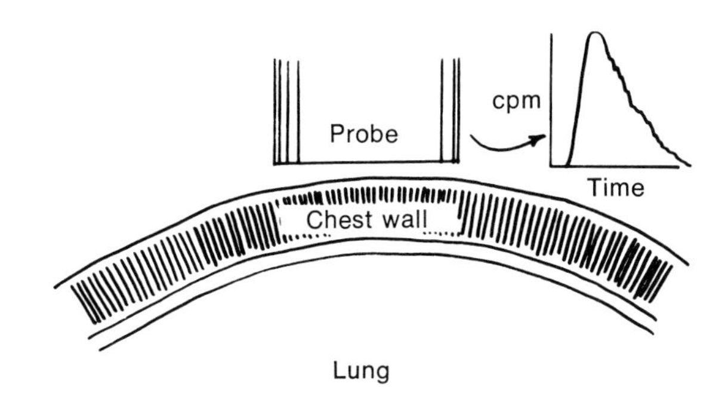

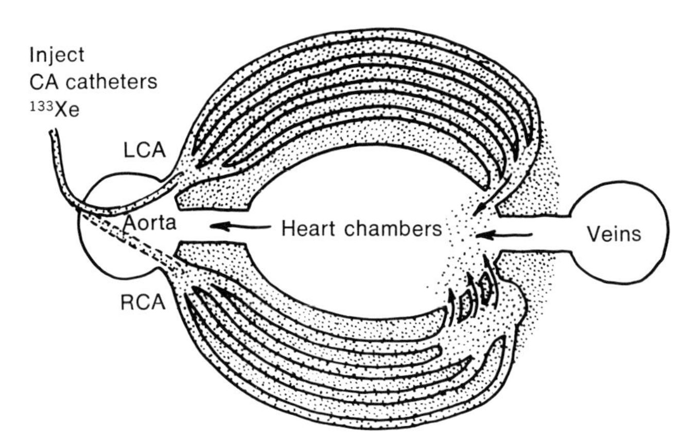

Fig. 8-4. Isolation of heart from remainder of circulation by collimation and selective injection.

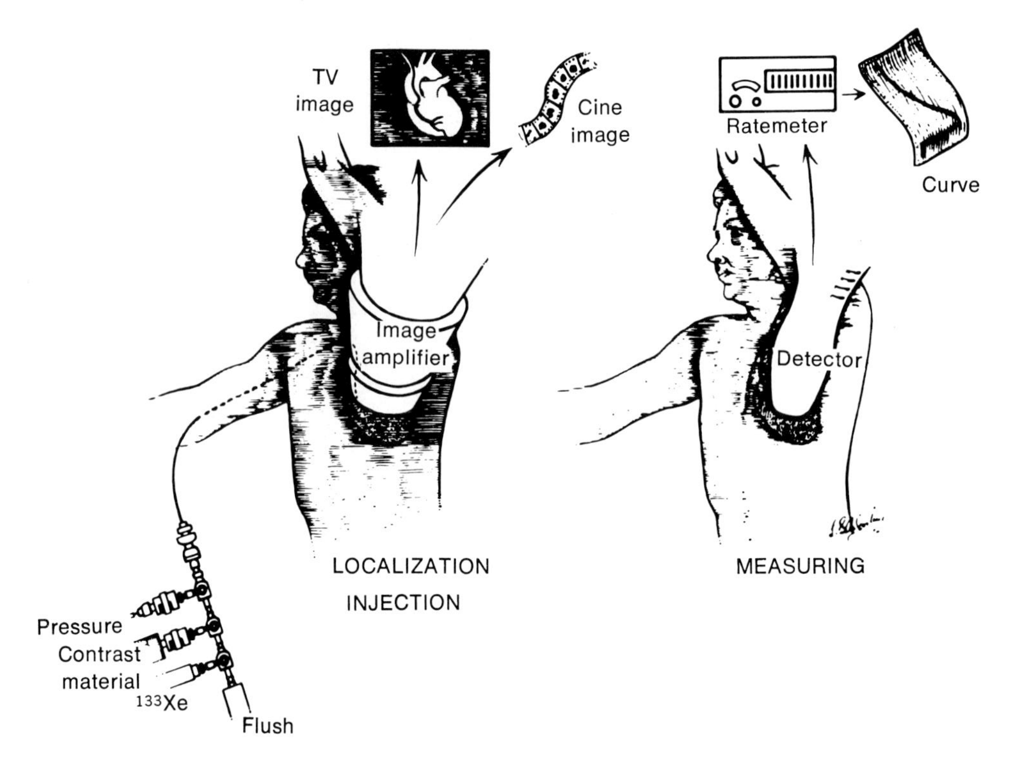

Fig. 8-5. Combined cineangiographic technique for detecting anatomic abnormalities and selective xenon injection for measuring perfusion.

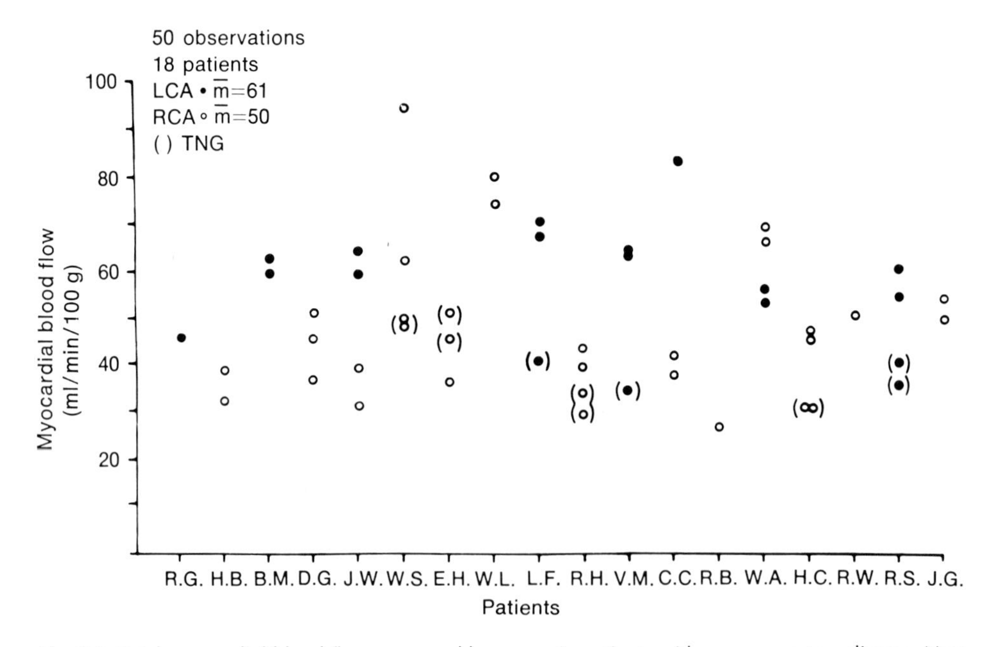

Fig. 8-6. Total myocardial blood flow measured by xenon in patients with coronary artery disease. Note wide scatter. There is no difference between patients with normal coronary flow and those with significant disease.

Fig. 8-6 also illustrates the principle that ischemic heart disease is a diffuse disease. This figure shows a set of myocardial blood flow measurements in a number of patients who were normal and also in a number of people who had disease. There should be a line between patients W. L. and L. F. The patients to the left of that line had normal coronary arteries, and the patients to the right had coronary artery disease. It may be fortuitous that the line is left out, since indeed there is absolutely no difference between the two populations. The point of this figure is that coronary artery disease is a diffuse disease; that there is normal tissue mixed with diseased tissue; and that if an indicator such as xenon is injected into the coronary artery, the washout reflects the state of the normal tissue. Coronary artery disease cannot be detected by this very gross technique, and these studies suggest that a more selective technique such as those discussed later in this section is needed.

The same information is displayed in another way in Fig. 8-7. The patients with normal coronary arteriograms and patients with abnormal coronary arteriograms are not different. The first study is designated as C_1 and the second as C_2, and there is no difference whatsoever between the normal and abnormal patients. This figure also shows that the method is reproducible and that it measures myocardial blood flow very consistently.

Fig. 8-8 illustrates the general principle in the evaluation of the coronary circulation that myocardial blood flow can be increased by stress.[5] The measurements in this figure were made in a set of dogs as part of another study in which the right ventricular myocardial blood flow was measured as a function of stroke work. The stroke work was changed by cranking down a clamp on the pulmonary artery and therefore making the right ventricle work harder; as it worked harder, the myocardial blood flow increased. This is a technique that is attractive for uncovering a deficit. It illustrates the importance of stress in increasing the myocardial blood flow.

The principle can be used to uncover

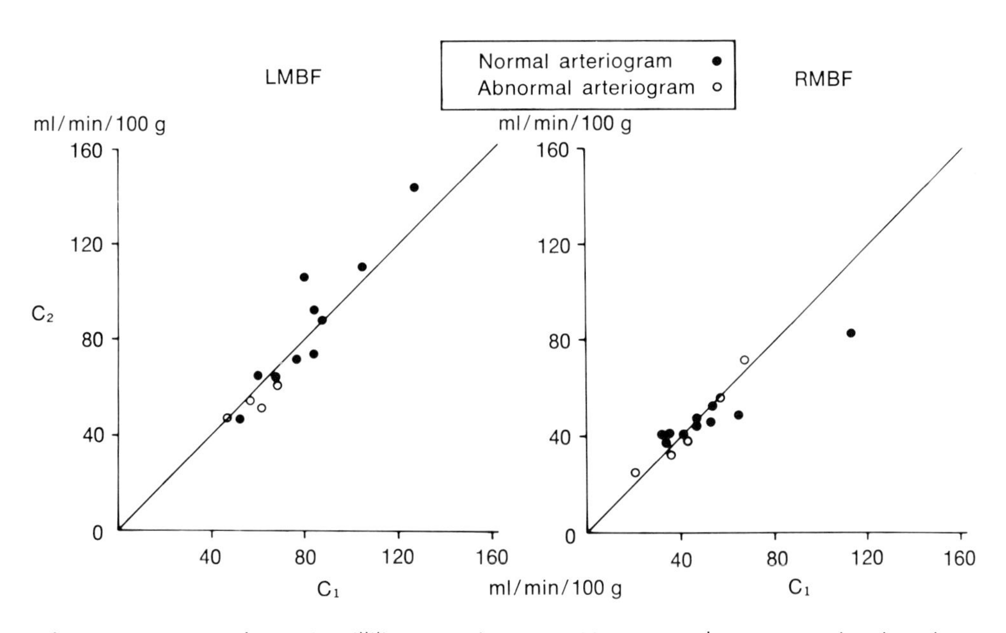

Fig. 8-7. Coronary perfusions in milliliters per minute per 100 g measured on two occasions in patients with normal coronary arteriograms *(dark circles)* and those with abnormal arteriograms *(light circles)* are compared. The method is reproducible but insensitive. LMBF = left myocardial blood flow; RMBF = right myocardial blood flow.

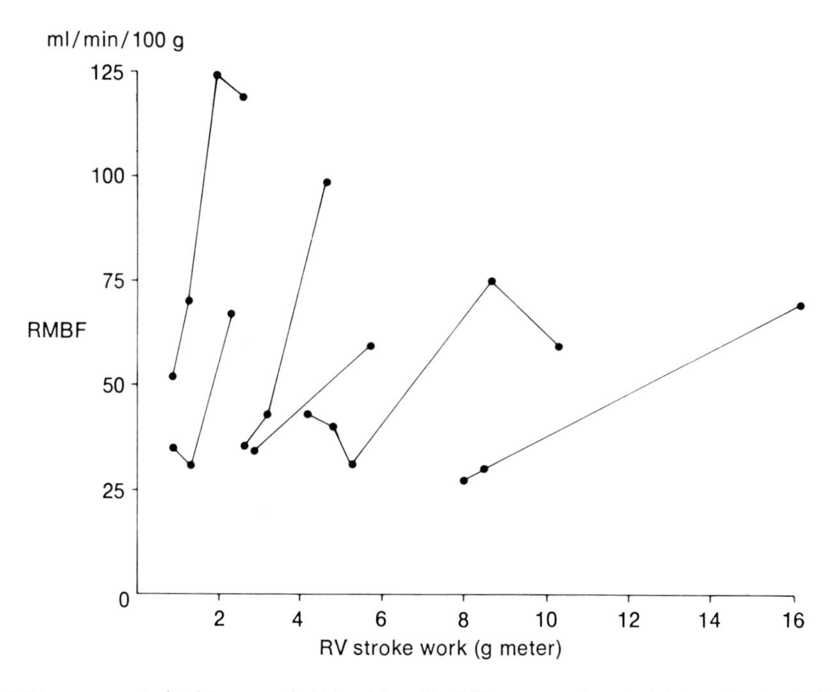

Fig. 8-8. Measurement of right myocardial blood flow (RMBF) in comparison to right ventricular (RV) stroke work.

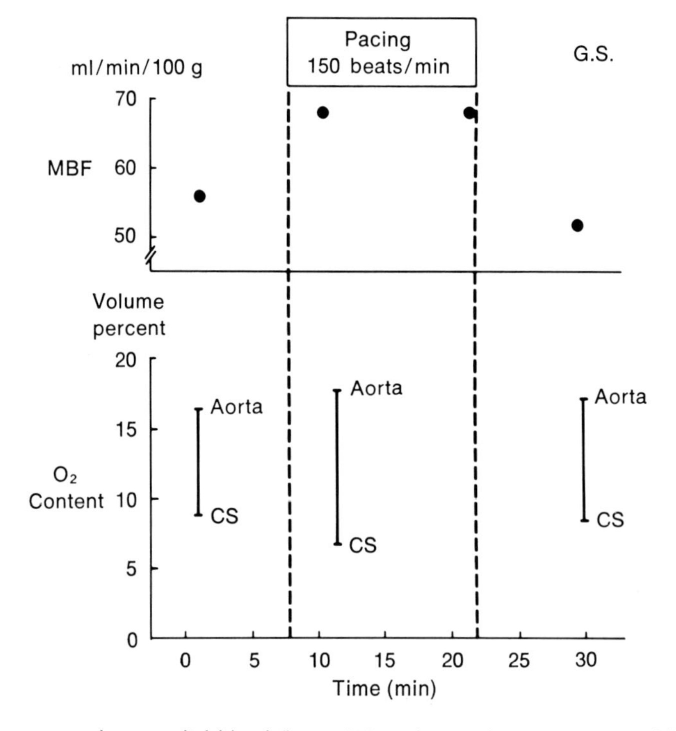

Fig. 8-9. Measurement of myocardial blood flow (MBF) and arterial venous oxygen difference in one patient at rest, during pacing, and after cessation of pacing.

disease, as demonstrated in Fig. 8-9. The patient was studied with the myocardial blood flow xenon method for measurements of myocardial blood flow, and at the same time measurements were made of aorta and coronary sinus oxygen saturation.[3] During the middle period, the patient had right atrial pacing, and his heart rate was accelerated to 150 beats/min. If the heart rate in a patient is artificially increased without exercise while the patient lies on a table, the cardiac output stays about the same, the heart rate increases, and therefore stroke volume must decrease. Because of the increase in heart rate, the oxygen consumption of the myocardium is increased. Therefore this is a method of increasing the oxygen consumption of the myocardium of the patient lying flat on the table. The atrioventricular oxygen difference gets wider and the myocardial blood flow increases in proportion to the increased oxygen demand. It could be reasoned that this technique could be used to uncover the deficit that exists in the patients with coronary ischemia.

It was hoped that the discrepancy could be uncovered by measurements of blood flow before, during, and after pacing, as shown in Fig. 8-9. These are the control observations of blood flow made before

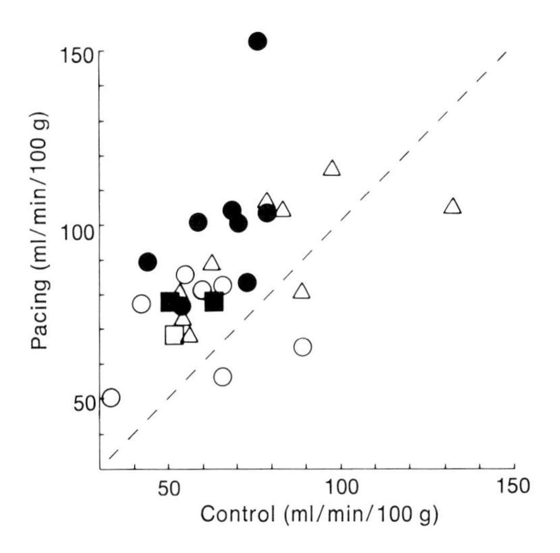

Fig. 8-10. Myocardial blood flow at rest and during pacing in patients with no ischemic response *(light symbols)* and those with definite response *(dark symbols)*.

pacing and during pacing-induced tachycardia with blood flow measured in millimeters per minute per hundred grams.

Fig. 8-10, contrary to what is expected, shows no great difference in the blood flow between the patients with no ischemic response to pacing, that is, who did not develop chest pain or S-T segment displacement, and those who did develop S-T segment changes or pain with pacing. This is a provocative finding. It may mean that the mass of tissue around the ischemic area for one reason or another has an increased flow. This may be the result of compensation for the ischemic area or possibly the result of the liberation of vasoactive substances. In this case, the technique of applying a stress to increase the demand did not separate patients with disease from patients without disease, probably because of the diffuse nature of the disease.

One other use for the xenon method is in the study of drugs.[2] Fig. 8-11 illustrates a study in which nitroglycerin was injected directly into the coronary circulation and myocardial blood flow measurements were made. The right-hand panel depicts observations in humans in the first minute after injection and then at 4 and 6 min. There are two patient populations studied: normal patients and patients with abnormal arteriograms. During the first minute after administration of nitroglycerin, there is a slight increase in myocardial blood flow, and it is possible that there is a greater increase in the blood flow in the patients with the normal vessels than in those who are abnormal. Unfortunately, this observation has not been confirmed, and it certainly does not achieve statistical significance; however, it is a technique that might be of value.

In the work of Fortuin and associates[4] and Becker and Pitt,[1] the problem of patchy disease and regional distribution of blood flow was explored with the use of radioactive microspheres. Microspheres are injected into the left atrium, where they become mixed with blood, and then are carried down into the anterior descending and circumflex branches of the coronary

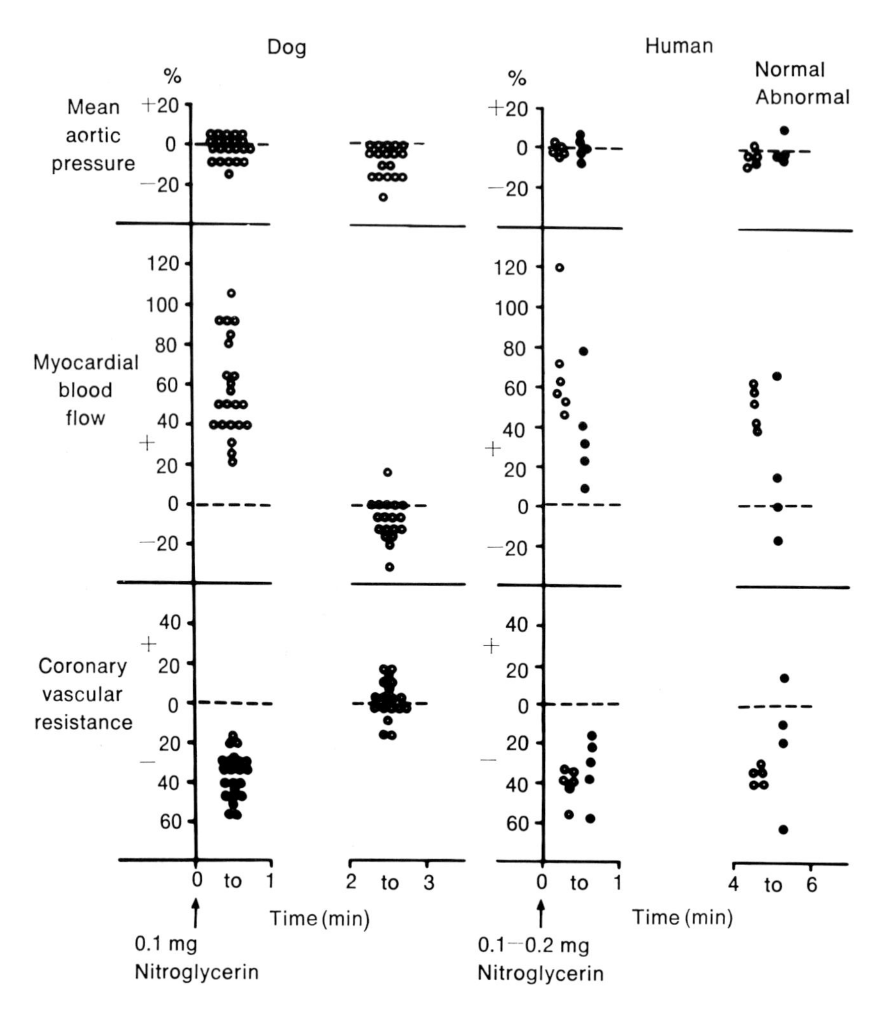

Fig. 8-11. Alterations in mean aortic pressure, myocardial blood flow, and coronary resistance after administration of nitroglycerin in dogs and humans. (Open circles represent normal patients and solid circles, abnormal patients.)

artery. By the procedure of counting the number of microspheres in a gram of tissue from one area and comparing it with the count in a gram of tissue from another area, it is possible to determine blood flow (Fig. 8-12).

Some of the principles of coronary heart disease have been illustrated. Ischemic heart disease is a diffuse disease with normal tissue next to diseased tissue. The heart is surrounded by blood-containing tissue and must be isolated; this can be accomplished by selective sampling, selective injection, selective uptake, collimation, or a combination of the last two. The rectilinear scanner, in which the tracer potassium is used, is a combination of these two; if it is desirable to refine it further, selective injections can be made. Flow may be normal at rest, and stress will uncover a deficit. Finally, it is a very important point that vascular lesions and flow deficit may not correspond. This principle has widespread application to the currently popular vein-bypass operation for angina. The logic that goes into that operative procedure is as follows: there is a lesion seen on the arteriogram, and it must be causing the trouble; therefore, if it is bypassed, the trouble will be cured. This logic is valid only if there is

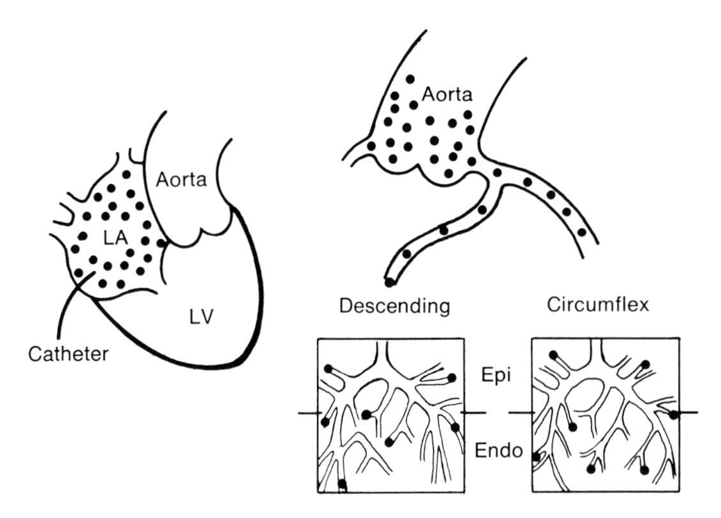

Fig. 8-12. Left atrial administration of microspheres and diagrammatic illustration of their distribution in precapillary arterioles.

a one-to-one correlation between the vascular lesions and the flow deficit, and I submit that this correlation does not exist.

Following are directions for the future. This is a list of areas we need to explore in evaluating the patient with coronary heart disease.

1. We need to know about the coronary artery lesions. The arteriographer can tell us where they are; however, this is not all we need to know.

2. We also need to know about myocardial blood flow. We want to know about total flow and the flow per unit mass. The flow volume ratio tells us something about the nutritional flow. Also, if we could measure it on a regional basis, it might be extremely important to know what the relative perfusion is. We need to know regional flow and the distribution between different areas such as the endocardium and epicardium.[6] Some of the most exciting physiologic information coming out now has to do with the reasons for endocardial ischemia; it is in this area that we are going to find the explanation for patients who have angina with normal coronary arteries. These peculiar states may well be explained on the basis of regional distribution. The problem of coronary spasm may also be explained on this basis. The importance of the response to stress should be reemphasized.

3. Then we need to know about the function of the heart as a pump. We need to know what the ejection fraction is and how the muscle moves in the distribution of a particular vessel.

4. We need to know what the relationship is between anatomic lesions and blood flow in the vessel. We need to know whether a severe lesion always indicates abnormal flow and a minor lesion, good flow.

In summary, it is a very complex situation; however, in looking back over the past fifteen years and comparing the state of knowledge of fifteen years ago with that of today, it can be seen that we have come a long way.

REFERENCES

1. Becker, L., and Pitt, B.: Collateral blood flow in conscious dogs with chronic coronary artery occlusion, Am. J. Physiol. **221**:1507, 1971.
2. Bernstein, L., Friesinger, G. C., Lichtlen, P. R., and Ross, R. S.: The effect of nitroglycerin on the systemic and coronary circulation in man and dogs, Circulation **33**:107, 1966.
3. Conti, C. R., Pitt, B., Gundel, W. D., Friesinger, G. C., and Ross, R. S.: Myocardial blood flow in pacing-induced angina, Circulation **42**:815, 1970.
4. Fortuin, N. J., Kaihara, S., Becker, L., and Pitt, B.: Regional myocardial blood flow in the dog studied

with radioactive microspheres, Cardiovasc. Res. **5:** 331, 1971.

5. Pitt, B., Friesinger, G. C., and Ross, R. S.: Measurement of blood flow in the right and left coronary artery beds in humans and dogs using the [133]xenon technique, Cardiovasc. Res. **3:**100, 1969.
6. Ross, R. S.: Pathophysiology of the coronary circulation: the Sir Thomas Lewis lecture of the British Cardiac Society, Br. Heart J. **33:**173, 1971.
7. Ross, R. S., and Friesinger, G. C.: Anatomical and physiological considerations in myocardial blood flow measurements. Symposium on The Detection of Coronary Artery Disease, Circulation **32:**630, 1965.
8. Ross, R. S., Ueda, K., Lichtlen, P. R., and Rees, J. R.: The measurement of myocardial blood flow in animals and man by selective injection of radioactive inert gas into the coronary arteries, Circ. Res. **15:**28, 1964.
9. Zaret, B. L., Strauss, W., Hurley, P. J., Natarajan, N. K., and Pitt, B.: A noninvasive scintiphotographic method for the detection of regional ventricular dysfunction in man, N. Engl. J. Med. **284:** 1165, 1971.

9 □ Importance of electrophysiologic, enzymatic, and tomographic estimation of infarct size

Burton E. Sobel
Michael J. Welch
Michel M. Ter-Pogossian

Estimation of the extent of myocardium undergoing irreversible injury induced by ischemia, which has been designated infarct size in this discussion, was stimulated initially by investigative rather than clinical considerations. As tools became available for estimating infarct size in experimental animals and in patients, it became clear that this parameter was one important determinant of prognosis and that, at least under some conditions, the evolution of infarction could be modified favorably by physiologic or pharmacologic interventions designed to improve the balance between myocardial oxygen requirements and myocardial oxygen supply or to facilitate washout of potentially deleterious metabolic products.[27] Accordingly, as clinicians became vitally interested in approaches that might exert favorable influences on evolving infarction, a pressing need developed to refine and implement practical and accurate methods for assessing infarct size in the clinical setting.[39]

In this discussion, material is considered from three points of view: (1) evolution of the concept that infarct size can be modified, (2) delineation of the importance of infarct size as a factor influencing prognosis, and (3) extent to which available techniques provide accurate assessments of infarct size in the clinical setting early after the onset of ischemia. More extensive discussion of many of the methods used to estimate infarct size is available in recent reviews[2,17,46,47] that focus on technical considerations, virtues and limitations of specific approaches, factors influencing results in the clinical setting, and aspects requiring continued development.

NATURE OF MYOCARDIAL INJURY INDUCED BY ISCHEMIA AND FACTORS INFLUENCING INFARCT SIZE

Prior to the establishment of myocardial infarction research units and specialized centers of research in ischemic heart disease supported by the National Heart, Lung, and Blood Institute, the extent of investigation directed toward characterizing reversible and irreversible ischemic injury to myocardium subjected to ischemia was relatively modest. Herdson, Sommers, and Jennings,[15] had been interested in morphologic criteria indicative of reversible injury and had developed an experimental model in which homogeneous infarction had been characterized at selected intervals after the onset of ischemia. Based on results of these studies, it became clear that heart muscle sustained irreversible injury after ischemia lasting from 20 min to 1 hr, with characteristic ultrastructural changes that served as criteria of irreversible injury.

In seeking to determine whether ischemic myocardium could be protected by interventions designed to improve the balance between myocardial oxygen supply and demand during the evolution of infarction, Maroko, Braunwald, and their associates needed some index of the severity of ischemic injury applicable to intact experimental animals that did not require destructive analysis of tissue.[27] With epicardial ST-segment recordings obtained serially during the evolution of ischemic injury in dogs, they demonstrated that regional, epicardial ST-segment elevation 15 min after the onset of ischemia presaged irreversible injury, reflected morphologically and biochemically in biopsies from corresponding sites 24 hr later. In addition, changed relationships between initial ST-segment elevation and subsequent biochemical and morphologic evidence of infarction occurred when interventions designed to protect the heart were implemented relatively soon after the onset of coronary occlusion.

From theoretical considerations[10,46] and the results of many studies, it is clear that ST-segment elevation in epicardial recordings is not always a measure of irreversible injury to myocardium. On the other hand, evaluation of ST-segment elevation in epicardial recordings has provided valuable information regarding the evolution of injury under a defined set of experimental conditions. The controversy surrounding the use of ST-segment elevation as an index of infarct size in patients should not obscure the importance of results of studies in which epicardial ST-segment elevation was employed, not as an absolute index of infarction but rather as a criterion of anticipated change, verified by subsequent morphologic and biochemical analysis of the heart. With the use of epicardial ST-segment recordings employed in this fashion, it became clear, for the first time, that in the experimental animal subjected to coronary occlusion, physiologic and pharmacologic interventions were able to reduce the rate of evolution or limit the ultimate extent of irreversible injury.

Several factors besides ischemia may influence ST-segment elevation, and changes in epicardial recordings may be relatively insensitive indices of ischemia deep within the ventricular wall.[21] This may account for some deviation from close correlations between ST-segment elevation in epicardial recordings and regional myocardial perfusion assessed with radioactively labeled microspheres. Since the magnitude of ST-segment elevation changes with the duration of ischemia and recordings obtained from sites distant from the heart are influenced by multiple electrical vectors occurring during the same portion of the cardiac cycle, it is not surprising that ST-segment elevation in precordial recordings is not an absolute index of the extent of irreversible injury. On the other hand, surface electrocardiographic mapping of ST-segments has shed some light on the progress of ischemic injury in serial studies in the same experimental animal or patient. Whether or not analysis of ST-segment elevation, combined with other markers of irreversible injury to myocardium, finds a permanent role in the assessment of infarct size in patients, electrocardiographic mapping provided cogent information suggesting that the evolution of ischemic injury could be influenced favorably by interventions implemented relatively soon after the onset of ischemia. Recent refinements in the surface electrocardiographic approach to evaluating the progress of ischemic injury have incorporated relationships between ST-segment elevation and subsequent loss of r wave amplitude or development of q waves in the same electrocardiographic sites.[26]

Biochemical markers of ischemic injury

Assessment of myocardial ischemic injury in intact organisms is formidable. Morphologic analysis of the heart, often cited as the "gold standard," entails sampling errors even in experimental animals in which the entire heart is available and is limited by the time required for the evolution of microscopic and ultrastructural changes after the onset of an ischemic insult. Furthermore, even partial, direct analysis of the heart is often not possible in clinical studies concerned with the evolution of myocardial infarction. Because of the recognized sensitivity of enzymes and other biochemical markers of ischemic injury appearing in the blood of intact experimental animals and patients, they are attractive as potentially useful quantitative indices of irreversible damage to the heart. For this purpose, several relationships between myocardial cell death and release of the biochemical marker from myocardium require definition, such as the extent of variation in concentration of the marker within normal and ischemic myocardium; the processes governing release of the marker into the circulation; the mode of transport of the marker and its behavior in compartments such as interstitial fluid and lymph; and the factors influencing distribution and removal of the marker from the circulation, including renal, hepatic, and cardiac function as well as the metabolic and pharmacologic milieu.* In a series of preliminary experiments from our laboratory,[22,49,50] work was focused on creatine kinase (CK) because of (1) the virtually exclusive distribution of CK in the heart within myocardial cells themselves (of particular importance because the heart contains approximately 50% myocytes and 50% nonmyocardial components) and (2) results in rabbits and dogs subjected to coronary occlusion, indicating that CK loss from the heart 24 hr after occlusion was related quantitatively to the severity of ischemia detected with radioactively labeled microspheres and late ultrastructural and histologic changes indicative of necrosis.[22,49] A one-compartment model was developed in which serial changes in plasma CK activity were used to estimate total CK loss from the heart and hence to infarct size in conscious dogs subjected to coronary occlusion.[49,52] In animals with hemodynamically uncomplicated infarction, CK loss from the heart estimated in this fashion correlated closely with myocardial CK depletion measured directly.[49]

Estimations of infarction based on analysis of plasma CK time-activity curves employ several parameters, including the rate of disappearance of CK from the circulation, assumed to be first order in our initial studies, and the ratio of CK recovered in blood compared to that lost from myocardium.[25,55,56] Improved enzymatic estimates of infarct size have been obtained by excluding noncardiac contributions to plasma time-activity curves based on analysis of the MB CK isoenzyme rather than total CK.[41] In humans, the MB isoenzyme is found almost exclusively in myocardium.[40,63] Thus changes in plasma MB CK activity reflect enzyme release from the heart rather than from other organs such as skeletal muscle or brain.[1] Additional improvements in enzymatic estimation of infarct size have come from clarification of CK disappearance constants,[55,56] and utilization of multicompartmental models that reflect the physiologic processes affecting release of enzyme from the heart,[25] transport in lymph,[24,44] and removal from blood, while the vascular space is being resupplied concomitantly with enzyme from an extravascular distribution space.[25] Despite the simplifications inherent in enzymatic estimates of infarct size obtained with one-compartment models, results obtained with this approach have indicated that infarct size is an important determinant of several sequelae of infarction, including alterations in electrical stability of the heart,[4,8,42] impairment of ventricular function,[3,12,23,45] alterations in ventricular compliance,[29] early clinical manifestations of severe infarction,[33] and the probability of death within the first six months after the acute episode.[54] Recently enzymatic estimates of infarct size have been shown to correlate with the mass of the left ventricle exhibiting dyskinesis, evaluated angiographically, and

*See references 1, 22, 25, 41, 43, 52, 55, 64, 70.

with the extent of infarction assessed histochemically in patients who die.[3]

Although estimates of infarct size may be possible with other biochemical markers such as myoglobin,[58] difficulties may be encountered with any marker of low molecular weight. This is true because clearance by the kidneys, likely to be variable in patients with acute myocardial infarction, will alter quantitative relationships between the time-activity curve of the marker in blood and loss of the marker from the heart itself.

Enzymatic estimates of infarct size have been employed from one other point of view. In conscious dogs, serial changes in plasma CK activity have been projected with the use of curve-fitting techniques on the basis of values obtained during the first five hr after coronary occlusion.[50] In animals with coronary occlusion alone, the correlation between observed and projected plasma CK values was close. In animals in whom ischemic injury was exacerbated (by acceleration of heart rate), observed values exceeded those projected prior to the deleterious intervention. In other animals in whom myocardial oxygen requirements were presumably decreased by administration of agents such as propranolol, observed values deviated below those projected, reflecting apparent salvage of myocardium. In the experimental animal, it was possible to verify directly the apparent salvage of myocardium induced by favorable interventions because the heart could be analyzed at the conclusion of each experiment.

Since relations between observed and projected CK values reflected directional changes, verified independently, in the severity of ischemic injury sustained in conscious dogs with experimental coronary occlusion, analysis of plasma CK time-activity curves in this fashion was undertaken in clinical studies as well. Among hypertensive patients with acute myocardial infarction, blood pressure was lowered cautiously with intravenous trimethaphan as soon as sufficient data were available to permit projection of plasma CK curves.[51] On the basis of comparison of observed to projected plasma CK values after implementation of the intervention, reduction of ventricular afterload to decrease myocardial oxygen requirements protected ischemic myocardium with an overall reduction in enzymatically estimated infarct size of approximately 20%, associated with a significant reduction in early mortality. These observations are presented not to extol a particular mode of therapy for all patients with acute myocardial infarction but rather to indicate that, based on enzymatic estimates of infarction, it appears likely that a favorable modification of the evolution of ischemic injury is attainable in at least some patients in a fashion analogous to that observed in experimental animals. Unfortunately, however, objective evaluation of potentially therapeutic interventions based on comparison of observed to projected plasma enzyme values requires substantial delay prior to implementation of the intervention, during which data required for curve-fitting must be acquired. Accordingly, techniques for more rapid estimation of the extent of myocardium in jeopardy would be helpful. One approach entails tomographic estimation of the zone in jeopardy soon after the onset of ischemia coupled with analysis of complete plasma CK time-activity curves to estimate the overall extent of infarction in control patients and those treated with an intervention implemented as soon as tomography has been completed.

INFARCT SIZE AND PROGNOSIS

In a prescient study of prognosis after myocardial infarction employing discriminant analysis,[32] Norris noted that several factors appeared to carry weight as indices of early prognosis, including the presence of congestive heart failure, hypotension, and electrocardiographic changes indicative of anterior transmural infarction, all compatible with the concept that extensive infarction was the common denominator relating these parameters to prognosis. Subsequently, extensive necrosis of myocardium was found to occur in patients who died of acute myocardial infarction associated with cardiogenic shock,[13,34] suggesting that the marked depression of left ventricular function characteristic of cardiogenic shock reflected loss of viable tissue. In a prospective study utilizing enzymatic estimates of infarct size, extensive infarction was associated with substantially higher mortality within the first six months after infarction compared to the mortality rate among patients with small infarcts, based on results of plasma CK time-activity curves.[54]

Nevertheless, one should not expect perfect correlations between infarct size and prognosis. The extent of underlying vascular disease will undoubtedly exert a major influence on the long-term mortality rate after acute myocardial infarction. Furthermore, the physiologic implications of an infarct affecting 10% of left ventricular mass will be different in a patient who has sustained previous remote infarction compared to one whose infarction is an initial one.[29,45] On the other hand, malignant dysrhythmia may not be associated with unequivocal evidence of infarction, and even though patients who survive episodes of ventricular fibrillation under these circumstances remain at extraordinarily high risk of sudden death, the risk is not due to extensive loss of viable myocardium.[48] Consideration of causal connections between the magnitude of infarct size and the outcome of infarction should help to avoid unrealistic expectations and inappropriate deprecation of established relationships between infarct size and prognosis.

Several potential causal connections appear to exist relating the mass of myocardium subjected to ischemia to ventricular dysrhythmia.[53] Extensive infarction could set the stage for reentry by increasing the availability of pathways for slow impulse conduction, particularly if changes within the ischemic zone such as accumulation of extracellular potassium potentiate slow current responses with decremental conduction.[9] Regional adrenergic stimulation of the heart appears to be associated with spontaneous ventricular fibrillation after infarction induced experimentally and may depend on the locus and extent of infarction sustained.[7,69] Thus, even in the absence of severe hemodynamic decompensation, deleterious effects of regional cardiac adrenergic stimulation may facilitate phase IV depolarization and repetitive responses associated with slow-current action potentials.[9] In view of these potential causal connections, it is not surprising that reduction of the ventricular fibrillation threshold[4] in animals with coronary occlusion and the severity and persistence of premature ventricular complexes in patients[8,42] appear to be directly related to infarct size.

Since impaired ventricular function may result in diminished coronary perfusion, further compromising the heart, extensive infarction may set the stage for progressive or recurrent ischemic injury. Mathey, Bleifeld, and colleagues[29] have demonstrated close correlations between the extent of infarction estimated enzymatically and impairment of ventricular function and compliance. Since pump failure continues to account for a great deal of the early deaths among patients hospitalized with acute myocardial infarction, it is therefore not surprising that infarct size is correlated with early mortality. The shape of plasma MB CK time-activity curves suggests that, in at least some patients with cardiogenic shock, a vicious cycle of progressive impairment of contractility and myocardial necrosis, giving rise to continual liberation of MB CK[11] into the circulation, underlies the relentless progression of the clinical syndrome to death.

If infarct size is indeed a determinant of early death, it will affect overall, long-term mortality in any cohort of patients. However, it may not be a determinant of late death, or affect the late mortality in the cohort. In fact, it appears that the mortality among patients with large as compared to that among patients with small infarcts becomes similar during the latter part of the first year of follow-up. Thus infarct size may have a great deal to do with whether or not a patient survives after acute myocardial infarction, but it may have little to do with the relative risk among survivors during the late follow-up interval. This interpretation is in keeping with the well-recognized dependence of the natural history of ischemic heart disease on the severity of underlying vascular disease, with yearly mortality figures for patients with one-, two-, and three-vessel disease in the range of 4%, 8%, and 12%.[20] Nevertheless, from the point of view of public health, even modestly successful efforts to reduce infarct size might pay large dividends in improving the quality of life and reducing overall mortality because of its impact on events during the interval early after infarction and the high incidence of infarction in the population.

ASSESSMENT OF INFARCT SIZE WITH RADIONUCLIDES

Progress in modifying the evolution of infarction in experimental animals and in delineating the effect of infarct size on prognosis in patients stimulated efforts to estimate infarct size in the clinical setting after administration of radionuclides intravenously. Imaging of myocardium is generally performed with a scintillation camera, a device well suited to many clinical applications but unfortunately limited for imaging of the heart by several intrinsic factors, including the following[60]:

1. Images obtained with the scintillation camera represent three-dimensional objects in a two-dimensional plane with superimposition reducing the image contrast and impairing quantitative evaluation of the distribution of the tracer.
2. Contrast and resolution characteristics of images obtained with the scintillation camera are influenced by the depth of tracer within the tissue.
3. Since gamma radiation, particularly at lower photon energies, is attenuated by tissues interposed between the tracer and the detector, quantitative relationships between the distribution of radionuclide within the organ of interest and its representation in the image are distorted.

Despite these intrinsic limitations, however, remarkable progress has been made in detecting the heterogeneity of radiopharmaceutical accumulation related to altered blood flow within the heart and in detecting myocardial infarction on the basis of infarct avidity for "bone-seeking" and other radiopharmaceuticals[14,18,35] (Chapters 14 to 16).

Myocardial perfusion scintigraphy has been performed in patients at rest and undergoing exercise. Thallium 201 (^{201}Tl)[59] and several other analogs of potassium have been employed for this purpose.[5,19] Diminished accumulation of the tracer occurs in zones of infarction and permits "cold-spot" imaging. On the other hand, transient ischemia leads to efflux of potassium from myocardial cells, and that, coupled with decreased delivery of tracer in zones with diminished perfusion, appears to account for decreased uptake seen in ischemic regions. Unfortunately, however, the fact that both ischemia and infarction can elicit decreased accumulation of potassium analogs limits perfusion scintigraphy as a tool for quantitative detection of the severity and distribution of myocardial ischemia in vivo.

Other techniques for assessment of perfusion, including inert gas washout methods after injection of xenon 133 (^{133}Xe) directly into the coronary artery, permit more effective quantitation of the distribution of regional myocardial perfusion,[6] particularly when multicrystal scintillation detection systems and deconvolution of washout curves are employed. However, this technique, as well as procedures utilizing intracoronary injection of radioactively labeled macroaggregated albumin or microspheres (Chapters 10 and 11), are limited by the need for injection of the radiotracer directly into the coronary circulation and the consequent difficulty entailed in performing studies serially.

Assessment of myocardial infarction with the use of radionuclides has progressed rapidly since Parkey, Bonte, Willerson, and their co-workers[35] recognized the avidity for myocardial infarcts of bone-seeking radiopharmaceuticals, injected intravenously. With the use of tracers such as [^{99m}Tc] pyrophosphate, "hot-spot" detection of myocardial infarcts is possible, based on accumulation of the tracer within 1 to 2 hr after intravenous injection during an interval from one to ten days after infarction. Although detection of infarction has been facilitated with this technique, several factors make it difficult to assess infarct size quantitatively with intravenously labeled [^{99m}Tc] pyrophosphate, or other bone-seeking radiotracers. In addition to limitations inherent in the gamma camera detection system for this purpose, accumulation of the tracer within regions of infarction may not correlate closely with the extent of necrosis within the same region. This is due in part at least to the dependence of tracer uptake on blood flow. In regions with markedly diminished perfusion, virtually no accumulation of the tracer is recognizable, probably because the delivery of tracer to the region is insufficient. On the other hand, even when some necrosis is present within a region, if overall perfusion within the region is substantial and the proportion of cells exhibiting necrosis is low, external detection of infarction based on accumulation of tracer may be difficult. Other lesions, such as calcified ventricular aneurysms or valves, may accumulate the tracer, and under some circumstances accumulation of the tracer in skeletal muscle or the heart after electrical cardioversion may impair diagnostic specificity.[17] Despite these limitations, however, sequestration of several radiotracers that appear to share the property of complexing with calcium (including diphosphonate, tetracycline, and glucoheptonate) is the basis for what has already become a widely utilized diagnostic procedure for external detection of infarction in the clinical setting.

ESTIMATION OF INFARCT SIZE WITH POSITRON EMISSION TRANSAXIAL TOMOGRAPHY (PETT)

The possibility that zones of myocardial ischemia and/or infarction could be visualized quantitatively with the use of positron-emitting radionuclides and instrumentation permitting tomography by computer reconstruction led us to undertake a series of studies with cyclotron-produced, positron-emitting [^{11}C] palmitate, a physiologic substrate of the heart.[57,61,65,66,68] The use of positron-emitting radionuclides for this purpose offers several advantages potentially useful in quantitative estimation of infarction with external detection systems.[67] The positron particles emitted by these tracers lose kinetic energy as they move through matter due to interactions similar to those affecting electrons, particles with similar mass but opposite charge. When the kinetic energy of the positron is almost exhausted, the particle undergoes annihilation by interacting with an electron. In the process, the mass of both particles is converted into two annihilation photons, each with an energy of 511 keV. Because annihilation photons are emitted in diametrically opposite directions (i.e., at an angle of 180 degrees), detection of each pair can be accomplished with crystal scintillation detectors 180 degrees apart and connected through a fast-coincidence circuit that will record detected radiation as an event only when both detectors respond simultaneously or within defined limits. The field of view of such a pair of detectors is limited to a cylindric volume between them providing "electronic collimation." In contrast to the fields of view of conventional gamma emission detectors, the field of view of electronic collimators is uniform over relatively large distances.

The use of coincidence detection obviates another difficulty encountered with gamma radiation detectors, namely, variation in detected counts due to attenuation dependent on the distance of the tracer from the detector. With single photon detectors, this problem is substantial. With positron detectors, the algebraic sum of attenuation of the two photons emitted at an angle of 180 degrees from each other and detected by a pair of detectors 180 degrees apart remains constant regardless of the distance of the radiation source from each of the detectors in the pair. In other words, when the source of radiation is moved closer to one member of the pair of detectors, decreased attenuation of radiation detected by that member of the pair will be offset by the increased attenuation of radiation perceived by the other member of the pair. These properties of positron emission, coupled with the high energy of the photons emitted, facilitate computer reconstruction of the distribution of positron-emitting radionuclides within a cross section of an organ of interest. In positron emission transaxial tomography, each transaxial cross-sectional image is reconstructed from a series of radiation profiles obtained from pairs of detectors rotated through selected angles around the organ of interest.

Because several positron-emitting radionuclides, including oxygen 15 (^{15}O), nitrogen 13 (^{13}N), and carbon 11 (^{11}C) are incorporated readily into organic molecules such as physiologic substrates of the heart, their use permits assessment of regional metabolism of normal and ischemic myocardium and detection of infarction reflected by inadequate utilization of substrate. The short half-lives of these radionuclides (^{15}O—2 min, ^{13}N—10 min, and ^{11}C—20 min) permit sequential studies within relatively brief intervals.[16,36,37,62] Although extensive utilization of these agents presently requires availability of a cyclotron, several radionuclide-generator systems have been fabricated to permit synthesis of selected agents without this constraint.

The distribution of $^{13}NH_3$[14] and rubidium 81 (^{81}Rb)[28] in reconstructed cross sections of normal and ischemic myocardium has been evaluated with several detection systems after intravenous injection of the radiotracers. However, in part for the same reasons applying to assessment of infarction with cold-spot images obtained with a gamma camera, decreased accumulation of potassium analogs, even when quantified more accurately with computer reconstruction tomography, is unlikely to reflect infarct size exclusively. This is because of the important effects on accumulation of these tracers of even transitory alterations in regional perfusion and because of the difficulty of differentiating of decreased accumulation due to infarction from that due to ischemia.

It is well known that ischemia affects metabolism of free fatty acids (FFA) in heart muscle by decreasing oxidation and enhancing conversion of substrate to triglycerides.[30,31] External detection of accumulation of radioactively labeled FFA in normal myocardium was undertaken early during the evolution of attempts to image the heart.[38] Since FFA is a primary substrate of normal myocardium, it appeared likely to us that free fatty acids labeled with positron-emitting radionuclides would be particularly suitable for quantitative studies of myocardial metabolism and assessment of infarct size with positron emission transaxial tomography. In studies with isolated perfused hearts, we found that decreased accumulation of [^{11}C]palmitate was readily detectable externally when perfusion was decreased from 20 to 5 ml/min, even though a transitory diminution of perfusion to produce the same reduction in residence time of tracer did not inhibit accumulation of palmitate.[66] Thus under conditions in which the molar ratio of FFA to albumin and the availability of oxygen could be controlled independently, reversible but consistent depression of extraction of FFA was detectable externally in hearts subject to decreased flow.

These observations were subsequently extended to intact dogs with reversible or irreversible coronary occlusion. After intravenous injection of 4 to 8 mCi of [^{11}C] palmitate, the distribution of the tracer was determined by positron emission transaxial tomography in selected cross sections of the heart from apex to base.[65] Zones with decreased [^{11}C] palmitate accumulation were evident in tomographic images and

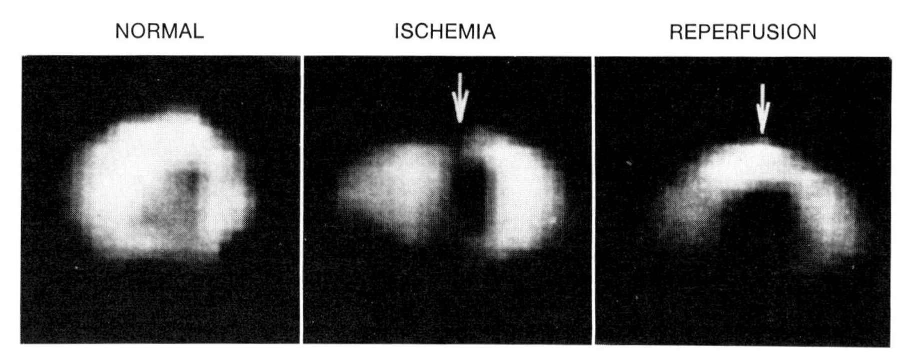

Fig. 9-1. Electrocardiographically gated diastolic images obtained with positron-emission transaxial tomograph after induction of transient myocardial ischemia by constriction of exteriorized coronary artery occlusive cuff in intact dog. Each image represents reconstructed cross-sectional slice through heart at ventricular level. Anterior, posterior, left, and right are represented in top, bottom, left, and right portions of each panel, respectively. *Left:* Homogeneous accumulation of [^{11}C] palmitic acid is evident in normal left ventricular myocardium. Tomogram was obtained during a 20 min interval after intravenous injection of tracer. *Center:* Transmural defect representing failure of [^{11}C] palmitate accumulation *(arrow)* is present anteriorly in image obtained after 30 min of myocardial ischemia. *Right:* Image was obtained during 20 min interval immediately following release of coronary artery occlusive cuff after an interval of ischemia of 30 min, hence insufficient to produce extensive infarction. As can be seen, after reperfusion myocardial metabolic integrity is demonstrable in area of previous defect *(arrow)* and, in fact, accumulation of tracer in this region exceeds that in adjacent and presumably normal myocardium. Only appreciable uptake of tracer within thorax is in region corresponding to heart. (From Weiss, E. S., et al.: Circ. Res. **39:**24, 1976. By permission of the American Heart Association, Inc.)

corresponded to the ischemic regions supplied by the transiently occluded coronary artery. When occlusion was maintained for less than 20 min, subsequent tomograms exhibited enhanced accumulation of [^{11}C] palmitate in zones that had been ischemic previously. In contrast, when ischemia was maintained for 1 hr or more and reperfusion permitted subsequently, inhibition of accumulation of [^{11}C] palmitate in zones of ischemia persisted, consistent with the time course of irreversible myocardial injury detectable by ultrastructural, histologic, and biochemical techniques (Fig. 9-1).

To determine whether positron emission transaxial tomography provided quantitative information regarding the distribution of myocardial necrosis in experimental animals subjected to coronary occlusion, we performed a series of studies in which [^{11}C] palmitate was injected intravenously in lightly anesthetized dogs that had been subjected to coronary occlusion 48 hr earlier.[65] The distribution of [^{11}C] palmitate in cross sections of the heart was determined by analysis of computer-reconstructed tomograms. In each cross section, the calculated percentage of infarction was compared to estimates of infarction from the corresponding cross section analyzed morphologically and biochemically after the animals were sacrificed. Morphologic analysis was conducted by reconstructing the entire cross section from histologic sections, planimetering the area of infarction recognizable by conventional histologic criteria of necrosis and verifying results by comparison with the planimetered area of infarction estimated from a photograph of the gross section. Biochemical analyses were performed by calculating depletion of myocardial CK in the entire cross section. In addition, in some animals [^{14}C] palmitate was injected intravenously at the same time that [^{11}C] palmitate was administered to permit comparison of the distribution of [^{14}C] palmitate within the myocardium to biochemical and morphologic criteria of necrosis in the same region. Estimates of infarct size obtained by positron emission transaxial tomography correlated closely

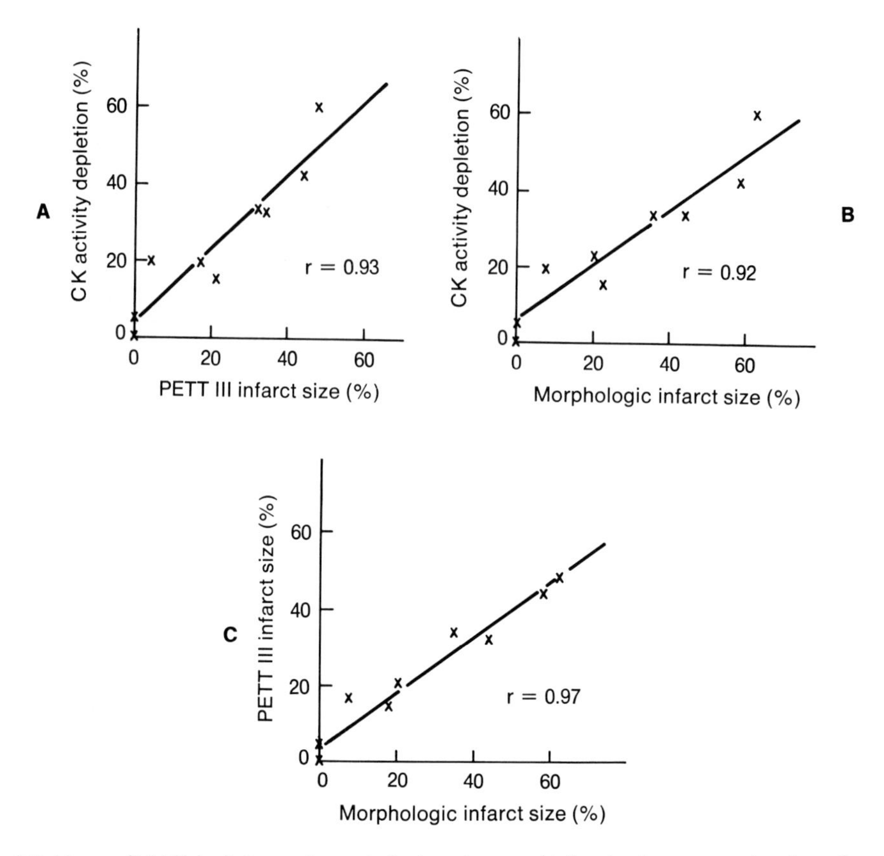

Fig. 9-2. Myocardial CK depletion and morphologic estimates of infarction in cross section through entire left ventricle, compared with percentage of infarction in corresponding section delineated by positron-emission transaxial tomography in vivo. Results from nine pairs of sections from six different dogs are shown. Correlation coefficients (least squares linear regression) indicate that estimates of infarction based on CK depletion (n = 9) and morphology correlated closely with estimates obtained in vivo by positron-emission transaxial tomography of corresponding cross section (n = 9) of left ventricle. Similarity of slopes in **B** and **C** is reflected by slope approaching 1 in **A.** Since 100% CK depletion does not occur within 48 hr (with approximately 10% residual activity—unpublished observations), slope in **B** would be anticipated to be less than 1. Slope in **C** is less than unity, probably because true image of normal ventricle is not congruent with concentric circle model used for calculations. (From Weiss, E. S., et al.: Circulation **55:**66, 1977. By permission of the American Heart Association, Inc.)

with morphologic estimates of infarction in corresponding cross sections (r = 0.93) and with biochemical estimates of infarction based on loss of CK activity (r = 0.92). In addition, the distribution of [^{14}C] palmitate correlated closely with myocardial CK depletion in the same 500 mg samples of myocardium from normal, ischemic, and necrotic myocardium (r = 0.97). These results indicate that necrosis can be estimated quantitatively and noninvasively by positron emission transaxial tomography after intravenous administration of [^{11}C] palmitate in animals with coronary occlusion (Fig. 9-2).

Recently these observations were extended to patients with acute myocardial infarction at least three months prior to study.[57] In normal subjects, the distribution of [^{11}C] palmitate in tomograms of the heart was homogeneous after intravenous administration of the tracer (Fig. 9-3). In patients with transmural myocardial infarction at least three months prior to study, decreased accumulation of palmitate was readily detectable in tomograms obtained by PETT

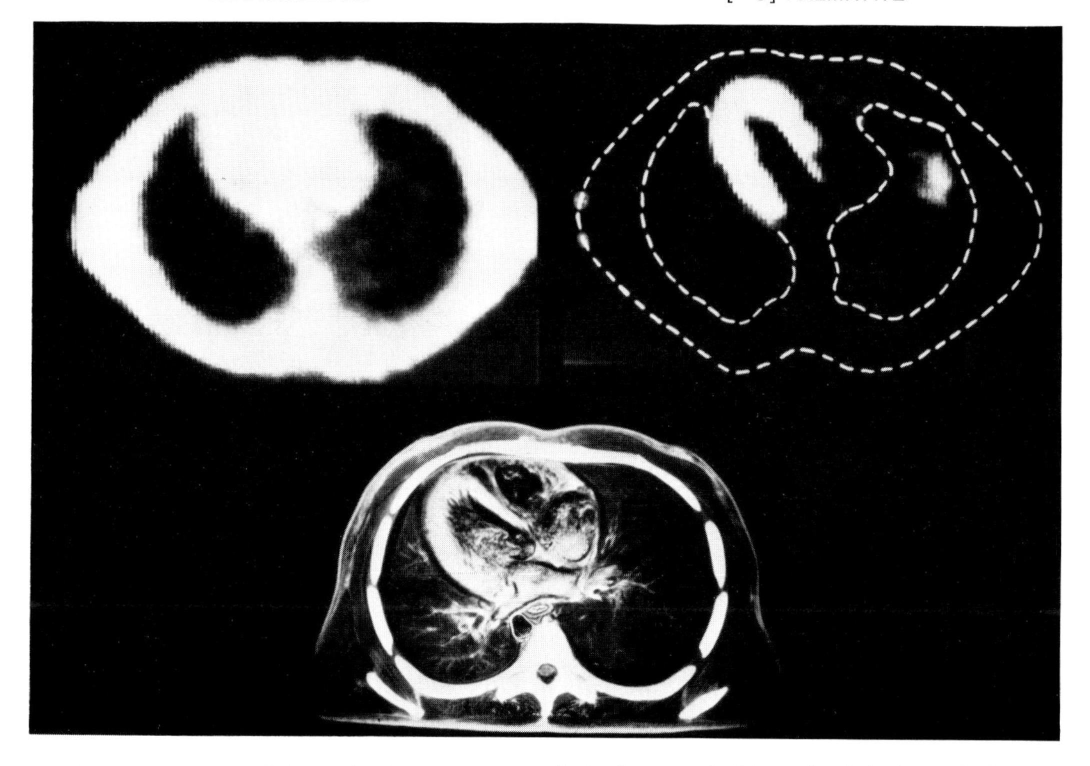

Fig. 9-3. *Top:* Transmission and emission tomograms obtained in normal subject at level of atrioventricular valves. Transmission image was obtained by placing a positron emitter, ^{64}Cu, around subject at selected level of thorax and was used for calculations of attenuation coefficients employed in reconstruction of emission tomograms. *Top right:* [^{11}C] palmitate emission tomogram from normal subject. Horseshoe-shaped image of left ventricular myocardial distribution of [^{11}C] palmitate is compared with photograph *(bottom)* of cross section of cadaver at corresponding level. The slight accumulation of tracer in right portion of emission tomogram is in dome of liver. (From Sobel, B. E., et al.: Circulation **55:**854, 1977. By permission of the American Heart Association, Inc.)

and corresponded to the electrocardiographic locus of infarction in every case (Fig. 9-4). Although these preliminary studies in patients were performed without electrocardiographic gating and by analyzing a cross section at only one selected level of the heart, other versions of positron-detecting tomographic systems, such as the MGH camera and PETT IV now operational in the Washington University Cardiac Care Unit, are capable of providing data needed to construct tomograms representing multiple cross sections of the heart simultaneously.

Differentiation of ischemic zones of myocardium from regions already necrotic can be accomplished with positron emission transaxial tomography, since depression of [^{11}C] palmitate accumulation in ischemic zones will be reversible in contrast to the irreversible depression associated with necrosis. On the basis of numerous studies concerned with the duration of ischemia required to produce irreversible injury, it appears likely that inhibition of [^{11}C] palmitate accumulation for more than 1 hr will be tantamount to irreversible injury, although this remains to be evaluated objectively. In addition, with the use of multiple positron-emitting tracers, such as those influenced primarily by perfusion in conjunction with those influenced primarily by metabolic integrity of myocardial cells, it should be possible to differentiate ischemic myocar-

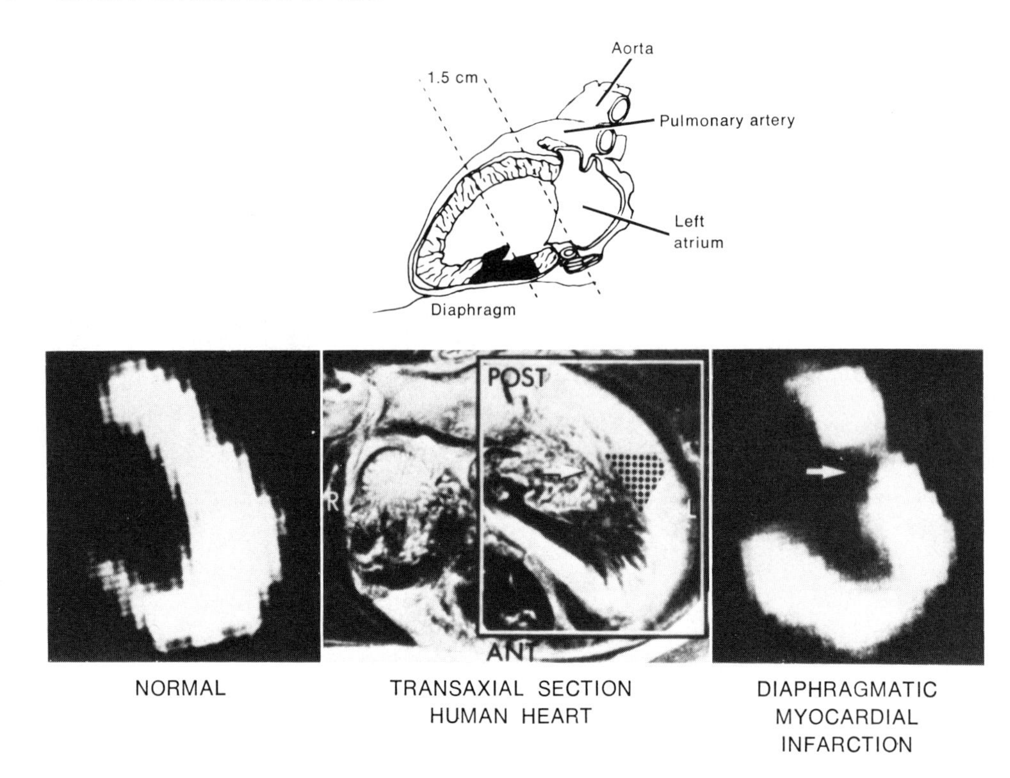

Fig. 9-4. Myocardial infarction involving posterior and diaphragmatic regions. *Left:* For comparison, normal emission tomogram at similar level. *Center:* Cross-sectional view of cadaver at level corresponding to level of emission tomogram with crosshatched marking indicating location of infarct. Myocardial infarction in 1.5 cm thick cross section of heart, depicted tomographically, is wedge shaped and localized in posterior and inferolateral portions of left ventricle. *Top:* Schematic drawing. Field of view encompasses left atrium not visualized in tomogram and normal myocardium extending posterior to infarct. (From Sobel, B. E., et al.: Circulation **55:**855, 1977. By permission of the American Heart Association, Inc.)

dium from tissue with impaired metabolic integrity, regardless of the prevailing regional perfusion.

CONCLUSIONS

Several techniques for estimation of infarct size have been developed and implemented in laboratory and clinical studies. Analysis of ST elevation in epicardial recordings obtained from experimental animals lent credence to the view that the evolution of ischemic injury sustained by the heart was potentially modifiable by physiologic and pharmacologic interventions favorably influencing the balance between myocardial oxygen supply and demand and facilitating washout of noxious metabolites. Enzymatic estimates of infarct size supported the hypothesis that the extent of infarction was an important determinant of prognosis and that it influenced the severity of impairment of ventricular function, diminution of ventricular compliance, and cardiac electrical instability soon after the onset of myocardial infarction as well as the probability of death during the first six months. Scintigraphic techniques have facilitated detection of heterogeneous perfusion in patients with ischemic heart disease and detection of infarction and its localization. Each of these techniques has limitations that may detract from universal application or impair accuracy. However, each has contributed to the growing recognition of the importance of infarct size as a factor influencing prognosis and the need to evaluate objectively interventions designed to protect ischemic myocardium.

External detection of jeopardized and blighted myocardium may be best achieved by techniques employing computer reconstruction to obtain tomograms, which represent the distribution of selected tracers within cross sections of the heart from apex to base. Labeling of physiologic substrates of the heart with positron-emitting radionuclides appears to be particularly promising in this regard because of several salutary physical characteristics of positron-emitting radionuclides and because of the extensive information already available regarding alterations in intermediary metabolism induced by ischemia and associated with infarction. Even a modest reduction in morbidity or mortality associated with reduction of infarct size in patients hospitalized with acute myocardial infarction would have a substantial impact on the population as a whole because of the high prevalence of coronary artery disease in our society. The need for definitive tools for objective and quantitative assessment of infarct size in patients has never been more clear. Fortunately progress in the development and validation of such tools has never been more encouraging.

ACKNOWLEDGMENT

Preparation of the manuscript by Ms. Carolyn Lohman is gratefully appreciated.

REFERENCES

1. Ahmed, S. A., Williamson, J. R., Roberts, R., Clark, R. E., and Sobel, B. E.: The association of increased plasma MB CPK activity and irreversible ischemic myocardial injury in the dog, Circulation **54:**187, 1976.
2. Ahumada, G., Roberts, R., and Sobel, B. E.: Evaluation of myocardial infarction with enzymatic indices, Prog. Cardiovasc. Dis. **18:**405, 1976.
3. Bleifeld, W., Mathey, D., Hanrath, P., Buss, H., and Effert, S.: Infarct size estimated from serial serum creatine phosphokinase in relation to left ventricular hemodynamics, Circulation **55:**303, 1977.
4. Bloor, C. M., Ehsani, A., White, F. C., and Sobel, B. E.: Ventricular fibrillation threshold in acute myocardial infarction and its relation to myocardial infarct size, Cardiovasc. Res. **9:**468, 1975.
5. Budinger, T. F., Yano, Y., and Hoop, B.: A comparison of $^{82}Rb^{+}$ and $^{13}NH_3$ for myocardial positron scintigraphy, J. Nucl. Med. **16:**429, 1974.
6. Cannon, P. J., Dell, R. B., and Dwyer, E. M., Jr.: Regional myocardial perfusion rates in patients with coronary artery disease, J. Clin. Invest. **51:** 978, 1972.
7. Corr, P. B., Pearle, D. L., Hinton, J. R., Roberts, W. C., and Gills, R. A.: Site of myocardial infarction: a determinant of the cardiovascular changes induced in the cat by coronary occlusion, Circ. Res. **39:**840, 1976.
8. Cox, J. R., Jr., Roberts, R., Ambos, H. D., Oliver, G. C., and Sobel, B. E.: Relations between enzymatically estimated myocardial infarct size and early ventricular dysrhythmia, Circulation **53** (suppl. 1):150, 1976.
9. Cranefield, P. F.: The conduction of the cardiac impulse, Mount Kisco, N.Y., 1975, Futura Publishing Co., Inc.
10. Fozzard, H. A., and DasGupta, D. S.: ST-segment potentials and mapping. Theory and experiments, Circulation **54:**533, 1976.
11. Gutovitz, A. L., Sobel, B. E., and Roberts, R.: Cardiogenic shock: a syndrome frequently due to slowly evolving myocardial injury, abstracted, Am. J. Cardiol. **39:**322, 1977.
12. Hanrath, P., Bleifeld, W., and Mathey, D.: Assessment of left ventricular hemodynamic reserve in acute myocardial infarction by volume loading and infarct size determination, Eur. J. Cardiol. **3:**99, 1975.
13. Harnarayan, C., Bennett, M. A., Pentecost, B. L., and Brewer, D. B.: Quantitative study of infarcted myocardium in cardiogenic shock, Br. Heart J. **32:**728, 1970.
14. Harper, P. V., Lathrop, K. A., Krizek, H., Lembares, N., Stark, V., and Hoffer, P. B.: Clinical feasibility of myocardial imaging with $^{13}NH_3$, J. Nucl. Med. **13:**278, 1972.
15. Herdson, P. B., Sommers, H. M., and Jennings, R. B.: A comparative study of the fine structure of normal and ischemic dog myocardium with special reference to early changes following temporary occlusion of a coronary artery, Am. J. Pathol. **46:**367, 1965.
16. Hoffman, E. J., Phelps, M. E., Mullani, N. A., Higgins, C. S., and Ter-Pogossian, M. M.: Design and performance characteristics of a whole-body positron transaxial tomograph, J. Nucl. Med. **17:** 493, 1976.
17. Holman, B. L.: Radionuclide methods in the evaluation of myocardial ischemia and infarction, Circulation **53** (suppl. 1):112, 1976.
18. Holman, B. L., Lesch, M., Zweiman, F. G., Temte, J., Lown, B., and Gorlin, R.: Detection and sizing of acute myocardial infarcts with ^{99m}Tc (Sn) tetracycline, N. Engl. J. Med. **291:**159, 1974.
19. Hoop, B., Jr., Smith, T. W., Burnham, C. A., Correll, J. E., Brownell, G. L., and Sanders, C. A.: Myocardial imaging with $^{13}NH_4^{+}$ and a multicrystal positron camera, J. Nucl. Med. **14:**181, 1973.
20. Humphries, J. O., Kuller, L., Ross, R. S., Frie-

singer, G. C., and Page, E. E.: Natural history of ischemic heart disease in relation to arteriographic findings. A twelve year study of 224 patients, Circulation **49**:489, 1974.

21. Kjekshus, J. K., Maroko, P. R., and Sobel, B. E.: Distribution of myocardial injury and its relation to epicardial ST-segment changes after coronary artery occlusion in the dog, Cardiovasc. Res. **6**: 490, 1972.
22. Kjekshus, J. K., and Sobel, B. E.: Depressed myocardial creatine phosphokinase activity following experimental myocardial infarction in rabbit, Circ. Res. **27**:403, 1970.
23. Kostuk, W. J., Ehsani, A. A., Karliner, J. S., Ashburn, W. L., Peterson, K. L., Ross, J., Jr., and Sobel, B. E.: Left ventricular performance after myocardial infarction assessed by radioisotope angiocardiography, Circulation **47**:242, 1973.
24. Malmberg, P.: Time course of enzyme escape via heart lymph following myocardial infarction in the dog, Scand. J. Clin. Lab. Invest. **30**:405, 1972.
25. Markham, J., Karlsberg, R. P., Roberts, R., and Sobel, B. E.: Mathematical characterization of kinetics of native and purified creatine kinase in plasma. In Computers in cardiology, Long Beach, Calif., 1976, IEEE Computer Society, p. 3.
26. Maroko, P. R., Davidson, D. M., Libby, P., Hagan, A. D., and Braunwald, E.: Effects of hyaluronidase administration on myocardial ischemic injury in acute infarction, Ann. Intern. Med. **82**:516, 1975.
27. Maroko, P. R., Kjekshus, J. K., Sobel, B. E., Watanabe, T., Covell, J. W., Ross, J., Jr., and Braunwald, E.: Factors influencing infarct size following experimental coronary artery occlusions, Circulation **43**:67, 1971.
28. Martin, N. D., Zaret, B. L., McGowan, R. L., Wells, H. P., Jr., and Flamm, M. D.: Rubidium-81: a new myocardial scanning agent, Radiology **111**: 651, 1974.
29. Mathey, D., Bleifeld, W., Hanrath, P., and Effert, S.: Attempt to quantitate relation between cardiac function and infarct size in acute myocardial infarction, Br. Heart J. **36**:271, 1974.
30. Neely, J. R., and Morgan, H. E.: Relationship between carbohydrate and lipid metabolism and the energy balance of heart muscle. In Comroe, J. H., Jr., Sonnenschein, R. R., and Zierler, K. L., editors: Annual review of physiology, vol. 36, Palo Alto, Calif., 1974, Annual Reviews Inc., p. 413.
31. Neely, J. R., Rovetto, M. J., Whitmer, J. T., and Morgan, H. E.: Effects of ischemia on function and metabolism of the isolated working rat heart, Am. J. Physiol. **225**:651, 1973.
32. Norris, R. M., Caughey, D. E., Mercer, C. J., Deeming, L. W., and Scott, P. J.: Coronary prognostic index for predicting survival after recovery from acute myocardial infarction, Lancet **2**:485, 1970.
33. Norris, R. M., Whitlock, R. M. L., Barratt-Boyes, C., and Small, C. W.: Clinical measurement of myocardial infarct size. Modification of a method for the estimation of total creatine phosphokinase release after myocardial infarction, Circulation **51**:614, 1975.
34. Page, D. L., Caulfield, J. B., Kastor, J. A., DeSanctis, R. W., and Sanders, C. A.: Myocardial changes associated with cardiogenic shock, N. Engl. J. Med. **285**:133, 1971.
35. Parkey, R. W., Bonte, F. J., Meyer, S. L., Atkins, J. M., Curry, G. L., Stokely, E. M., and Willerson, J. T.: A new method for radionuclide imaging of acute myocardial infarction in humans, Circulation **50**:540, 1974.
36. Phelps, M. E., Hoffman, E. J., Higgins, C., Mullani, N., and Ter-Pogossian, M. M.: Performance analysis of a positron transaxial tomograph (PETT III), part II. In Ter-Pogossian, M. M., Phelps, M. E., Brownell, G. L., Cox, J. R., Jr., Davis, D. O., and Evens, R. E., editors: Reconstructive tomography in diagnostic radiology and nuclear medicine, Baltimore, 1977, University Park Press, p. 371.
37. Phelps, M. E., Hoffman, E. J., Coleman, R. E., Welch, M. J., Raichle, M. E., Weiss, E. S., Sobel, B. E., and Ter-Pogossian, M. M.: Tomographic images of blood pool and perfusion in brain and heart, J. Nucl. Med. **17**:603, 1976.
38. Poe, N. D.: A critical evaluation of myocardial imaging. In Subramanian, G., Rhodes, B. A., Cooper, J. F., et al., editors: Radiopharmaceuticals, New York, 1975, Society of Nuclear Medicine, p. 359.
39. Protection of the ischemic myocardium, Circulation **53**(suppl. I), 1976.
40. Roberts, R., Gowda, K. S., Ludbrook, P. A., and Sobel, B. E.: Specificity of elevated serum MB creatine phosphokinase activity in the diagnosis of acute myocardial infarction, Am. J. Cardiol. **36**: 433, 1975.
41. Roberts, R., Henry, P. D., and Sobel, B. E.: An improved basis for enzymatic estimation of infarct size, Circulation **52**:743, 1975.
42. Roberts, R., Husain, A., Ambos, H. D., Oliver, G. C., Cox, J., Jr., and Sobel, B. E.: Relation between infarct size and ventricular arrhythmia, Br. Heart J. **37**:1169, 1975.
43. Roberts, R., and Sobel, B. E.: The effect of selected drugs and myocardial infarction on the disappearance of creatine kinase from the circulation in conscious dogs, Cardiovasc. Res. **11**:103, 1977.
44. Robison, A. K., Gnepp, D. R., and Sobel, B. E.: Inactivation of CPK in lymph, abstracted, Circulation **52**(suppl. II):5, 1975.
45. Rogers, W. J., McDaniel, H. G., Smith, L. R., Mantle, J. A., Russell, R. O., Jr., and Rackley, C. E.: Correlation of CPK-MB and angiographic estimates of infarct size in man, abstracted, Circulation **54**(suppl. II):28, 1976.
46. Ross, J., Jr.: Electrocardiographic ST-segment analysis in the characterization of myocardial

ischemia and infarction, Circulation **53**(suppl. I): 73, 1976.

47. Ross, J., Jr., and Franklin, D.: Analysis of regional myocardial function, dimensions, and wall thickness in the characterization of myocardial ischemia and infarction, Circulation **53**(suppl. I):88, 1976.
48. Schaffer, W. A., and Cobb, L. H.: Recurrent ventricular fibrillation and modes of death in survivors of out-of-hospital ventricular fibrillation, N. Engl. J. Med. **293**:259, 1975.
49. Shell, W. E., Kjekshus, J. K., and Sobel, B. E.: Quantitative assessment of the extent of myocardial infarction in the conscious dog by means of analysis of serial changes in serum creatine phosphokinase activity, J. Clin. Invest. **50**:2614, 1971.
50. Shell, W. E., Lavelle, J. F., Covell, J. W., and Sobel, B. E.: Early estimation of myocardial damage in conscious dogs and patients with evolving acute myocardial infarction, J. Clin. Invest. **52**: 2579, 1973.
51. Shell, W. E., and Sobel, B. E.: Protection of jeopardized ischemic myocardium by reduction of ventricular afterload, N. Engl. J. Med. **291**:481, 1974.
52. Shell, W. E., and Sobel, B. E.: Biochemical markers of ischemic injury, Circulation **53**(suppl. I):98, 1976.
53. Sobel, B. E.: Infarct size, prognosis, and causal contiguity, Circulation **53**(suppl. I):146, 1976.
54. Sobel, B. E., Bresnahan, G. F., Shell, W. E., and Yoder, R. D.: Estimation of infarct size in man and its relation to prognosis, Circulation **46**:640, 1972.
55. Sobel, B. E., Larson, K. B., Markham, J., and Cox, J. R., Jr.: Empirical and physiological models of enzyme release from ischemic myocardium. In Computers in cardiology, Long Beach, Calif., 1974, IEEE Computer Society, p. 189.
56. Sobel, B. E., Markham, J., and Roberts, R.: Factors influencing enzymatic estimates of infarct size, Am. J. Cardiol. **39**:130, 1977.
57. Sobel, B. E., Weiss, E. S., Welch, M. J., and Ter-Pogossian, M. M.: Detection of remote myocardial infarction in patients with positron emission transaxial tomography and intravenous ^{11}C-palmitate, Circulation **55**:853, 1977.
58. Stone, M. J., Waterman, M. R., Murray, G., Harimoto, D., Platt, M. R., Blomqvist, G., and Willerson, J. T.: The serum myoglobin level as a diagnostic test in patients with acute myocardial infarction, Br. Heart J. **39**:375, 1977.
59. Strauss, H. W., Harrison, K., Langan, J. K., Lebowitz, E., and Pitt, B.: Thallium-201 for myocardial imaging. Relation of thallium-201 to regional myocardial perfusion, Circulation **51**:641, 1975.
60. Ter-Pogossian, M. M.: Limitations of present radionuclide methods in the evaluation of myocardial ischemia and infarction, Circulation **53**(suppl. I):119, 1976.
61. Ter-Pogossian, M. M., Hoffman, E. J., Weiss, E. S., Coleman, R. E., Phelps, M. E., Welch, M. J., and Sobel, B. E.: Positron emission reconstruction tomography for the assessment of regional myocardial metabolism by the administration of substrates labeled with cyclotron produced radionuclides. In Harrison, D. C., Sandler, H., and Miller, H. A., editors: Proceedings of the Conference on Cardiovascular Imaging and Image Processing Theory and Practice, vol. 72, Palos Verdes Estates, Calif., 1975, Society of Photo-Optical Instrumentation Engineers, p. 277.
62. Ter-Pogossian, M. M., Phelps, M. E., Hoffman, E. J., and Coleman, R. E.: Performance analysis of a positron transaxial tomograph (PETT III), part 1. In Ter-Pogossian, M. M., Phelps, M. E., Brownell, G. L., Cox, J. R., Jr., Davis, D. O., and Evens, R. E., editors: Reconstructive tomography in diagnostic radiology and nuclear medicine, Baltimore, 1977, University Park Press, p. 359.
63. Van der Veen, K. J., and Willebrands, A. F.: Isoenzymes of creatine phosphokinase in tissue extracts and in normal and pathological sera, Clin. Chim. Acta **13**:312, 1966.
64. Wakim, K. G., and Fleisher, G. A.: The fate of enzymes in body fluids—an experimental study, IV. Relationship of the reticuloendothelial system to activities and disappearance rates of various enzymes, J. Lab. Clin. Med. **61**:107, 1963.
65. Weiss, E. S., Ahmed, S. A., Welch, M. J., Williamson, J. R., Ter-Pogossian, M. M., and Sobel, B. E.: Quantification of infarction in cross sections of canine myocardium in vivo with positron emission transaxial tomography and ^{11}C-palmitate, Circulation **55**:66, 1977.
66. Weiss, E. S., Hoffman, E. J., Phelps, M. E., Welch, M. J., Henry, P. D., Ter-Pogossian, M. M., and Sobel, B. E.: External detection and visualization of myocardial ischemia with ^{11}C-substrates in vitro and in vivo, Circ. Res. **39**:24, 1976.
67. Weiss, E. S., Siegel, B. A., Sobel, B. E., Welch, M. J., and Ter-Pogossian, M. M.: Evaluation of myocardial metabolism and perfusion with positron-emitting radionuclides, Prog. Cardiovasc. Dis. **20**:191, 1977.
68. Weiss, E. S., Welch, M. J., Ter-Pogossian, M. M., Higgins, C. S., and Sobel, B. E.: Non-invasive quantification of myocardial infarction with positron emission transaxial tomography. In Computers in cardiology, Long Beach, Calif., 1976, IEEE Computer Society, p. 41.
69. Witkowski, F. X., Sobel, B. E., and Corr, P. B.: Characterization of regional myocardial electrograms with an automated system for pulse analysis. In Computers in cardiology, Long Beach, Calif., 1976, IEEE Computer Society, p. 19.
70. Witteveen, S. A. G. J., Hemker, H. C., Hollaar, L., and Hermens, W. T.: Quantitation of infarct size in man by means of plasma enzyme levels, Br. Heart J. **37**:795, 1975.

10 □ Regional myocardial blood flow in the human

Attilio Maseri
Antonio L'Abbate

The main field of application of myocardial blood flow measurements in the human is ischemic heart disease. The interest in the behavior of myocardial perfusion in ischemic heart disease stems both from the need for extension of knowledge about the pathophysiologic aspects of the disease and from the desire for improvement of the functional diagnostic capability for the individual patient.[22]

Measurements of total or average myocardial blood flow obtained in humans by different methods are, in general, conflicting and inconclusive. This is essentially related to the fact that even though pathologic, coronarographic, and electrocardiographic studies indicated a patchy distribution of coronary artery lesions, perfusion alterations were sought with methods that measure average or total myocardial flow; thus changes in opposite directions in adjacent zones could not be detected.

In recent years this fact has stimulated the development of techniques capable of providing information on myocardial perfusion abnormalities on a regional basis. Nuclear medical external counting principles and instruments have made this significant leap forward possible. In this chapter we will discuss the methods with which we have had most direct experience, without any pretention of covering homogeneously the entire field, which is in continuous, rapid expansion.

PATHOPHYSIOLOGIC PROBLEMS

The necessity for extending knowledge of the pathophysiologic mechanisms resulting in ischemic heart disease stems from the consideration that, although it is universally accepted that myocardial ischemia results from a deficiency of perfusion in relation to the metabolically generated demand for perfusion, the mechanisms responsible for the imbalance between demand and supply, for its reversibility or irreversibility, and for its extension to surrounding areas are not yet clear. Pathologic and clinical data suggest that the traditional concept that organic arterial obstructions are responsible, on purely hydraulic terms, for the impairment of blood supply leading to myocardial ischemia might be rather simplistic. Therefore functional coronary or myocardial mechanisms may intervene and be involved to a significant extent in the genesis, duration, and extension of myocardial ischemia. It is the possible importance of these functional factors that calls for a pathophysiologic study on the behavior of the coronary circulation in ischemic heart disease. Indeed, if myocardial blood flow alterations were simply and only related to anatomic lesions in what could be defined as the plumbing system of the heart, which is controlled strictly by the fixed-resistance supply vessels, the only reason for studying myocardial perfusion in ischemic heart disease would be a purely diagnostic one. On the other hand, although animal studies have provided some basic elements for the understanding of the experimental ischemia and infarction produced by the sudden occlusion of a single vessel, the sudden occlusion of a major coronary branch, as produced in animal models, seems rather remote from the pathogenesis of angina pectoris or of human myocardial infarction, since in the human, angina and infarction generally occur in the presence of slowly developing stenotic lesions, single or multiple, and of collateral vessels and often in the absence of complete occlusions.

The study of regional myocardial blood flow should provide insight into the sequence of alterations occurring during the onset, development, and disappearance of acute ischemia (in relation to a quantification of the metabolically generated demands) so that appropriate forms of therapy can be developed.

With the purpose of obtaining this type of pathophysiologic information, it is possible to select the patients who are the more typical cases and who appear to be suitable for the available methods of study. The method selected for this measurement should be capable of detecting and recording transients or giving frequent, repeated estimates of flow.

DIAGNOSTIC PROBLEMS

Clinical symptoms and findings may be useful in the diagnosis of ischemic heart disease, but they are of little value in the evaluation of the location, severity, and extension of ischemia. Accordingly, electrocardiography, which is an objective means of assessing ischemia, is largely based on indirect empiric signs.

Direct evaluation of myocardial perfusion alterations offers an objective means of detecting the site and extent of ischemia and could provide the cardiologist with the functional counterpart of the anatomic information provided by coronary arteriography. The methods must be capable of detecting regional differences in flow or regional changes of perfusion during effort tests that induce acute ischemia to give an objective indication of the location, extension, and severity of ischemia.

Although noninvasive techniques are essential for screening and follow-up studies, more complex procedures, which are associated with coronary arteriography, may be acceptable for a thorough functional evaluation of patients who are possible candidates for surgery.

METHODOLOGY

Two different approaches are currently possible: static imaging and dynamic recording of the regional indicator dilution curves (Table 10-1).

Static imaging

Static imaging can be used only when it can be assumed that the myocardial distribution of the indicator remains constant up to the completion of the measurement. The following types of indicators can be used: (1) biodegradable particles (labeled with suitable nuclides) that do not pass the capillary filter and are arrested in the myocardium and that must be injected into the arterial side of the circulation (left side of the heart or coronary arteries) (Chapter 11); (2) diffusible indicators (with intracoronary or aortic injection only) when the measurement can be completed within 5 to 10 sec after injection, or they can be infused at a constant rate; and (3) tracers selectively taken up by the myocardium, such as potassium tracers, that can be injected into the venous side of the circulation (Chapter 12).

Underlying principle. The underlying principle is that the distribution of the indicator in the myocardium is proportional to that of the myocardial blood flow. For particulate and diffusible indicators, this condition is certainly met if they are uniformly mixed with blood before they enter the coronary circulation (or before the first branching point with intracoronary injection). With potassium tracers, for which the extraction is not complete with the first circulation, this condition is met only when the amount of indicator leaving the organ before the

Table 10-1. Methods for the study of regional myocardial perfusion

Principle	Indicators: Type	Indicators: Administration	Instruments	Possibilities
Static imaging (of intramyocardial distribution of radioactivity)	Biodegradable particles (human albumin macroaggregates or microspheres labeled with short-lived nuclides)	Left side of heart or coronary arteries	Scanner/scinticamera	Qualitative evaluation of relative changes of the regional distribution of perfusion (with successive injections) in the solid angle of each field
	Diffusible indicators (only for a short time after injection of them)	Intracoronary Aortic root	Scinticamera	
	Potassium tracers selectively taken up by the myocardium	Intravenous	Scanner/scinticamera	
Dynamic recording (of regional time-activity curves)	Diffusible tracers	Intracoronary Aortic root	Scinticamera and computer	Semiquantitative evaluation of regional myocardial perfusion (under appropriate conditions)

completion of the measurement equals the amount that reenters during the same time. This occurs when average myocardial and total body extraction are equal.[9] However, when this condition is met, the equilibrium is only transiently maintained (Fig. 10-1).[24,35] Semiquantitative information on the relative regional distribution of perfusion can also be obtained at later times under appropriate conditions.

Methodologic problems. The problems of static imaging are those of external counting and are also related to the type of indicator and the way it is administered.

With the external counting systems currently available, three-dimensional information is condensed into two-dimensional information that constitutes the scintigram in a given projection. The detected activity for each area of the scintigram depends on the amount of muscle in the solid angle outlined by the area, on the geometric efficiency with which it is seen, and on its perfusion per unit mass. Only when the first two factors can be adequately determined can the perfusion of each area be estimated in absolute terms. The flow value obtained would correspond to the average perfusion of the muscle in the solid angle included in the area.

Coincidence counting with positron emitters and computerized longitudinal transaxial tomography at present offers the most advanced solution to the problem of uniform geometric counting efficiency in the counting field (Chapter 9).[5,45]

Available indicators and techniques. The available indicators are macroaggregates or microspheres of human albumin labeled with ^{99m}Tc, ^{113m}In, or ^{131}I; diffusible indicators such as ^{133}Xe and ^{81m}Kr; and radioactive potassium tracers such as ^{43}K, ^{131}Cs, ^{129}Cs, ^{201}Tl, and ^{81}Rb.

The labeled macroaggregates or microspheres must be injected into the arterial side of the circulation. When they are injected into the ventricle or atrium as a bolus, information can be obtained about the regional distribution of the indicator, and also the fraction of cardiac output perfusing the myocardium can be estimated.[29] A further advantage is the distribution of the indicator to the various zones of the myocardium in exact proportion to their blood flow and independent of the anatomic origin of their blood supply, since it appears reasonable to assume that the indicator is thoroughly mixed with blood at the aortic outlet. In this instance, the main difficulties stem from the subtraction of pericardiac

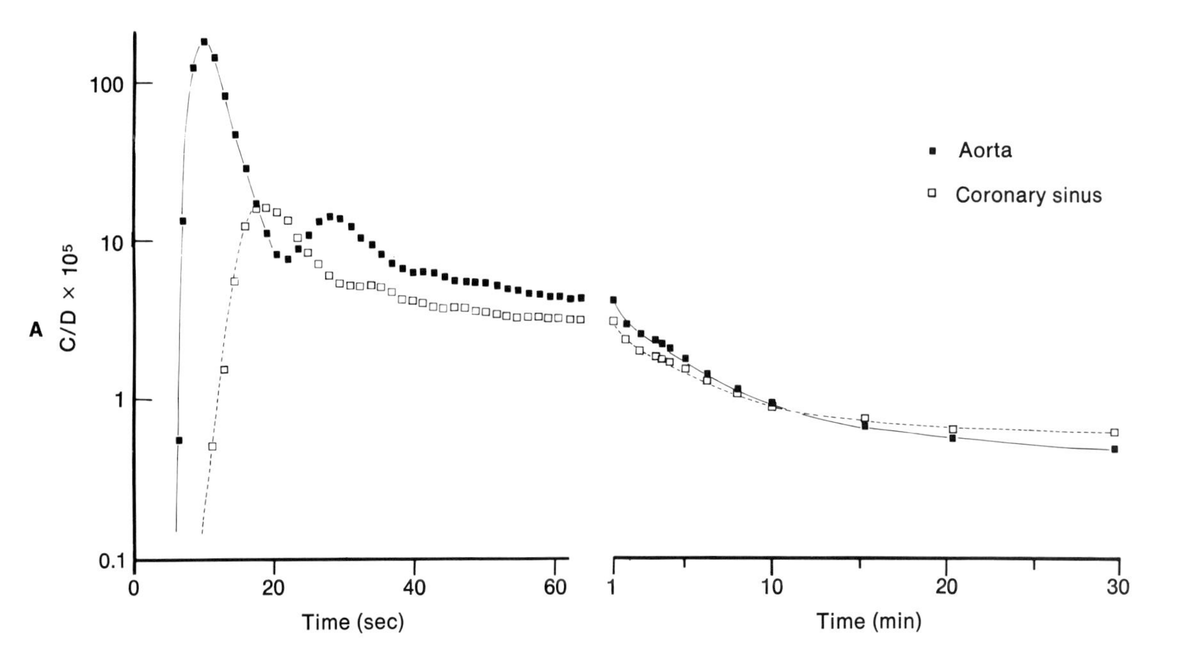

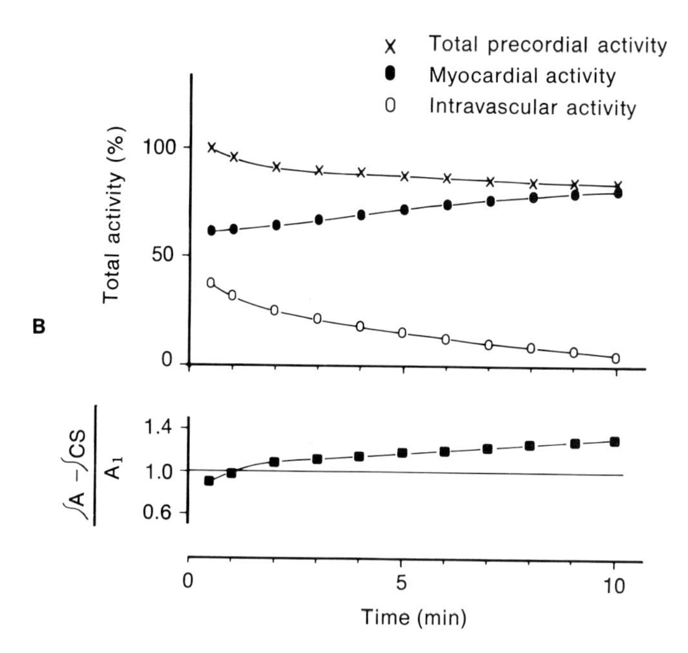

Fig. 10-1. A, Aortic and coronary sinus ^{201}Tl indicator dilution curves following pulmonary artery injection of indicator. Automated sampling at 1.5 sec intervals for initial forty-seven samples, manual sampling afterwards. **B,** Time course of total precordial ^{201}Tl activity over heart, intravascular activity and its contribution to precordial counting, and myocardial uptake as fraction of activity that entered coronary arteries with first circulation.

tissue contribution and from the relatively low fraction of total tracer distributed to the myocardium (about 5% to 10% of the amount injected).

By contrast, after direct intracoronary injection, adequate mixing can be assumed only when the coronary main stem is long and the injection is rapid. Furthermore, in the presence of intracoronary collateral circulation, the indicator will reveal the perfusion pattern at the moment of the injection, which might be modified by the speed of injection itself. An additional problem that arises when the same or two different indicators are injected into the two coronary arteries is the requirement that the amount injected into each vessel should be exactly proportional to its flow. Only under these conditions is a comparable, flow-dependent indicator distribution obtained. Thus, although this form of application may retain some usefulness for the qualitative demonstration of nonperfused areas, it does not appear to be particularly suitable for the evaluation of changes in the relative distribution of perfusion with successive injections into the left or right coronaries except in particular cases in which there is a long main-stem coronary artery and absence of collateral circulation.

Potassium tracers present the same general problems when used to measure total blood flow by nonimaging systems as when used to measure regional blood flow. These problems have been discussed in other articles[4,9,19,24,35] and have not been completely solved.

^{201}Tl is at present the most widely used potassium tracer for qualitative or semiquantitative scintigraphic studies of regional myocardium perfusion. Knowledge of the myocardial and blood kinetics of this tracer is essential for the appropriate timing and interpretation of the scintigraphic studies. In a human at rest, maximal uptake occurs from 14 to 20 min after intravenous injection[15a] and is 1.2 to 1.5 times greater than the fraction that entered the myocardium with the first circulation (Fig. 10-1). The intracardiac contribution is still appreciable 5 min after the injection[3,25] (Fig. 10-1). Therefore, ideally, scintigraphy at rest should be performed 5 to 10 min after the injection to avoid intracardiac blood activity contribution and to be closer to the time of maximal uptake, which occurs shortly after the injection and is followed by a very slow decrease of myocardial activity. The presence of regional ischemia will be more easily detected if it is maintained from the moment of the injection to the time of scintigraphy. Precise knowledge of the accuracy with which this technique allows the detection of ischemia is still lacking; although the fraction of the indicator reaching the ischemic zones is reduced in exact proportion to blood flow, its extraction should tend to be increased because of low flow[17,38] and decreased because of ischemic impairment of the membrane Na^+,K^+ pump—back diffusion should also be faster because of the reduced potassium pool of ischemic cells. However, comparison with the microsphere technique indicates that it may reflect ischemia satisfactorily for clinical purposes.[2,39,43] When regional ischemia persists only a few minutes after the injection, the redistribution of the tracer occurs in a rather short time (given the still high arterial concentration of the tracer) and tends to abolish rapidly the regional differences in uptake. This process occurs more gradually when ischemia lasts some minutes after the ^{201}Tl injection,[41] as shown in Fig. 10-2. (See also Chapter 12.)

For quantitative measurements a difficulty, which is common to particulate indicators injected into the left side of the heart, is the separation of the heart from the spleen and liver, which normally have an uptake per gram equal to about half that of the heart.[26] The relative uptake of these organs with respect to the myocardium can vary significantly under conditions such as exercise; thus their relative contribution to the precordial image can also vary appreciably. This contribution results both from direct radiations due to partial overlapping of these organs with the heart and from scattered radiations.

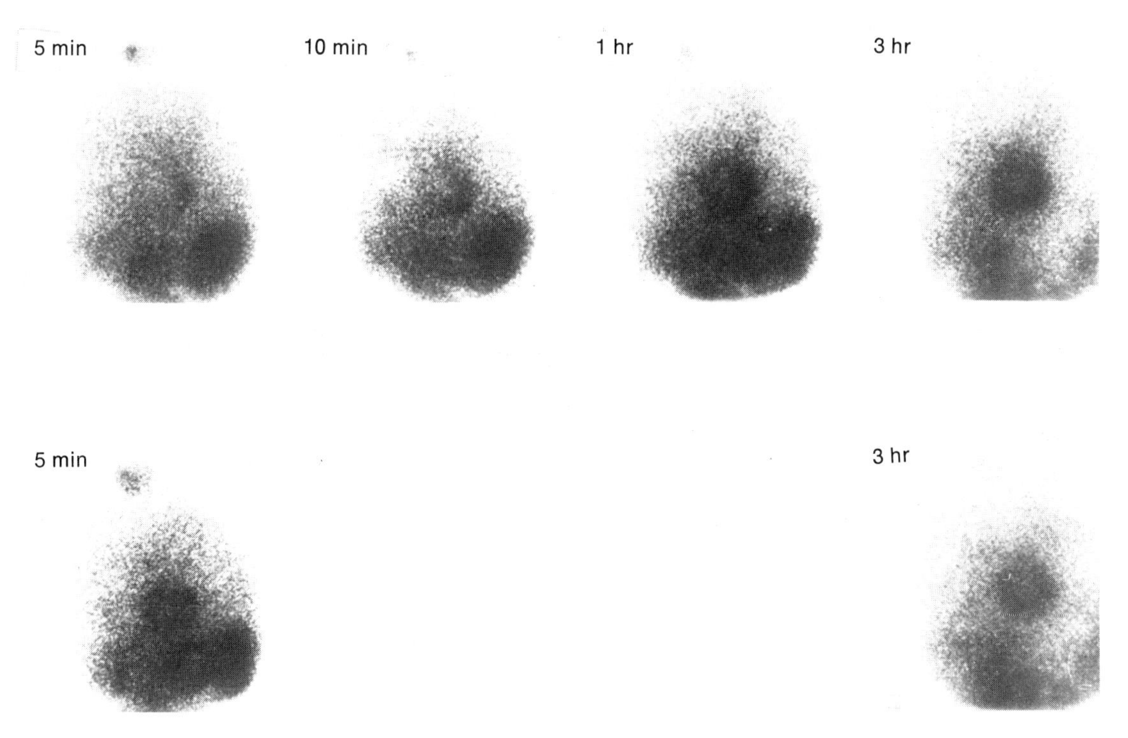

Fig. 10-2. Myocardial ^{201}Tl scintigrams in 30-degree LAO projection in patient with "variant" angina. *Top:* Four scintigrams obtained at successive intervals following injection during an anginal attack with S-T segment elevation in anterior leads. *Bottom:* Two scintigrams obtained one week later following an injection in absence of symptoms. Progressive redistribution of myocardial tracer activity occurs with time and abolishes defect of perfusion seen immediately after injection performed in presence of ischemia.

Of particular interest is the method based on the use of a continuous infusion of ^{81m}Kr into the aortic root for the evaluation of changes in the distribution of regional myocardial perfusion.[10,14,46] This technique is based on the fact that this nuclide has a 13 sec half-life. It assumes that its myocardial distribution is proportional to the regional blood flow rates and that its decay is completed before leaving the myocardium but is not too short to also reach regions with low perfusion. Under these circumstances the amount of tracer present in each region at any given time is a function of regional perfusion. The practical problems in the application of this technique stem from (1) the impossibility of ascertaining that the amount of tracer that enters the right and left coronary artery is proportional to the fraction of cardiac output respectively perfusing these vessels; (2) the fact that, given the relatively long half-life of ^{81m}Kr (which requires 43 sec for a 90% decay), at high flow rates an appreciable amount of ^{81m}Kr leaves the myocardium before its decay. For example, at 200 ml/min/100 g the mean transit time of ^{81m}Kr through the myocardium is 20.8 sec, which is short relative to the half-life of the nuclide and would result in an underestimate of flow by about 40%. Indeed, since the frequency function of transit times across the myocardium for diffusible indicators can be approximated by a monoexponential or a multiexponential function with a time lag of about 5 sec (appearance time) at rest, it is conceivable that the decay of ^{81m}Kr may not be complete for the shorter transit times, whereas it may be complete before reaching the tissues with longer transit times. Thus this isotope can be reasonably used in the most favorable conditions for quantitative studies of regional perfusion redistribution in the range of normal resting myocardial blood flows, but it can also provide valuable qualitative information about low or relatively high flow rates.

Possibilities of static imaging. The possibilities of static imaging are limited so far to the qualitative or semiquantitative evaluation of relative changes in the regional flow distribution to different areas of the myocardium with successive injections when the spatial distribution of the myocardium, the geometric efficiency with which it is seen, and the possible contribution of activity from pericardiac tissues remain constant.

The ^{201}Tl technique offers the advantage of the possibility of comparing the flow distribution (in scintigrams taken very early after the injection) with the distribution of the cellular mass (in scintigrams taken 3 to 4 hr after the injection) (Fig. 10-2).

Scintigrams in multiple projections, gating techniques, control of the constancy of counting geometry during scintigraphy, and finally objective quantification of tracer distribution will contribute to the detection of localized areas of ischemia.

Undoubtedly considerable advances will be made possible by the use of computerized longitudinal transaxial tomography in the field of the methodology and by the availability of ^{81m}Kr in the field of nuclides.

Dynamic recording of washout curves

Computer analysis of scinticamera data allows recording of regional indicator dilution curves in selected areas of the myocardial scintigram. Different types of indicators and forms of administration are theoretically possible, as in the case of single external counters.[20,48] Washout curves obtained after intracoronary bolus injection of diffusible indicators recorded over different areas of the heart have been the most widely used. Only recently has continuous infusion of ^{81m}Kr into the aortic root been proposed and used in dogs, as discussed in the section on static imaging (pp. 189 to 194).

Underlying principle. The principle of the washout technique is that the mean transit time (actually the mean washout time) of the diffusible indicator in different areas of the myocardium can be calculated from the time-activity curves recorded in the individual areas in which the myocardial scintigram can be subdivided. If the distribution space of the indicator corresponds to and is confined to the myocardial tissue, multiplication of the reciprocal of the mean transit time by the blood-tissue partition coefficient of the indicator gives the average flow per unit volume of tissue in the solid angle outlined by each area.

Methodologic problems. The problems of dynamic recording are essentially those of single-counter precordial detection, which have been discussed in another article.[20] However, we will describe the essential points that condition the interpretation of regional washout curves and the choice of the most suitable indicators, forms of administration, and analysis of the curves.

First, within the solid angle of any given area of the scintigram, perfusion may be markedly inhomogeneous; therefore a correct calculation of the mean washout time can be obtained only by stochastic analysis (area-peak ratio of the complete curve).[37,48]

Theoretic models indicate that following bolus injections, initial monoexponential extrapolation over the first 50% of the curve gives reasonable estimates of average flow until the difference between better perfused and ischemic myocardium varies by a factor of about 3 in the field of view for each area of interest. It markedly overestimates average flow when the flow differs by a factor greater than 5. Under these conditions the monoexponential extrapolation of the initial slope of the washout curve leads to a systematic underestimate of the true mean time of the washout curve for the following reasons: (1) it neglects the tail of the curve, which has a less steep slope, and (2) it overweights the behavior of the better perfused zones, which receive the larger fraction of the indicator, because after bolus injection, the initial indicator distribution is flow dependent. In the presence of inhomogeneous perfusion, a further critical point is the geometric efficiency of the counting system, which overweights the tracer behavior in the tissue closer to the counter head; thus an ischemic area can be more easily detected when located in the heart wall closer to the counter head.

Second, the indicator should be removed from the myocardium into which it diffuses only by blood and not by diffusion into surrounding tissues; it should not recirculate, or recirculation should be subtracted by appropriate procedures.[16,21]

Finally, it is essential that the geometric relationships between the heart and the scinticamera remain the same during the whole course of the measurement. If a second indicator fixed in the myocardium, such as ^{99m}Tc microspheres, is used, a continuous check can be made to ensure that this condition is met.

Available diffusible indicator. The indicator with suitable emission for washout scinticamera studies is ^{133}Xe. Other short-lived cyclotron-produced radionuclides can be used only on the spot where they are produced and will not be dealt with here. Since ^{133}Xe is eliminated through the lung, it recirculates to a small extent. However, this relatively small amount of recirculation appears to influence significantly the washout curve and to lead to a systematic underesti-

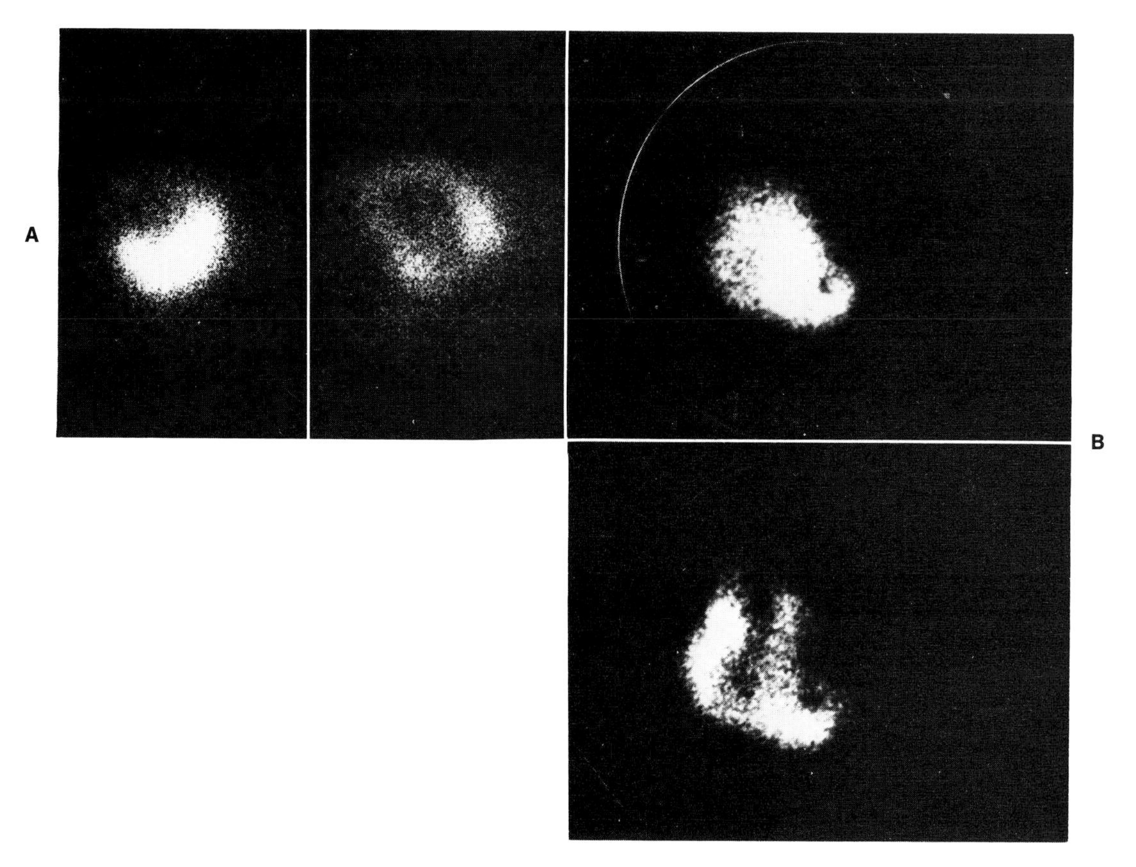

Fig. 10-3. A, Scintigrams obtained in left lateral position in patient with large postinfarction anterior aneurysm. *Left:* Scintigram obtained immediately after injection of 3 mCi of ^{133}Xe into the left coronary artery (LCA). Note large superoanterior filling defect. *Right:* Image obtained 25 min after injection of ^{133}Xe. Distribution of ^{133}Xe 25 min after intracoronary injection is markedly different from initial one (being distributed at base and along interventricular groove). **B,** Scintigrams obtained in LAO position in patient with collateral vessels to LAD branch from right coronary artery. *Top:* Scintigram obtained immediately after injection of 3 mCi of ^{133}Xe into right coronary artery. *Bottom:* Image obtained 25 min later.

mate of the actual washout rate from the myocardium, which becomes considerable at high flow rates.[37] This is most likely to represent the cause of the large underestimate of flow observed with external washout of ^{133}Xe in the presence of recirculation at high flow rates relative to the reference method.[12]

For a given amount of recirculation the initial slope of the myocardial washout curve will be more affected when the initial fractional distribution of indication to this area was smaller than that to other areas.

Furthermore, because of its high lipid solubility and the very large surface of contact between myocardium and epicardial fat, a sort of tonometer effect occurs, that is, ^{133}Xe rapidly diffuses from epicardial muscle into fat during the initial phase of the washout and is practically held up there. This phenomenon has been demonstrated in dogs[42] and is quite evident in humans through serial pictures taken with a scinticamera after intracoronary ^{133}Xe injection (Fig. 10-3).[20] Simultaneous myocardial washout curves of ^{133}Xe and [^{125}I] iodoantipyrine obtained in humans with a single precordial counter and subtraction of recirculation indicate that ^{133}Xe diffusion into fat alters the curves from the very beginning (Fig. 10-4)[21,37]; therefore regional differences in ^{133}Xe washout curves in different zones of the scintigram may be related also to varying degrees of epicardial fat. Prelimi-

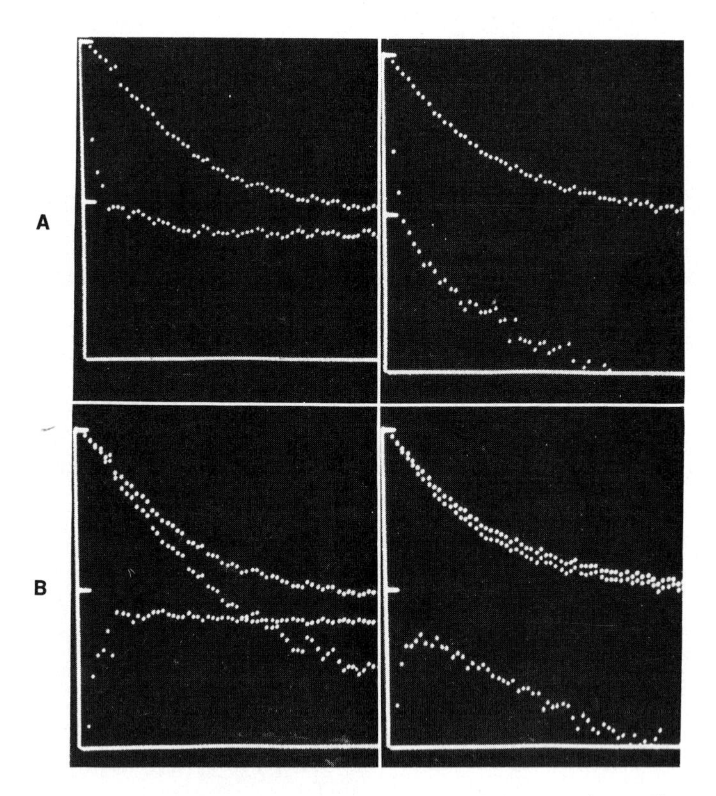

Fig. 10-4. Typical semilogarithmic plot (two log cycles) of precordial curves obtained by single precordial counter in patient with mixture of [^{125}I] iodoantipyrine and ^{133}Xe injected as bolus. **A,** Experimental [^{125}I] iodoantipyrine *(left)* and ^{133}Xe *(right)* curves obtained after left coronary (*higher curves*) and right atrial (*lower curves*) injections. **B,** Lower curves are computed recirculation curves, upper curves are experimental myocardial washout curves (reported from top graphs), and intermediate curves are obtained by subtraction of recirculation. $\Delta t = 5$ sec. Although myocardial-blood partition coefficient is smaller for ^{133}Xe (0.72 versus 1), Xe washout is similar to or slower than that of [^{125}I] iodoantipyrine from the beginning of washout.

nary studies with ^{133}Xe and [^{125}I] iodoantipyrine suggest that this may be the case in certain patients. Indeed, ^{133}Xe curves do not decrease below 5% to 10% even 5 min after the injection; thus it is impossible to calculate the mean time by the stochastic approach (peak-area ratio). In an attempt to circumvent this problem, the mean washout time of the curve has been estimated from the monoexponential that fits the initial part of the curves.[6,31] This type of analysis, however, causes serious errors when there is a large inhomogeneous perfusion in the solid angle outlined by each area[15]; the errors lead to overestimation of the average flow per unit volume of myocardium, which because of counting geometry also varies with the relative position of the ischemic tissue with respect to the counter. The effect of the progressive ^{133}Xe diffusion into the epicardial fat tends to compensate for this error. These considerations clearly indicate that it is by no means warranted to express the result obtained from initial ^{133}Xe washout slopes as average myocardial blood flow per gram or per 100 g of tissue in the solid angle of the area.

The use of iodoantipyrine or other indicators that are not fat soluble could avoid these difficulties if adequate correction were made for recirculation.[16,21] Unfortunately, the emission and half-life of ^{131}I and ^{125}I, which are commercially available, are not suitable for scinticamera studies.

Possibilities of regional washout curves. Even with several limitations, the possibilities of regional washout curves are remarkable as long as one is aware of the potential pitfalls. In fact, they can provide valuable semiquantitative evaluations both in single measurements and in studies of regional distribution changes.

In single measurements, scinticamera pictures taken 5 min after intracoronary injection of ^{133}Xe indicate the areas in which significant diffusion holdup in fat occurs, which are usually the base, apex, and interventricular groove. Thus regional reductions or myocardial perfusion can be identified with certainty when the initial washout is slower in areas corresponding to zones with little final ^{133}Xe accumulation, but no definite conclusions can be drawn when the washout is slower in an area with final accumulation of ^{133}Xe unless the difference is greater than about 30% or unless zones with apparently equal final ^{133}Xe accumulation have obviously different slopes. Accordingly, the absence of appreciable differences in the initial regional washout slopes does not exclude the existence of inhomogeneous perfusion even of a severe degree, since initial slopes reflect predominantly the washout from the well-perfused muscle to which the indicator is mainly distributed.

Thus it can be concluded that, in single measurements, regional ^{133}Xe washout curves can be used for detecting and quantifying ischemia when the reduction of perfusion is relatively uniform in the myocardium included in the solid angle of a given zone (1) if this relative reduction of perfusion occurs in areas in which there is no significant tracer diffusion holdup in epicardial fat and (2) if recirculation is shown to be negligible or is corrected for.

By contrast, for the areas with considerable final ^{133}Xe content at 15 min, even after correction of recirculation, one is not entitled to derive the theoretic flow values. In fact, the back extrapolation approach and "peeling off" the final exponential, proposed by Holman,[13] also do not appear warranted because in the initial part of the curve the back extrapolation would subtract a much larger activity than that present in fat (which takes time to accumulate by diffusion from the muscle). Under these conditions a qualitative and semiquantative assessment of differences in regional flow, in which perfusion may be expected to be inhomogeneous within each area of interest, can be approached by selecting areas with similar ^{133}Xe accumulation at 15 min and comparing the residual ^{133}Xe activity at the time 90% of the injectate has been washed out.[27,28] The evaluation of the slope (and also of the residual activity) at a fixed percentage of the peak appears by far preferable to the evaluation at a fixed time after the peak of the curves, as performed by Cannon.[6,7] In fact, computation of the initial slope over 30 sec after the peak implies that different portions of the curve will be included in the calculation, depending on the values of flow (at 50 ml/min/100 g, the curve will be included only down to about 70% of the peak; at 200 ml/min/100 g it will be included down to about 25% of the peak).

In studies of changes in the regional distribution of myocardial perfusion, either of the following approaches can be followed: (1) a second injection of ^{133}Xe can be performed after the change or (2) the change can be induced during the course of the washout, when it can be assumed that the changes in the flow distribution take place within a short time. The first approach has the disadvantage of the bolus injection, which delivers the indicator to the various zones of the myocardium in proportion to their blood flow; thus if the change markedly reduces the flow to one fraction of the muscle and increases it to another within a given solid angle, the initial slope may become faster even if mean flow actually decreases. On the other hand, if the main-stem coronary artery is sufficiently long to ensure adequate indicator mixing, this approach may provide information about the changes in the initial indicator distribution, which is obtained immediately after the injection; the changes in the distribution may provide information complementary to that given by the washout curves. In both cases, only a continuous check of the constancy of the counting geometry conditions allows the assumption that differences in initial distribution, residual activity, and slopes reflect actual changes in tracer distribution or washout.

Dynamic recording of myocardial time-activity curves after intracoronary administration appears to be particularly suited for pathophysiologic studies that demand the evaluation of transients of regional myocardial flow. This method will acquire greater potential when diffusable indicators with appropriate physical characteristics, such as [^{123}I] iodoantipyrine, become available and different forms of tracer administration other than bolus injection, such as continuous infusion of ^{81m}Kr, become practicable.

SURVEY OF RESULTS

The practical significance of the methods actually in use for the evaluation of regional myocardial blood flow can be gathered from the consideration of the results that they have so far produced.

Static imaging

Myocardial scintigraphic studies using a gamma camera computer system following selective intracoronary injection of xenon during pacing angina have demonstrated that perfusion is considerably lower in ischemic areas than in the surrounding healthy myocardium (Fig. 10-5).[27,28,32,33] The use of a reference indicator fixed in the myocardium (^{99m}Tc-labeled human albumin microspheres) allowed quantitation of these regional differences (Fig. 10-6).

Successive left intracoronary injection of microspheres labeled with ^{99m}Tc and ^{131}I has been used to study the impairment of coronary vasodilatation in poststenotic myocardium[11,40] and the alterations of regional myocardial perfusion during an attack of "variant" angina.

Potassium analogs have been used to study the alterations of regional myocardial perfusion during exercise tests[44,47] and angina at rest.[23,24] The physical characteristics of ^{201}Tl made it a more popular nuclide than ^{43}K. Experience at our clinic in the study of angina at rest using ^{201}Tl has been extremely positive, in particular for the so-called "variant" type, which is associated with a sizable transmural deficit of myocardial perfusion, which results in an obvious cold area in myocardial scintigrams taken early after the injection (Figs. 10-2, 10-7, and 10-8).[23,24] This cold area tends to disappear very early when ischemia persists only 1 or 2 min after ^{201}Tl injection; it remains barely visible in scintigrams taken 20 to 30 min after the injection and is no longer evident 2 to 4 hr later (Fig. 10-2).

As to the diagnostic use of ^{201}Tl in exer-

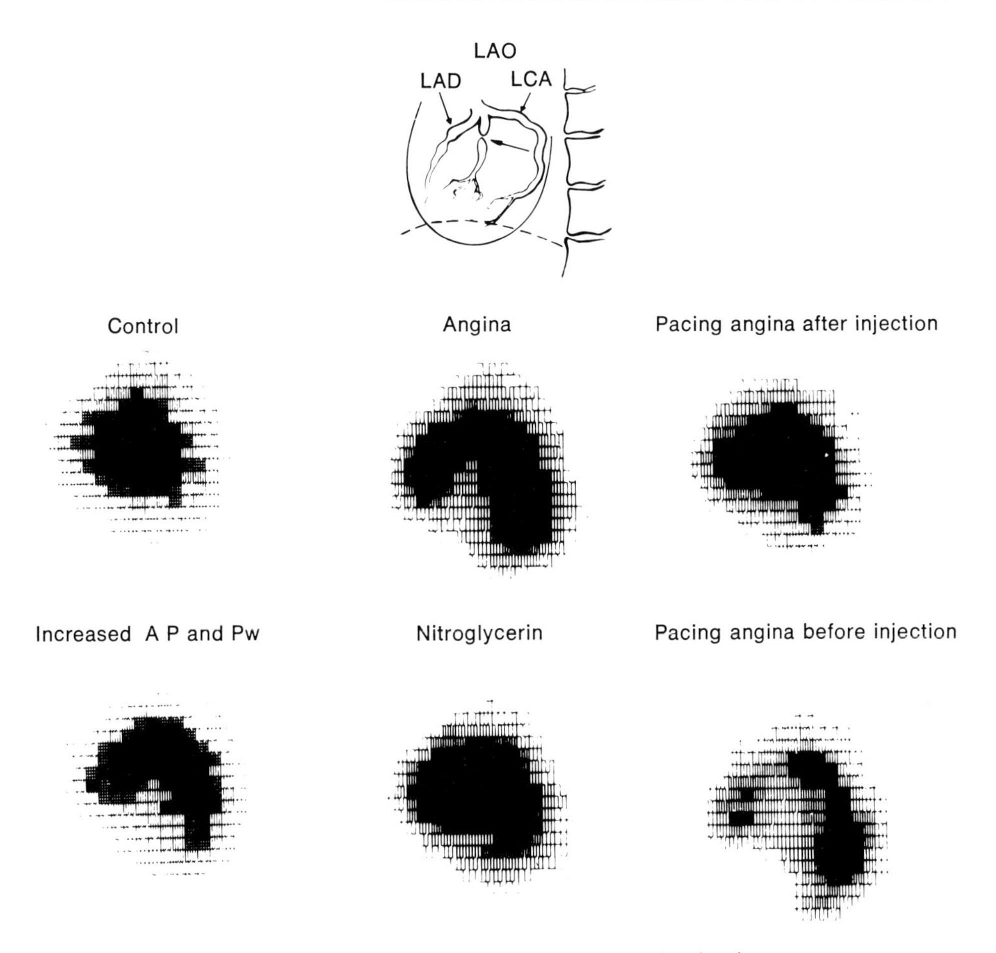

Fig. 10-5. Myocardial scintigrams in LAO position, obtained immediately after (10 sec) intracoronary injections of ^{133}Xe in different hemodynamic conditions in patient with complete occlusion of left diagonal branch and 60% stenosis of LAD branch. Before appearance of angina *(bottom left)*, during spontaneous angina *(top center)*, and during pacing-induced angina *(bottom right)*, an obvious deficit of perfusion is apparent and disappears after administration of nitroglycerin *(bottom center* and *top right)*.

cise stress testing, some studies report a good sensitivity and a correlation with electrocardiographic signs of myocardial ischemia and/or with the presence of atherosclerotic obstructions.[1] However, our experience indicates that visual assessment of the deficits of perfusion on ^{201}Tl exercise scintigrams usually is far more disputable than that on scintigrams in "variant" angina with transmural ischemia. Indeed, from a theoretic point of view, if ischemia is only distributed in a subendocardial area, its detection by conventional scintigraphy will be difficult. Thus the detection of smaller degrees of ischemia seems to require further progress both in the acquisition technique and in the objective image interpretation. The appropriate timing for the imaging technique has not always been used. A check of the constancy of the counting geometry during the time of scintigraphy, such as repeated scintigrams in the same projection at subsequent times, has not been performed. The analysis of the scintigrams often was not performed by objective means (with standardized automatic procedures). Furthermore, no attempt has been made to establish whether the cold areas of the myocardial scintigram result from an inadequate increase in perfusion or from an actual decrease that has relevant pathophysiologic implications.

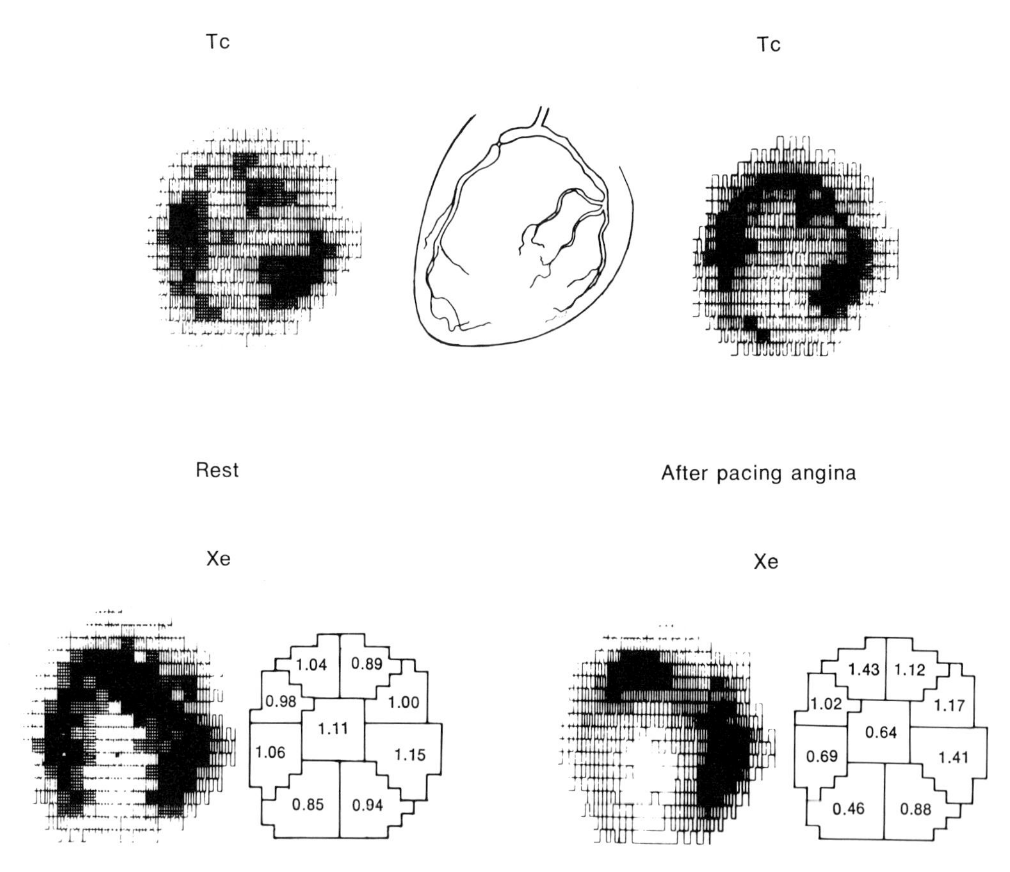

Fig. 10-6. ^{99m}Tc and ^{133}Xe initial distribution at rest and during pacing-induced angina in patient with proximal stenosis on LAD coronary studied in LAO projection. Severe reduction in the ^{133}Xe regional distribution to anterior, apical, and central areas relative to basal and circumflex areas is apparent (*bottom right* scintigram). Initial regional distribution is quantified from fractional average activity per element of matrix in each area, which, by contrast, showed negligible differences in ^{99m}Tc scintigrams.

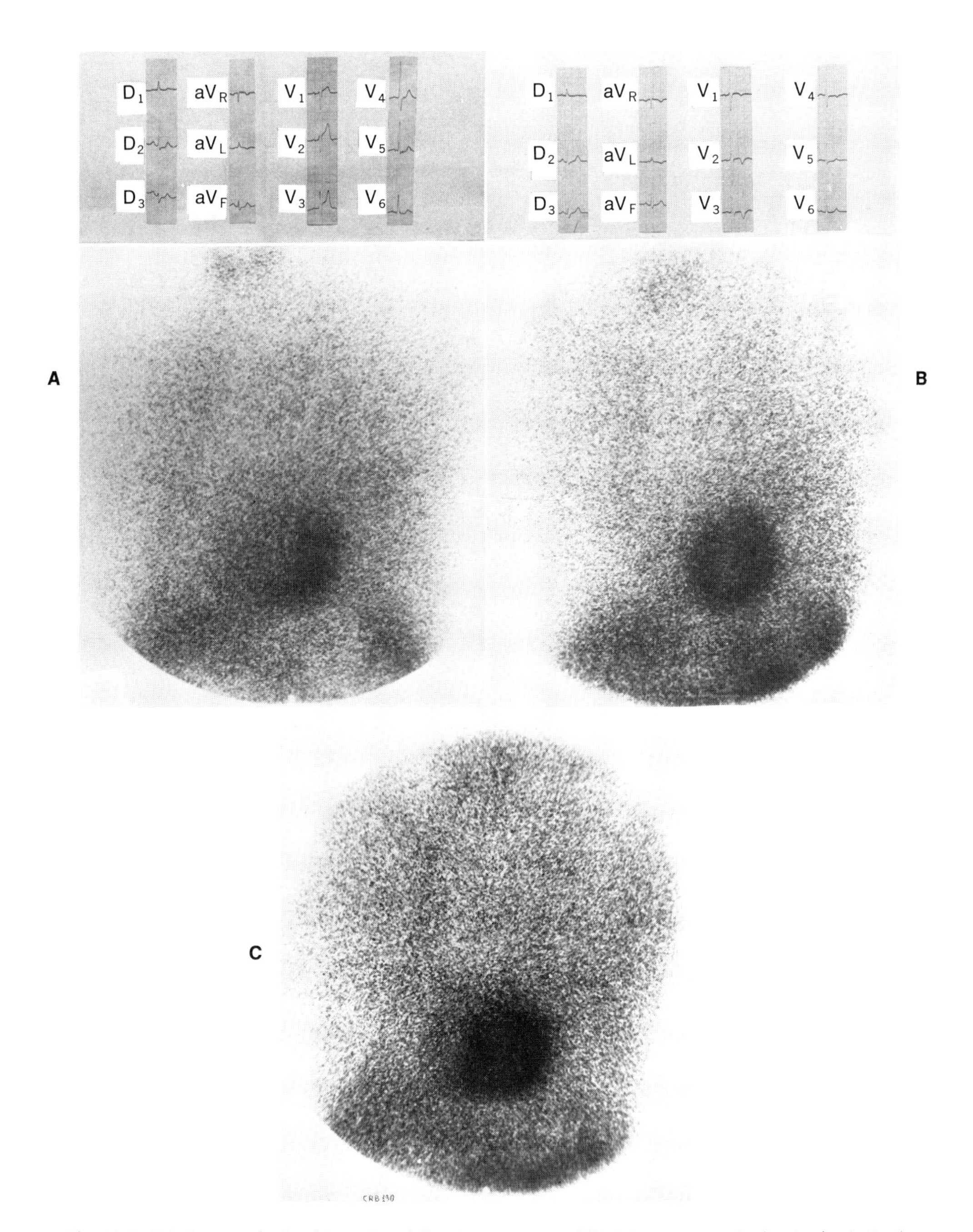

Fig. 10-7. Scintigrams obtained in patient following injection of ^{201}Tl during an anginal episode. **A,** 7 min after injection; **B,** 2 hr after injection; **C,** a week later. Massive transmural reduction of tracer uptake in anterior wall and septum is apparent in **A.**

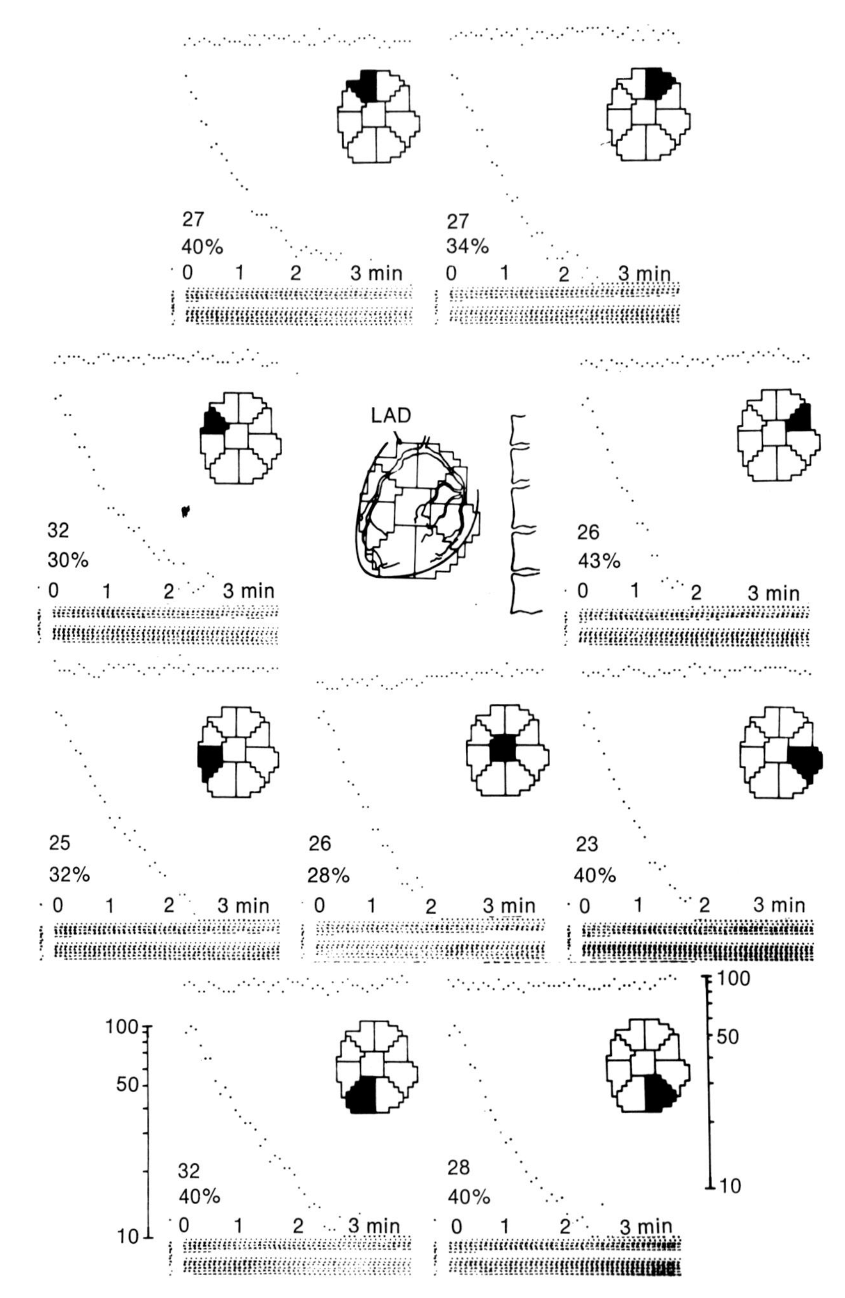

Fig. 10-8. Simultaneous time-activity curves of ^{99m}Tc and ^{133}Xe in semilogarithmic scale for each of nine areas of interest at rest in same patient as Fig. 10-6. Each dot represents activity in 5 sec interval. ^{99m}Tc activity remains constant over whole time of ^{133}Xe washout. For each panel, $t_{0.5}$ and percentage of peak at which ^{133}Xe washout deviates from monoexponential course are indicated. Washout is slower at base, where greater final ^{133}Xe accumulation was observed.

Dynamic regional recording at rest

Dynamic regional recording at rest of myocardial ^{133}Xe washout curves obtained after intracoronary injection (Fig. 10-8) has demonstrated a significantly larger variability of the regional initial slopes in patients with coronary artery disease than in patients without obstructions.[7] This observation is certainly suggestive of topographic differences of perfusion in this condition. With an appropriate projection it was possible to separate clearly territories perfused by normal vessels from others distal to severe obstructions or occlusions; in this series of patients, the regional differences of initial slope were not correlated with the severity of the arterial lesions as judged by arteriography. Only distally to occluded vessels were lower perfusion values frequently found.[8,13] We noted a correlation between slower initial ^{133}Xe washouts and hypokinesia in the same region at ventriculography[36] (Figs. 10-9 and 10-10). It is conceivable that at rest in the absence of symptoms of ischemia, local metabolic demands are essentially satisfied by variations of perfusion rather than by a varying extraction of oxygen and metabolites; thus low regional perfusion and hypokinesia are locally matched by regulatory mechanisms. In fact, the resting level of flow in hypokinetic areas can increase during induced stimulations, although to a lesser extent than in healthy zones.

Text continued on p. 208.

A

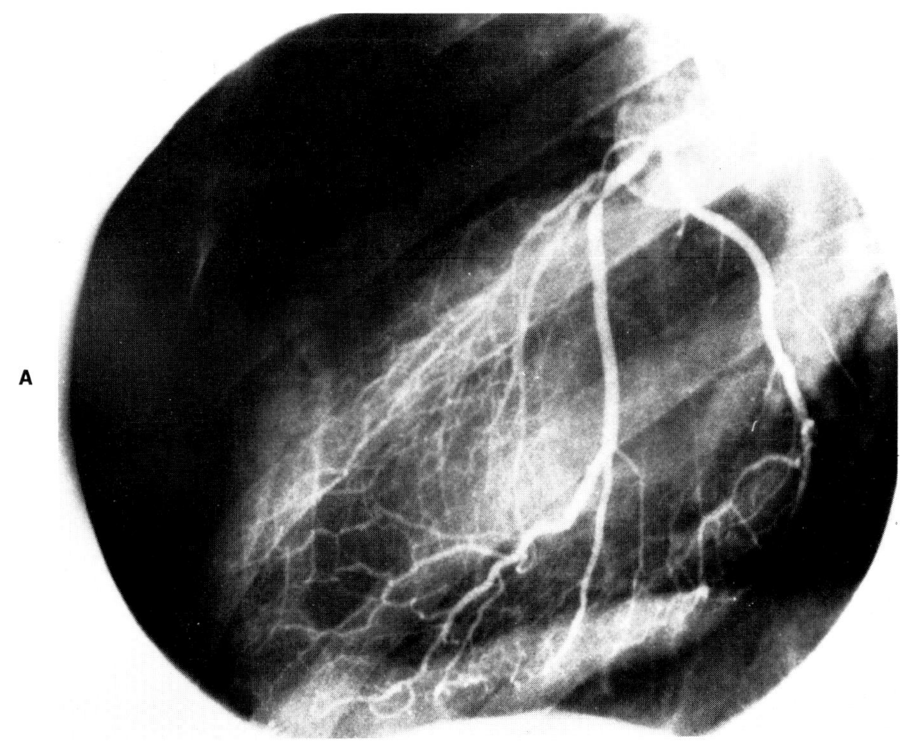

Continued.

Fig. 10-9. A, Coronary angiogram in left lateral (L Lat) projection in patient with atypical chest pain and doubtful ECG signs of ischemia. LAD branch is occluded, diagonal branch is severely stenotic, and circumflex branch is normal. Ventriculogram was normal. **B,** Computer plot of initial 90% of washout curves of ^{133}Xe in semilogarithmic scale with patient at rest for various areas (in black) in which scintigram was subdivided. Each point is 5 sec interval. The half-times of curves and percent value up to which an exponential was fitted by X^2 method are also reported. Initial 90% of washout in areas distal to obstructed vessels is slightly faster than that distal to normal circumflex artery. ^{133}Xe scintigram at 10 min showed basal ^{133}Xe accumulation.

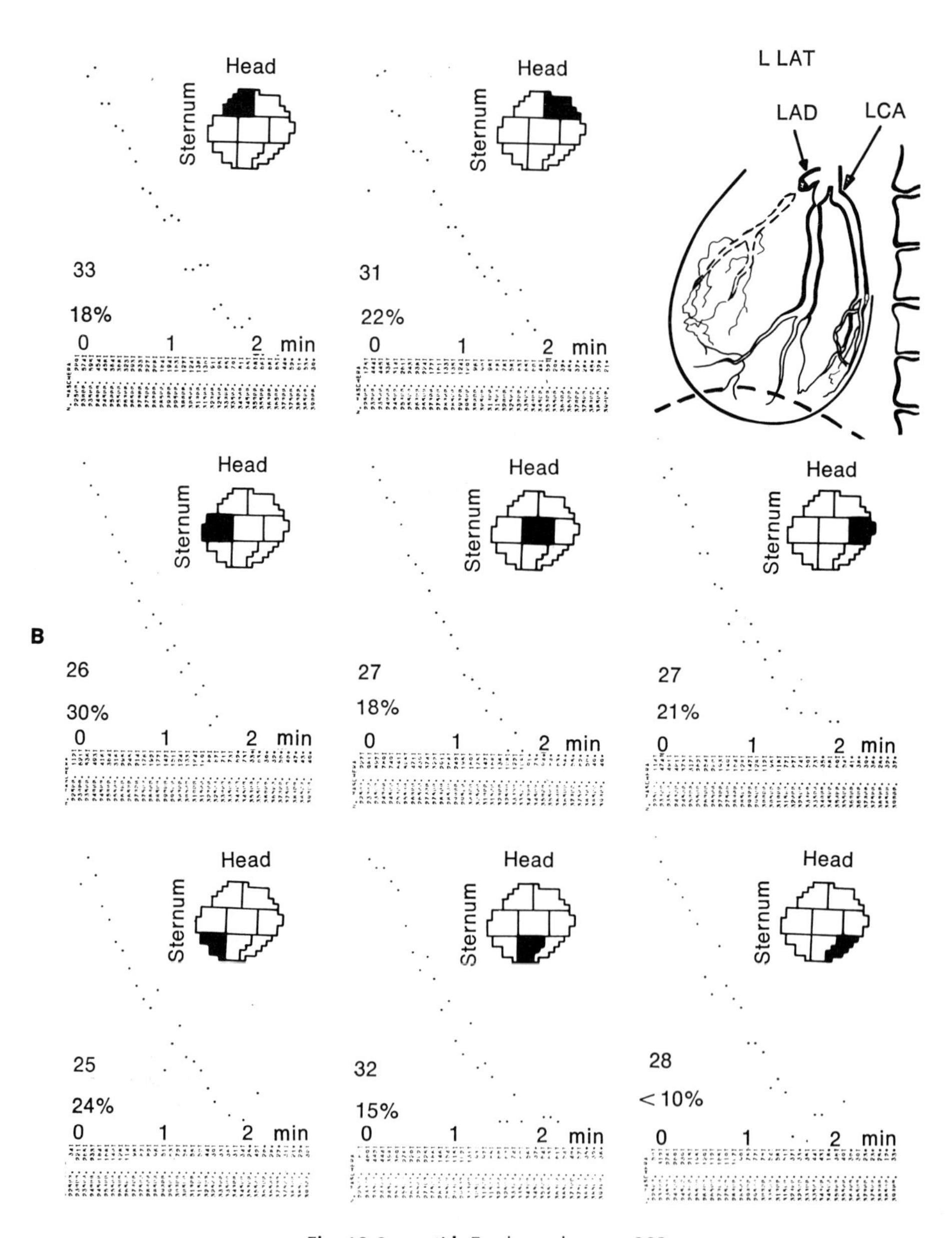

Fig. 10-9, cont'd. For legend see p. 203.

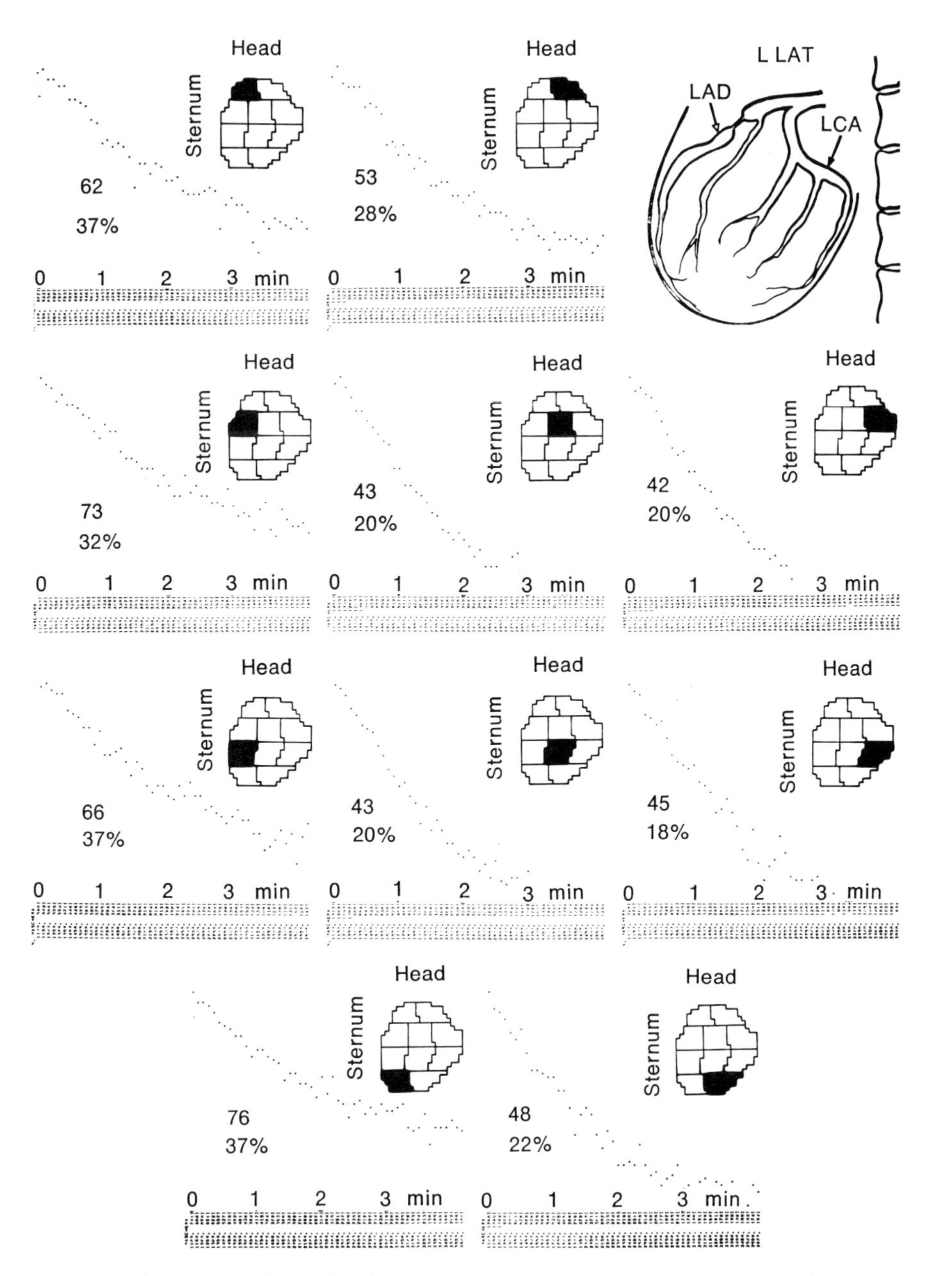

Fig. 10-10. Washout curves obtained with patient at rest in L Lat projection in patient with old anterior infarction, subocclusion of LAD branch, and large anterior wall akinetic area on ventriculogram. Washout is significantly slower in all areas corresponding to anterior wall. ^{133}Xe scintigram at 10 min showed prevalent ^{133}Xe accumulation at base of heart. (Computer plot of curves is described in Fig. 10-9, *B*.)

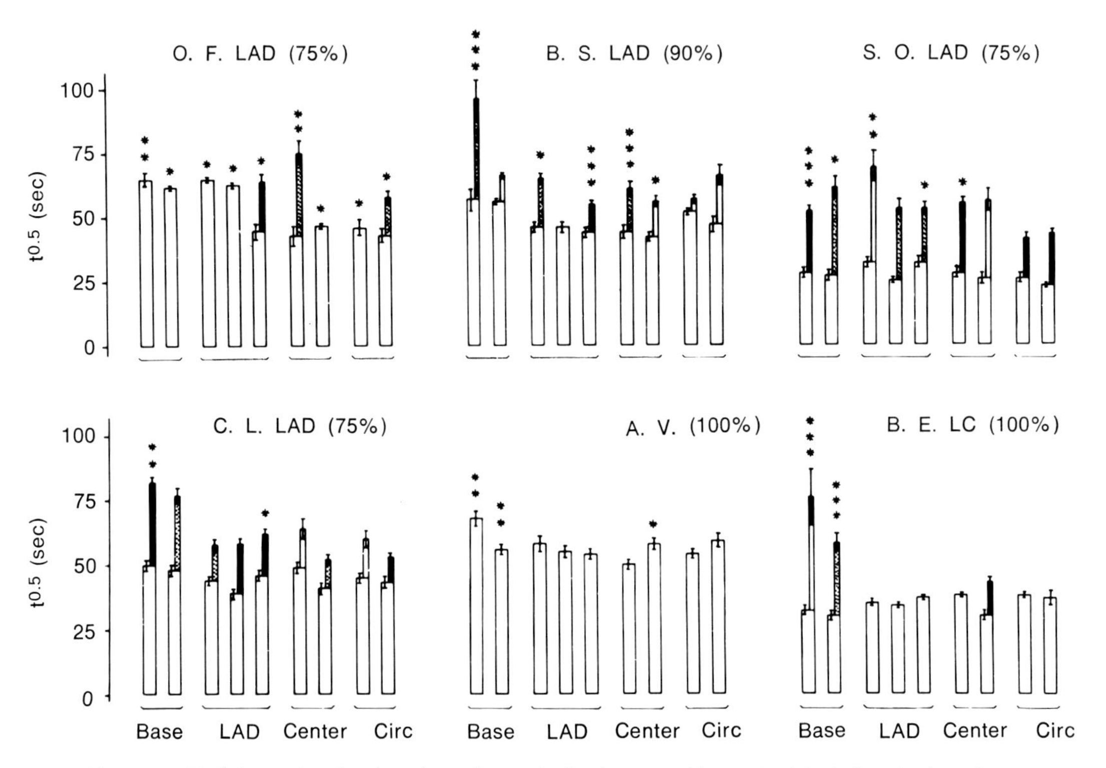

Fig. 10-11. Half-times of regional washout slopes obtained at rest with standard deviations in six patients with critical single-vessel stenosis in whom reference ^{99m}Tc scintigram in LAO projection allowed continuous check on counting geometry. Asterisks indicate extent of ^{133}Xe accumulation in 15 min scintigram. Narrow bars above indicate half-time of second exponential when present: black when appearing between 50% and 40% of peak, hatched when appearing between 40% and 30% of peak, and white between 30% and 20% of peak. No consistent differences in relation to site of stenosis are noticeable.

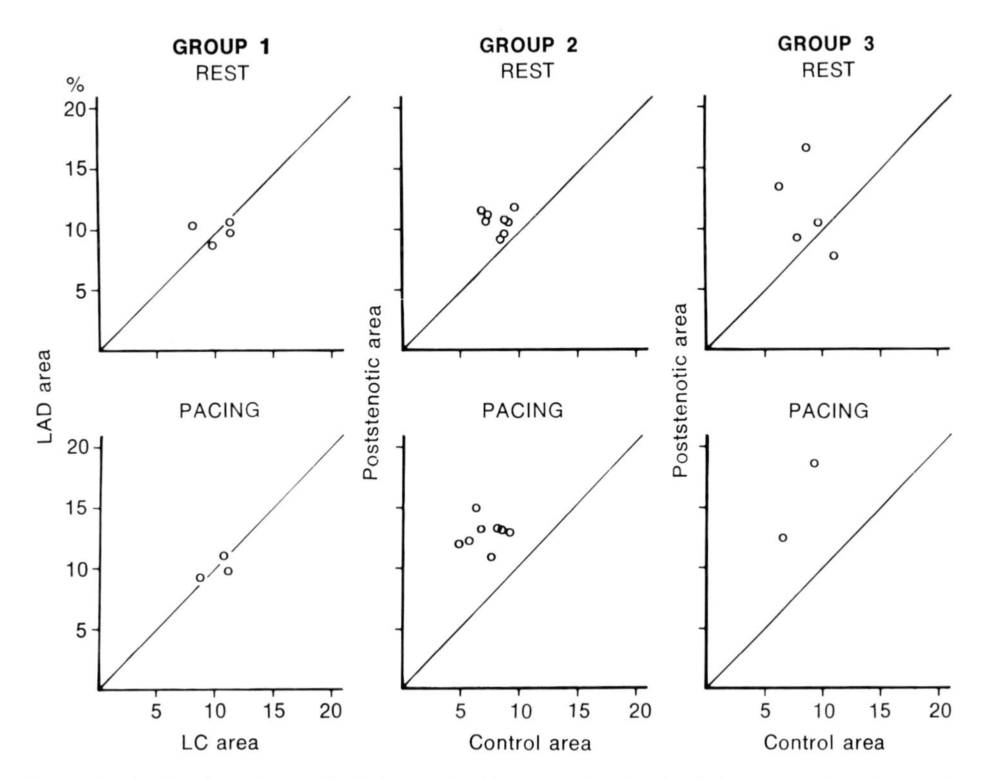

Fig. 10-12. Residual activity at 90% total washout in areas with similar ^{133}Xe accumulation in 15 min scintigram. Activity is consistently larger in poststenotic area at rest. Difference approximately doubles during pacing-induced angina. ^{133}Xe was administered after pacing was initiated.

Distally to a critically obstructed vessel in the absence of hypokinesis we did not observe any consistent difference in initial ^{133}Xe washout slope (Fig. 10-11), but a significantly higher residual activity relative to control areas (for a comparable amount of tracer accumulation in the 15 min scintigram) when 90% of the injectate was washed out (Fig. 10-12). This finding suggests the presence of small areas of reduced perfusion distally to critical obstructions.

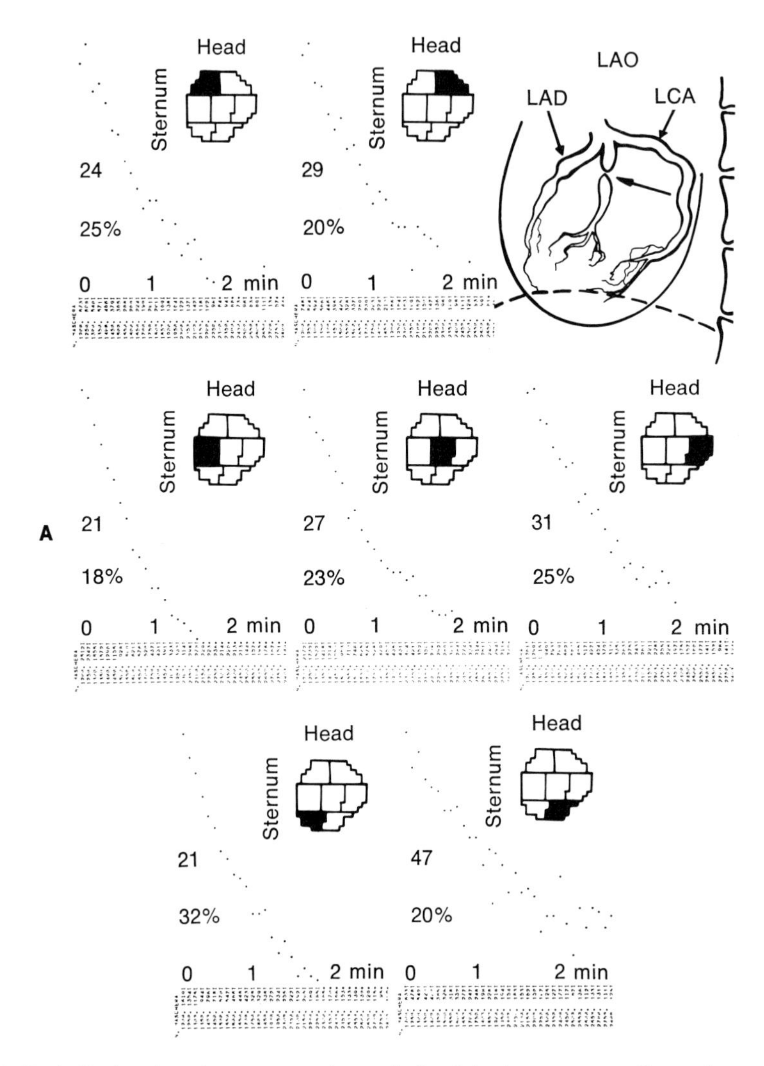

Fig. 10-13. A, Regional washout curves observed after injection corresponding to bottom right initial scintigram of Fig. 10-5. The apical area *(bottom right)* shows slow washout, but washouts corresponding to LAD territory appear to be faster than those corresponding to circumflex territory. **B,** Changes induced by angina on regional washout rates of indicator corresponding to top right scintigram of Fig. 10-5 (same time scale and type of representation as in preceding figures). First arrow indicates beginning of pacing, which was followed within 10 sec by marked ECG ischemic changes in V_3 to V_5, and second arrow represents appearance of pain. Although washout continued regularly at base and in left circumflex territory, it was markedly slowed at apex and in center after beginning of pacing; this indicated that perfusion to this area at apex became markedly impaired. This area indeed corresponds to that showing deficit of indicator when injection was made during angina, that is, when perfusion was already impaired before injection of indicator (Fig. 10-5). Therefore washout curves of **A** reflect washout from better perfused myocardium, and those of **B** are composite of washout from areas with different perfusion.

Dynamic regional recording during stress

Dynamic regional recording during stress has provided valuable pathophysiologic information about the behavior of myocardial perfusion in ischemic areas during acute coronary insufficiency. Studies were performed in patients with isolated obstructions in the left anterior descending or circumflex coronary artery at rest and during pacing angina. The induction of angina immediately after injection of ^{133}Xe into the left coronary artery was accompanied by an increase of the slope of the washout curves in the myocardium perfused by normal vessels and by a reduction of the washout rate in the territory beyond the obstructed vessel[13] (Figs. 10-13 and 10-14). The degree of reduction of the washout rate was greater in the patients with more severe pain and ECG changes.[19,28,30,32] This information is complementary to that obtained by static imaging during angina because it indicates that the relative changes of the initial distribution are related not to the inability of flow to increase in the ischemic areas but mainly to decreased perfusion in these areas, which is probably caused by functional factors triggered by the imbalance between de-

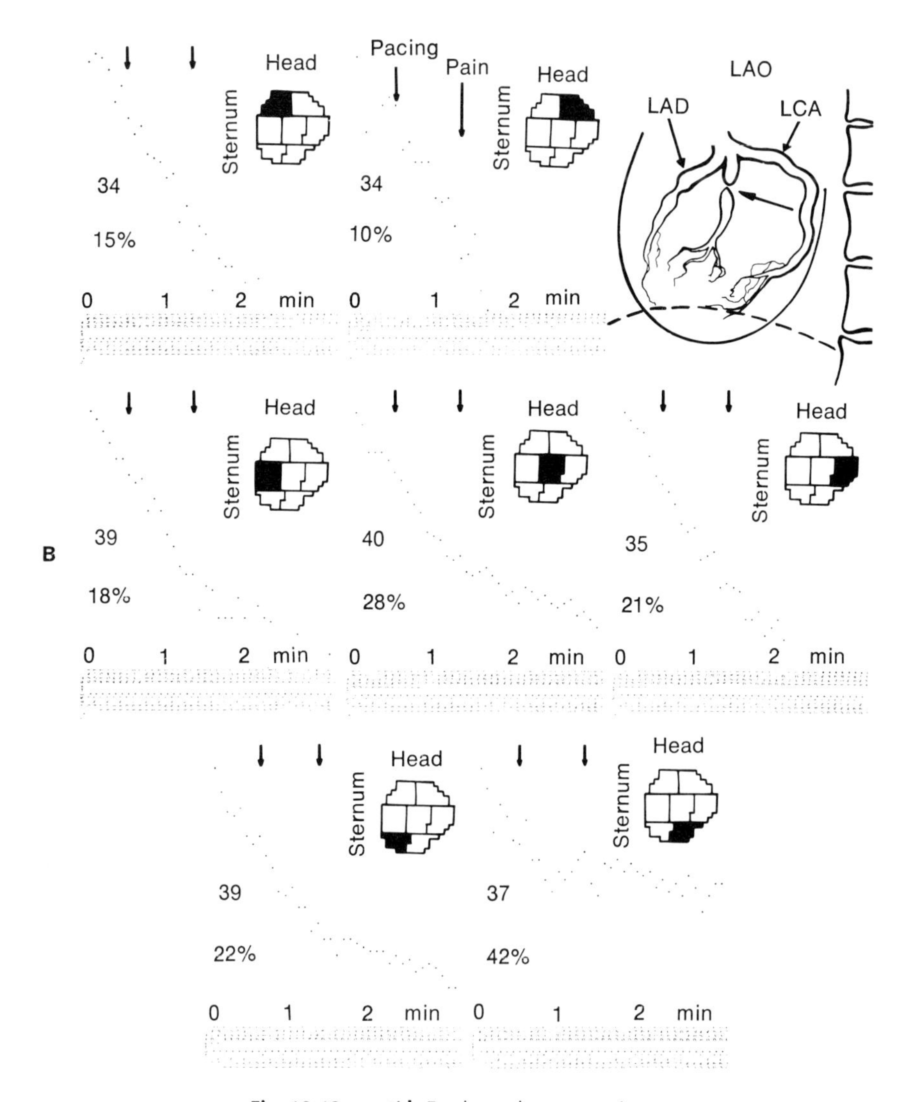

Fig. 10-13, cont'd. For legend see opposite page.

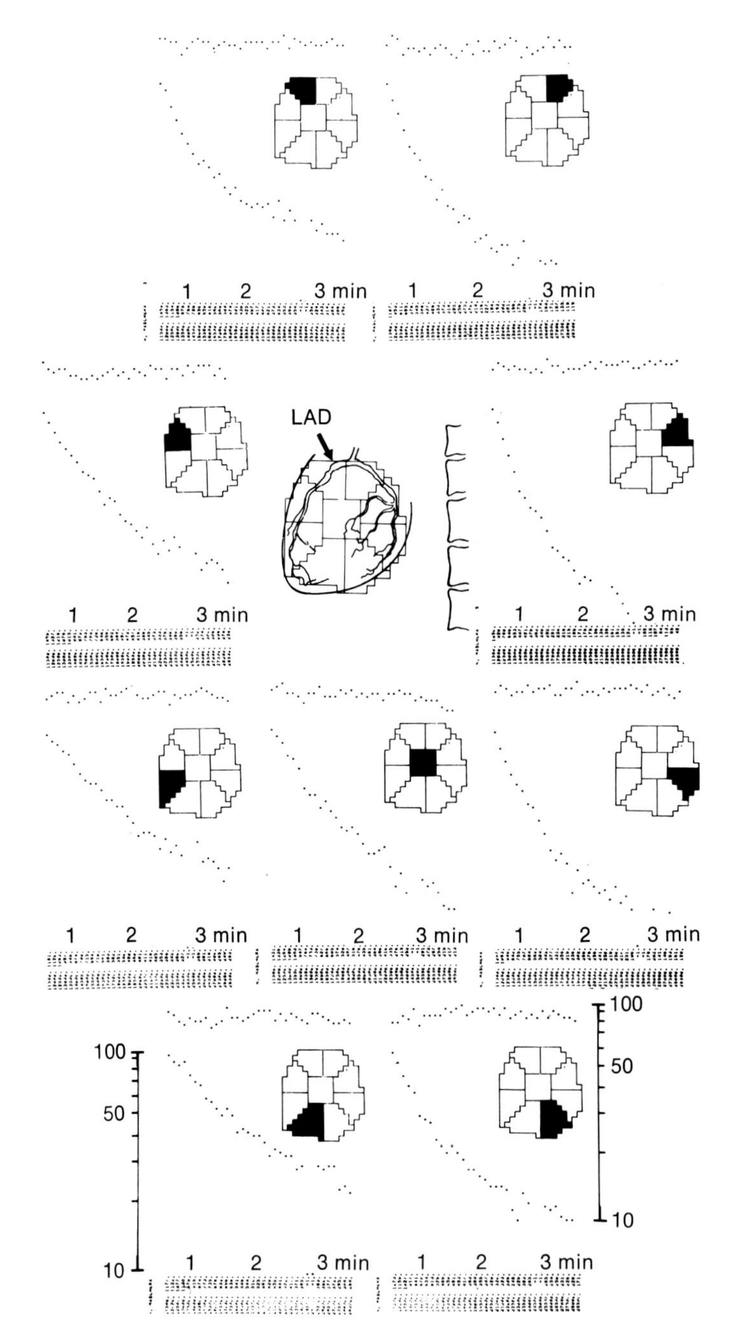

Fig. 10-14. Alterations of regional washout slopes induced by pacing above anginal threshold immediately after injection of ^{133}Xe in patient with critical stenosis of LAD coronary artery as unique lesion. Counting geometry remains constant. Rate of washout becomes much slower than during rest in areas distal to obstructed vessel.

mand and supply. The degree of inhomogeneity of flow between well-perfused and ischemic myocardium in the visual field of each area of interest must be much greater than 3:1. In fact, the extent of flow reduction is considerably underestimated from initial monoexponential extrapolation of xenon following injection during ischemia. Thus these differences in perfusion are most likely to occur in opposite cardiac walls, rather than transmurally.

CONCLUSIONS

The methods so far available for the study of regional myocardial blood flow in humans can provide only qualitative or semiquantitative information on the distribution of perfusion to the myocardium because of inadequacy of the available indicators and imaging devices. However, the results so far produced by the application of these techniques to the study of ischemic heart disease represent a breakthrough both in the understanding of the disease and in the diagnostic evaluation of patients and can provide more diagnostic and pathophysiologic information than an exact measurement of total or average myocardial blood flow. Therefore further development of instruments and tracers for a wider application of this approach appear at the present time to be a fruitful investment.

REFERENCES

1. Baily, I. K., Griffith, L. S. C., Strauss, H. W., and Pitt, B.: Detection of diffuse coronary disease and myocardial ischemia by electrocardiography and myocardial perfusion scanning with ^{201}Tl, Am. J. Cardiol. **37**:118, 1976.
2. Becker, L., Ferreira, R., and Thomas, M.: Comparison of ^{86}Rb and microsphere estimates of left ventricular blood flow distribution, J. Nucl. Med. **15**:969, 1974.
3. Biagini, A., L'Abbate, A., Michelassi, C., and Maseri, A.: Study of myocardial kinetics of ^{201}Tl in man at rest. Comparison with ^{125}I RIHSA, THO and ^{42}K, J. Nucl. Med. All. Sci. **21**:59, 1977.
4. Botti, R. E., MacIntyre, W. J., and Pritchard, W. A.: Identification of ischaemic area of left ventricle by visualization of ^{43}K myocardial deposition, Circulation **47**:486, 1973.
5. Burnham, C. A., and Brownell, G. L.: Quantitative imaging with the MGH positron camera. In Proceedings of Conference on Quantitative Organ Visualization in Nuclear Medicine, University of Miami School of Medicine, Division of Nuclear Medicine, May 1970.
6. Cannon, P. J., Dell, R. B., and Dwyer, E. M., Jr.: Measurement of regional myocardial perfusion in man with ^{133}Xenon and a scintillation camera, J. Clin. Invest. **51**:964, 1972.
7. Cannon, P. J., Dell, R. B., and Dwyer, E. M., Jr.: Regional myocardial perfusion rates in patient with coronary artery disease, J. Clin. Invest. **51**:978, 1972.
8. Cannon, P. J., Schmidt, D. H., Weiss, M. B., Fowler, D. L., Sciacca, R. R., Ellis, K., and Casarella, W. J.: The relationship between regional myocardial perfusion at rest and arteriographic lesions in patients with coronary atherosclerosis, J. Clin. Invest. **56**:1442, 1975.
9. Donato, L., Bartolomei, G., and Giordani, R.: Evaluation of myocardial perfusion in man with radioactive potassium or rubidium and precordial counting, Circulation **29**:195, 1964.
10. Fazio, F., and Jones, T.: Assessment of regional ventilation by continuous inhalation of radioactive Krypton-81m, Br. Med. J. **3**:673, 1975.
11. Gould, G. M., Lipscomb, K., and Hamilton, G. W.: Physiologic basis for assessing critical coronary stenosis: instantaneous flow response and regional distribution during coronary hyperemia as measures of coronary flow reserve, Am. J. Cardiol. **33**:87, 1974.
12. Hirzel H. O., and Krayenbuehl H. P.: Validity of the ^{133}Xenon method for measuring coronary blood flow. Comparison with coronary sinus outflow determined by an electromagnetic Flowprobe, Plüegers Arch. **349**:159, 1974.
13. Holman B. L., Adams D. F., Jevitt, D., et al.: Measuring regional myocardial blood flow with ^{133}Xe and anger camera, Radiology **112**:99, 1974.
14. Kaplan, E., Mayron, L. W., Friedman, A. M., Gindler, J. E., Frazin, L., Moran, J. M., Loeb, H., and Gunnar, R. M.: Definition of myocardial perfusion by continuous infusion of krypton-81m, Am. J. Cardiol. **37**:878, 1976.
15. Klocke, F. J., and Whittenberg, S. M.: Heterogeneity of coronary flow in human coronary artery disease and experimental myocardial infarction, Am. J. Cardiol. **24**:782, 1969.

15a. L'Abbate, A., Biagini, A., Michelassi, C., and Maseri, A.: Myocardial kinetics of thallium and potassium in man, Circulation (in press).

16. Larson, K. B., and Snyder, D. L.: Measurement of relative blood flow, transit-time distributions and transport-model parameters by residue detection when radiotracer recirculates, J. Theor. Biol. **37**:503, 1972.
17. Love, W. D., Ishihara, Y., Lyon, L. D., and Smith, R. O.: Differences in the relationship between coronary blood flow and myocardium clearance of isotopes of potassium, rubidium, and cesium, Am. Heart J. **76**:353, 1968.

18. MacIntyre, W. J., Ishii, Y., Pritchard, W. H., and Eckstein, W.: Measurement of total myocardial blood flow with ^{43}K and the scintillation camera. In Maseri, A., editor: Myocardial blood flow in man. Methods and significance in coronary disease, Proceedings of international symposium, Pisa, Italy, June 1971, Turin, Italy, 1972, Minerva Medica.
19. Maseri, A.: Methods of studying regional myocardial flow and alterations during angina. In Oliver, M. F., Julian, D. G., and Donald, K. W., editors: Effect of acute ischaemia on myocardial function, Edinburgh, 1972, Churchill Livingstone.
20. Maseri, A.: Myocardial flow by precordial residue detection following intracoronary slug injection of radioactive diffusible indicators. In Maseri, A., editor: Myocardial blood flow in man. Methods and significance in coronary disease, Proceedings of international symposium, Pisa, Italy, June 1971, Turin, Italy, 1972, Minerva Medica.
21. Maseri, A.: Correction of recirculation in regional blood flow studies by residue detection, J. Appl. Physiol. **36**:375, 1974.
22. Maseri, A.: Myocardial blood flow and measurements in acute ischaemia. In Oliver, M. F., editor: Modern trends in cardiology, ed. 3, London, 1976, Butterworth & Co. Ltd., p. 115.
23. Maseri, A.: Pathophysiologic studies of the pulmonary and coronary circulations in man, Am. J. Cardiol. **38**:751, 1976.
24. Maseri, A.: Radioactive tracer technique for evaluating coronary flow. In Yu, P. N., and Goodwin, J. F., editors: Progress in cardiology, Philadelphia, 1976, Lea & Febiger, p. 141.
25. Maseri, A., Biagini, A., Distante, A., L'Abbate, A., and Guzzardi, R.: Methodological study of ^{201}Tl for the evaluation of regional myocardial blood flow in man, J. Nucl. Biol. Med. **20**:105, 1976.
26. Maseri, A., Kushnir, E., Duce, T., Pesola, A., and Donato, L.: The effect of counting geometry on the significance of myocardial blood flow measurements by the injection method, J. Nucl. Biol. Med. **13**:140, 1969.
27. Maseri, A., L'Abbate, A., Michelassi, C., Pesola, A., Pisani, P., Marzilli, M., De Nes, M., and Mancini P.: Possibilities, limitations and techniques for the study of regional myocardial perfusion in man by xenon-133, Cardiovasc. Res. **11**:277, 1977.
28. Maseri, A., L'Abbate, A., Pesola, A., Michelassi, C., Marzilli, M., and De Nes, M.: Regional myocardial perfusion in patients with atherosclerotic coronary artery disease at rest and during angina pectoris induced by tachycardia, Circulation, **55**:423, 1977.
29. Maseri, A., Mancini, P., Contini, C., Pesola, A., and Donato, L.: Method for the estimate of total coronary flow by ^{99}Tc tagged albumin microspheres, J. Nucl. Biol. Med. **15**:58, 1971.
30. Maseri, A., Mancini, P., L'Abbate, A., Contini, C., and Pesola, A.: Alterations of regional myocardial perfusion during angina. In Maseri, A., editor: Myocardial blood flow in man. Methods and significance in coronary disease, Proceedings of international symposium, Pisa, Italy, June 1971, Turin, Italy, 1972, Minerva Medica.
31. Maseri, A., Mancini, P., L'Abbate, A., and Magini, G.: Method for regional dynamic study of myocardial blood flow in man, J. Nucl. Biol. Med. **15**:54, 1971.
32. Maseri, A., Mancini, P., L'Abbate, A., Pesola, A., and Contini, C.: Risultati preliminari sulla misura del flusso miocardico regionale durante angina pectoris, Boll. Soc. Ital. Cardiol. **16**:531, 1971.
33. Maseri, A., Mancini, P., Pesola, A., L'Abbate, A., Bedini, R., Pisani, C., Michelassi, C., Contini, C., Marzilli, M., and De Nes, D. M.: Method for the study of regional myocardial perfusion in patients with atherosclerotic coronary artery disease, Nuklearmedizin **15**:1, 1976.
34. Maseri, A., Parodi, O., Severi, S., and Pesola, A.: Transient transmural reduction of myocardial blood flow, demonstrated by ^{201}Tl scintigraphy, as a cause of variant angina, Circulation **54**:280, 1976.
35. Maseri, A., Pesola, A., Duce, T., and Donato, L.: Myocardial blood flow measurements by radioactive rubidium and potassium, J. Nucl. Biol. Med. **15**:87, 1971.
36. Maseri, A., Pesola, A., L'Abbate, A., and Contini, C.: Regional myocardial perfusion in ischaemic heart disease. In Roskamm, H., and Reindell, editors: Das chronische Kranke Herz, Proceedings of International Symposium on Chronic Diseases of the Heart, Bad Krozingen, Oct. 1972, Stuttgart, West Germany, 1974, F. K. Schattauer Verlag.
37. Maseri, A., Pesola, A., L'Abbate, A., Contini, C., Michelassi, C., and D'Angelo, T.: Contribution of recirculation and fat diffusion to myocardial washout curves obtained by external counting in man. Stochastic versus monoexponential analysis, Circ. Res. **35**:826, 1976.
38. Moir, T. W.: Measurement of coronary blood flow in dogs with normal and abnormal myocardial oxygenation and function, Circ. Res. **19**:695, 1966.
39. Prokop, E. K., Strauss, H. W., Shaw, J., Pitt, B., and Wagner, H. N., Jr.: Comparison of regional perfusion determined by ionic potassium-43 to that determined by microspheres, Circulation **50**:978, 1974.
40. Ritchie, J. L., Hamilton, G. W., Gould, K. L., Allen, D., Kennedy, J. W., and Hammermeister, K. E.: Myocardial imaging with indium-113m and technetium-99m-macroaggregated albumin: new procedure for identification of stress-induced regional ischemia, Am. J. Cardiol. **35**:380, 1975.
41. Severi, S., Parodi, O., Maseri, A., and Solfanelli, S.: Possibilities and limitations of ^{201}Tl myocardial scintigraphy in angina pectoris, J. Nucl. Med. All. Sci. **21**:81, 1977.
42. Shaw, D. J., Pitt, A., and Friesinger, G. C.: Auto-

radiographic study of the ^{133}Xenon disappearance method for measurements of myocardial blood flow, Cardiovasc. Res. **6**:268, 1972.
43. Strauss, H. W., Harrison, K., Langan, J. K., Lebowitz, E., and Pitt, B.: Thallium-201 for myocardial imaging. Relation of thallium-201 to regional myocardial perfusion, Circulation **51**:641, 1975.
44. Strauss, H. W., Zaret, B. L., Martin, N. D., Wells, H. P., and Flamm, M. D.: Noninvasive evaluation of regional myocardial perfusion with potassium 43, Radiology **108**:85, 1973.
45. Ter-Pogossian, M. M., Phelps, M. E., Hoffman, E. J., and Mullani, N. A.: A positron emission transaxial tomograph for nuclear imaging (PETT), Radiology **114**:89, 1975.
46. Turner, J. H., Selwyn, A. P., Jones, T., Evans, T. R., Raphael, M. J., and Lavender, J. P.: Continuous imaging of regional myocardial blood flow in dogs using ^{81m}Kr, Cardiov. Res. **10**:398, 1976.
47. Zaret, B. L., Strauss, H. W., Martin, N. D., Wells, H. P., Jr., and Flamm, M. D., Jr.: Noninvasive regional myocardial perfusion with radioactive potassium, N. Engl. J. Med. **388**:809, 1973.
48. Zierler, K. L.: Theory of measurement of myocardial blood flow in man by use of indicators and tracers, with consideration of assumptions and their violation in practical problems arising in special cases. In Maseri, A., editor: Myocardial blood flow in man. Methods and significance in coronary disease, Proceedings of international symposium, Pisa, Italy, June 1971, Turin, Italy, 1972, Minerva Medica.

11 □ Myocardial imaging with radioactive particles

Glen W. Hamilton

Historically, radioactive particles have been used to study the coronary circulation since 1965. Shortly after the introduction of the lung scan in 1964[14,17] Ueda[15] reported the use of ^{131}I-labeled macroaggregates of albumin (MAA) to measure regional myocardial blood flow, and Quinn[9] demonstrated the feasibility of external imaging in dogs. The potential danger of capillary blockage in the myocardium delayed clinical myocardial imaging until 1970, when Endo[3] and Ashburn[1] reported the use of ^{131}I-labeled MAA and ^{99m}Tc-labeled MAA in humans. Nearly simultaneously, Schelbert,[13] Poe,[8] and Weller[18] reported extensive studies of the effects of MAA and human albumin microspheres (HAM) when injected into the canine coronary circulation. Based on measurements of coronary flow, contractile force, left ventricular pressure, ECG, enzymes, and histologic sections, all these authors concluded that coronary particle injection was safe if the size and number of particles were adequately controlled. During the past five years, experience at several centers[6,7,12] has confirmed the safety in humans through well over 5000 studies.

PHYSIOLOGIC BASIS OF IMAGING WITH PARTICLES

Small particles injected into the circulation are distributed in proportion to regional blood flow, if they are of the appropriate size and are injected in a manner to ensure complete mixing prior to vascular branching. If the particle size exceeds 15 to 20 μm, they will not be accurately distributed in terms of epicardial-endocardial blood flow[2,19]; this is important for measurements employing in vitro tissue counting but of little importance for clinical external imaging, in which particles in the 30 to 60 μm range are satisfactory. Ensuring adequate mixing prior to arterial branching is a definite problem with coronary injections because of the short length of the left main coronary artery.[4] Careful injection techniques are thus critical to prevent erroneous studies (p. 217).

In the normal heart, the resting myocardial blood flow is about 100 ml/min/100 g. Regional myocardial blood flow is normally quite uniform throughout the left ventricular muscle mass. Blood flow to the right ventricular and atrial myocardium is slightly less—approximately 75% to 80% of left ventricular flow on a milliliters per gram basis. During periods of increased oxygen demand, coronary flow can increase to levels four or five times the resting flow rate. This ability to increase coronary flow is almost entirely due to diminished coronary vascular resistance and is commonly termed coronary reserve.[5] Experimental production of a progressive coronary constriction or stenosis causes a progressive pressure drop across the stenosis, but the flow remains normal until the stenosis exceeds 85% to 90% by diameter.[5] The pressure drop across the stenosis may, however, cause decreased flow in the endo-

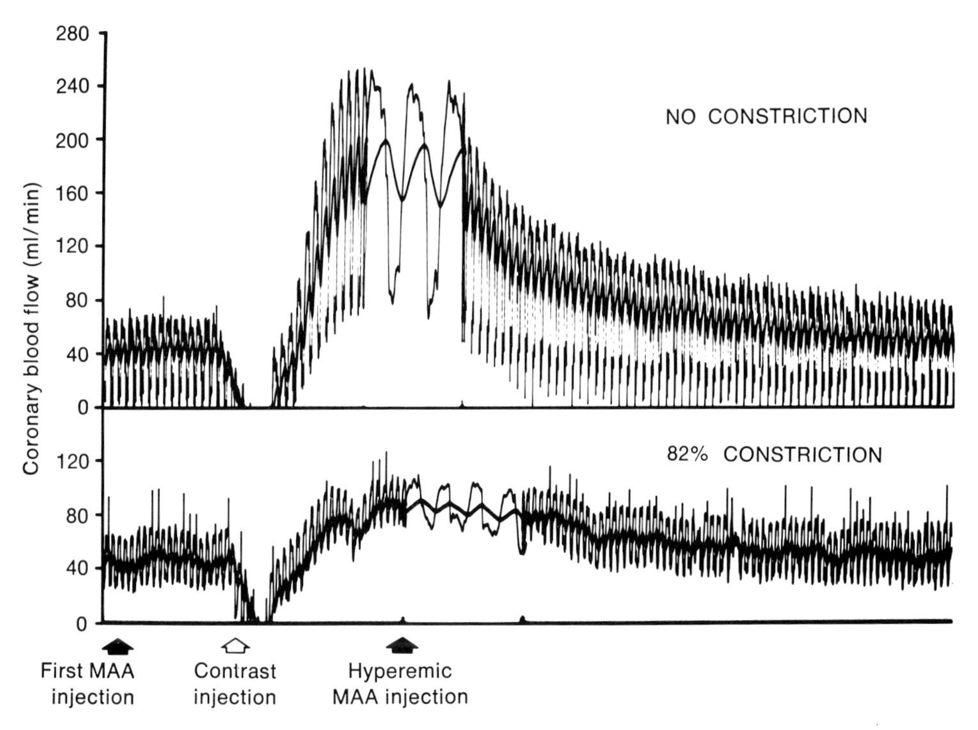

Fig. 11-1. Flow response to contrast injection in normal artery and artery with 82% stenosis. Study was performed in dog, with electromagnetic flowmeter and variable coronary constrictor on coronary artery. *Top:* No stenosis, and flow increases to four times resting level immediately following contrast injection. *Bottom:* An 82% (by diameter) stenosis has been induced; resting flow remains normal, but flow response following contrast injection is markedly impaired. This is fundamental abnormality in patients with angina pectoris. Note that MAA injected at rest will be equally distributed to normal and constricted vessels because resting flow is essentially equal in both. MAA injected during peak hyperemia following contrast injection will be distributed preferentially to normal vessel (due to flow differential), creating image defect in myocardium supplied by constricted vessel. (From Ritchie, J. L., Hamilton, G. W., and Gould, K. L.: Am. J. Cardiol. **35:**380, 1975.)

cardial myocardium, even though total coronary flow remains normal. Despite the normal resting flow, coronary stenosis exceeding 40% to 50% impairs coronary reserve, or the capacity to increase flow during stress. This is graphically shown in Fig. 11-1, which demonstrates the resting flow and the flow response to angiographic contrast material—a potent coronary vasodilator. The vessel above has no stenosis, and flow increases to a level four times the resting flow following contrast injection. The artery below (Fig. 11-1) has an 82% diameter constriction; resting flow is quite normal, but the response following contrast injection is markedly impaired. This is the fundamental abnormality in patients with coronary artery disease manifested by exercise-induced angina pectoris; resting myocardial blood flow remains normal but is unable to increase adequately during the period of increased oxygen need accompanying exercise. The relationship between resting flow, contrast-induced hyperemia, and coronary stenosis is shown in Fig. 11-2.

The situation in patients with previous myocardial infarction is quite different. Fibrous regions of scar tissue that develop after infarction have much less flow per unit mass (usually less than 10% to 15%) than normal myocardium. External imaging following particle injection will thus demonstrate regions that appear virtually devoid of radioactivity. Often, old myocardial infarction is manifested by less discrete regions of patchy myocardial scarring inter-

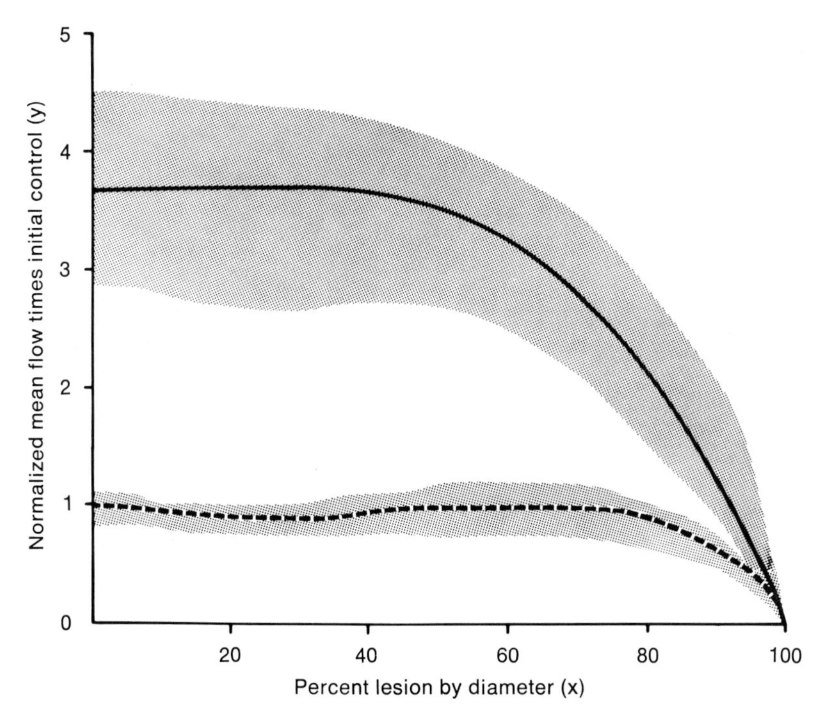

Fig. 11-2. Flow response at rest and during contrast-induced hyperemia. Data represent series of stenoses ranging from 0% to 100% in each of eight dogs. Resting blood flow remained normal until coronary constriction exceeded 80% to 90%. Hyperemic flow response remained normal with stenoses up to 40% to 50%. Stenoses greater than 50% caused diminution in flow response, which became marked with stenoses in 70% to 80% range. Note that with stenoses of 85% to 90% and normal resting flow, hyperemic flow response was virtually eliminated. (From Gould, K. L., Lipscomb, K., and Hamilton, G. W.: Am. J. Cardiol. **33:**87, 1974.)

mixed with normal myocardium, and external imaging will show areas of relatively decreased radioactivity due to the mixture of fibrous tissue and normal myocardium.

Based on these relationships, particle injections are made during the resting state to detect regional scarring due to previous infarction or during induced coronary hyperemia to detect regions of myocardium with impaired coronary reserve.

RADIOPHARMACEUTICAL SELECTION

Either MAA or HAM tagged with ^{99m}Tc, ^{113m}In, ^{131}I, or ^{123}I can be used. Satisfactory preparations of both MAA and HAM with particle sizes in the 20 to 50 μm range are available and can be labeled with 1.0 to 4.0 mCi of either ^{113m}In or ^{99m}Tc in such a way that the final preparation contains *less than 50,000 particles*. Commercially available ^{131}I-labeled MAA containing 100 μCi on less than 800,000 particles has also been used extensively[7] but is less desirable because of the low photon yield and lessened margin of safety. For the resting study, 100 μCi of ^{131}I or 1.5 mCi of ^{113m}In particles is injected into the right coronary artery (RCA) and 1.5 to 3.0 mCi of ^{99m}Tc particles into the left coronary artery (LCA). ^{99m}Tc-labeled MAA or HAM may be used to inject both the right and left coronary arteries (one third into the RCA; two thirds into the LCA), although it is difficult to ensure that both arteries receive equal amounts of activity on a millicurie per gram of myocardium basis because of the variable sizes of the right and left coronary artery distributions. If each arterial distribution does not receive equal activity per gram of myocardium, activity will not be proportional to blood flow throughout the myocardium, and erroneous image interpretation can occur.[12]

For resting hyperemic studies, ^{131}I or ^{113m}In particles are injected into the LCA at rest and ^{99m}Tc particles injected immediately following contrast injection. I have attempted to perform resting hyperemic studies by dividing both isotopes between the RCA and LCA just discussed. This has not proved to be satisfactory, and the technique is currently limited primarily to study of the LCA alone.

Since particle preparations with either large particles (exceeding 100 μm) or an excessive number of particles cause diminished coronary flow and contractile force,[8] strict quality control of the size and number of particles is imperative.

INJECTION AND IMAGING TECHNIQUES

Particle injections are made by the angiographer immediately following coronary arteriography. The operator should be certain that the catheter position provides equal visualization of both the anterior descending and circumflex coronary arteries prior to injection. For the resting study, 2 min should be allowed from the last contrast injection to the particle injection. Prior to injection the particles are agitated by hand, then injected through the coronary catheter and flushed in with saline solution. The same procedure is repeated for the contralateral artery. For studies of flow distribution during contrast-induced coronary hyperemia, the resting study is made as just discussed. From 5 to 10 ml of contrast material is then injected at a rate identical to that used for arteriography. The second particle (labeled with another radionuclide) injection is made 6 to 10 sec after the contrast injection. Injection of saphenous vein bypass grafts is made using similar techniques.

Following catheterization, imaging is performed with a scintillation camera. The ideal system is a camera-computer system with dual spectroscopy, so that images of two radionuclides (right and left injections or resting and hyperemic injections) can be collected simultaneously. Images (100,000 to 200,000 counts) in the anterior, 45-degree LAO, and left lateral positions are adequate for most cases. ECG gating of the images to end of diastole improves resolution slightly but increases imaging time greatly and is not routinely used. Collimation depends on the particular isotopes used; the highest resolution collimator consistent with the emission energy is the most satisfactory. In each position, images of both radionuclides should be obtained with care to avoid patient motion (if dual spectroscopy is not available), so that images of the right and left circulation can be directly added (by computer or photographically) to form a composite image of flow distribution. The same applies during rest-hyperemia studies if one wishes to use computer subtraction of the hyperemic image from the rest image in an effort to accentuate the regions of changing flow distribution.

RESTING MYOCARDIAL IMAGE

The normal resting myocardial image is shown in Fig. 11-3. Images of the left coronary artery injection alone visualize only the left ventricle. All views will show a variable diminution in activity inferiorly in the region of the myocardium supplied by the posterior descending branch of the RCA. In 10% to 20% of the cases, this inferior area of decreased activity is inconspicuous or absent because of the dominance of the LCA. Injection of the RCA usually has a "ball and tail" pattern, with the "ball" approximately twice as intense as the "tail" (Fig. 11-3). The tail represents flow to the right ventricular muscle and has less activity, primarily because of less muscle mass and to a lesser extent the lower right ventricular blood flow per gram of muscle. The more intense "ball," or wedge of activity, represents the inferior left ventricular muscle supplied by the posterior decending coronary artery. In all views, the size and shape of this region should correspond to the inferior region of diminished activity demonstrated on the LCA injec-

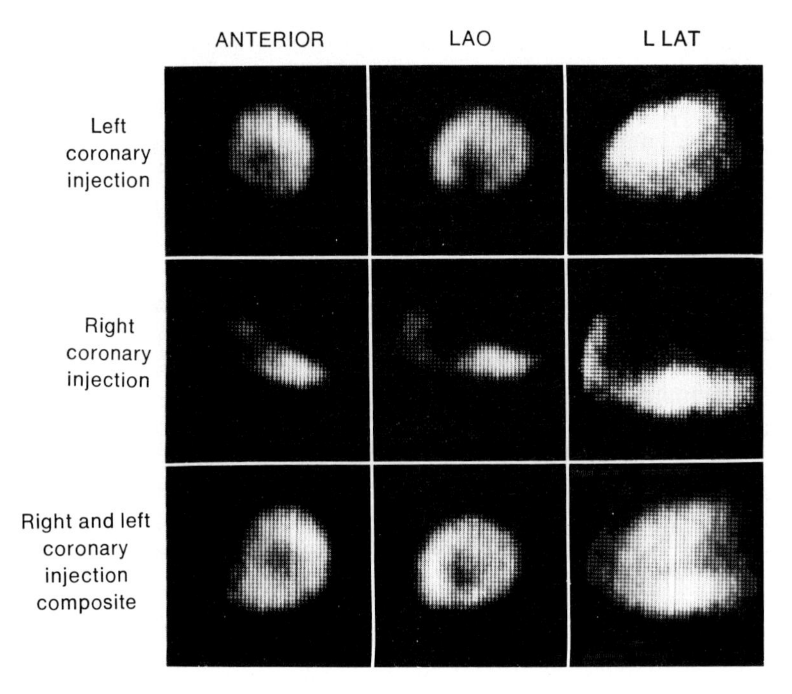

Fig. 11-3. Normal myocardial image following injection of right and left coronary arteries. *Bottom:* Photographic composite of right and left injections. Inferior region of myocardium shows decreased activity when left coronary is injected; this is area of myocardium normally supplied by posterior descending branch of right coronary artery. When right and left injections are superimposed, entire left ventricular myocardium shows equal activity. (From Ritchie, J. L., Hamilton, G. W., Williams, D. L., and Kennedy, J. W.: Radiology **121:**131, 1976.)

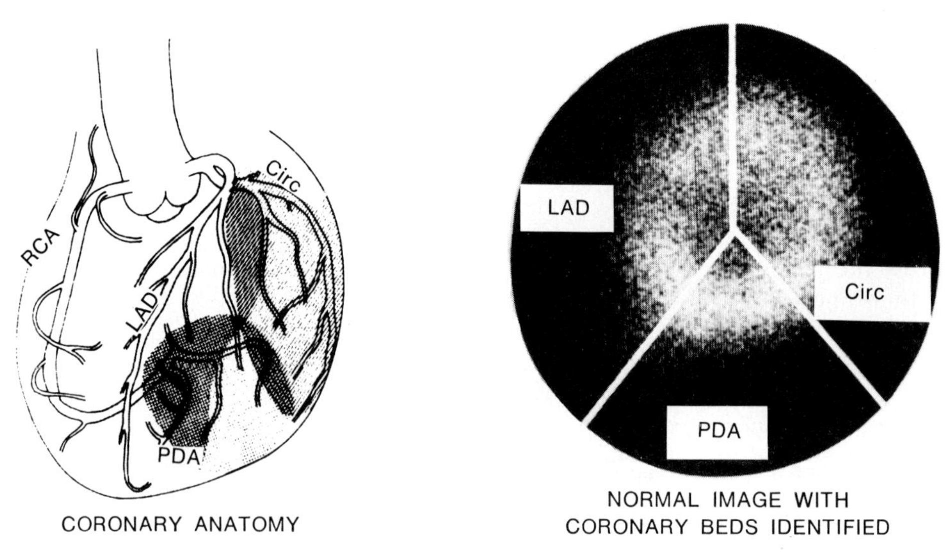

Fig. 11-4. Coronary arterial supply and myocardial image in LAO view. This view provides least overlap of major arterial distributions. Anterior myocardium is supplied by LAD, posterior myocardium by circumflex, and inferior myocardium by posterior descending. (From Ritchie, J. L., Hamilton, G. W., Williams, D. L., and Kennedy, J. W.: Radiology **121:**131, 1976.)

tion. This can be confirmed by adding or superimposing the RCA and LCA images to form a composite, as shown at the bottom of Fig. 11-3.

Regions of diminished activity are abnormal and usually visualized on several views. The LAO view is most satisfactory for determining the location of the abnormality because this position presents the three major arterial distributions (anterior myocardium supplied by the left anterior descending; inferior myocardium by the posterior descending artery, and posterior myocardium by the circumflex coronary artery) with the least overlap (Fig. 11-4). Because of the variable central area of decreased activity (corresponding to the left ventricular cavity), apical lesions are often not appreciated on the LAO view but can be readily seen on the anterior or left lateral views. Fig. 11-5 demonstrates the four basic locations of abnormalities: anterior, inferior, posterior, and apical. Three typical LAO images with anterior, inferior, and posterior infarction are shown in Fig. 11-6.

Studies in our laboratory revealed the following relationships between the resting myocardial image, coronary angiogram, and ECG (Table 11-1). Of 127 patients studied for suspected or known coronary disease, fifty-six (44%) had resting image defects. Fifty-three of these fifty-six (95%) had evidence of previous myocardial infarction in the area of the defect by ECG, ventriculogram, or direct inspection and biopsy. Conversely, only two patients with ECG Q waves did not have image defects. The presence of a resting defect is thus quite specific for infarction and significantly more sensitive than the ECG. The group of patients with the preinfarction syndrome merit special attention. Six patients in this series were in that category and three of

Table 11-1. Relationship of coronary stenosis, ECG Q wave, and the resting myocardial image

Coronary stenosis (%)	No. of patients	ECG Q waves	Rest defects
0-50	29	0	1 (3%)
51-75	12	1 (8%)	3 (25%)
76-99	32	6 (19%)	11 (34%)
100	54	32 (59%)	41 (76%)
TOTAL	127	39 (31%)	56 (44%)

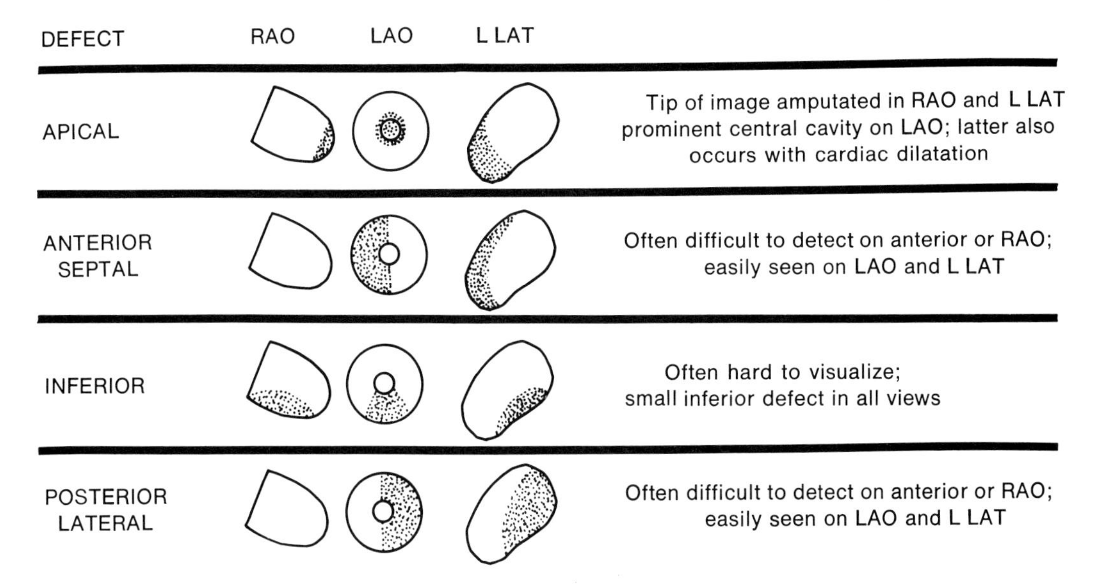

Fig. 11-5. Location of four major types of defects seen with myocardial imaging. Most abnormalities are noted in two views. Apical defects are most commonly associated with lesions in LAD, but can also occur with circumflex and posterior descending lesions. (From Ritchie, J. L., Hamilton, G. W., Williams, D. L., and Kennedy, J. W.: Radiology **121:**131, 1976.)

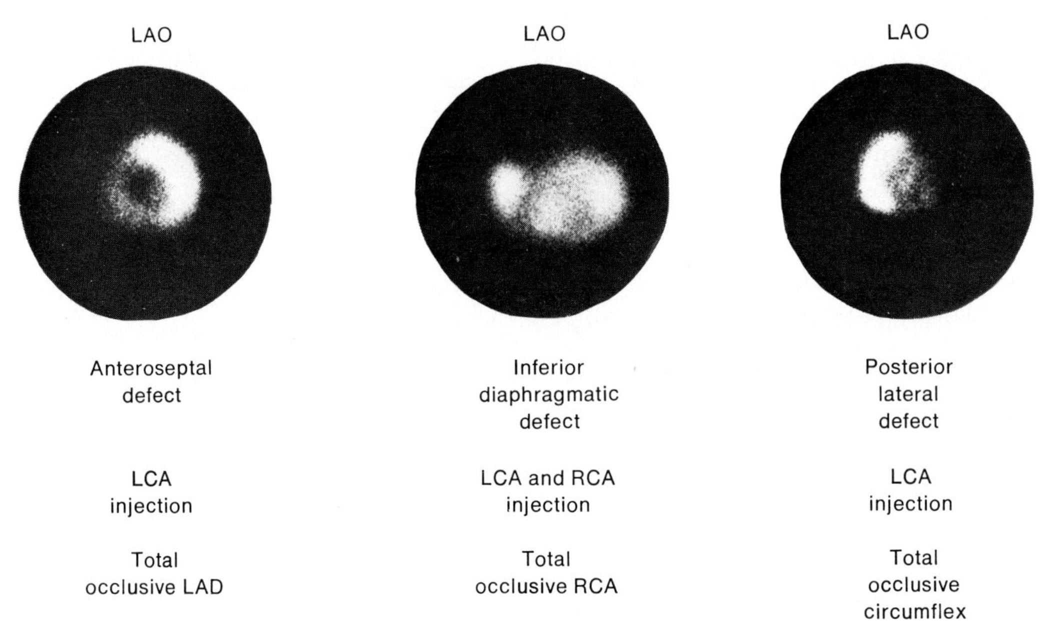

Fig. 11-6. Myocardial images from three patients with documented myocardial infarction. All images are from LAO view. (From Ritchie, J. L., Hamiltion, G., W., Williams, D. L., and Kennedy, J. W.: Radiology **121:**131, 1976.)

these (50%) had resting defects without evidence of infarction by other parameters (this has also been noted by Wackers and co-workers[16] in studies using ^{201}Tl). Therefore I do not believe that defects in this group can be considered evidence of previous infarction.

The relation between coronary anatomy and resting defects is less clear (Table 11-1). Few patients will demonstrate defects in the absence of significant coronary stenosis (≥75%); these presumably represent patients who have sustained infarction due to an embolus or a previous stenosis that has recanalized. About half of the patients at our laboratory with greater than 75% stenosis have demonstrated defects due to infarction. However, many patients (including thirteen out of fifty-four with total occlusion) with severe coronary disease have completely normal resting images. The resting image is thus not a sensitive method for detecting coronary disease in patients without previous myocardial infarction.

IMAGING DURING INDUCED CORONARY HYPEREMIA

Immediately following contrast injection, coronary flow increases markedly (four to five times resting flow). Coronary arteries with significant stenosis (>75%) have an impaired response, resulting in maldistribution of coronary flow during the period of hyperemia. Particles injected during this period will be maldistributed, the degree of maldistribution being directly related to the degree of flow impairment in the diseased vessel.[19] When combined with the resting study (using ^{113m}In or ^{131}I at rest and ^{99m}Tc during exercise), four patterns will result. Patients with no coronary disease will have normal images at rest and during hyperemia (Fig. 11-7). If the hyperemic image is subtracted from the resting image, the remaining activity will be minimal, since both activity distributions are equal. Patients with significant coronary disease but without previous infarction will have normal resting studies. During hyperemia, maldistribution of flow will occur as shown in Fig. 11-8. The

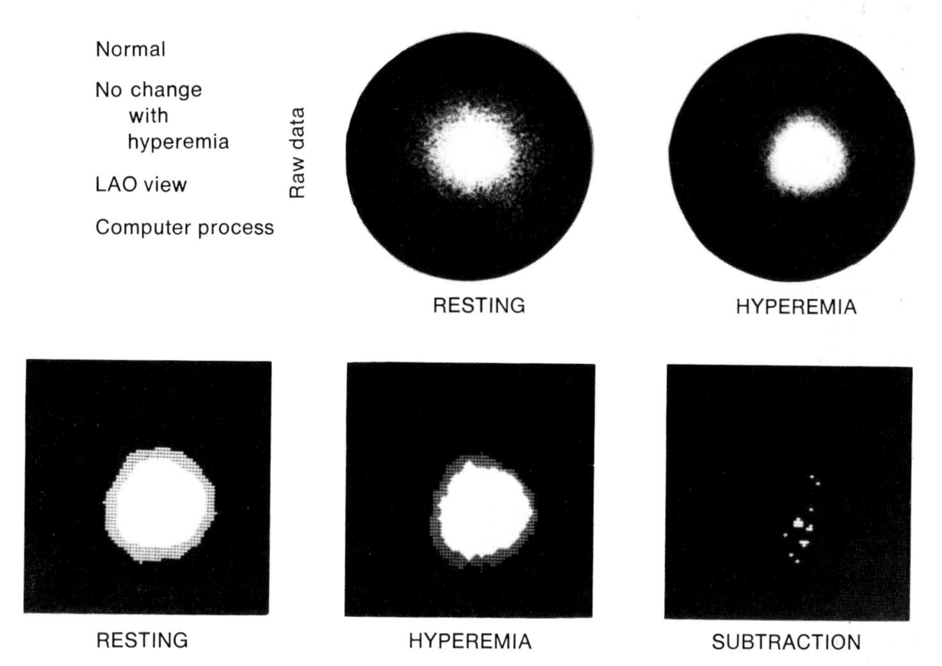

Fig. 11-7. Myocardial images from patient with no coronary disease. Images are all LAO view. Pattern of isotope distribution is essentially identical at rest and during contrast-induced hyperemia. Subtraction of hyperemic image from resting image shows minimal residual activity, documenting that distribution of activity was similar with both injections. (From Ritchie, J. L., Hamilton, G. W., and Gould, K. L.: Am. J. Cardiol. **35:**380, 1975.)

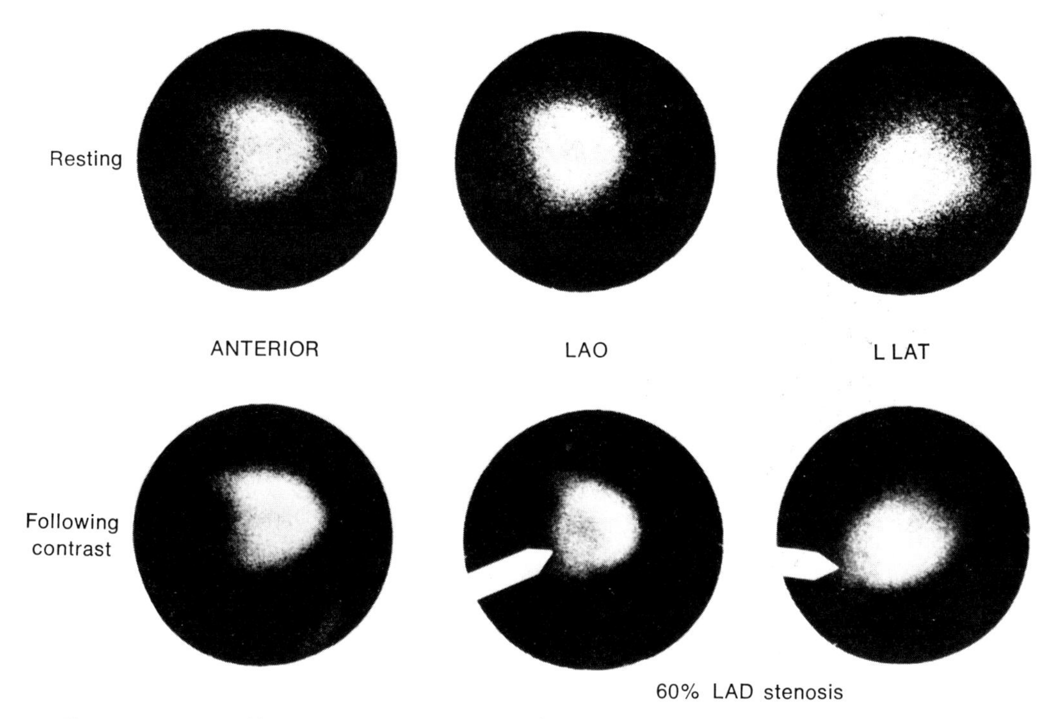

Fig. 11-8. Resting and hyperemic images in patient with coronary disease. Resting image is normal despite stenosis in LAD artery. During hyperemia, there is decreased activity in anterior myocardium compared to posterior myocardial region, best seen on LAO and lateral views.

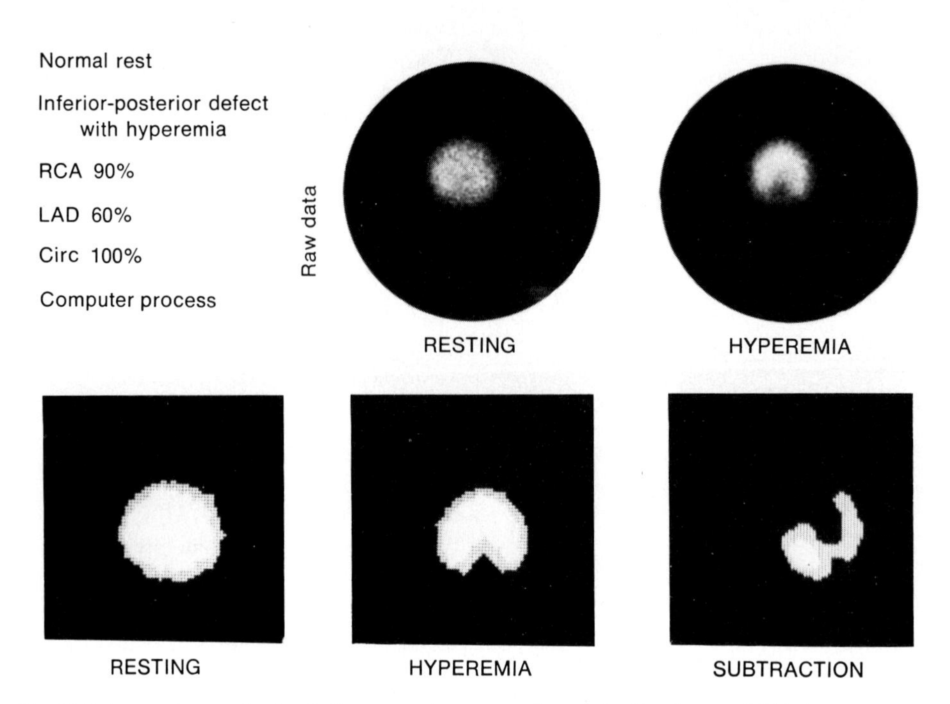

Fig. 11-9. Images from patient with three-vessel coronary disease. Resting images (LAO view) are normal. During hyperemia, inferior region of decreased activity is readily apparent on both raw data and computer-processed image. Computer subtraction shows area of decreased activity in circumflex region as well. Circumflex defect can be appreciated retrospectively on initial images, but shape and location tend to obscure difference. Changes of this type are easily recognized following computer subtraction of hyperemic image from resting image. (From Ritchie, J. L., Hamilton, G. W., and Gould, K. L.: Am. J. Cardiol. **35:**380, 1975.)

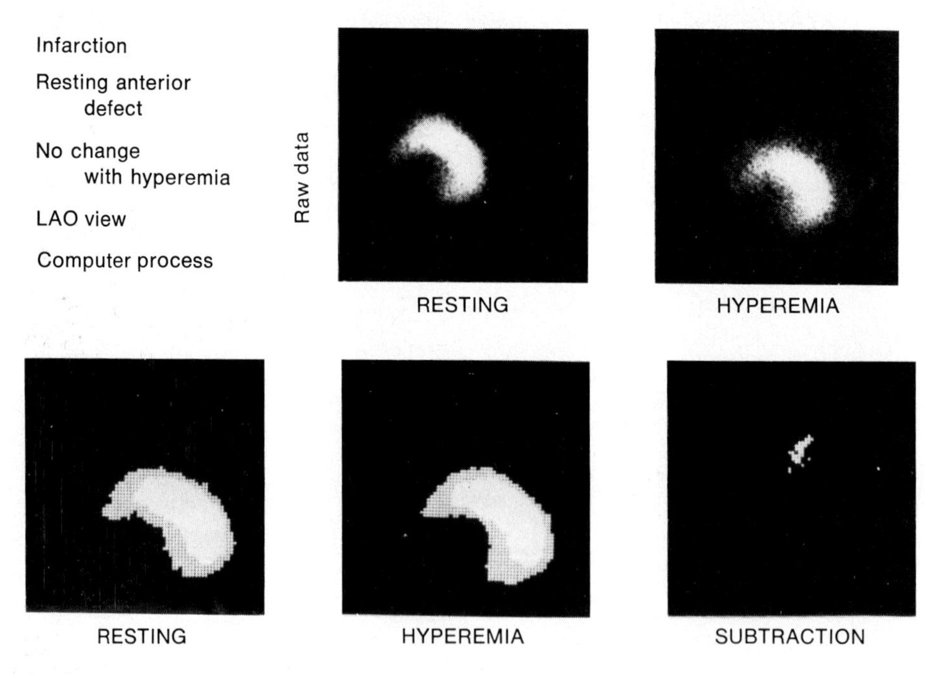

Fig. 11-10. Images from patient with large anterior defect at rest. Coronary arteriography demonstrated total occlusion of LAD artery with insignificant minor lesions of circumflex and right coronary arteries. During contrast hyperemia, relative distribution of activity remains unchanged. (From Ritchie, J. L., Hamilton, G. W., and Gould, K. L.: Am. J. Cardiol. **35:**380, 1975.)

change in distribution between rest and hyperemia is usually easily visible; however, occasionally computer subtraction brings out regions not immediately apparent, as illustrated in Fig. 11-9. Similarly, patients with resting defects due to infarction can show further maldistribution during hyperemia or remain unchanged (Fig. 11-10.)

In studies at our laboratory, the rest-hyperemia images were more sensitive than ^{201}Tl rest-exercise imaging for detecting coronary disease, particle studies being positive in 95% of the cases and ^{201}Tl positive in 80%.[10,11] Particle imaging, however, has the following serious limitations:

1. Only the LCA system can be studied with the rest-hyperemia technique.
2. The particle study can only be done at the time of catheterization.
3. Equal lesions in the left anterior descending and circumflex arteries will not cause maldistribution because the flow response is impaired in both vessels. Although this is also a problem with ^{201}Tl or ^{43}K, these agents visualize both right and left systems, and the likelihood of balanced lesions in three vessels is considerably less.

The major advantages are the improved resolution, short imaging time, and low cost afforded by the direct injection technique. Currently, the rest-hyperemia particle study is the single most useful test for determining the significance of lesions in the LCA system that are equivocal by arteriography.

ROLE OF PARTICLE IMAGING COMPARED TO ^{201}Tl

The recent introduction of ^{43}K and ^{201}Tl has lessened the need for particle studies to some extent. ^{201}Tl in particular provides excellent images using standard gamma camera–collimator systems and has the definite advantage of being noninvasive. A resting particle study during catheterization is clearly redundant if a resting thallium study has been or will be performed. Conversely, the resting particle study provides a satisfactory, relatively inexpensive alternative if a ^{201}Tl study is not performed. Similarly, the rest-hyperemia study can be replaced by the rest-exercise ^{201}Tl study in many cases. This is particularly true in cases that demonstrate exercise-induced thallium defects in the left anterior descending or circumflex distributions. These cases will simply show defects during hyperemia in the corresponding area, and no additional information will be gained. However, about 20% to 30% of patients with coronary disease do not develop exercise-induced ^{201}Tl defects; in these cases, the rest-hyperemia particle study may aid in documenting regional maldistribution of blood flow due to coronary stenoses. I have found this most helpful in patients with exertional angina and coronary lesions of questionable significance (i.e., those judged to be in the 40% to 65% range), particle maldistribution during hyperemia providing strong evidence that the lesion in question is physiologically significant.

SURGICAL IMPLICATIONS

Initial reports suggest that regional distribution by particle injection is useful in surgical selection. It is currently reasonable to restrict surgery in patients with large- or multiple-image defects indicating extensive scarring due to infarction, since it seems improbable that saphenous vein grafts to regions of scarring can accomplish much. Alternatively, regions with normal resting flow, which demonstrate decreased flow during hyperemia, would appear to benefit most from grafting.

REFERENCES

1. Ashburn, W. L., Braundwald, E., Simon, A. L., and Gault, J. H.: Myocardial perfusion imaging in man using ^{99m}Tc MAA, J. Nucl. Med. **11**:618, 1970.
2. Buckberg, G. D., Luck, J. C., Payne, D. B., Hoffman, J. I. E., Archie, J. P., and Fixler, D. E.: Some sources of error in measuring regional blood flow with radioactive microspheres, J. Appl. Physiol. **31**:598, 1971.
3. Endo, M., Yamazaki, T., Konno, S., Hiratsuka, H., Akimoto, T., Tanaka, T., and Sakakibara, S.: The direct diagnosis of human myocardial ischemia using ^{131}I-MAA via the selective coronary catheter, Am. Heart J. **80**:498, 1970.

4. Fox, C., Davies, M. J., and Webb-Peploe, M. M.: Length of left main coronary, Br. Heart J. **35:**796, 1973.
5. Gould, K. L., Lipscomb, K., and Hamilton, G. W.: Physiologic basis for assessing critical coronary stenosis, Am. J. Cardiol. **33:**87, 1974.
6. Hamilton, G. W., Ritchie, J. L., Allen, D. R., Lapin, E., and Murray, J. A.: Myocardial perfusion imaging with ^{99m}Tc or ^{113m}In macroaggregated albumin: correlation of the perfusion image with clinical, angiographic, surgical and histologic findings, Am. Heart J. **89:**708, 1975.
7. Jansen, C., Judkins, M. P., Grames, G. M., Gander, M. P., and Adams, R.: Myocardial perfusion color scintigraphy with MAA, Radiology **109:**369, 1973.
8. Poe, N.: The effects of coronary arterial injection of radio-albumin macroaggregates on coronary hemodynamics and myocardial function, J. Nucl. Med. **12:**724, 1971.
9. Quinn, J. L., Serratto, M., and Kezdi, P.: Coronary artery bed photoscanning using radioiodine albumin macroaggregates (RAMA), J. Nucl. Med. **7:**107, 1966.
10. Ritchie, J. L., Hamilton, G. W., Gould, K. L., Allen, D., Kennedy, J. W., and Hammermeister, K. E.: Myocardial imaging with indium-113m- and technetium-99m- macroaggregated albumin. New procedure for identification of stress-induced regional ischemia, Am. J. Cardiol. **35:**380, 1975.
11. Ritchie, J. L., Hamilton, G. W., Williams, D. L., English, M. T., and Leibowitz, E.: Myocardial imaging with ^{201}thallium—correlation with intracoronary macroaggregated albumin imaging, abstracted, Circulation **II-52:**231, 1975.
12. Ritchie, J. L., Hamilton, G. W., Williams, D. L., and Kennedy, J. W.: Myocardial imaging with radionuclide-labeled particles, Radiology **121:**131, 1976.
13. Schelbert, H., Ashburn, W., Covell, J., Simon, A., Braunwald, E., and Ross, J.: Feasibility and hazards of the intracoronary injection of radioactive serum albumin macroaggregates for external myocardial perfusion imaging, Invest. Radiol. **6:**379, 1971.
14. Taplin, G. V., Johnson, D. E., Dore, E. K., and Kaplan, H. S.: Lung photoscans with macroaggregates of human serum radioalbumin. Experimental basis and initial clinical trials, Health Phys. **10:**1219, 1964.
15. Ueda, H., Kaihara, S., Ueda, K., Sugishita, Y., Sasaki, Y., and Iio, M.: Regional myocardial blood flow measured by I-131 labeled macroaggregated albumin (I-131 MAA), Jpn. Heart J. **6:**534, 1965.
16. Wackers, F. J. Th., Sokole, E. B., Samson, G., Schoot, J. B., Lie, K. I., Liem, K. L., and Wellens, H. J. J.: Value and limitations of thallium-201 scintigraphy in the acute phase of myocardial infarction, N. Engl. J. Med. **295:**1, 1976.
17. Wagner, H. N., Jr., Sabiston, D. C., Iio, M., McAfee, J. G., Meyer, J. K., and Langan, J. K.: Regional pulmonary blood flow in man by radioisotope scanning, J.A.M.A. **187:**601, 1964.
18. Weller, D., Adolph, R., Wellman, H., Carroll, R., and Kim, O.: Myocardial perfusion scintigraphy after intracoronary injection of ^{99m}Tc-labeled human albumin microspheres, Circulation **46:**963, 1972.
19. Yipintsoi, T., Dobbs, W. A., Jr., Scanlon, P. D., Knopp, T. J., and Bassingthwaighte, J. B.: Regional distribution of diffusible tracers and carbonized microspheres in the left ventricle of isolated dog hearts, Circ. Res. **33:**573, 1973.

12 □ Kinetics of thallium distribution and redistribution: clinical applications in sequential myocardial imaging

George A. Beller
Denny D. Watson
Gerald M. Pohost

CHARACTERISTICS OF THALLIUM

Thallium is a metallic element in group III-A of the periodic table with properties similar to potassium. Like K^+ and Rb,$^+$ $^{201}Tl^+$ produces an inhibitory effect on cardiac Na^+,K^+-ATPase activity in the presence of Na^+ and K^+ and a dose-dependent increase in cardiac contractile force.[18] $^{201}Tl^+$ has highly polarizable outer electron shells and readily complexes with water, thus rendering it highly diffusible across cell membranes. It has been postulated that intracellular accumulation of $^{201}Tl^+$ is caused by a passive transport mechanism, which is driven by the transmembrane electropotential gradient.[27]

Thallium is biologically similar to potassium in terms of organ distribution.[7] The physical-chemical explanation for the biologic similarity of $^{201}Tl^+$ and K^+ is that the hydrated ionic radius of $^{201}Tl^+$ is between K^+ and Rb^+ in size, and this radius has been suggested as the property that determines passive penetration through a lipid-layered membrane.

The use of radiothallium in nuclear medicine was first suggested by Kawana and associates in 1970.[16] $^{201}Tl^+$ decays by electron capture with a 73 hr half-life. Its principal photopeaks are at 135 and 167 keV, and it emits mercury x rays of 69 to 83 keV with 98% abundance. The 80 keV mercury x ray is at the low end of the energy scale for resolution with a gamma scintillation camera. One can therefore employ a high-resolution low-energy collimator for myocardial imaging with a scintillation camera. Several other properties of $^{201}Tl^+$ make it a very attractive tracer for imaging the myocardium. Compared to K^+ and Rb,$^+$ a greater percentage of the total dose of $^{201}Tl^+$ concentrates in the myocardium.[28] There is also relatively lower hepatic and gastric uptake of $^{201}Tl^+$ compared to ^{43}K and ^{81}Rb.[28]

INITIAL DISTRIBUTION OF THALLIUM

The myocardial uptake of $^{201}Tl^+$ immediately following intravenous injection is the result of both the blood flow delivery of the tracer to the heart and the extraction of the tracer by the myocardium. Weich and co-workers[33] found the mean extraction fraction of $^{201}Tl^+$ by the heart in dogs in the basal resting state to be 88% ± 2%. The myocardial extraction fraction for $^{201}Tl^+$ remains relatively unaltered under various physiologic and metabolic conditions. Weich and colleagues also reported that an increase in heart rate from 106 to 195 in anesthetized dogs resulted in no significant change in $^{201}Tl^{+}$ extraction. The production of acidosis (pH=7.02) resulted in a slight diminution in extraction fraction from 88% to 79%, and with hypoxemia the extraction fraction decreased to 78%. Propranolol and acetyl strophanthidin had no significant effect on the extraction fraction. The relationship between regional myocardial $^{201}Tl^{+}$ concentration and regional blood flow as determined by the microsphere technique in experimental canine studies was shown to be nearly linear within the physiologic range of myocardial blood flow.[24,28] The initial regional concentration of $^{201}Tl^{+}$ following intravenous injection may thus be considered equivalent to regional myocardial blood flow.

REDISTRIBUTION OF THALLIUM

It was originally expected that $^{201}Tl^{+}$ would remain fixed in myocardial cells at least for several hours after initial uptake. This assumption was based on the observation that following injection, the net myocardial $^{201}Tl^{+}$ clearance rate was on the order of 7 hr.[7] However, resolution of initial perfusion defects with redistribution of $^{201}Tl^{+}$ has been observed in delayed images following exercise stress in patients.[24]

Further studies revealed rapid redistribution following resting injections in humans with severe fixed stenotic lesions and without evidence of transient changes in myocardial perfusion.[5,12] Studies in animal models have confirmed $^{201}Tl^{+}$ redistribution in myocardial regions with fixed perfusion defects[23] as well as reperfusion after transient coronary occlusion.[2] Disappearance of $^{201}Tl^{+}$ defects over a 4 hr period of reperfusion after transient coronary occlusion in anesthetized dogs was found to result from accumulation of $^{201}Tl^{+}$ in the previously ischemic zone as well as from washout from the normal myocardium.[2] Fig. 12-1

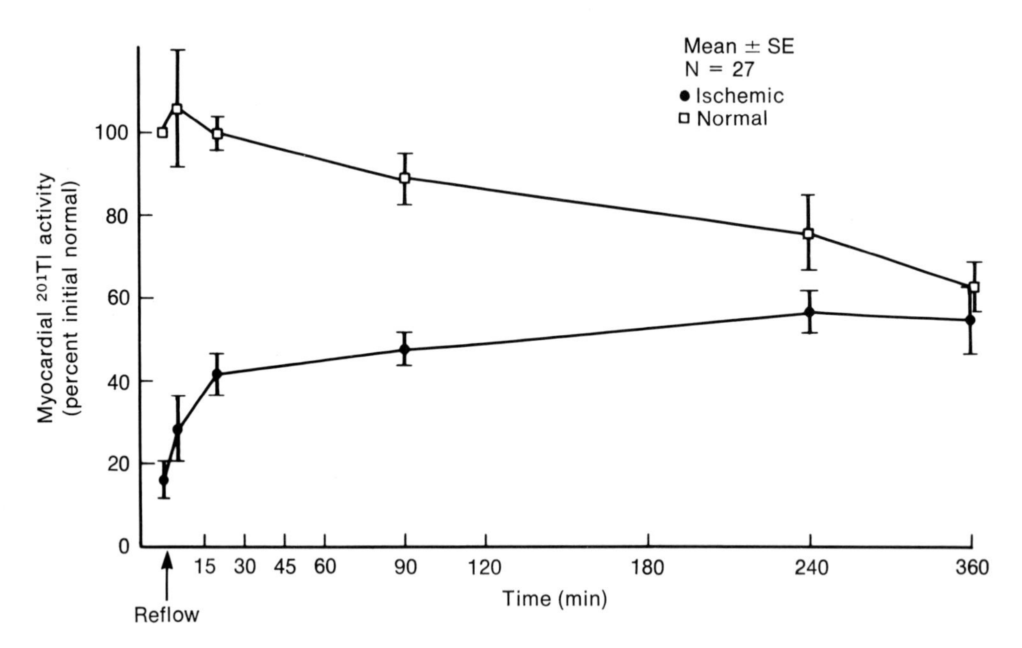

Fig. 12-1. Serial determinations of myocardial $^{201}Tl^{+}$ activity expressed as percentage of initial normal $^{201}Tl^{+}$ activity in five groups of dogs occluded for 20 min and reperfused for 5, 20, 90, 240, and 360 min. Initial values for $^{201}Tl^{+}$ activity in ischemic and normal regions were obtained prior to reflow and are mean values for all twenty-seven dogs. By four hours of reperfusion gradient in $^{201}Tl^{+}$ activity between normal and previously ischemic regions is markedly decreased. Scintigraphically, this would correspond to redistribution with filling in of initial defect.

summarizes the time course of the $^{201}Tl^{+}$ redistribution observed in this study. Redistribution occurred even in the presence of a fixed stenosis and did not require complete restoration of flow.

There are two conclusions that can be drawn from these studies: (1) $^{201}Tl^{+}$ redistribution can occur even when there is no change in coronary blood flow between the initial and delayed images, and (2) the distribution of $^{201}Tl^{+}$ in delayed images is not equivalent to the distribution obtained initially after a resting injection.

To understand the mechanism of redistribution and the conditions required for it to occur, it should be recognized that $^{201}Tl^{+}$ does exchange between the myocardium and the extracardiac $^{201}Tl^{+}$ pool. A simplified compartmental model is useful in explaining the redistribution phenomenon.

In this model there are three major compartments within which $^{201}Tl^{+}$ is distributed:

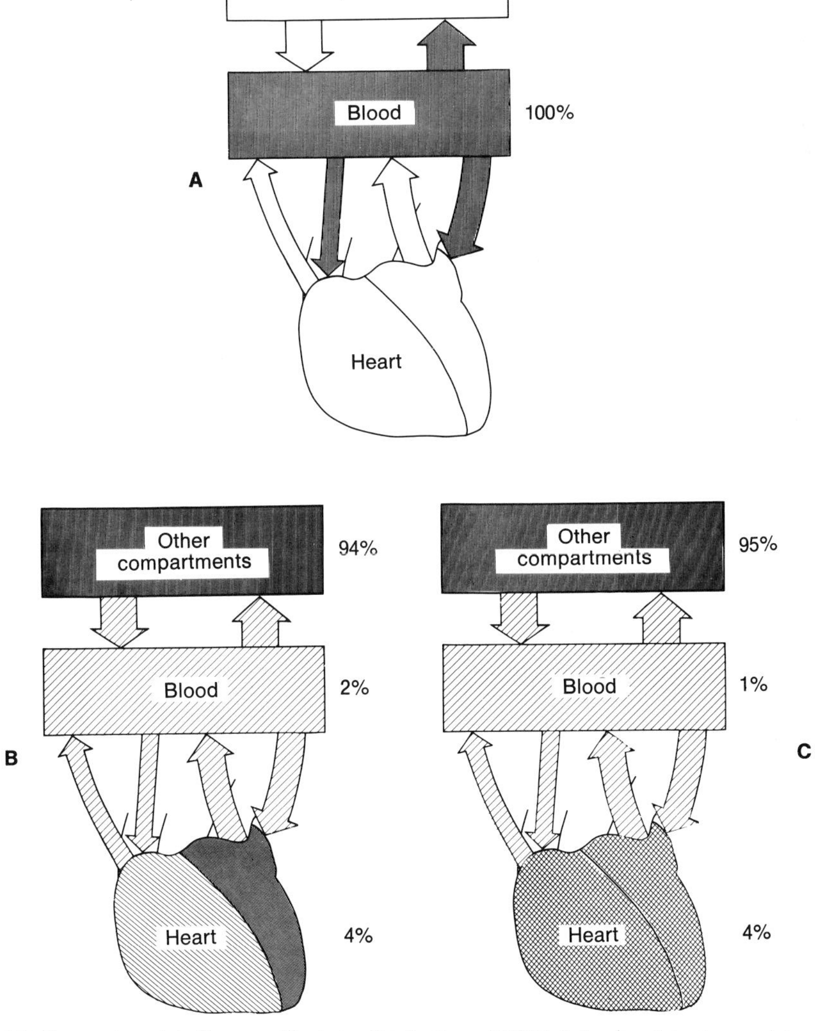

Fig. 12-2. Compartmental diagrams illustrate distribution of $^{201}Tl^{+}$ following injection and in delayed equilibrium phase. (See text for discussion.)

blood, myocardium, and all extracardiac organs. The latter contain all of the extravascular $^{201}Tl^+$ except that which is in the myocardium. Fig. 12-2 depicts this model to help visualize the compartmental exchange of $^{201}Tl.^+$ In Fig. 12-2, *A*, the dark-shaded region indicates the concentration of $^{201}Tl^+$ just after injection into the blood. $^{201}Tl^+$ is extracted avidly by most organs in the body and within a few minutes will be distributed roughly in proportion to the distribution of cardiac output. This distribution is illustrated in Fig. 12-2, *B*. The light-shaded region of the heart is presumed to have diminished blood flow in comparison to the dark-shaded region of the heart and consequently to have taken up proportionately less $^{201}Tl.^+$ The actual concentration of $^{201}Tl^+$ at this time would be proportional to the product of perfusion and extraction fraction. As long as myocardial cellular integrity remains intact with maintenance of the electrochemical gradient for $^{201}Tl^+$, the myocardial extraction fraction is approximately 85% and the total myocardial concentration of $^{201}Tl^+$ would then be very nearly proportional to the blood flow. Approximately 4% of the $^{201}Tl^+$ that was injected is now in the heart, with more than 90% in other body organs. Thus there is a very large systemic reservoir of $^{201}Tl^+$ outside the heart. $^{201}Tl^+$ is in equilibrium with the systemic pool and the cardiac pool, while the blood acts as the transport vehicle. Thus a portion of $^{201}Tl^+$ will enter the myocardium from the exchangeable pool. This is referred to as $^{201}Tl^+$ being "imported." On the other hand, Tl^+ may wash out of the myocardium and be deposited in the systemic pool. This washout is referred to as Tl^+ being "exported" from the myocardium.

In the type of exchange process just described, in which $^{201}Tl^+$ is both entering and leaving a myocardial region, the net concentration of $^{201}Tl^+$ in that region may change if there is an imbalance between the rate of import and the rate of export. For example, in Fig. 12-2, *B*, the area of myocardium with initially high concentration might be exporting $^{201}Tl^+$ to the systemic pool more rapidly than it can import $^{201}Tl^+$ from the systemic pool, resulting in a decrease in net $^{201}Tl^+$ concentration. In contrast, the area of myocardium with initially low concentration might tend to import more $^{201}Tl^+$ than it exports and consequently increase its net concentration over time. Thus the local concentration of $^{201}Tl^+$ will continue to change until a static equilibrium condition is achieved so that $^{201}Tl^+$ import is equal to $^{201}Tl^+$ export. Ultimately the total amount of $^{201}Tl^+$ extracted will be related to the ability of the cell to maintain the electrochemical $^{201}Tl^+$ gradient. Under the condition of static equilibrium, no further net change in concentration occurs. This condition is illustrated in Fig. 12-2, *C*.

The myocardial distribution of $^{201}Tl^+$ may therefore be different in the delayed or equilbrium images than it was in the initial images. This change in concentration over time is referred to as "$^{201}Tl^+$ redistribution." Redistribution can occur even when absolute blood flow to both areas (Fig. 12-2) remains unchanged. This condition is not a true equilibrium (no further change in $^{201}Tl^+$ concentration), since there is a continued net loss of $^{201}Tl^+$ due to systemic excretion. However, the total body loss of $^{201}Tl^+$ is quite slow compared to the rate of systemic exchange. Thus the exchange and the equilibrium will be unaffected by systemic excretion except for a gradual net loss of total circulating $^{201}Tl^+$.

The rate at which $^{201}Tl^+$ exchanges between myocardium and the systemic pool is largely determined by the *intrinsic* myocardial washout rate. This intrinsic rate reflects the rate of $^{201}Tl^+$ output from the myocardium in the absence of continuous reintroduction of $^{201}Tl^+$ to the myocardium from the systemic pool. In contrast, the *net* myocardial washout rate represents the net difference between the rate of input and the rate of output. It is essential to consider the two washout rates independently. The intrinsic washout rate depends solely on factors intrinsic to the myocardium, whereas the net washout rate depends on how rapidly the tracer is being recirculated back into the heart from all other body organs and will therefore depend on factors extrinsic to the myocardium.

The intrinsic washout rate has been experimentally determined by serial measurements of myocardial $^{201}Tl^+$ activity after direct intracoronary injection of $^{201}Tl^+$ in closed chest anesthetized dogs.[32] With intracoronary administration, approximately 85% of the injected dose is initially distributed in the heart, compared to 4% in the instance of an intravenous injection. Under these conditions the $^{201}Tl^+$ that is being recirculated back to the heart from the systemic pool becomes insignificant. Fig. 12-3 shows the intrinsic myocardial $^{201}Tl^+$ washout curve measured in a representative dog from this group. The average value obtained for the entire group was $T_{1/2} = 75$ min.[32] The intrinsic washout curve is monoexponential over the time range of 5 min to 2 hr after infection. This finding implies that the intrinsic washout rate is proportional to the myocardial $^{201}Tl^+$ concentration. That is, $\frac{dC(t)}{dt} = kC(t)$. This relationship is directly implied from the experimental data, indicating that $C(t) = C_0e^{-kt}$, where C(t) is the myocardial $^{201}Tl^+$ concentration at time t and K is the intrinsic clearance coefficient $\left(\frac{0.693}{T_{1/2}}\right)$.

In contrast, the other curve in Fig. 12-3 shows the net washout of $^{201}Tl^+$ from the myocardium in the same dog after intravenous injection. The value obtained for the half life for net washout was approximately 8 hr, which is in agreement with previously reported values.[7] The explanation for the slower net washout rate is that after intravenous injection, as $^{201}Tl^+$ leaves the myocardium, it is continuously replaced by new $^{201}Tl^+$ from the systemic pool at almost the same rate; thus there is little change in $^{201}Tl^+$ concentration. Even though there is little decrease in net $^{201}Tl^+$ concentration over time, $^{201}Tl^+$ is being continously exchanged between the myocardial and systemic compartments. In fact, following intravenous injection, half the myocardial $^{201}Tl^+$ will be exchanged every 75 min. This dynamic exchange allows local redistribution of $^{201}Tl^+$ to occur.

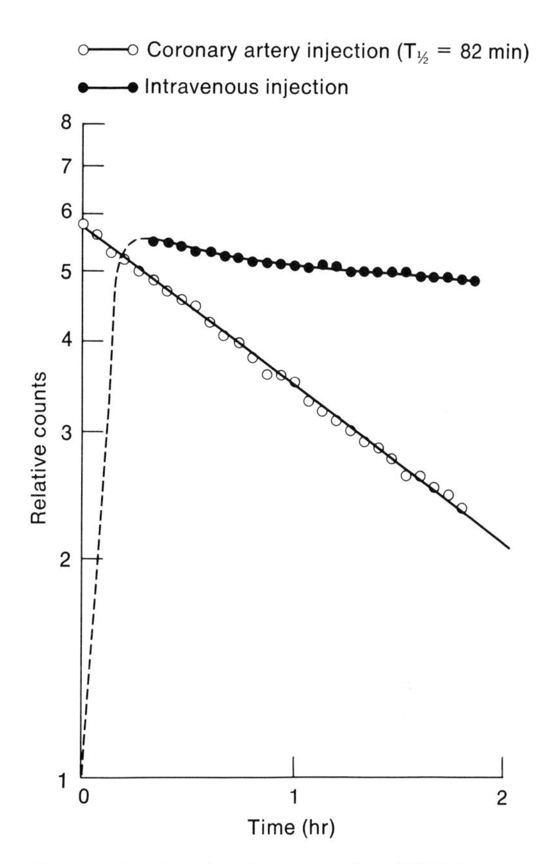

Fig. 12-3. Semilog plot of concentration of $^{201}Tl^+$ in myocardium following intracoronary injection *(open circles)*, indicating intrinsic washout. Concentration after intravenous injection *(closed circles)* indicates net myocardial washout.

DELAYED DISTRIBUTION OF THALLIUM

The term "redistribution" means that the myocardial distribution in the delayed images is identifiably different from the myocardial distribution in the initial images. The initial distribution is predominantly related to regional blood flow. A mechanism was just discussed by which the transition is made from an initial $^{201}Tl^+$ distribution to a potentially different distribution in the delayed images, which is determined by an equilibrium between local input and output.

It is essential when redistribution is observed to understand what factors have caused the delayed equilibrium image to be different from the initial image. In the delayed images, an equilibrium state prevails; thus the rate of $^{201}Tl^+$ input to each local myocardial region will equal the intrinsic rate of output. In a specified myocardial region the rate of $^{201}Tl^+$ input is expressed as input = PC_B. Output (as just shown) is computed as output = kCm. At equilibrium, when input equals output, $PC_B = kCm$, or, $C_m = C_B \frac{P}{k}$. P is the effective perfusion (perfusion × extraction fraction); C_m is the concentration of $^{201}Tl^+$ in a specified myocardial region; k is the intrinsic myocardial washout constant; and C_B is the concentration of $^{201}Tl^+$ in the arterial blood supply.[4]

Thus the local concentration of $^{201}Tl^+$ *at equilibrium* depends directly on effective perfusion and inversely on the intrinsic washout constant. It should be pointed out that "equilibrium" will occur only for an instant, that is, at the time when the myocardial cell has achieved its peak activity. The time at which ischemic cells achieve "equilibrium" may be considerably delayed. Uptake continues until the membrane-determined equilibrium.

This relationship can perhaps be better visualized in Fig. 12-4. This figure diagrammatically depicts three myocardial segments in equilibrium. The first segment is assumed to have normal perfusion and thus a normal $^{201}Tl^+$ input and a normal intrinsic washout rate. The second is assumed to have reduced perfusion and consequently a decreased rate of $^{201}Tl^+$ input. The intrinsic washout rate remains normal. This region at equilibrium would have a *persistently lower* $^{201}Tl^+$ concentration compared to the normal segment and would appear as a persistent defect. The third segment is assumed to have a reduced intrinsic washout rate as well as reduced input of $^{201}Tl^+$ and would accumulate $^{201}Tl^+$ in the delayed equilibrium phase equal to or greater than the concentration of $^{201}Tl^+$ in the normal region. The dependence on the washout rate is the new factor, which causes the delayed equilibrium image to be potentially different from the initial postinjection image. If the washout rates are different between two myocardial regions, then redistribution will necessarily occur, even when there is no change in perfusion between the initial and delayed images. Recent data from Gewirtz and associates[12] suggest that the loss of thallium does not depend on flow; in their studies of infarcted dogs, the rate of thallium loss from both normal and ischemic tissue was identical. However, more recent data have shown that the intrinsic rate of thallium washout may be markedly pro-

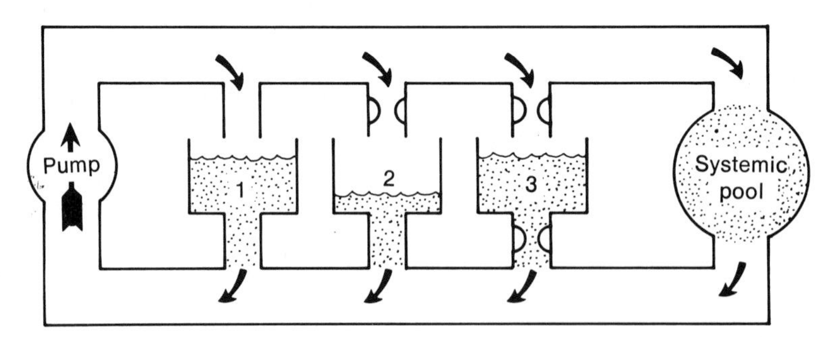

Fig. 12-4. Schematic representation of $^{201}Tl^+$ concentration at equilibrium. Containers represent three myocardial segments. Fluid levels represent relative myocardial $^{201}Tl^+$ concentration. Input and output of each region represent rates at which $^{201}Tl^+$ is entering and leaving myocardial cells. Regions *2* and *3* have reduced $^{201}Tl^+$ input (secondary to reduced blood flow or cellular extraction). Region *3* is assumed to also have reduced intrinsic washout of $^{201}Tl^+$.

longed during chronic diminution of myocardial perfusion.

In summary, from experimental data obtained in animal studies and from clinical experience in patients, a better understanding of $^{201}Tl^{+}$ kinetics has evolved. It is clear that the initial distribution of $^{201}Tl^{+}$ in the myocardium is proportional to regional flow at the time of intravenous injection under resting or stress conditions. Redistribution begins immediately and proceeds continuously until an equilibrium distribution is reached, which is determined by a net balance between input and output. Redistribution results from a continuous exchange of $^{201}Tl^{+}$ between myocardium and all other extracardiac compartments.

CLINICAL APPLICATIONS

Diminished myocardial uptake of $^{201}Tl^{+}$ has been observed clinically with intravenous administration of the radionuclide during exercise stress,[1,9] during pain accompanying spontaneous angina of the Prinzmetal-variant type,[21,22] in patients with acute or prior myocardial infarction,[30] and in resting patients with severe coronary artery disease.[5,12] In each of these situations, serial imaging after a single intravenous dose of $^{201}Tl^{+}$ can be employed to assess initial distribution and redistribution of $^{201}Tl^{+}$ to obtain noninvasive scintigraphic data relevant to myocardial perfusion and cellular viability.

The most clinically utilized application of myocardial imaging with $^{201}Tl^{+}$ has been in conjunction with exercise stress testing in patients with either suspected or documented coronary artery disease. When $^{201}Tl^{+}$ is injected at the peak of exercise, images obtained shortly thereafter demonstrate the regional perfusion pattern at the time of exercise. Regions of diminished activity on these images may represent areas of transient relative underperfusion or ischemia as well as myocardial scar. To differentiate between transient relative underperfusion or ischemia and scar, it has been customary to repeat the imaging study approximately one week later with the patient in the resting state. Regions of diminished activity corresponding to old infarction will appear on both studies, whereas regions representing exercise-induced relative underperfusion or ischemia will not be evident on the rest study.

An alternative method to differentiate transient underperfusion or ischemia from scar employs a sequential imaging technique after a single intravenous dose of $^{201}Tl^{+}$.[24] The rationale for this method is based on knowledge of $^{201}Tl^{+}$ distribution and redistribution kinetics summarized earlier in this chapter. The technique is based on the observation that there is redistribution of $^{201}Tl^{+}$ over time into areas of transient relative underperfusion or ischemia, whereas areas of scar or infarction demonstrate persistent defects in $^{201}Tl^{+}$ uptake. The explanation for this phenomenon is that initially at peak exercise normally perfused myocardial segments acquire an increased fraction of the intravenously injected $^{201}Tl^{+}$, which is attributed to the exercise-induced increase in coronary blood flow. Myocardial segments supplied by stenotic coronary arteries may receive adequate resting flow but not an adequate increase in flow during exercise. Thus in the initial distribution phase, focal defects in $^{201}Tl^{+}$ uptake will be observed in these relatively underperfused segments. During the 2 to 3 hr postexercise resting phase, underperfused segments with initial $^{201}Tl^{+}$ defects may accumulate $^{201}Tl^{+}$, whereas normal segments will have lost more $^{201}Tl^{+}$ than abnormally perfused segments. In the equilibrium phase, normally perfused and underperfused myocardial segments will approach equal $^{201}Tl^{+}$ concentration, and images obtained at that time (2 to 3 hr postexercise) will demonstrate relatively uniform $^{201}Tl^{+}$ concentration in the left ventricular myocardium.

Clinical examples of the abnormal initial distribution during exercise and subsequent redistribution patterns of $^{201}Tl^{+}$ with delayed filling in of defects are shown in Figs. 12-5 and 12-6. The serial images shown in Fig. 12-5 demonstrate redistribution in in-

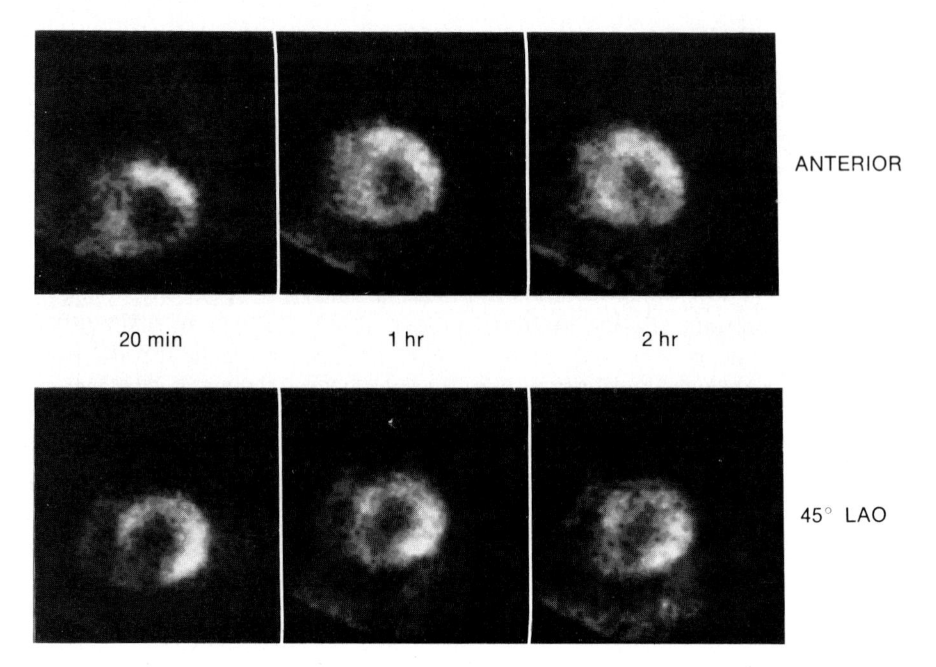

Fig. 12-5. Sequential myocardial $^{201}Tl^{+}$ images during and after exercise in anterior and 50-degree LAO projections, demonstrating $^{201}Tl^{+}$ redistribution in inferior, apical, and lower septal regions on delayed views.

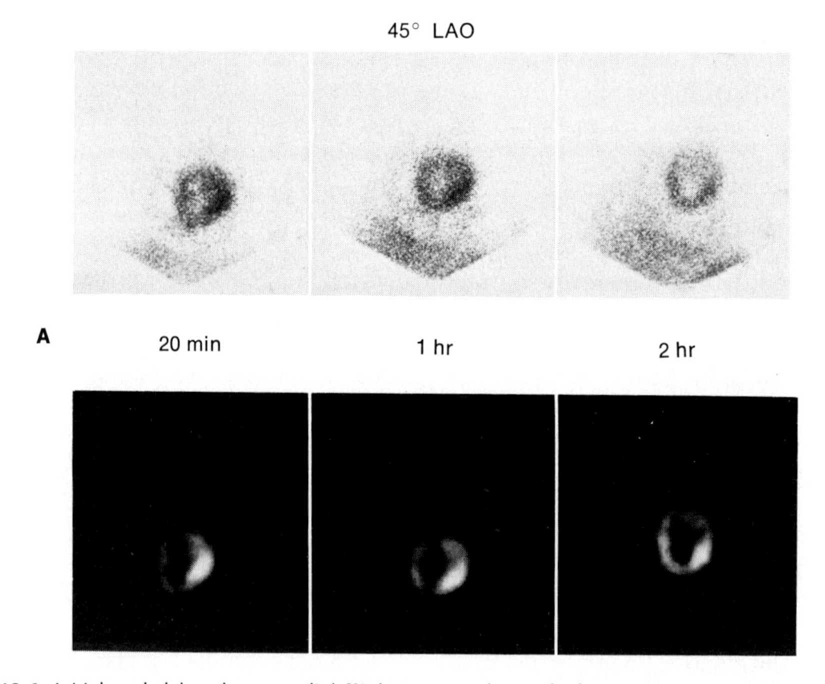

Fig. 12-6. Initial and delayed myocardial $^{201}Tl^{+}$ images obtained after exercise stress in patient with coronary artery disease. **A,** Initial image. *Top:* Unprocessed scintiphotographs. *Bottom:* Computer-processed images with background subtraction. Septal defect in LAO projection. **B,** Delayed image. $^{201}Tl^{+}$ redistribution into region with washout in high posterolateral wall region. Time-activity curves obtained from septal and posterolateral segments demonstrate quantitatively diminished early uptake with delayed equalization of $^{201}Tl^{+}$ activity, indicating redistribution.

ferior, apical, and septal segments in the postexercise period. Fig. 12-7 shows the early postexercise and delayed images in a patient with persistent apical and septal defects secondary to prior myocardial infarction. No redistribution is apparent in these images. Myocardial segments demonstrating redistribution after exercise have been shown by coronary angiography to be supplied by stenotic coronary arteries but to exhibit either normal or hypokinetic wall motion.[24,35] Persistent defects correlated with sites of previous myocardial infarction and more severe segmental wall motion abnormality. By utilizing the sequential imaging technique after a single intravenous dose of $^{201}Tl,^{+}$ sensitivity and specificity for detecting coronary artery disease are comparable to values obtained when stress and rest images are performed one week apart.

The redistribution images obtained 2 to 3 hr after exercise in patients undergoing stress testing can provide information similar to the images obtained at rest one week later. Initial postexercise defects showing redistribution usually revert to normal after successful coronary artery bypass surgery. Fig. 12-8 shows the preoperative and postoperative exercise $^{201}Tl^{+}$ sequential imaging studies in a patient who underwent coronary artery bypass surgery for intractable angina pectoris. Disappearance postoperatively of the initial inferoapical defect at a higher rate-pressure product is consistent with the normalization of perfusion.

Serial imaging after a single intravenous dose of $^{201}Tl^{+}$ has been employed to differentiate false from true positive electrocardiographic responses to exercise stress in asymptomatic patients or patients with atypical chest pain. In a study by Guiney and associates,[14] thirty-five patients referred for coronary angiography because of an exercise test interpreted as positive for ischemic ST-segment changes (> 1.0 mm or more reversible horizontal or down sloping ST depression) but who were asymptomatic or had a history of atypical chest pain un-

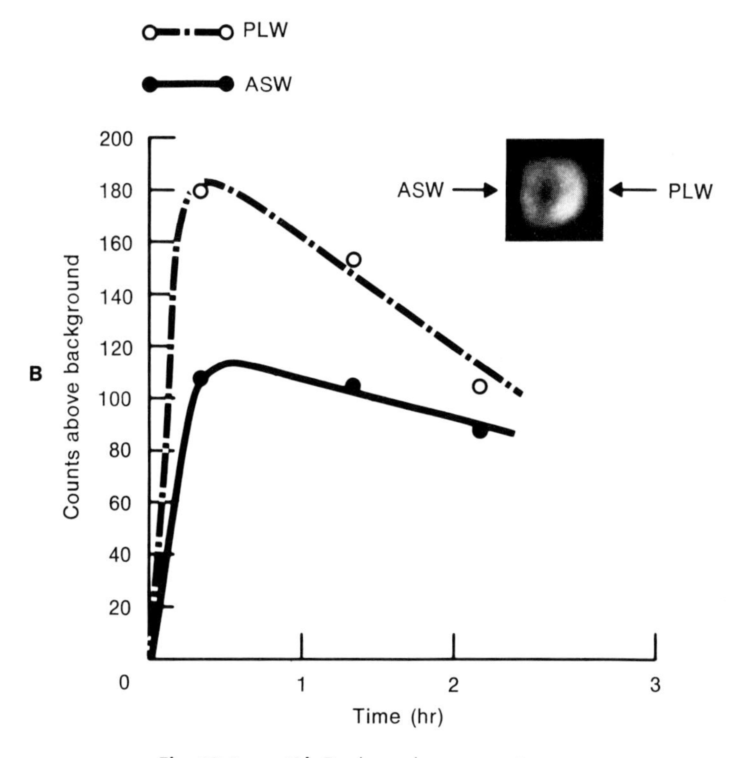

Fig. 12-6, cont'd. For legend see opposite page.

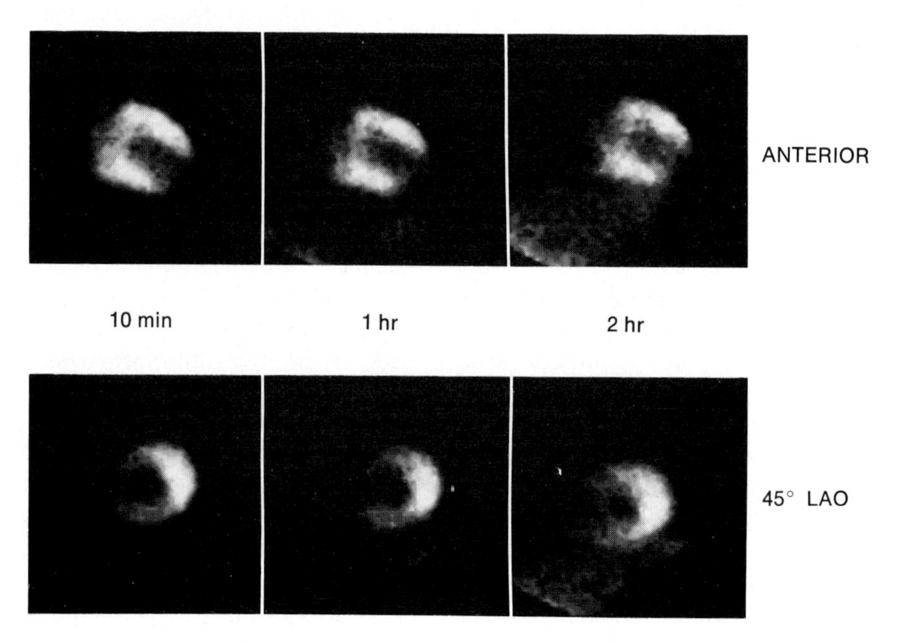

Fig. 12-7. Initial and delayed postexercise $^{201}Tl^{+}$ images in patient with prior anterior wall infarction. Persistent apical and septal defects are observed. No redistribution is apparent.

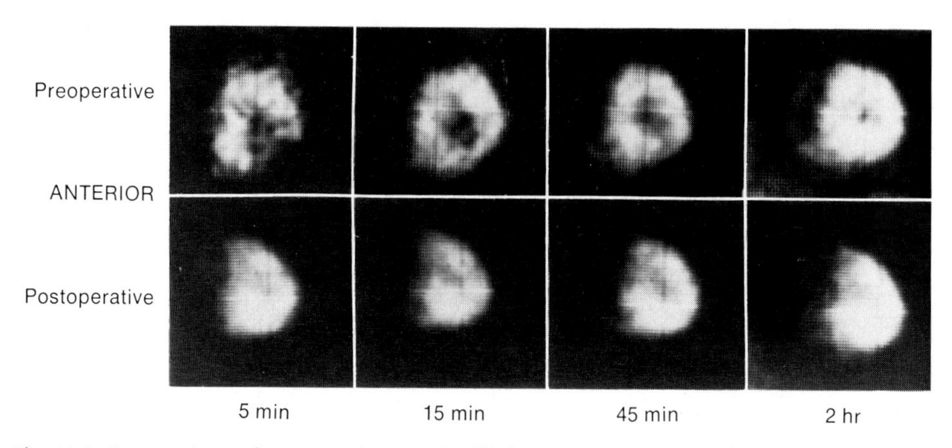

Fig. 12-8. Preoperative and postoperative exercise $^{201}Tl^{+}$ images in patient undergoing coronary artery bypass surgery. On preoperative scans, redistribution is observed in apical region on delayed images. Postoperatively, uniform $^{201}Tl^{+}$ uptake is observed in initial postexercise image.

derwent repeat exercise testing in conjunction with $^{201}Tl^{+}$ myocardial imaging. Of these thirty-five patients, twenty-four (69%) had normal postexercise $^{201}Tl^{+}$ scans and eleven (31%) demonstrated segmental defects in $^{201}Tl^{+}$ uptake on initial postexercise images; eight of the latter also demonstrated $^{201}Tl^{+}$ redistribution. Of the twenty-four with a positive ST-segment response and normal stress $^{201}Tl^{+}$ scan, twenty-three had no significant coronary stenoses at the time of angiography. Of the eleven patients with a positive ST-segment response and abnormal $^{201}Tl^{+}$ scans, ten had coronary artery stenoses at the time of angiography.

Although it was generally thought that resting coronary blood flow is normal in patients with coronary artery disease, recent work by Klocke[17] and Cannon and coworkers[8] has shown, by using inert gas

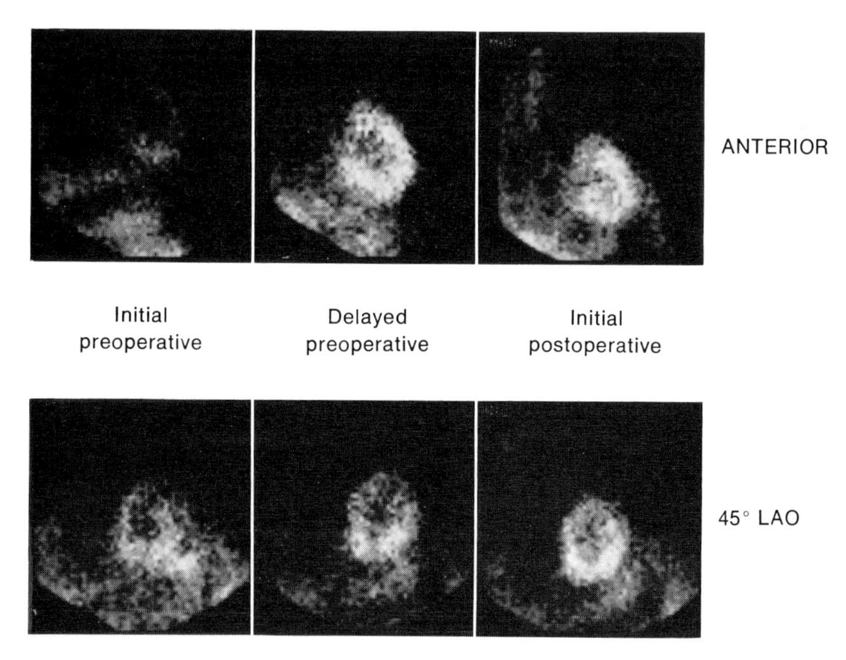

Fig. 12-9. Preoperative and postoperative resting $^{201}Tl^+$ images in patient with three-vessel disease who underwent coronary artery bypass surgery. Preoperative images show initial defects in multiple myocardial segments with redistribution on delayed images. Postoperative initial images show uniform uptake of $^{201}Tl^+$ in left ventricular myocardium. The initial postoperative images are similar to the delayed preoperative images, indicating improvement in early $^{201}Tl^+$ uptake in hypoperfused zones.

washout techniques, that some asymptomatic patients with severe coronary artery disease may have decreased myocardial blood flow at rest. It was previously thought that resting coronary blood flow is not diminished until the coronary arterial diameter is reduced by 75% to 85%. Recent experimental work in animals suggests that 40% to 60% narrowing, if sufficiently prolonged, can reduce resting regional blood flow.[11] More recently, Gewirtz and colleagues,[12] Berger and associates,[5] and Wackers and co-workers[29] have demonstrated myocardial defects on resting $^{201}Tl^+$ images in the absence of pain or electrocardiographic evidence of acute ischemia or recent or old infarction. The majority of these defects, seen 20 min after injection of $^{201}Tl^+$ at rest, fill in with $^{201}Tl^+$ activity on delayed images. Myocardial segments demonstrating rest redistribution are usually perfused by severely stenotic coronary arteries (> 70% narrowing) but demonstrate normal or only hypokinetic wall motion. Berger and colleagues[5] studied fifteen patients with stable and fourteen with unstable angina pectoris and correlated perfusion patterns observed on resting $^{201}Tl^+$ scintigrams with coronary anatomic changes detected on angiograms and segmental wall motion assessed by ventriculography. These investigators showed that fifty-two of sixty-three segments demonstrating redistribution at rest correlated with normal or hypokinetic wall motion. Twenty-two patients in this study had repeat resting $^{201}Tl^+$ scintigrams performed seven to ten days after coronary artery bypass surgery. Of forty-eight myocardial segments demonstrating redistribution preoperatively, thirty-seven became normal, ten continued to show redistribution, and one became a persistent defect. Fig. 12-9 shows preoperative and postoperative resting $^{201}Tl^+$ scintigrams in a patient with three-vessel coronary artery disease (CAD) undergoing coronary artery bypass graft (CABG) surgery. In the preoperative study, diminished early $^{201}Tl^+$ distribution 20 min after injection is apparent in multiple myocardial segments on anterior and LAO views. Delayed images show $^{201}Tl^+$ redistribution into these segments. Postoperatively the early images show uniform uptake throughout the left ventricular myocardium. These early images are identical to delayed images obtained preoperatively. Thus defects in $^{201}Tl^+$ uptake observed at rest 20 min after injection may not represent myocardial scar and irreversible myocardial cellular injury. Rest redistribution is related to severely diminished resting regional blood flow to myocardial segments with normal or hypokinetic wall motion. It is important to note that a defect in initial images obtained after injection of $^{201}Tl^+$ in the resting state may represent viable myocardium, not myocardial scar. Viable and scarred myocardial segments can be differentiated by serial imaging. Those segments which fill in are viable, whereas those which persist are most likely to be scar. Defects in $^{201}Tl^+$ uptake that are demonstrable in the resting state are probably associated with more advanced coronary artery disease than are defects appearing only after exercise.

Sequential myocardial imaging at rest with $^{201}Tl^+$ appears to be useful in the evaluation of patients with variant angina. Maseri and co-workers[21] demonstrated redistribution at rest in patients with variant angina when $^{201}Tl^+$ was injected during an episode of chest pain accompanied by ST-segment elevation. Large transmural defects were demonstrated in images performed within 10 to 20 min of tracer administration. These defects correspond in location to leads showing ST elevation. These defects disappeared 2 hr later, after relief of angina. In contrast to patients with stable or unstable angina pectoris, disappearance of defects in patients with variant angina is most likely secondary to redistribution after restoration of normal or near-normal blood flow with reversal of the spasm by nitroglycerin.

Finally, serial imaging after $^{201}Tl^+$ administration has been employed to distinguish between myocardial ischemia and infarction.[24] Defects that resolve over time appear to represent ischemia. Defects that persist appear to be related to infarction.

IMAGING TECHNIQUES

The sensitivity and specificity of detecting the presence and extent of coronary artery disease by sequential imaging after a single $^{201}Tl^+$ dose will greatly depend on imaging techniques. There are several factors that influence the diagnostic quality of $^{201}Tl^+$ images. The specifications of gamma cameras and collimators vary. Resolution, sensitivity, and field uniformity are all important in $^{201}Tl^+$ imaging. Unfortunately, these factors are interrelated; thus improvement in one is frequently at the expense of another. The best system performance will thus result from a good compromise between different components of the entire system, including the collimator and the imaging system. Most recent vintage cameras will provide adequate $^{201}Tl^+$ images. Some older models, which are still in use and adequate for many radionuclide procedures, may not provide $^{201}Tl^+$ images of diagnostic quality. Many of the newer cameras provide field-nonuniformity correction capability. Since field uniformity is essential, this feature is desirable, although care must be taken in performing correction floods on cameras with automatic correction because errors can introduce artifacts, which will then be impressed into all the subsequent images.

Compton scattering is a major problem associated with the low-energy photons from $^{201}Tl^+$, since 80 keV photons can scatter through an angle greater than 90 degrees with loss of only 10 keV. For example, an energy window of 25% centered on the photopeak will accept photons that have been scattered up to 90 degrees. Using a narrower window to limit the acceptance of scattered photons, would, however, reduce the detection efficiency to such a degree that a loss in image quality would result. A window width of 25% is close to an optimum compromise between loss of statistical definition from a narrow window and loss of image contrast from a wide window because of acceptance of excessive Compton scatter. For the same reason, the $^{201}Tl^{+}$ source must not be placed on the floor or around any massive object when adjusting the camera and setting its window. This will cause excessive backscatter and result in erroneous camera adjustments. The $^{201}Tl^{+}$ energy spectrum recorded from a patient study will also have a greatly distorted photopeak due to low-angle Compton scattering. The automatic peak-tracking feature found in some commercial gamma cameras will be misguided by the distorted photopeak and caused to incorrectly lower the window setting. Automatic peaking should not be used with $^{201}Tl^{+}$.

A medium-resolution (general all-purpose) collimator may be preferable to a high-resolution collimator, since it allows higher count rates and shorter imaging times, which are important in detecting transient perfusion defects; these defects rapidly fill in by redistribution. The general all-purpose collimator has nearly the same resolution as the high-resolution collimator but loses resolution more rapidly with increasing distance from the collimator face. This appears to be an acceptable compromise to reduce imaging time and achieve a more rapid image sequence. Images are acquired using a constant time per image (6 to 10 min after a standard $^{201}Tl^{+}$ dose of 1.5 mCi). Since one is interested in what fraction of the injected $^{201}Tl^{+}$ is accumulated regionally in the heart, imaging for a preset time is a more direct way of obtaining that information than recording for constant counts or for constant count densities. On the other hand, imaging for a preset count or count density may lead to more comparable images, but duration of image acquisition should be recorded to determine the relative change in the actual counts that are in each myocardial segment.

Adequate stress and well-timed injection of $^{201}Tl^{+}$ are important to the exercise stress study. It must be assumed that the $^{201}Tl^{+}$ images only demonstrate perfusion defects that exist at the time of injection and will not reveal potential perfusion defects, which would occur at higher levels of exercise stress. Exercise should optimally be continued for at least 1 full min after the injection, if necessary, at a reduced level. Myocardial cells in an ischemic region are known to accumulate a potassium debt during exercise and to begin very rapid potassium extraction almost immediately on cessation of exercise. Although a similar effect has not been demonstrated with ^{201}Tl,$^{+}$ it is reasonable to assume that it does occur and that the injection should be timed so most of the $^{201}Tl^{+}$ will be extracted prior to the complete cessation of exercise. Additionally, to obtain optimum studies, patients should fast prior to $^{201}Tl^{+}$ scintigraphy. Food will increase mesenteric splanchnic blood flow, causing higher background activity adjacent to the heart, particularly in resting studies. This will degrade the diagnostic quality of the study.

Images should be recorded from the computer on transparency film. Although adequate images can be recorded on Polaroid film, it is much more difficult to maintain reproducible and optimal gray scale.

All $^{201}Tl^{+}$ studies should be recorded on a computer for standardized image formation and display of relative myocardial distribution and delayed redistribution. A modified exponential gray scale for use with computer videotape displays has been designed to facilitate visual perception of small defects within the region of highest activity.[31] This nonlinear intensity scale provides adequate visualization of raw $^{201}Tl^{+}$ images and avoids the arbitrary and sometimes misleading use of conventional background cutoff and contrast enhancement. The purpose of this display is to produce clear visual perception of small brightness decrements (representing possible perfusion defects) within regions of otherwise intense myocardial uptake. A similar scale can be used with a color videotape system but with a color scale, that maintains a constant hue (e.g., green) and adds increments of white to enhance the perception of small intensity changes near the high-intensity portion of the color scale.[31] This monohue scale allows an apparent expansion of the gray scale but preserves the intuitive sense of relative intensities, which is lost in the conventional multicolor formats. With this image presentation, background suppression is not required to see minimal defects. All sequential $^{201}Tl^+$ images are then formed on the same scale with no adjustable parameters.

Quantitative evaluation of $^{201}Tl^+$ images cannot be accomplished without adequate background subtraction techniques. Background activity is variable and has an obviously nonuniform distribution. Moreover, the distribution of background activity changes quite significantly over the time interval between initial and delayed images. One satisfactory background subtraction method is the bilinear interpolation scheme of Goris and associates.[13] Slightly more complex weighting functions[31] have been introduced, which differ from the original formula. These functions provide proximity weighting near the edges of the background-defining region, which also causes more rapid fall-off of the generated background because it is extrapolated beneath the myocardial rim in the proximity of intense background regions, such as the liver and stomach. This latter method of background subtraction has proved to be highly reproducible and independent of the computer operator.

After background subtraction is performed, residual myocardial activity can be easily determined either by setting small regions of interest in selected myocardial segments or by observing profiles generated across the selected regions. The image matrix is 64 by 64 channels and nine-point smoothing is probably useful prior to quantitation. Myocardial segments are typically chosen in the inferior, apical, and anterolateral segments on anterior projections and in upper and lower septal, upper and lower posterior, and inferoapical segments of LAO projections. The use of image smoothing and rather coarse selection of regions of interest provides adequate statistical accuracy but is consistent with the limited resolution of this technique.

Utilizing quantitative techniques for assessing $^{201}Tl^+$ distribution and redistribution may allow detection of underperfused or ischemic zones that visually do not appear as "defects" on scintigrams. For example, the most intense region of any image is conventionally considered to be the "normal" area to which other segments are compared. Certain patients with severe three-vessel disease have diffusely decreased $^{201}Tl^+$ uptake after stress or at rest. Areas that are more severely underperfused will appear as defects in relation to another segment with a lesser perfusion deficit. However, on sequential images, a global net accumulation of $^{201}Tl^+$ into all myocardial regions may be demonstrated, indicating redistribution. Fig. 12-10 illustrates this phenomenon in a patient with multivessel coronary disease. Quantitation of serial images appears to increase the detection sensitivity of diminished perfusion to multiple segments when no single region is perfused by a normal vessel.[3]

We can now describe at least five possible patterns of $^{201}Tl^+$ uptake in the myocardium (Fig. 12-11). Type I is characterized by persistently diminished $^{201}Tl^+$ uptake in a myocardial segment. This region has decreased initial uptake, which remains decreased by a constant fraction relative to the normal area. There is reduced uptake but not concomitant reduced washout in the affected segment. This may be because one is looking at normal areas in the solid angle through the infarct region. This pattern may be observed in the presence of myocardial scar (Fig. 12-11). The type II pattern is the prototype for redistribution following exercise stress. The normally perfused myocardial segment receives an increased fraction of the injected dose of $^{201}Tl^+$. The abnormally perfused segment receives less $^{201}Tl^+$ during exercise because it is supplied by a stenotic artery, which can deliver adequate resting blood flow but cannot adequately increase blood flow during exercise.

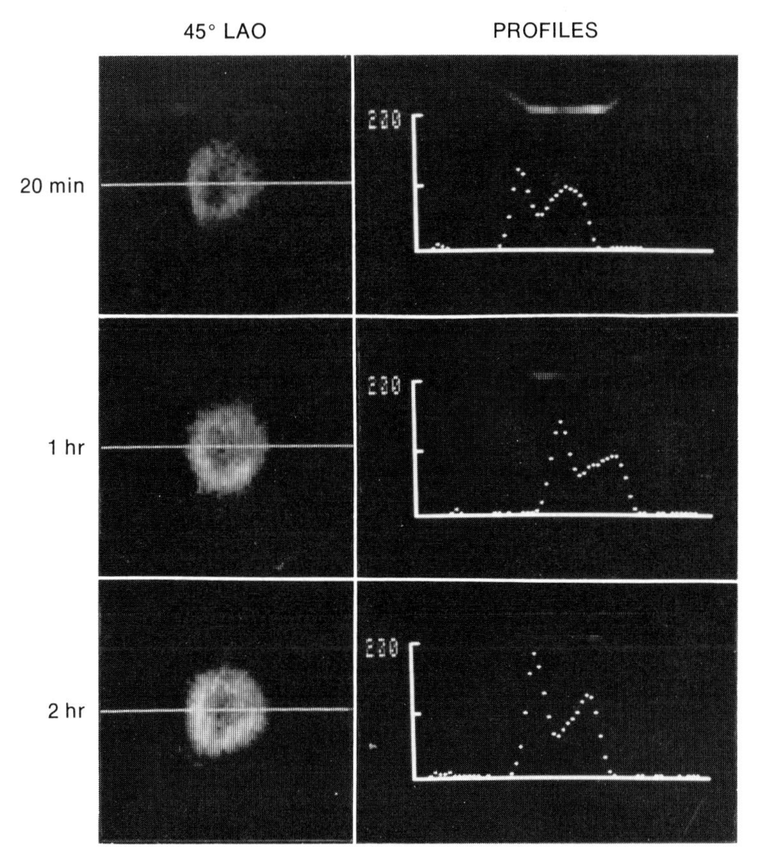

Fig. 12-10. Serial $^{201}Tl^+$ images at rest in LAO projection demonstrating multiple areas of redistribution. Intensity scale is same for all images. Count profiles for each image are on right. Although septal region has higher regional activity in initial image and "defect" is present in posterolateral region, delayed image shows further accumulation of $^{201}Tl^+$ in septum. Persistent posterolateral defect is observed. Postoperatively, initial image shows improved perfusion in posterolateral region and septum with almost uniform early $^{201}Tl^+$ uptake.

After the initial distribution phase, the normal area washes out more rapidly than the abnormal region. In the equilibrium phase, both normal and abnormal segments will approach equal $^{201}Tl^+$ concentration, since delayed accumulation of $^{201}Tl^+$ occurs in the previously ischemic region. Redistribution in this case is caused entirely by a change in input rate, reflecting transiently increased blood flow at the time of exercise to segments supplied by normal coronary arteries.

A similar image sequence can also be observed when $^{201}Tl^+$ is injected during a spasm of a normal coronary artery. In this instance the initial defect in $^{201}Tl^+$ uptake occurs because the flow was transiently decreased in the distribution of the occluded vessel. When the spasm is relieved, redistribution occurs because of a rapid accumulation of $^{201}Tl^+$ in the involved segment, since total flow has been restored.

The type IIIA pattern occurs without a change in perfusion after $^{201}Tl^+$ injection at rest. This pattern is characterized by a focal decrease in initial $^{201}Tl^+$ uptake secondary to chronic underperfusion in an area of persistently diminished blood flow. Redistribution of $^{201}Tl^+$ occurs in the underperfused zone because of delayed equilibration and possibly a decreased intrinsic washout. In this case, washout from the normally per-

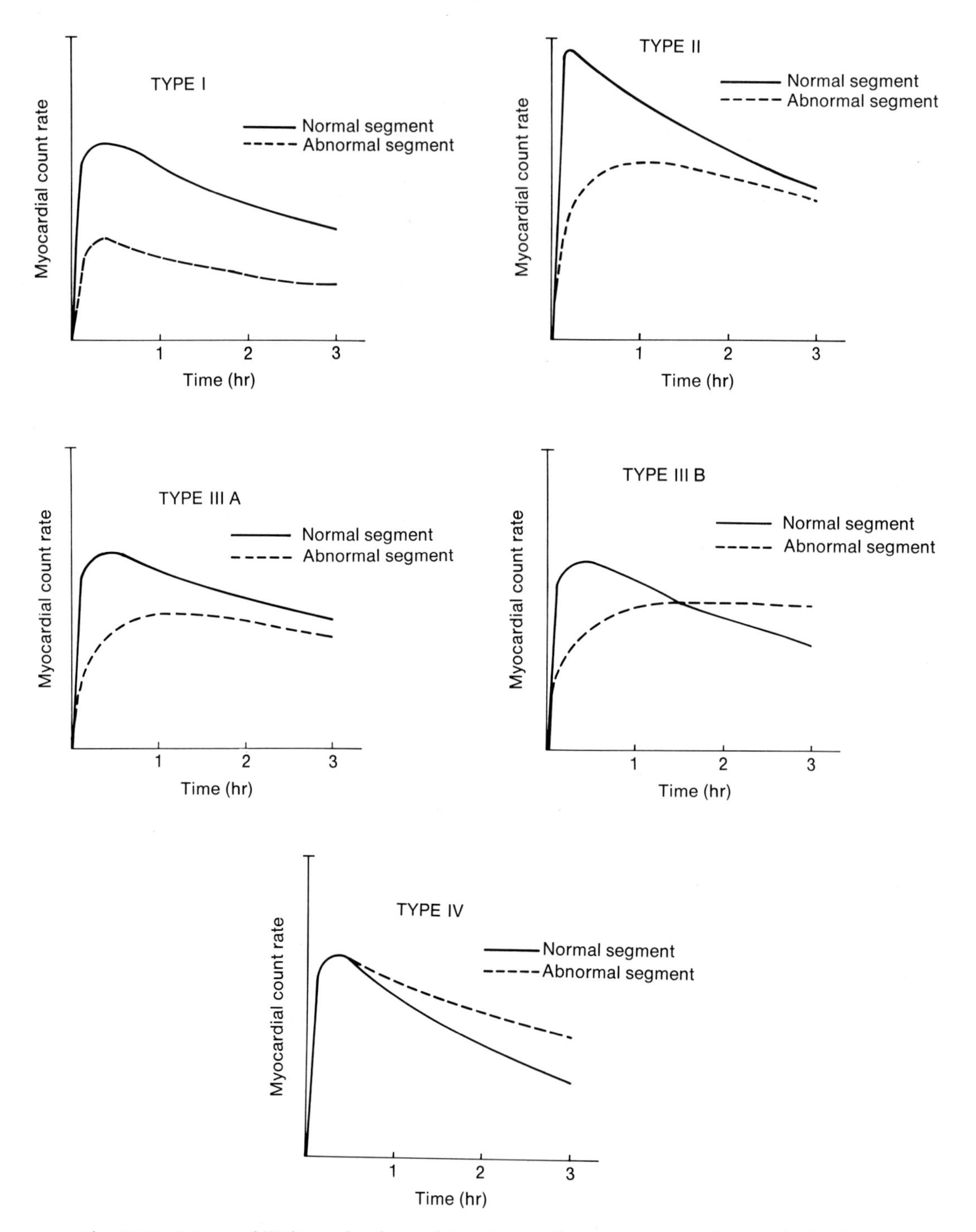

Fig. 12-11. Patterns of $^{201}Tl^{+}$ uptake observed in patients with coronary artery disease. Idealized time-activity curves are shown for normal and ischemic segments.

fused zones contributes more importantly to apparent filling in of defects than in the patterns previously described. The type IIIB pattern is similar to type IIIA except that intrinsic washout is markedly decreased in abnormally perfused regions. Thus on late images, "defect reversal" may be observed because $^{201}Tl^{+}$ concentration in normal myocardial segments ultimately falls below $^{201}Tl^{+}$ concentration in the abnormal segment.

There have been anecdotal reports of another pattern of $^{201}Tl^{+}$ uptake for which no easy explanation is apparent. This pattern (type IV) is characterized by normal uniform $^{201}Tl^{+}$ uptake on initial images with a defect appearing for the first time on delayed images. This may imply that a myocardial region with normal perfusion and extraction of $^{201}Tl^{+}$ had an altered clearance rate of the radionuclide. Another possibility is that one is actually comparing two abnormal myocardial regions, both with altered perfusion. The time course of $^{201}Tl^{+}$ redistribution might be asynchronous, with one segment accumulating less $^{201}Tl^{+}$ than the other over time. On a delayed image, this segment could be visualized as a defect in comparison to the segment that accumulated more $^{201}Tl^{+}$ over time. The development of a defect in delayed images is uncertain at the present time and further studies are warranted to investigate the significance of the pattern. Suffice it to say that this pattern has been found in a number of patients without change in coronary arterial or myocardial disease.

In conclusion, sequential myocardial imaging after a single intravenous dose of $^{201}Tl^{+}$ requires an understanding of $^{201}Tl^{+}$ kinetics and knowledge of the factors and variables that determine the initial myocardial distribution of $^{201}Tl^{+}$ and the change in this distribution over time. The clinical utilization of the sequential imaging method requires adequate instrumentation, including computer technology, high-quality display systems, technical expertise in acquiring and processing of $^{201}Tl^{+}$ imaging data, and clinical experience in interpreting the studies. The sensitivity and specificity of $^{201}Tl^{+}$ imaging may continue to improve with further technical advances. The recent introduction of three-dimensional reconstruction techniques may be helpful in this regard (Chapter 2).

REFERENCES

1. Bailey, I. K., Griffith, I. S. C., Rouleau, J., et al.: Thallium-201 myocardial perfusion imaging at rest and during exercise. Comparative sensitivity to electrocardiography in coronary artery disease, Circulation **55:**80, 1977.
2. Beller, G. A., and Pohost, G. M.: Mechanism for thallium-201 redistribution after transient myocardial ischemia, abstracted, Circulation **56:**141, 1977.
3. Beller, G. A., Watson, D. D., Berger, B. C., et al.: Sensitivity of exercise scintigraphy for coronary disease detection using quantitation of thallium-201 activity (submitted for publication).
4. Bennett, K. R., Smith, R. O., Lehan, P. H., et al.: Correlation of myocardial ^{42}K uptake with coronary arteriography, Radiology **102:**117, 1972.
5. Berger, B. C., Watson, D. D., Sipes, J. N., et al.: Redistribution of thallium at rest in patients with coronary artery disease, abstracted, J. Nucl. Med. **19:**680, 1978.
6. Botvinick, E. H., Taradosh, M. R., Shames, D. M., et al.: Thallium-201 myocardial perfusion scintigraphy for the clinical clarification of normal, abnormal and equivocal electrocardiographic stress tests, Am. J. Cardiol. **41:**43, 1978.
7. Bradley-Moore, P. R., Lebowitz, E., Greene, M. W., et al.: Thallium-201 for medical use, II. Biologic behavior, J. Nucl. Med. **16:**156, 1975.
8. Cannon, P. U., Dell, R. B., and Dwyer E. M., Jr.: Regional myocardial perfusion rates in patients with coronary artery disease, J. Clin. Invest. **51:**978, 1972.
9. Carello, A. P., Marks, D. S., Pickard, S. D., et al.: Correlation of exercise ^{201}thallium myocardial scan with coronary arteriograms and the maximal exercise test, Chest **73:**321, 1978.
10. Carr, E. A., Jr., Beierwaltes, W. H., Wegst, A. V., et al.: Myocardial scanning with rubidium-86, J. Nucl. Med. **3:**77, 1962.
11. Felman, R. L., Nichols, W. W., Pepine, C. J., et al.: Hemodynamic significance of the length of a coronary arterial narrowing, Am. J. Cardiol. **41:** 865, 1978.
12. Gewirtz, H., Beller, G. A., Strauss, H. W., et al.: Transient defects on resting thallium scans in patients with coronary artery disease, Circulation (in press).
13. Goris, M. L., Daspit, S. G., McLauglin, P., et al.: Interpolative background subtraction, J. Nucl. Med. **17:**744, 1976.
14. Guiney, T. E., Pohost, G. M., McKusick, K. A.,

et al.: Differentiation of false from true positive electrocardiographic responses to exercise stress by single dose of thallium-201 perfusion imaging (submitted for publication).

15. Harper, P. V., Schwartz, J., Beck, R. N., et al.: Clinical myocardial imaging with nitrogen-13 ammonia, Radiology **108:**613, 1973.
16. Kawana, M., Krizek, H., Porter, J., et al.: Use of ^{199}Tl as a potassium analog in scanning, J. Nucl. Med. **11:**333, 1970.
17. Klocke, F. J.: Coronary blood flow in man, Prog. Cardiovasc. Dis. **19:**117, 1976.
18. Ku, D., Akera, T., Tobin, T., et al.: Effects of monovalent cations on cardiac Na^+, K^+-ATP-ase activity and on contractile force, Arch. Pharm. **290:**113, 1975.
19. Lebowitz, E., Greene, M. W., Fairchild, R., et al.: Thallium-201 for medical use, I, J. Nucl. Med. **16:**151, 1975.
20. Martin, N. D., Zaret, B. L., McGowan, R. L., et al.: Rubidium-81: a new myocardial scanning agent, Radiology **111:**651, 1974.
21. Maseri, A., Parodi, O., Severi, S., et al.: Transient transmural reduction of myocardial blood flow, demonstrated by thallium-201 scintigraphy, as a cause of variant angina, Circulation **54:**280, 1976.
22. McLaughlin, P. R., Doherty, P. W., Martin, R. P., et al.: Myocardial imaging in a patient with reproducible variant angina, Am. J. Cardiol. **39:**126, 1977.
23. Pohost, G. M., O'Keefe, D. D., Gewirtz, H., et al.: Thallium redistribution in the presence of severe fixed coronary stenosis, abstracted, J. Nucl. Med. **19:**680, 1978.
24. Pohost, G. M., Zir, L. M., Moore, R. H., et al.: Differentiation of transiently ischemic from infarcted myocardium by serial imaging after a single dose of thallium-201, Circulation **55:**294, 1977.
25. Ritchie, J. L., Trobaugh, G. B., Hamilton, G. W., et al.: Myocardial imaging with thallium-201 at rest and during exercise. Comparison with coronary arteriography and resting and stress electrocardiography, Circulation **56:**66, 1977.
26. Romhilt, D. W., Adolph, R. J., Sodd, U. C., et al.: Cesium-129 myocardial scintigraphy to detect myocardial infarction, Circulation **48:**1242, 1973.
27. Skulskii, A., Manninen, V., and Jarnfelt, J.: Interaction of thallous scans with the cation transport mechanism in erythrocytes, Biochim. Biophys. Acta **298:**702, 1973.
28. Strauss, H. W., Harrison, K., Langan, J. K., et al.: Thallium-201 for myocardial imaging. Relation of thallium-201 to regional myocardial perfusion, Circulation **51:**641, 1975.
29. Wackers, F. J., Lie, K. L., Liem, K. L., et al.: Thallium-201 scintigraphy in unstable angina pectoris, Circulation **57:**738, 1978.
30. Wackers, F. J., Sokole, E. B., Samson, G., et al.: Value and limitations of thallium-201 scintigraphy in the acute phase of myocardial infarction, N. Engl. J. Med. **295:**1, 1976.
31. Watson, D. D., Gridley, D., and Beller, G. A.: The brightness increment display (submitted for publication).
32. Watson, D. D., Irving, J. F., Berger, B. C., et al.: A mechanism for thallium redistribution (submitted for publication).
33. Weich, H., Strauss, H. W., and Pitt, B.: The extraction of thallium-201 by the myocardium, Circulation **56:**188, 1977.
34. Zaret, B. L., Strauss, H. W., Martin, N. D., et al.: Noninvasive regional myocardial perfusion with radioactive potassium, N. Engl. J. Med. **288:**809, 1973.
35. Zir, I. M., Beller, G. A., Strauss, H. W., et al.: Thallium-201 stress testing. Experience utilizing the redistribution technique (submitted for publication).

13 □ Clinical application of myocardial imaging with thallium

Bertram Pitt
H. William Strauss

Myocardial imaging with ^{201}Tl has been found to be widely applicable in the clinical evaluation of patients with both ischemic and nonischemic cardiac disease.[21,29]

Over a wide physiologic range of myocardial blood flows, the uptake of ^{201}Tl by the heart is proportional to the myocardial blood flow.[27] Within the normal range of myocardial blood flow the extraction rate for ^{201}Tl is approximately 88%.[32] At the extremes of the myocardial blood flow range, the extraction of ^{201}Tl by the myocardium is altered so that at low blood flow rates the extraction is greater and at high rates, less. The relationship between myocardial blood flow and ^{201}Tl uptake may also be altered by a number of pharmacologic and physiologic interventions that affect $Na,^{+}K^{+}$-ATPase activity.[17,32]

In normal individuals, ^{201}Tl is relatively homogeneously distributed throughout the myocardium. There is often a relative diminution of tracer uptake at the apex in the anterior view because the apex is physiologically thinner than the left ventricular free wall.[14] Uptake of ^{201}Tl by the right ventricle at rest is usually not well defined in the LAO. When ^{201}Tl is taken up by the right ventricular free wall, its mass and blood flow per gram are much less than in the left ventricle. During tachycardia, exercise, and conditions resulting in increased right ventricular mass, the right ventricular free wall is, however, easily detectable on the LAO ^{201}Tl image.[12]

There is at present no general agreement as to the best way to display ^{201}Tl myocardial images for analysis. Adequate results have been obtained by viewing the unprocessed image recorded on Polaroid or 35 mm film from the CRT of the scintillation camera. More recently the ^{201}Tl image has been subjected to various forms of computer processing and enhancement. Programs have been developed to extract the myocardium from surrounding background areas, such as the pneumonic and hepatic-gastric areas. The extracted myocardial image can then be further processed by simple background subtraction or more complex computer programs. Attempts have also been made to analyze the distribution of ^{201}Tl within the myocardium by computer processing. An early attempt was made to correlate relative myocardial ^{201}Tl activity to background. The ratio of myocardial ^{201}Tl activity (RMTA) to background was then compared between individuals and those with infarction or ischemia.[26] Although initial studies showed good separation of values in normal individuals and those with ischemia or infarction, further experience with a larger group of patients revealed the difficulties with this ratio approach. The RMTA value could be altered by either a decrease in the myocardial concentration of ^{201}Tl or an increase in the pulmonary concentration. Recent work by Bingham[6] indicates that the pulmonary extraction of ^{201}Tl is inversely related to the transit time through the lungs. A rapid transit time is associated with a low ^{201}Tl extraction, whereas a prolonged transit time is associated with an increased extraction. In patients with congestive cardiac failure due to the prolonged transit time through the lungs, a high ^{201}Tl lung concentration is commonly found. Therefore a low RMTA value could be due either to an intrinsic decrease in myocardial ^{201}Tl concentration or to a increase in lung concentration. The ambiguity in the cause of a low RMTA value makes this ratio approach less useful for quantifying ^{201}Tl scans than was originally hoped.

More recently, computer programs have been generated to detect the center of activity within the myocardial image.[5] Radii are constructed from the center of activity to the edge of the myocardial image, and the activity along each radus is determined. The activity for the entire 360 degrees of the outlined edge of the myocardial image is then displayed on a linear graph. Images from patients with myocardial ischemia or infarction will have radii with decreased activity compared to those of normal patients. Although attractive and of possible clinical value, caution should be used in the application of any currently available computer-assisted program for analysis and interpretation of ^{201}Tl activity in view of the heterogeneous and variable nature of the background activity and the individual differences in attentuation coefficient. Although many of the difficulties may be overcome with further experience or by newer imaging techniques, such as planar or time-lapse tomographic imaging, it is essential that for the immediate future each institution establish criteria for interpretation of normal and abnormal images on its own equipment. If computer facilities are available, normal volunteers should be imaged and their data stored on magnetic tape or disk. The images should be subject to whatever processing is decided on, for example, 20% background subtraction, and all subsequent patients interpreted relative to the normal studies. It is unacceptable to view patient studies at high contrast or with 30% to 50% background subtraction and to diagnose the presence of myocardial ischemia or infarction if normal images have not been processed in an identical manner and found to have uniform ^{201}Tl uptake. It is relatively easy with current computer and image display systems to obtain false-positive defects in tracer uptake. This is especially true if the images are viewed on a color display. Color tables can be constructed in which a 5% reduction in activity appears as a major defect in tracer uptake. For example, by choosing a color scale so that 5% increments in activity result in a difference between red and white, a 5% difference may appear to be a defect in tracer uptake, leading to a false diagnosis of myocardial ischemia.

One of the major advantages of a computer system is that the images can be viewed at several different contrast and brightness settings. It is best to evaluate the study for the presence of pulmonary background and right ventricular activity using a low-contrast setting, whereas small changes in the thickness of the myocardium and the regional activity in one zone compared to another are best defined with higher levels of contrast enhancement.

RIGHT VENTRICULAR MYOCARDIUM

Although, as mentioned previously, the right ventricle is not well defined on the resting LAO ^{201}Tl myocardial image, it is easily seen under conditions that increase right ventricular myocardial flow, such as exercise, excitement, or increased pressure or volume work leading to right ventricular hypertrophy. Right ventricular hypertrophy is often difficult to detect by standard electrocardiography, particularly if concomitant left ventricular hypertrophy is present. The LAO ^{201}Tl image, however, allows independent assessment of right and left ventricular myocardium. In patients with pulmonary hypertension and resultant right ventricular hypertrophy, the ventricular free wall is easily visualized on the resting LAO ^{201}Tl myocardial image.[12] The ability to independently examine the intraventricular septum as well as the left and right ventricular free walls with the ^{201}Tl myocardial image has been found to be useful in the evaluation of patients with cyanotic congenital heart disease.[20] The detection of single ventricles is often difficult clinically, and occasionally, even with contrast ventriculography, it is difficult to determine the exact diagnosis and anatomic relationship. In a patient suffering from cyanotic congenital heart disease with a ventricular septal defect, such as with the tetralogy of Fallot, the LAO ^{201}Tl image reveals the presence of an intraventricular septum with clear separation of both the right and left ventricular free walls. In a patient with a single ventricle, however, a single ventricular cavity or a large chamber with a small outlet chamber is seen without the presence of an interventricular septum, no matter how the patient is rotated. Although the right atrium has been detected in a patient with cyanotic congenital heart disease,[15] this is a very rare occurrence.

CORONARY ARTERY DISEASE

Ischemia

^{201}Tl myocardial imaging has been found of the greatest use in the evaluation of patients with suspected ischemic heart disease.[4] Results from several research centers and a recent multicenter trial have shown that the sensitivity and specificity of exercise ^{201}Tl myocardial imaging for the detection of myocardial ischemia are greater than with conventional exercise electrocardiography.[4,7,25] Although exact figures vary from center to center, depending on patient selection and technique, the sensitivity and specificity for exercise ^{201}Tl myocardial imaging appear to be in the range of 85% to 90%, compared to 65% to 70% for exercise electrocardiography.

Exercise ^{201}Tl myocardial imaging is normally performed in conjunction with exercise electrocardiography. Resting supine and upright electrocardiograms are obtained, and the patient then exercised on a bicycle ergometer or treadmill to maximum capacity. At that point 1.5 to 2.0 mCi of ^{201}Tl is injected through an indwelling intravenous catheter and the patient asked to continue exercising for an additional 20 to 30 sec at a lower level to allow tracer to be taken up over several circulations to the tissues in proportion to blood flow. From 2 to 5 min after injection of tracer, patient is placed beneath the detector of the scintillation camera and imaging is begun in the anterior, LAO and left lateral positions. It is important to begin imaging within 5 to 10 min after injection of the tracer because redistribution of ^{201}Tl into an ischemic area is relatively rapid, resulting in a loss of definition and sensitivity for detection of ischemia. A focal defect in tracer uptake seen on the postexercise myocardial image suggests the presence of myocardial ischemia or infarction. Ischemia can be diagnosed if the initial defect found at exercise is no longer seen on reimaging after several hours (to observe the redistribution of ^{201}Tl[22]) or imaging at rest after a second injection of the tracer several days later.[4] Myocardial infarction can be diagnosed if the tracer uptake defect seen on the postexercise image persists unchanged at rest. The ability to reimage hours later without additional doses of the tracer has the advantage that only a single dose of the tracer is required, thereby resulting in considerable savings in cost and reduction of radiation exposure. If, however, the initial defect is smaller but persistent on reimaging several hours later, the distinction between ischemia and infarction becomes uncertain, and the defect should be reimaged sometime within the next 12 hr or a second dose of ^{201}Tl administered at rest several days later. The variability in the time of redistribution for ^{201}Tl has in part eliminated the clinical utility of redistribution ^{201}Tl imaging. In a patient being evaluated for suspected ischemic heart disease, it may, however, be sufficient to detect a defect in tracer uptake during stress, thereby establishing the presence of ischemic heart disease.

Redistribution of ^{201}Tl into an ischemic region over time has been shown to occur independently of any change in myocardial blood flow. In a patient with a persistent area of myocardial ischemia, the initial ^{201}Tl myocardial image will show a defect in tracer uptake. This defect in tracer uptake depends to a large extent on the initial delivery of ^{201}Tl. Since myocardial flow to the ischemic region is less than normal, regional tracer uptake will be less and therefore a tracer defect will be appreciated on imaging. Over time, however, the ischemic area continues to extract ^{201}Tl from residual blood-flow activity. Eventually the intramyocardial ^{201}Tl concentration in the ischemic region approaches that in the normal area, the final concentration depending on the myocardial mass and intracellular cation content.

Exercise ^{201}Tl myocardial imaging is of particular value under conditions in which the exercise electrocardiogram is difficult to interpret, such as in the presence of an intraventricular conduction defect, old myocardial infarction, left ventricular hypertrophy, and electrolyte abnormalities. In patients with a normal resting electrocardiogram, the sensitivity of ^{201}Tl to detect ischemia is approximately equal to or only slightly greater than that of exercise electrocardiography. Exercise ^{201}Tl myocardial imaging is, however, of special value in the evaluation of asymptomatic patients for ischemic heart disease.[11] In asymptomatic individuals, an exercise electrocardiogram showing "an ischemic ST-segment" response may often be falsely positive. Since the prevalence of significant coronary artery disease and myocardial ischemia is relatively low in asymptomatic individuals and the specificity of exercise electrocardiography is greater than 90%, a high proportion of positive electrocardiographic responses on exercise testing will be false. An independent test of myocardial ischemia such as ^{201}Tl imaging adds increased certainty to the diagnosis. If both the exercise electrocardiogram and the exercise ^{201}Tl myocardial image are suggestive of ischemia, significant obstructive coronary artery disease will be detected in a high percentage of patients. If, however, the exercise ^{201}Tl myocardial image is negative in an asymptomatic individual with a positive exercise electrocardiogram, the certainty of significant anatomic coronary artery disease diminishes greatly. If both the exercise electrocardiogram and exercise ^{201}Tl myocardial image are negative, myocardial ischemia is unlikely. Significant coronary artery disease may, however, be present even in the absence of a positive exercise electrocardiogram or exercise ^{201}Tl myocardial image. This may be because functional collateral vessels are present or that the functional significance of a given anatomic coronary artery lesion may have been overestimated by conventional cineangiography.

Exercise ^{201}Tl myocardial imaging is also of value in the evaluation and follow-up studies of patients being considered for CABG surgery.[13,23] As mentioned earlier, the location of the defect with the ^{201}Tl-labeled tracer can be of value in localizing the area of significant coronary artery narrowing. A number of instances have arisen after coronary arteriography, in which there was uncertainty as to the functional significance of a given 40% to 60% coronary artery lesion. During exercise, the appearance of a defect in tracer uptake within the distribution of the vessel under question is evidence of the functional significance of the lesion.

The distribution of myocardial blood flow within the region of the three major coronary arteries can be assembled from the anterior, 40- and 60-degree LAO ^{201}Tl myocardial images. On the basis of cineangiographic studies by Bailey and associates in patients with single-, double-, and triple-vessel disease, the distribution of the left anterior descending coronary artery was shown to be reflected in the anterior apical portion of the anterior ^{201}Tl myocardial image and the septal portion of the 40- and 60-degree LAO image.[3] The right and left circumflex coronary artery distribution appears to overlap in the inferior border of the anterior image and the lateral border of the 60-degree LAO image. The high lateral border of the 40-degree LAO image appears, however, to be relatively specific for the left circumflex coronary artery. Defects in the apex of the 40- and 60-degree LAO images have relatively little localizing value because they may occur with lesions of any of the major coronary arteries. Significant coronary artery lesions of the left anterior descending coronary artery above the first septal perforator can be distinguished from those below it by the finding of a defect involving the anterior apical area of the anterior image and the septal area of the 40- and 60-degree LAO images. In patients with lesions distal to the first septal perforator, tracer uptake defects are found more distally on the anterior view involving the apex and do not involve the septum in the 40- and 60-degree LAO image. This distinction could be of clinical importance because lesions of the left anterior descending coronary artery above the first septal perforator are associated with a higher mortality than those below the first septal perforator.

Although useful in locating the area of ischemia, ^{201}Tl has only limited utility in determining whether or not single- or multiple-vessel disease is present. The appearance of a defect in two or more vascular areas is strong evidence for the presence of double- or triple-vessel disease. The appearance of a defect in tracer uptake in a single vascular area cannot, however, be used to exclude multiple-vessel disease because one vascular area may become ischemic before another, and exercise may have been terminated before the other vascular areas became ischemic. In some instances, a significant coronary artery lesion may not become apparent during exercise because of the presence of functional collateral vessels. ^{201}Tl myocardial imaging may also be of value in determining whether a segment of myocardium is viable or contains only scar tissue. Absence of tracer uptake at rest with an RMTA to background of approximately 1:1 and a failure to see increased ^{201}Tl uptake on redistribution imaging suggests the presence of scar tissue. Persistent, although diminished, uptake at rest with an increase on redistribution imaging suggests viable myocardium. This differentiation is often of importance in the decision whether or not a bypass graft to a particular region should be attempted.

Perhaps the most important use of ^{201}Tl myocardial imaging in patients undergoing CABG surgery is in the assessment of the results of the operation. Symptomatic relief has been found to occur in over 85% of the patients undergoing CABG surgery. Relief of symptoms after CABG surgery may, however, be due to a number of causes other than relief of myocardial ischemia, including the placebo effect, destruction of sensory coronary artery nerves, or perioperative infarction with resultant loss of previously ischemic myocardium. In a patient with an exercise-induced ^{201}Tl uptake defect preoperatively, the finding of a completely normal rest and exercise ^{201}Tl image postoperatively is evidence of graft patency and successful relief of ischemia. Persistence of an exercise-induced tracer defect postoperatively despite relief of symptoms suggests a placebo effect, often with an occluded graft, whereas the appearance of a new resting ^{201}Tl uptake defect postoperatively suggests perioperative infarction as the cause for relief of symptoms.

Although exercise has been of great value in provoking myocardial ischemia in patients with suspected ischemic heart disease, because of peripheral vascular or pulmonary disease, not every patient is able to achieve sufficient exercise to determine whether or not significant coronary artery disease is present. Another approach involves the use of maximum coronary vasodilation.[28] In a patient with a subcritically narrowed coronary artery, for example, resting coronary blood flow and hence tracer uptake will be similar in the distribution of the narrowed left anterior descending and normal right and left circumflex coronary artery vascular regions. After administration of a maximal coronary artery vasodilator such as ethyl adenosine or dipyridamole, coronary flow will increase greater than twofold in the normal vascular regions, while flow in the distribution of the subcritically narrowed left anterior descending coronary artery will increase only slightly. Administration of ^{201}Tl during this period will reveal a relative defect in tracer uptake because flow and hence tracer uptake will be relatively greater in the distribution of the normal coronary arteries. This technique has proved clinically useful for the detection of ischemic heart disease in patients for whom exercise was not feasible or desirable.[1]

In patients with myocardial ischemia due to a coronary artery spasm, ^{201}Tl myocardial imaging has also been useful. Injection of ^{201}Tl during a spontaneous episode of Prinzmetal-variant angina reveals a defect in tracer uptake, which, on redistribution imaging several hours later, fills in.[18] The finding that the RMTA to background falls to 1:1 during the episode of spontaneous variant angina lends indirect support to the premise that transmural ischemia caused by a coronary spasm is an important mechanism in these patients. ^{201}Tl myocardial imaging may also be of value in provocative testing for coronary artery spasms. Ergonovine has been used to provoke coronary artery spasms during arteriography and has also been used with ^{201}Tl myocardial imaging. In a patient with suspected coronary artery spasms, ergonovine is injected in increments, with the patient beneath the detector of the scintillation camera. At the onset of precordial chest pain or after the maximum dose of ergonovine has been administered, ^{201}Tl should be injected and imaging begun within 5 min. If a defect in ^{201}Tl myocardial uptake is detected, imaging should be repeated approximately 3 hr later to determine whether the initial tracer defect was because of a coronary artery spasm as manifested by transient myocardial ischemia or because of previous myocardial infarction.

Infarction

^{201}Tl myocardial imaging has been found of wide use in the detection and evaluation of acute myocardial infarction.[31] If the patient is evaluated within the first 6 hr of onset of symptoms, about 100% of those with transmural and nontransmural infarction can be detected. With increasing time from onset of symptoms there is a moderate reduction in the sensitivity for detection of transmural infarction and a marked reduction in detection of nontransmural infarction. Wackers and colleagues[31] detected infarctions in less than half of their patients with nontransmural infarction that were studied 24 hr or more after the onset of their symptoms. Despite this loss of sensitivity with time for detection of myocardial infarction, the sensitivity of ^{201}Tl myocardial imaging is greater than that of standard electrocardiography. In a study of 101 patients with angiographically significant coronary artery disease and a history of myocardial infarction, Bailey[2] found a resting myocardial ^{201}Tl uptake defect in a significantly greater number of patients than those with definite electrocardiographic evidence of infarction. Although more sensitive than the electrocardiogram for the detection of old myocardial infarction, small infarcts of less than 7g may be missed by myocardial imaging.[19]

The clinical value of an independent technique to detect myocardial infarction is evident in patients entering the coronary care unit (CCU) with a history suggestive of myocardial infarction but in whom the electrocardiogram is difficult or impossible to interpret because of the presence of left bundle-branch block. The location and extent of the initial defect in ^{201}Tl uptake may also be of value in predicting the likelihood of subsequent congestive heart failure, complex ventricular arrhythmias, and death. In a recent study, ^{201}Tl myocardial imaging was used to assess the effect of therapeutic interventions during the early stages of infarction.[5] In a double-blind study of the effect of nitroglycerin versus a placebo in patients with acute myocardial infarction, patients receiving nitroglycerin had a significantly greater reduction in the extent of their defect in ^{201}Tl uptake compared to those receiving placebo. This finding was determined by comparing their initial images on entry into the CCU to images obtained one week later. However, caution should be used in the interpretation of serial ^{201}Tl myocardial imaging to assess the effect of therapeutic interventions in patients with acute myocardial infarction, since, as mentioned earlier, the extent of the initial ^{201}Tl uptake defect reflects the extent of ischemia as well as infarction.

Myocardial perfusion imaging is difficult to use for the detection of acute changes in myocardial perfusion resulting from therapy. Recent studies by Gewirtz[16] reveal that the rate of ^{201}Tl loss from lesions made worse by the intervention is similar to that from normal tissue. In addition, the thickness of the myocardium and size of the chamber, which are frequently altered by interventions, will markedly alter the image. For these reasons, conclusions drawn from acute changes in the myocardial scan must be interpreted with caution.

^{201}Tl myocardial imaging is also of value in assessing patients with cardiogenic shock. In the patient with anterior infarction and cardiogenic shock due to massive infarction of the left ventricle, the ^{201}Tl myocardial image will reveal a large defect in tracer uptake, which corresponds to the location and extent of the infarction. In patients with cardiogenic shock and electrocardiographic evidence of inferior or posterior infarction, ^{201}Tl myocardial imaging may reveal a number of clinically useful patterns. Patients with predominantly left ventricular damage will be found to have a defect in tracer uptake on the inferior or posterolateral portion of the myocardium. Those with predominantly right ventricular infarction may have a normal ^{201}Tl myocardial image because in most instances only the left ventricular myocardium is visualized on the ^{201} Tl image. In those with right ventricular infarction, increased right ventricular activity may be occasionally detected with a defect in tracer uptake. The presence of right ventricular infarction can be confirmed by hemodynamic monitoring, [^{99m}Tc] pyrophosphate imaging, or gated cardiac blood pool imaging. In patients with cardiogenic shock or severe left ventricular dysfunction due to papillary muscle rupture or rupture of the interventricular septum, ^{201}Tl myocardial imaging alone or in combination with gated blood pool imaging is useful in decisions concerning surgical intervention. The prognosis for surgical repair is good in those with a relatively small defect in tracer uptake, whereas in those with an extensive defect in tracer uptake, which suggests massive myocardial infarction, the prognosis is poor, even if the patient initially survives the operation.

Initially it was thought that ^{201}Tl myocardial imaging would not be of value in distinguishing acute from old infarction. Recent experience suggests that this distinction may be possible. Pohost and coworkers[22] have shown that redistribution of ^{201}Tl into ischemic areas occurs over time. A defect in tracer uptake within an ischemic areas occurs because the initial delivery to the ischemic area and extraction of tracer by the area is diminished. Over several hours, the ischemic myocardium continues to extract ^{201}Tl from residual blood pool activity until the intracellular cation space and transport system are saturated. On reimaging of the initial defect in tracer uptake several hours later, uniform distribution of the tracer may be encountered. This technique has been used to evaluate patients entering the CCU with suspected acute myocardial infarction in whom the initial electrocardiograms and serum enzyme measurements were nondiagnostic.[23] Patients with unstable angina pectoris studied within 6 hr of the onset of symptoms who did not show evidence of subsequent infarction by serial electrocardiograms and serum enzyme measurements were found to have an initial defect in tracer uptake, which on reimaging 3 hr later could not be detected. In contrast, those with unstable angina pectoris who subsequently were found to have acute infarction on serial enzyme study were also found to have a changing but persistent defect on reimaging. In patients with a history of old infarction and chest pain not associated with an episode of acute infarction, an initial defect in tracer uptake could be detected, which on reimaging several hours later was unchanged. Patients with chest pain of noncardiac origin were found to have a normal distribution of ^{201}Tl both on their initial and on redistribution images. Although experience with this approach is as yet still limited, the early results appear promising. Application of this approach to the triage of patients with suspected acute myocardial infarction could result in considerable savings of high-cost intensive care unit beds and a more rapid application of appropriate diagnostic and therapeutic interventions to those with infarctions.

Cardiomyopathy

Myocardial imaging with ^{201}Tl has also been found to be of value in the evaluation of patients with ischemic cardiomyopathy.[8] The diagnosis of ischemic cardiomyopathy is often difficult on clinical grounds alone. Patients with ischemic cardiomyopathy may have congestive heart failure without a typical history of electrocardiographic changes suggestive of previous myocardial infarction. These patients are often suspected of having idiopathic congestive cardiomyopathy. The presence of significant coronary artery disease as a cause for chronic congestive heart failure in these patients is recognized in many instances only at postmortem examination. Conversely, patients with idiopathic congestive cardiomyopathy may have a history of precordial chest discomfort and loss of R-wave voltage across the anterior precordium, which is suggestive of previous myocardial infarction. ^{201}Tl myocardial imaging may be useful in detecting patients with ischemic cardiomyopathy and distinguishing them from those with idiopathic congestive cardiomyopathy. In those with ischemic cardiomyopathy, a defect in tracer uptake greater than 40% of the image circumference can be detected, reflecting massive myocardial replacement by fibrosis, whereas in those with idiopathic congestive cardiomyopathy, ^{201}Tl uptake is relatively uniform. Small scattered defects in tracer uptake may, however, be found in patients with idiopathic congestive cardiomyopathy, but the extent of these defects is usually less than 20% of the circumference of the image and does not account for the extent of left ventricular dysfunction. In one study of thirteen patients with ischemic cardiomyopathy and eight patients with idiopathic congestive cardiomyopathy, ^{201}Tl myocardial imaging allowed the correct diagnosis in twenty out of twenty-one patients. However, caution should be used in assigning an etiologic diagnosis to a given tracer uptake defect. Although in the United States the most likely explanation for a defect in myocardial ^{201}Tl uptake is ischemic heart disease, in other parts of the world other causes may be of equal or greater importance. For example, a defect in ^{201}Tl uptake in patients in Argentina, Venezuela, and Brazil may be due to Chagas' disease of the myocardium. Any process that results in loss of viable myocardium, including tumors, infiltrates, or granuloma formation, causes a defect in ^{201}Tl uptake. ^{201}Tl myocardial imaging has been found to be of value in the assessment of patients with sarcoidosis.[10] Patients with pulmonary sarcoidosis have been found to have large defects in tracer uptake of the left ventricle resulting from granuloma formation. Many of these patients were asymptomatic and were not suspected of having myocardial involvement on clinical grounds alone. Myocardial involvement may subsequently become evident by the occurrence of an intraventricular conduction defect, dysrhythmia, congestive heart failure, or sudden death. In other patients with pulmonary sarcoidosis, ^{201}Tl myocardial imaging has suggested right ventricular hypertrophy and dilatation compatible with the diagnosis of cor pulmonale.

^{201}Tl myocardial imaging has also been found to be helpful in the evaluation of patients with suspected hypertrophic obstructive cardiomyopathy.[9] In our initial experience with ^{201}Tl myocardial imaging in patients with suspected hypertrophic obstructive cardiomyopathy, a number of patients were encountered with both a false-positive and a false-negative diagnosis by conventional M-mode echocardiography. Asymmetric septal hypertrophy (ASH), a diagnostic clue for hypertrophic obstructive cardiomyopathy, also occurs in patients with right ventricular hypertrophy. Right ventricular hypertrophy may not be appreciated on standard M-mode echocardiography but is obvious on a LAO ^{201}Tl myocardial image, in which both the right and left ventricles can be independently assessed. Conversely, ASH is occasionally missed by M-mode echocardiography, depending on the technique and angulation of the transducer, but may be detected on ^{201}Tl myocardial imaging by the finding of a thickened intraventricular septum and apex in comparison to the left ventricular wall. Gated ^{201}Tl myocardial imaging should be used in assessing the relative thickness of the myocardium, since with ungated imaging myocardial wall motion will blur the outline of the myocardial image in such a way that hyperkinetic regions will appear relatively thicker than akinetic or hypokinetic areas.

From the experience previously referred to, it can be seen that ^{201}Tl myocardial imaging is being applied to a wide variety of clinical situations. The clinical utility of ^{201}Tl myocardial imaging should increase even more with the further development and application of tomographic myocardial imaging. Several approaches are being explored. The most promising for widespread clinical application is the use of new collimators with computer reconstruction of tomographic slices through the myocardium. Initial experience with a multiple-pinhole collimator and tomographic reconstruction has shown increased sensitivity for detection of myocardial ischemia and infarction.[30] The use of ^{201}Tl myocardial imaging has shown increased sensitivity for detection of myocardial ischemia and infarction.[30] Application of ^{201}Tl myocardial imaging alone or in combination with other radioactive tracer studies, such as infarct-avid imaging with [^{99m}Tc] pyrophosphate and/or gated cardiac blood pool imaging, adds a new dimension to cardiovascular diagnosis, which at this relatively early stage of development and application can only be partially appreciated.

REFERENCES

1. Albro, P. C., Gould, K. L., Westcott, R. J., and Hamilton, G. W.: Noninvasive assessment of coronary disease in man by myocardial imaging during pharmacologic vasodilatation, abstracted, J. Nucl. Med. **19:**743, 1978.
2. Bailey, I. K.: Unpublished observations.
3. Bailey, I., Burow, R., Griffith, L. S. C., and Pitt, B.: Localizing value of thallium-201 myocardial perfusion imaging in coronary artery disease, abstracted, Am. J. Cardiol. **39:**320, 1977,
4. Bailey, I. K., Rouleau, J. R., Griffith, L. S. C., Strauss, H. W., and Pitt, B.: Myocardial perfusion imaging to detect patients with single and multivessel disease, Herz. **2:**135, 1977.
5. Becker, L. C., Bulkley, B. H., Pitt, B., Flaherty, J. T., Weiss, J. L., Gerstenblith, G., Rehn, T., Pond, M., Mason, S., Silverman, K., Wang, D. G., and Weisfeldt, M. L.: Enhanced reduction of thallium-201 defects in acute myocardial infarction by nitroglycerin treatment: initial results of a prospective randomized trial, abstracted, Clin. Res. **26:**219A, 1978.
6. Bingham, J. B., Strauss, H. W., Pohost, G. M., and McKusick, K. A.: Mechanisms of lung uptake of Tl-201, abstracted, Circulation **58:**62, 1978.
7. Botvinick, E. H., Taradash, M. R., Shames, D. M., and Parmley, W. W.: Thallium-201 myocardial perfusion scintigraphy for the clinical clarification of normal, abnormal and equivocal electrocardiographic stress test, Am. J. Cardiol. **41:**43, 1978.
8. Bulkley, B. H., Hutchins, G. M., Bailey, I., Strauss, H. W., and Pitt, B.: Thallium-201 imaging and gated cardiac blood pool scans in patients with ischemic and idiopathic cardiomyopathy: a clinical and pathologic study, Circulation **55:**753, 1977.
9. Bulkley, B. H., Rouleau, J. R., Strauss, H. W., and Pitt, B.: Idiopathic hypertrophic subaortic stenosis: detection by thallium-201 myocardial perfusion imaging, N. Engl. J. Med. **293:**1113, 1975.

10. Bulkley, B. H., Rouleau, J. R., Whitaker, J. Q., Strauss, H. W., and Pitt, B.: The use of thallium-201 for myocardial perfusion imaging in sarcoid heart disease, Chest **72**:27, 1977.
11. Caralis, D. G., Kennedy, H. L., Bailey, I. K., and Pitt, B.: Thallium-201 myocardial perfusion scanning in the evaluation of asymptomatic patients with ischemic ST segment depression, abstracted, Am. J. Cardiol. **39**:320, 1977.
12. Cohen, H. A., Baird, M. G., Rouleau, J. R., Fuhrmann, C. F., Bailey, I. K., Summer, W. R., Strauss, H. W., and Pitt, B.: Thallium-201 myocardial imaging in patients with pulmonary hypertension, Circulation **54**:790, 1976.
13. Cohen, H., Rouleau, J. Griffith, L., Gott, V., Brawley, R., Strauss, H. W., and Pitt, B.: Myocardial perfusion and wall motion pre- and post-coronary bypass surgery, abstracted, Circulation **51&52**:170, 1975.
14. Cook, D. J., Bailey, I., Strauss, H. W., Rouleau, J., Wagner, H. N., and Pitt, B.: Thallium-201 for myocardial imaging: appearance of the normal heart, J. Nucl. Med. **17**:184, 1976.
15. Cowley, M. J., Coghlan, H. C., and Logic, J. R.: Visualization of atrial myocardium with thallium-201, J. Nucl. Med. **18**:984, 1977.
16. Gewirtz, H., Beller, G. A., Strauss, H. W., Dinsmore, R. E., Zir, L. M., McKusick, K. A., and Pohost, G. M.: Transient defects of resting thallium scans in patients with coronary artery disease, Circulation **59**:707, 1979.
17. Hamilton, G. W., Narahara, K. A., Yee, H., Ritchie, J. L., Williams, D. L., and Gould, K. L.: Myocardial imaging with thallium-201: effect of cardiac drugs on myocardial images and absolute tissue distribution, J. Nucl. Med. **19**:10, 1978.
18. Maseri, A., Parodi, O., Severi, S., and Pesola, A.: Transient transmural reduction of myocardial blood flow demonstrated by thallium-201 scintigraphy as a cause of variant angina pectoris, Circulation **54**:280, 1976.
19. Mueller, T. M., Marcus, M. L., Ehrhardt, J. C., Chowdhuri, T., and Abboud, F. M.: Limitations of thallium-201 myocardial perfusion scintigrams, Circulation **54**:640, 1976.
20. Neill, C., Kelly, D., Bailey, I., White, R., Strauss, H. W., and Pitt, B.: Thallium-201 myocardial scintigraphy in single ventricle, abstracted, Circulation **53&54**:174, 1976.
21. Pitt, B., and Strauss, H. W.: Myocardial imaging in the noninvasive evaluation of patients with suspected ischemic heart disease, Am. J. Cardiol. **37**:797, 1976.
22. Pohost, G. M., Zir, L. M., Moore, R. H., McKusick, K. A., Guiney, T. E., and Beller, G. A.: Differentiation of transiently ischemic from infarcted myocardium by serial imaging after a single dose of thallium-201, Circulation **55**:294, 1977.
23. Pond, M., Rehn, T., Burow, R., and Pitt, B.: Early detection of myocardial infarction by serial thallium-201 imaging, abstracted, Circulation **56**: 893, 1977.
24. Ritchie, J. L., Narahara, K. A., Trobaugh, G. B., Williams, D. L., and Hamilton, G. W.: Thallium-201 myocardial imaging before and after coronary revascularization, Circulation **56**:830, 1977.
25. Ritchie, J. L., Zaret, B. L., Strauss, H. W., Pitt, B., Berman, D. S., Schelbert, H. R., Ashburn, W. L., and Hamilton, G. W.: Myocardial imaging with thallium-201 at rest and exercise—a multicenter study: coronary angiographic and electrocardiographic correlations, abstracted, J. Nucl. Med. **18**:642, 1977.
26. Rouleau, J., Griffith, L., Strauss, H. W., and Pitt, B.: Detection of diffuse coronary artery disease by quantification of thallium-201 myocardial images, abstracted, Circulation **51&52**:111, 1975.
27. Strauss, H. W., Harrison, K., Langan, J., Lebowitz, E., and Pitt, B.: Thallium-201 for myocardial imaging: relation of thallium-201 to regional myocardial perfusion, Circulation **51**:641, 1975.
28. Strauss, H. W., and Pitt, B.: Noninvasive detection of subcritical coronary arterial narrowings with a coronary vasodilator and myocardial perfusion imaging, Am. J. Cardiol. **39**:403, 1977.
29. Strauss, H. W., and Pitt, B.: Thallium-201 as a myocardial imaging agent, Semin. Nucl. Med. **7**: 49, 1977.
30. Vogel, R. A., Kirch, D. L., LeFree, M. T., and Steele, P. P., Improved diagnostic results of myocardial perfusion tomography using a new rapid inexpensive technique, abstracted, J. Nucl. Med. **19**:731, 1978.
31. Wackers, F. J., Sokole, E. B., Samson, G., Schoot, J. B., Lie, K. I., Liem, K. L., and Wellens, H. J. J.: Value and limitations of thallium-201 scintigraphy in the acute phase of myocardial infarction, N. Engl. J. Med. **295**:1, 1976.
32. Weich, H. F., Strauss, H. W., and Pitt, B.: The extraction of thallium-201 by the myocardium, Circulation **56**:188, 1977.

SECTION THREE

INFARCT-AVID IMAGING

14 □ Infarct-avid radiopharmaceuticals: biologic and structural characteristics

B. Leonard Holman

Utilization of the scintigraphic appearance of infarcted myocardium as an area of increased activity has considerable appeal, since present techniques, including serum enzyme tests, electrocardiography, and vectorcardiography, provide only indirect evidence of the presence, size, and location of infarcted myocardium. Although these techniques are usually accurate in detecting infarction, techniques to precisely locate the site of damaged myocardium and assess the extent of damage are limited. Radiopharmaceuticals that are extracted by normal myocardium, such as potassium and its analogs[24] and radio-labeled fatty acids,[18] offer additional help by outlining poorly perfused tissue as regions of decreased tracer concentration in myocardial scans.[43] However, these radiopharmaceuticals do not permit differentiation between acute infarction and fibrotic or previously infarcted tissue. To overcome these difficulties, agents that accumulate selectively in damaged myocardium have been sought.

INFARCT DETECTION WITH [^{203}Hg] CHLORMERODRIN AND [^{203}Hg] FLUORESCEIN

The earliest attempts at infarct imaging were with [^{203}Hg] chlormerodrin.[10] It was possible to demonstrate the acutely infarcted myocardium as an area of increased radioactivity after scanning both a living dog and then its excised heart. The concentration of [^{203}Hg] chlormerodrin was fifteen times greater in the infarcted than in the normal myocardium. This work was later confirmed in pigs, a species with a coronary circulation more similar to that of humans.[20]

The quantity of [^{203}Hg] chlormerodrin injected in these studies was substantially greater than would be possible in clinical trials. However, when smaller quantities (100 to 300 μCi) were injected into dogs with ligated coronary vessels, external imaging of the infarct was not possible.[33]

Attempts to apply this work to humans were unsuccessful.[11] Carr and co-workers detected the infarct in one of thirteen patients with acute myocardial infarction. These poor results may have been due to the small amount of [^{203}Hg] chlormerodrin injected and the fact that the patients were studied four to eight days after infarction.

Radiomercurifluorescein analogs also sequester in acutely damaged muscle. Malek and associates observed large concentration ratios between damaged and normal muscle after injection of ^{203}Hg-labeled hydroxymercurifluorescein.[32] In experimental acute myocardial infarction, the damaged myocardium could be imaged in the infarcted animal with a number of mercurifluorescein analogs after temporary ligation of the anterior descending artery. With this reperfusion model, infarct-to-normal myocardium concentration ratios ranged from 15:1 to 100:1.[28,33]

When the anterior descending artery was ligated permanently, the ratio of [^{203}Hg] hydroxymercurifluorescein in infarcted tissue to that in normal tissue was approximately 10:1. Although the concentration in infarcted myocardium was ten times greater when the coronary artery ligature was released than when it was permanently ligated, external imaging detected the infarct in all cases.

When the time course of fluorescein accumulation in the canine heart was studied, activity overlying the infarcted tissue was detectably greater than over surrounding tissue as early as 24 hr after occlusion.[30] By the third day, infarct avidity peaked at three times the 24 hr level. Tracer concentration in the lesion fell gradually, reaching normal levels by the end of the third week.

The high ratio of concentration in infarcted tissue to that in normal tissue has not been obtained universally with labeled fluorescein.[38] Ramanathan and colleagues[38] failed to elicit ratios of greater than 2:1 to 3:1 after temporary (1 hr to ten-day) coronary ligature in dogs injected with [^{203}Hg] hydroxymercurifluorescein.

In contrast to these results in animals, clinical investigations with [^{203}Hg] fluorescein in humans have been disappointing. Since ^{203}Hg has a long physical half-life (47 days), nonpenetrating radiation, and prolonged retention in the kidneys, only a small dose can be administered to the patient who is suspected of having myocardial infarction, a dose substantially smaller that that used for successful imaging in dogs. Although ^{197}Hg alleviates some of these problems, successful imaging with this radionuclide in humans has yet to be reported.

TETRACYCLINE

Noting that tetracycline analogs concentrate in pathologically altered tissues, Malek and associates demonstrated accumulation of this antibiotic in infarcted myocardium.[31] The dynamics of tetracycline fluorescence was divided into the following four phases:

1. Immediately after tetracycline injection, the infarcted focus was free of fluorescence, whereas the normal muscle showed yellow fluorescence.
2. By 3 hr after injection, the fluorescence had intensified along the borders of the infarct and disappeared from normal myocardium.
3. From 3 hr to four days after injection, the intense fluorescence was limited to the margins of the infarct. Initially the fluorescent border was continuous; however, after four days the areas of fluorescence broke up into clumps, still limited to the margin of the infarct.
4. By seven days after infarction, tetracycline had accumulated in foci not only along the borders but also throughout the infarct.

Calcium ion concentration, determined by von Kossa staining, correlated well with tetracycline fluorescence.[31] Although von Kossa staining was slightly positive throughout the entire infarct, intense staining occurred in clumps that demonstrated intense tetracycline fluorescence.

Although the potential for infarct detection with gamma-emitting tetracycline analogs was recognized, attempts to label tetracycline with radioiodine were unsuccessful because of the instability of the label.[4,16,17] Furthermore, [^{131}I] tetracycline was difficult to prepare.

The development of a method for production of stable ^{99m}Tc-labeled tetracycline[15] permitted evaluation of this agent for the detection, localization, and sizing of experimental infarcts.[23] Initial results demonstrated a high degree of accuracy in the detection of acute myocardial infarction 24 hr after the injection of [^{99m}Tc] tetracycline. A study of twenty-eight patients was done 24 hr after the injection of 20 mCi of [^{99m}Tc] tetracycline.[25] There were nine control patients who had evidence of cardiac disease with normal scans. By clinical criteria, fourteen patients had sustained acute myocardial infarction (ten transmural, four nontransmural). The scan was abnormal in all fourteen patients. The optimal time for imaging was within the first three days after the onset of symptoms. Abnormal scans returned to normal within one to two weeks in all six patients scanned serially. Scintigraphic and electrocardiographic localizations correlated well. The size of the infarct determined by scintigraphy correlated well with the peak serum creatine-phosphokinase level.

The major limitation to [^{99m}Tc] tetracycline for clinical use is the slow clearance rate from the blood, necessitating a 24 hr delay after injection before imaging can be performed. Thus, at the present time, [^{99m}Tc] pyrophosphate is the radiopharmaceutical of choice for the detection of acute myocardial infarction in humans.[26]

In dogs, increased concentrations of [^{99m}Tc] tetracycline are found in acutely infarcted myocardium but not in ischemic uninfarcted tissue.[45] Thus the tracer separates viable from irreversibly damaged tissue. There is a minimal concentration of the tracer in nonischemic myocardium and in ischemic tissue until the blood flow has fallen to 45% of normal. At this point, there

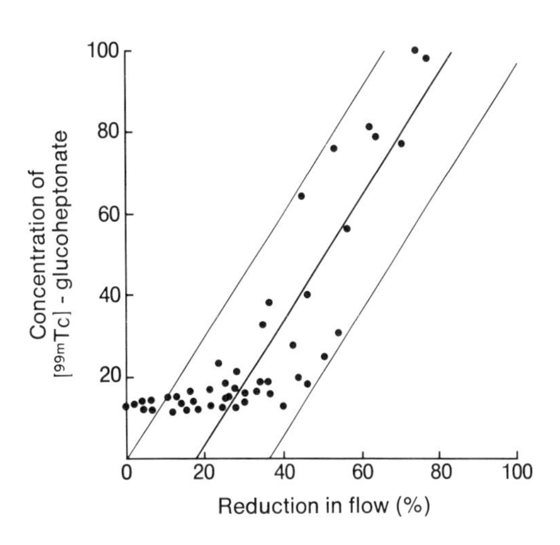

Fig. 14-1. Correlation between percentage of blood flow reduction and uptake of [^{99m}Tc] glucoheptonate. Regression line *(middle line)* was fitted from threshold blood flow reduction value to maximum blood flow reduction (r = 0.889). Lines bordering regression line are 95% confidence limits. (From Zweiman, F. G., et al.: J. Nucl. Med. **16**:975, 1975.)

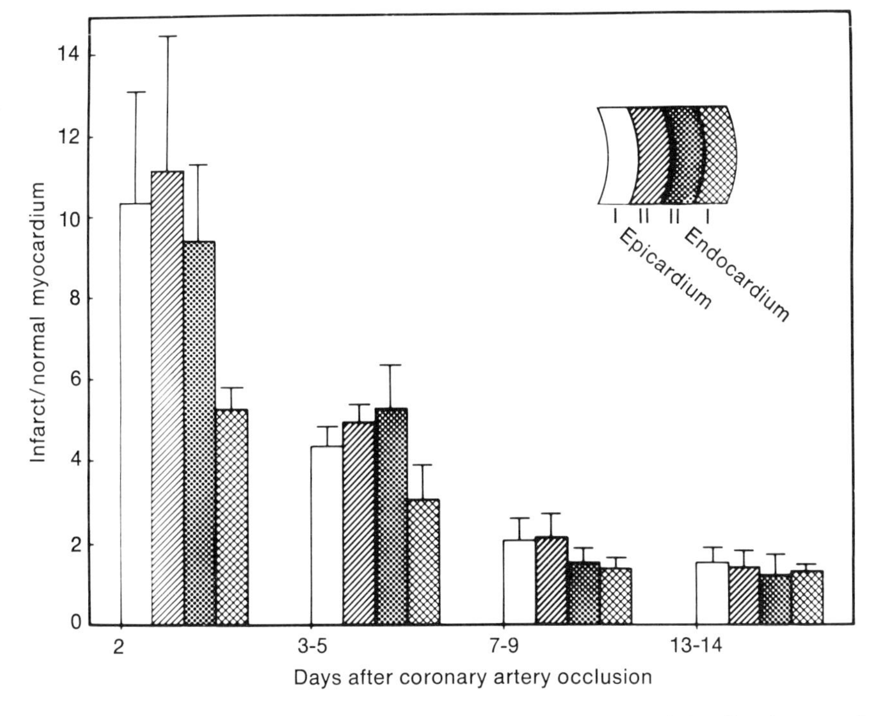

Fig. 14-2. Gradient of [^{99m}Tc] tetracycline concentration ratio across myocardial wall as function of time after coronary artery occlusion (± sem). (From Holman, B. L., et al.: J. Nucl. Med. **16**:1144, 1975.)

is a direct correlation between reduction in blood flow and increase in tracer concentration. This linear relationship persists even at very low blood flow rates. This relationship between blood flow reduction and tracer uptake is also seen with [^{99m}Tc] glucoheptonate but not with the ^{99m}Tc-labeled phosphates (Fig. 14-1).

When one assesses the transmural distribution of [^{99m}Tc] tetracycline, the radiotracer is relatively evenly divided within the endocardial and innermost portion of the epicardial layers[27] (Fig. 14-2). The concentration in the outer epicardial section is about half that in the endocardium two days after occlusion. Thus the greatest concentration of [^{99m}Tc] tetracycline is found in that portion of the wall with the greatest degree of tissue necrosis, indicating that the intensity of [^{99m}Tc] tetracycline uptake is a better indicator of tissue necrosis than that of [^{99m}Tc] pyrophosphate, the uptake of which becomes flow limited in regions of poor perfusion.

There is an excellent correlation between the size of the infarct when measured with either serum enzyme activity or electrophysiologic techniques and when measured with [^{99m}Tc] tetracycline.[12] In animals, there is also a direct correlation between the increase in [^{99m}Tc] tetracycline concentration and the size of the infarct. Infarcts involving greater than 40% of the myocardium have a sevenfold increase in infarcted tissue compared to normal myocardium. In infarcts involving less than 15% of the myocardium, the increase in [^{99m}Tc] tetracycline concentration is less than two times that in normal tissue.

[^{99m}Tc] GLUCOHEPTONATE

Other radiotracers have also been used for acute infarct scintigraphy. Accumulation of [^{99m}Tc] glucoheptonate within the infarcted focus has been observed in a number of animal models.[19,39,45] After acute myocardial infarction in dogs, concentration ratios between infarcted and normal myocardium have ranged from 11:1 to 20:1.

As with [^{99m}Tc] tetracycline, the threshold phenomenon has been observed; a reduction in flow of approximately 40% appears to be necessary before significant increases in the concentration of the labeled glucoheptonate are observed.

Infarct uptake occurs fairly rapidly after coronary artery occlusion. Concentrations of approximately 21:1 can be detected in the canine model 6 hr after infarction (4 hr after the tracer injection). Concentrations of approximately 6:1 can be recorded as early as 3 to 4 hr after coronary occlusion.

Increased uptake of purified radiolabeled antibody to cardiac myosin has also been demonstrated in regions of acute infarction.[29] Ratios of 6.1 ± 0.6 and 3.3 ± 0.4 between infarcted and normal myocardium have been obtained in the epicardium and endocardium, respectively. There was an inverse relationship between the blood flow and antimyosin uptake, even at low flow rates. Unfortunately the concentration of tracer within the infarct is too low to permit external detection of the infarct within the first 24 hr after coronary occlusion.

[^{99m}Tc] PYROPHOSPHATE

[^{99m}Tc] pyrophosphate is the agent most commonly employed for acute myocardial infarct scintigraphy.[26,35,36] [^{99m}Tc] pyrophosphate uptake depends on three factors: local blood flow, calcification, and tissue damage.

After acute coronary occlusion, increased concentrations of [^{99m}Tc] pyrophosphate are found in regions with only minimal reductions in blood flow.[44] The highest concentration ratios between damaged and normal myocardium occur when local blood flow is 20% to 30% of normal. As flow is reduced further, the concentration ratios begin to fall until, in regions of minimal flow (0% to 5% of normal), [^{99m}Tc] pyrophosphate concentration may be normal.

Buja and co-workers[8] found the highest [^{99m}Tc] pyrophosphate concentration ratios in the outer periphery of the infarct. In seventeen dogs that underwent left anterior descending artery ligation and were injected 18 to 48 hr later with [^{99m}Tc] pyrophosphate, the concentration of the pyrophosphate in the outer periphery zone of the infarction was 24.5 times that of normal myocardium, whereas the concentration in the inner periphery was 7.8 times normal. The concentration in the center of the infarct was only 1.7 times normal. Thus the status of the regional myocardial perfusion after occlusion appears to be a key determinant for the scintigraphic detection of such infarcts with pyrophosphate.

There was also a coincident correlation between the presence of contraction bands and calcium deposits and the uptake of labeled pyrophosphate. Thus numerous irreversibly injured muscle cells with contraction bands and calcium deposits were found in the outer periphery of the infarct, where the pyrophosphate uptake was high; the reverse was true in the central zones of the infarct. This led the investigators to hypothesize that pyrophosphate uptake also depends on the presence of calcium.

A close correlation was found between the subcellular distribution of [^{99m}Tc] pyrophosphate and that of calcium.[9] Thus abnormal-to-normal ratios for [^{99m}Tc] pyrophosphate and calcium of 11.0 ± 0.4 and 3.1 ± 0.9, respectively, were found in the border zones of the acute myocardial infarct with focal necrosis, 17.4 ± 1.9 and 16.7 ± 6.4 in peripheral zones with extensive necrosis and 6.5 ± 1.5 and 6.2 ± 1.0 in central zones with confluent necrosis. There was a similar correlation between the uptake of pyrophosphate and calcium within subcellular fractions. Thus these studies suggest that elevated calcium level plays a key role in myocardial scintigraphy by establishing a chemical milieu that induces extraction and concentration of [^{99m}Tc] pyrophosphate.

Observing that hydroxyapatite crystals were formed within mitochondria in the peripheral zones of the acute infarct, Bonte and associates[5] hypothesized that [^{99m}Tc] pyrophosphate, a calcium chelate, is sequestered by these crystals during the acute phase of myocardial infarction. Mitochondrial calcification has been detected as early as 2 to 4 hr after permanent occlusion and becomes marked by 12 to 24 hr.[41] The calcification occurred primarily in irreversibly damaged tissue.

Recent experimental evidence would suggest that, although the mechanism for localization may be related to the presence of calcium, the site of sequestration of the radiopharmaceutical is not the mitochondria. In studies carried out in a tissue-culture model, Dewanjee[14] found uptake primarily in the soluble and nuclear fractions after subcellular separation. Only a small percentage (8.5% in the live and 12.3% in the dead cells) were found in the mitochondria. Other investigators have also found the highest levels in the supernate and in membrane and cell debris and the lowest concentrations in microsomes and mitochondria.[9] Although pure fractionation of subcellular organelles is extremely difficult, particularly with myocardial tissue, this evidence is strongly suggestive that the method of localization is not mitochondrial sequestration.

Recent studies have cast further doubt on the precise role of calcification in [^{99m}Tc] pyrophosphate sequestration. Using a fetal mouse heart tissue-culture model, Schelbert and colleagues[40] showed a poor correlation between calcium and [^{99m}Tc] pyrophosphate uptake during oxygen and glucose metabolism.

Concentration of [^{99m}Tc] pyrophosphate is not proportional to the extent of infarction. Thus, in a canine model of acute infarction, there was a negative correlation between maximum radioactivity recorded in the precordial image and the extent of infarction. The variability of individual measurements about the regression line was such that the relationship was not useful in precisely predicting the extent of infarction in any given animal. There was, however, a strong direct correlation between the maximum radioactivity recorded in the precordial image and the extent of infarction in reperfused dogs. There was still a good deal of variability of individual measurements about the regression line. These workers concluded that, although increased radioactivity was a sensitive marker of tissue necrosis, the quantitated radioactivity in a given tissue sample was not proportional to the amount of myocardial infarction.

Other investigators have also found that, although pyrophosphate uptake appeared limited to histologically necrotic tissue, pyrophosphate ratios between normally perfused tissue adjacent to the ischemic segment and distant normal tissue were elevated (6.72 ± 0.4).[34,44] Also, these investigators did not find any linear relationship between the extent of necrosis and pyrophosphate accumulation. Because the intensity of the pyrophosphate accumulation on the infarcted scintigrams is related to the degree of perfusion as well as the extent of the necrosis, these investigators concluded that the intensity of uptake on the myocardial scintigram cannot be used as a quantitative index of infarct size.

Several studies have shown increased myocardial uptake of [^{99m}Tc] pyrophosphate in patients without other evidence of acute myocardial infarction. Thus eleven of fifteen patients with previous infarction and without evidence of acute myocardial infarction had abnormal scintigrams with uptake localized to angiographically visualized sites of left ventricular wall motion abnormality.[3] In addition, patients with unstable angina but without clinical evidence of infarction have been reported to have abnormal scintigrams.[1] Thus, in seven patients with unstable angina, five had abnormal and one had equivocal scintigrams. In another study, all twenty-four patients with unstable arteriosclerotic heart disease but without acute infarction had nonfocal, ill-defined accumulations of [^{99m}Tc] pyrophosphate.[37] In addition, six patients with myocardiopathy also showed diffuse uptake. It has been argued that [^{99m}Tc] pyrophosphate is being sequestered by reversibly ischemic myocardium or, conversely, that [^{99m}Tc] pyrophosphate is being sequestered in small nests of necrotic tissue in patients with insufficient irreversible myocardial damage to be detected with currently available laboratory tests for myocardial necrosis.

These studies would suggest that, although [^{99m}Tc] pyrophosphate is a sensitive detector of the presence of infarction, its usefulness in determining changes in the size of infarction is limited by (1) the low uptake in regions of low blood flow, (2) the poor correlation between tracer concentration and the degree of tissue necrosis, and (3) the low uptake in the early phase of acute myocardial infarction. Nevertheless, there have been promising preliminary results using the area of uptake as an index of infarct size 24 hr or more after occlusion.[6] In a canine model, computer-estimated infarct size and gross infarct area correlated well by linear regression analysis. There was good correlation between the computer-estimated in vivo infarct area and gross infarct area in the RAO and LAO projections. The lateral projection showed a poor correlation ($r = 0.68$). This work was corroborated by another animal study, in which good correlation was found between the scintigraphic infarct area and histologically determined infarct weight ($r = 0.92$).[42] Although these results suggested that pyrophosphate scintigrams provide a useful, noninvasive method for measuring infarct size, these studies were limited to left anterior descending artery occlusions, and hence the accuracy of estimating diaphragmatic infarct size was not assessed. Other investigators have found a poorer correlation between scintigraphy and infarct size in inferior wall infarction.[22]

STRUCTURE-ACTIVITY RELATIONSHIPS

Using the heat-damaged rat heart model described by Adler and associates,[2] it has been possible to demonstrate that a large number of radiopharmaceuticals have the property of being sequestered by acutely damaged myocardium[13] (Table 14-1). Of the commonly available radiopharmaceuticals, the bone-seeking agents, including [^{99m}Tc] pyrophosphate, [^{99m}Tc] diphosphonate, and [^{99m}Tc] methylene diphosphonate, have the

Table 14-1. Sequestration of radiopharmaceuticals in damaged rat myocardium

	Agent	Injected dose/ g of infarct (%)	Infarct-to-normal ratio
Common radiopharmaceuticals	[^{99m}Tc] pyrophosphate	2.2 ± 0.4	25.2 ± 9.2
	[^{99m}Tc] 1,1-hydroxyethylidine diphosphonate	0.8 ± 0.2	26.7 ± 8.9
	[^{99m}Tc] methylene diphosphonate	0.9 ± 0.3	30.2 ± 12.3
	[^{99m}Tc] glucoheptonate	0.7 ± 0.5	20.2 ± 13.2
	[^{99m}Tc] tetracycline	0.9 ± 0.3	13.9 ± 5.6
	[^{99m}Tc] oxytetracycline	1.6 ± 1.0	12.9 ± 5.1
	[^{99m}Tc] diethylenetriamine pentaacetate	0.7 ± 0.2	10.9 ± 1.9
	[^{131}I] rose bengal	0.7 ± 0.2	7.8 ± 1.0
Sulfhydryl-containing radiopharmaceuticals	[^{99m}Tc] thioglycerol	1.7 ± 1.3	11.3 ± 4.7
	[^{99m}Tc] 2,3-dimercaptosuccinic acid	2.4 ± 1.1	9.7 ± 3.1
	[^{99m}Tc] 2-mercaptoisobutyric acid	1.2 ± 0.9	7.0 ± 3.0
	[^{99m}Tc] penicillamine	0.6 ± 0.1	6.1 ± 1.0
	[^{99m}Tc] dihydrothioctic acid	0.8 ± 0.5	2.5 ± 0.9
Mercury-containing radiopharmaceuticals	[^{125}I] diiodohydroxymercurifluorescein	12.4 ± 3.4	31.7 ± 9.7
	[^{203}Hg] diiodohydroxymercurifluorescein		
	[^{203}Hg] chlormerodrin	1.5 ± 0.5	26.8 ± 6.6
	[^{203}Hg] mercuric nitrate	7.1 ± 3.4	15.5 ± 4.6
	[^{203}Hg] phenylmercuric acetate	3.7 ± 1.8	5.4 ± 2.3
	[^{203}Hg] bromomercurihydroxypropane	1.2 ± 0.2	2.5 ± 0.5

Table 14-2. Target-to-nontarget ratios of the labeled fluorescein compounds in rats

Agent	Injected dose/ g of infarct (%)	Infarct to normal ratio
[^{203}Hg] hydroxymercurifluorescein	5.0 ± 1.5	51.5 ± 13.5
[^{203}Hg] bishydroxymercurifluorescein	9.4 ± 1.5	42.8 ± 9.4
[^{203}Hg] 4′,5′-diiodohydroxymercurifluorescein	8.6 ± 2.2	25.4 ± 6.8
[^{203}Hg] 4′,5′-diiodobishydroxymercurifluorescein	12.6 ± 3.3	18.4 ± 4.1
[^{203}Hg] 2′,7′-dibromohydroxymercurifluorescein	11.0 ± 3.5	24.7 ± 6.4
[^{203}Hg] 2′,7′-dibromobishydroxymercuriflourescein	9.9 ± 3.7	26.5 ± 8.9
[^{131}I] rose bengal (3,4,5,6-tetrachloro-2′,4′,5′,7′-tetraiodofluorescein)	0.7 ± 0.2	7.8 ± 1.0
[^{203}Hg] mercuric acetate	7.1 ± 3.4	15.5 ± 4.6
[^{3}H] fluorescein	0.15 ± .03	2.4 ± 0.5

highest infarct-to-normal myocardium and infarct-to-blood concentration ratios. Of these, the percentage of injected dose per gram of damaged tissue is highest with [^{99m}Tc] pyrophosphate. [^{99m}Tc] glucoheptonate and [^{99m}Tc] tetracycline have concentration ratios approximately half to two thirds that of the bone-seeking tracers. A mercury-containing compound, diiodohydroxymercurifluorescein, has resulted in a concentration ratio between damaged and normal myocardium higher than any other radiopharmaceutical tested. In addition, the percentage of injected dose per gram is almost six times that of the bone-seeking radiotracers.

Indeed, the presence of mercury in the radiocompound has correlated strongly with infarct concentration.[13] The damaged-to-normal myocardium distribution of various fluorescein derivatives is presented in Table 14-2.

Table 14-3. Target-to-nontarget ratios in rats

Agent	Injected dose/g of infarct		Infarct-to-normal ratio	
	1 hr	3 hr	1 hr	3 hr
^{203}Hg 4′,5′-diiodohydroxymercurifluorescein	8.6 ± 2.2	5.4 ± 1.3	25.4 ± 6.8	45.3 ± 14.8
^{203}Hg 2′,7′-dibromohydroxymercurifluorescein	11.0 ± 3.5	5.2 ± 1.4	4.7 ± 6.4	38.6 ± 8.8

The evidence very strongly indicates that both the polycyclic aromatic moiety and the hydroxymercury group are necessary for the high selectivity that is required for imaging.[21] All of the hydroxymercurifluoresceins are vastly superior to [^{131}I] rose bengal (a nonmercurated fluorescein) and [^{3}H] fluorescein and significantly better than [^{203}Hg] mercuric acetate. The effect of introducing the halogen substituents onto the tricyclic portion of the molecule is the reduction of the agent's damaged tissue selectivity. There is no difference between the monohydroxymercurated and bishydroxymercurated fluoresceins, which indicates that only one binding site may be necessary for activity and a separation of the two components may not be necessary.

The results of this work suggest a two-step process for the localization of these damaged tissue–seeking agents, to which both the organic portion and the mercuric group contribute. The ability of the agent to cross the cell membrane in ischemic or necrotic tissue is enhanced by the large, somewhat lipophilic organic component. The second step is binding the agent by way of the mercuric functionality to an increased (relative to normal cells) population of sulfhydryl groups and preventing the diffusion of the agent back into the blood. The presence of the diiodo- or dibromo- substituents results in an increased acidity of the phenolic groups, thereby increasing the solubility in the blood and surrounding tissue and lowering the infarct-to-normal myocardium ratio. When the diiodo- and dibromo- agents were allowed to clear the blood to a greater extent (3 hr), the infarct-to-normal myocardium ratios approached that of hydroxymercurifluorescein (Table 14-3).

These results indicate that the fluorescein group enhances the uptake of the agent into the damaged myocardium. The hydroxymercury group then binds it to intracellular sulfhydryl groups.

As a result of the studies conducted with the hydroxymercurifluoresceins, my coworkers and I have concluded the following[2]:

1. Both the polycyclic aromatic organic system and the hydroxymercury group are required for high uptake and selectivity.
2. There is little difference in selectivity between the monohydroxymercurated and the bishydroxymercurated derivatives, which indicates that only one binding site per molecule is involved.
3. The oxygen bridge present in fluorescein is not required for selectivity.
4. Introduction of electron-withdrawing substituents such as iodine or bromine onto the phenolic rings tends to decrease the selectivity for damaged tissue.
5. The substitution of carboxylic acid for sulfonic acid results in little change in selectivity.

Thus it appears that structural features that augment the specificity of a radiopharmaceutical for acutely infarcted myocardium can be predicted. Based on structure-activity relationships, such as the presence of mercury and the configuration of the organic carrier, it should be possible to synthesize compounds with improved biologic and physical properties for the estimation of acute myocardial infarct size in animal models and in humans.

REFERENCES

1. Abdulla, A. M., Canedo, M. I., Cortez, B. C., et al.: Detection of unstable angina by ^{99m}technetium pyrophosphate myocardial scintigraphy, Chest **69:** 168, 1976.
2. Adler, N., Camin, L. L., and Shulkin, P.: Rat model for acute myocardial infarction: application to technetium-labeled glucoheptonate, tetracycline, and polyphosphate, J. Nucl. Med. **17:**203, 1976.
3. Ahmad, M., Dubiel, J. P., Verdon, T. A., et al.: Technetium 99m stannous pyrophosphate myocardial imaging in patients with and without left ventricular aneurysm, Circulation **53:**833, 1976.
4. Anghileri, L. J.: Absorption and excretion of radioiodinated tetracycline, Nucl. Med. **3:**368, 1963.
5. Bonte, F. J., Parkey, R. W., Graham, K. D., et al.: A new method for radionuclide imaging of myocardial infarcts, Radiology **110:**473, 1974.
6. Botvinick, E. H., Shames, D., Lappin, H., et al.: Noninvasive quantitation of myocardial infarction with technetium 99m pyrophosphate, Circulation **52:**909, 1975.
7. Bruno, F. P., Cobb, F. R., Rivas, F., et al.: Evaluation of ^{99m}technetium stannous pyrophosphate as an imaging agent in acute myocardial infarction, Circulation **54:**71, 1976.
8. Buja, L. M., Parkey, R. W., Stokely, E. M., et al.: Pathophysiology of technetium-99m stannous pyrophosphate and thallium-201 scintigraphy of acute anterior myocardial infarcts in dogs, J. Clin. Invest. **57:**1508, 1976.
9. Buja, L. M., Tofe, A. J., Mukherjee, A., et al.: Role of elevated tissue calcium in myocardial infarct scintigraphy with technetium phosphorus radiopharmaceuticals, abstracted, Circulation **54** (suppl. 2):219, 1976.
10. Carr, E. A., Jr., Beierwaltes, W. H., Patno, M. E., et al.: The detection of experimental myocardial infarcts by photoscanning, Am. Heart J. **64:**650, 1962.
11. Carr, E. A., Jr., Cafruny, E. J., and Bartlett, J. D.: Evaluation of ^{203}Hg chlormerodrin in the demonstration of human myocardial infarcts by scanning, Univ. Mich. Med. Bull. **29:**27, 1963.
12. Cook, G. A., Holman, B. L., Westmoreland, N., et al.: Localization, detection and sizing of myocardial infarction with technetium-99m-tetracycline and electrophysiologic techniques, abstracted, Am. J. Cardiol. **33:**131, 1974.
13. Davis, M. A., Holman, B. L., and Carmel, A. N.: Evaluation of radiopharmaceuticals sequestered by acutely damaged myocardium, J. Nucl. Med. **17:**911, 1976.
14. Dewanjee, M. K.: Localization of skeletal-imaging ^{99m}Tc chelates in dead cells in tissue culture: concise communication, J. Nucl. Med. **17:**993, 1976.
15. Dewanjee, M. K., Fliegel, C., Treves, S., et al.: ^{99m}Tc-labeled tetracycline: a new radiopharmaceutical for renal imaging, J. Nucl. Med. **13:**427, 1972.
16. Dunn, A. L., Eskelson, C. D., McLeay, J. F., et al.: Preliminary study of radioactive product obtained from iodinating tetracycline, Proc. Soc. Exp. Biol. Med. **104:**12, 1960.
17. Eskelson, C. D., Dunn, A. L., Ogborn, R. E., et al.: Distribution of some radioiodinated tetracycline in animals, J. Nucl. Med. **4:**382, 1963.
18. Evans, J. R., Gunter, R. W., Baker, R. G., et al.: Use of radioiodinated fatty acids for photoscans of the heart, Circ. Res. **16:**1, 1965.
19. Fink-Bennett, D., Dworkin, H. J., and Lee, Y.-H.: Myocardial imaging of the acute infarct, Radiology **113:**449, 1974.
20. Gorten, R. J., Hardy, L. B., McCraw, B. H., et al.: The selective uptake of ^{203}Hg-chlormerodrin in experimentally produced myocardial infarct, Am. Heart J. **72:**71, 1966.
21. Hanson, R. N., Davis, M. A., and Holman, B. L.: Myocardial infarct imaging agents. Synthesis and evaluation of ^{203}Hg-hydroxymercurifluoresceins, J. Nucl. Med. **18:**803, 1977.
22. Henning, H., Schelbert, H., O'Rourke, R. A., et al.: Dual myocardial imaging with Tc-99m-pyrophosphate and thallium 201 for diagnosing and sizing acute myocardial infarction, abstracted, J. Nucl. Med. **17:**524, 1976.
23. Holman, B. L., Dewanjee, M. K., Idoine, J., et al.: Detection and localization of experimental myocardial infarction with ^{99m}Tc-tetracycline, J. Nucl. Med. **14:**595, 1973.
24. Holman, B. L., Eldh, P., Adams, D. F., et al.: Evaluation of myocardial perfusion after intracoronary injection of radiopotassium, J. Nucl. Med. **14:**274, 1973.
25. Holman, B. L., Lesch, M., Zweiman, F. G., et al.: Detection and sizing of acute myocardial infarcts with ^{99m}Tc (Sn) tetracycline, N. Engl. J. Med. **291:**159, 1974.
26. Holman, B. L., Tanaka, T. T., and Lesch, M.: Evaluation of radiopharmaceuticals for the detection of acute myocardial infarction in man, Radiology **121:**427, 1976.
27. Holman, B. L., and Zweiman, F. G.: Time course of ^{99m}Tc (Sn)-tetracycline uptake in experimental acute myocardial infarction, J. Nucl. Med. **16:** 1144, 1975.
28. Hubner, P. J. B.: Radioisotopic detection of experimental myocardial infarction using mercury derivatives of fluorescein, Cardiovasc. Res. **4:**509, 1970.
29. Khaw, B. A., Beller, G. A., Haber, E., et al.: Localization of cardiac myosin-specific antibody in myocardial infarction, J. Clin. Invest. **58:**439, 1976.
30. Malek, P., Kolc, J., Vavrejn, B., et al.: Mercurascan test in ischaemia of the heart muscle caused by permanent ligature of the coronary artery, Rev. Czech. Med. **17:**11, 1971.
31. Malek, P., Kolc, J., Zastava, V. L., et al.:

Fluorescence of tetracycline analogues fixed in myocardial infarction, Cardiologia **42:**303, 1963.
32. Malek, P., Ratusky, J., Vavrejn, B., et al.: Ischaemia detecting radioactive substances for scanning cardiac and skeletal muscle, Nature **214:**1130, 1967.
33. Malek, P., Vavrejn, B., Ratusky, J., et al.: Detection of myocardial infarction by in vivo scanning, Cardiologia **51:**22, 1967.
34. Marcus, M. L., Tomanek, R. J., Ehrhardt, J. C., et al.: Relationships between myocardial perfusion, myocardial necrosis, and technetium-99m pyrophosphate uptake in dogs subjected to sudden coronary occlusion, Circulation **54:**647, 1976.
35. Parkey, R. W., Bonte, F. J., Meyer, S. L., et al.: A new method for radionuclide imaging of acute myocardial infarction in humans, Circulation **50:** 540, 1974.
36. Parkey, R. W., Bonte, F. J., Stokely, E. M., et al.: Analysis of Tc-99m stannous pyrophosphate myocardial scintigrams in 242 patients, abstracted, J. Nucl. Med. **16:**556, 1975.
37. Perez, L. A., Hayt, D. B., and Freeman, L. M.: Localization of myocardial disorders other than infarction with ^{99m}Tc-labeled phosphate agents, J. Nucl. Med. **17:**241, 1976.
38. Ramanathan, P., Ganatra, R. D., Daulatram, K., et al.: Uptake of ^{203}Hg-hydroxy-mercury-fluorescein in myocardial infarcts, J. Nucl. Med. **12:**641, 1971.
39. Rossman, D. J., Strauss, H. W., Siegel, M. E., et al.: Accumulation of ^{99m}Tc-glucoheptonate in acutely infarcted myocardium, J. Nucl. Med. **16:** 875, 1975.
40. Schelbert, H., Ingwall, J., Sybers, H., et al.: Uptake of Tc-99m pyrophosphate and calcium in irreversibly damaged myocardium, abstracted, J. Nucl. Med. **17:**534, 1976.
41. Shen, A. C., and Jennings, R. B.: Myocardial calcium and magnesium in acute ischemic injury, Am. J. Pathol. **67:**417, 1972.
42. Stokely, E. M., Buja, L. M., Lewis, S. E., et al.: Measurement of acute myocardial infarcts in dogs with ^{99m}Tc-stannous pyrophosphate scintigrams, J. Nucl. Med. **17:**1, 1976.
43. Strauss, H. W., Zaret, B. L., Martin, N. D., et al.: Noninvasive evaluation of regional myocardial perfusion with potassium-43. Technique in patients with exercise-induced transient myocardial ischemia, Radiology **108:**85, 1973.
44. Zaret, B. L., DiCola, V. C., Donabedian, R. K., et al.: Dual radionuclide study of myocardial infarction. Relationships between myocardial uptake of potassium-43, technetium-99m stannous pyrophosphate, regional myocardial blood flow and creatine phosphokinase depletion, Circulation **53:** 422, 1976.
45. Zweiman, F. G., Holman, B. L., O'Keefe, A., et al.: Selective uptake of ^{99m}Tc-complexes and ^{67}Ga in acutely infarcted myocardium, J. Nucl. Med. **16:**975, 1975.

15 □ Myocardial imaging with technetium phosphates

Robert W. Parkey
Frederick J. Bonte
L. Maximilian Buja
Ernest M. Stokely
James T. Willerson

Early in 1974, Bonte and associates[3] found that ^{99m}Tc-labeled phosphate compounds localized in acutely infarcted animal myocardium. By the spring of 1974, it was apparent in human volunteers with documented infarction that the technique was equally sensitive in identifying the disease in humans. Also apparent was a more variable pattern of infarct labeling in patients, probably due to a more complex pattern of residual blood flow. In the past three years much work has been done in the animal model to better understand the pathophysiology of the labeling process* and with patients to define its clinical usefulness.†

TECHNIQUE

Radiopharmaceuticals

[^{99m}Tc] stannous pyrophosphate ([^{99m}Tc] PYP) was originally used to image acute myocardial infarcts in both animals[3] and humans.[51] Since then, other ^{99m}Tc-tagged phosphate compounds have been shown to label acutely infarcted myocardium.[4,29,72] [^{99m}Tc] pyrophosphate continues to be the most widely used radiopharmaceutical. Any ^{99m}Tc-labeled phosphate or phosphonate agent that is stable, has a high tagging efficiency, has good blood clearance properties, and generally gives good bone images should work equally well. [^{99m}Tc] Sn complexed with 1-hydroxyethylidene-1, diphosphonic acid (HEDP) or disodium methylene diphosphonate (MDP) has given good quality myocardial images in animals. Acute myocardial infarction imaging agents may develop from the group of new bone-imaging radiopharmaceuticals being tested, such as multidentate phosphonates and imidodiphosphate.[20]

Poor labeling of the phosphate compounds with ^{99m}Tc or rapid breakdown of the label in the vial or syringe leads to poor clearance of the tracer from the blood pool. If this is not recognized, it can be misinterpreted as diffuse tracer concentration in the myocardium and be given a false-positive interpretation. Persistence of activity in the blood pool is one of the most common causes of false-positive interpretations. Each batch of the radiopharmaceutical must be tested for labeling efficiency. Sta-

This work was supported, in part, by the Harry S. Moss Heart Fund, the National Institutes of Health Ischemic Heart Disease Specialized Center of Research (SCOR) grant no. HL 17669, the National Institutes of Health grant no. HL 17777, and the Southwestern Medical Foundation.

*See references 5, 6, 9-12, 22, 23, 29, 47, 56a, 57, 58, 60, 62a, 63, 66, 70, and 72.

†See references 1, 13, 16, 24, 26, 31, 34-36, 39, 44, 48, and 52-56.

bility of the agent in the vial or syringe has been improved over the past three years by the manufacturers, with most claiming at least 3 to 4 hr of stability after preparation. Injection within 1 hr after preparation helps ensure a minimum of free $^{99m}TcO_4^-$.

Large amounts of tin are used by many manufacturers to stabilize the ^{99m}Tc-labeled phosphate compound. This leads to an interesting phenomenon in whole blood, in which the excess stannous ions reduce the $^{99m}TcO_4^-$, resulting in labeling of the red blood cells with the reduced ^{99m}Tc ion.[15,49] Although this in vivo labeling can be used to advantage in performing gated cardiac blood pool imaging,[64,65] it prevents visualization of acute myocardial infarcts. Activity in the cardiac blood pool disappears rapidly as $^{99m}TcO_4^-$ is removed from the blood by the kidneys, thyroid gland, and gastrointestinal tract. However, $^{99m}TcO_4^-$-labeled red blood cells remain in the blood pool for hours, making even delayed myocardial imaging difficult, if not impossible. This potential problem is only encountered when there is poor labeling efficiency or unstable ^{99m}Tc-labeled phosphate compounds are employed.

Dosages of 15 mCi ^{99m}Tc tagged to 5 mg of stannous pyrophosphate is most commonly used for acute myocardial necrosis imaging in patients. In general, 15 mCi of ^{99m}Tc can be labeled to the standard amount of phosphate or phosphonate compound used in routine bone imaging.

Equipment

Equipment requirements to perform myocardial imaging vary, depending on the patient's condition and location and the degree of image processing required. The minimum requirement is a gamma camera. There have been some attempts to perform [^{99m}Tc] phosphate myocardial imaging on rectilinear scanners, but increased imaging times, positioning difficulty, and a tomographic effect caused by the focal plane of the rectilinear collimator are limiting factors. The increased resolution of the newer gamma cameras is a *definite advantage* in myocardial imaging but not a necessity. Much of the imaging to date has been done with the older generation nineteen-tube gamma cameras and has been satisfactory. More important than a camera's resolution is its field uniformity, since warm areas in the field can be erroneously interpreted as a positive myocardial scintigram.

Patients suspected of having acute myocardial infarction can be brought to the nuclear medicine section of the hospital for imaging if they are believed to be clinically stable. This requires a portable electrocardiographic monitor, defibrillator, and emergency drugs. The whole procedure is simplified by using a portable gamma camera that can be taken to the patient's bedside. Critically ill patients who cannot be transported can only be imaged at their bedsides, and even cardiac patients who appear stable are subjected to less risk if imaging is done in the ICU. When coupled with a computer system, the portable gamma camera becomes a powerful tool in evaluating cardiac patients.

Computer systems are not necessary to do basic [^{99m}Tc] phosphate imaging. Image processing is helpful in 10% to 15% of the images and can be used to better delineate the lesion, but lack of a data-processing system should not prevent smaller hospitals from utilizing this technique.

Positioning

Patients are routinely imaged in the anterior, left lateral, and one or more LAO projections. All three views are usually obtained with the patient in the supine position. RAO projections have not been used routinely because of the increased distance between the camera face and the left ventricle but may be useful, particularly to image septal or posterior lesions. Lateral views in the supine projection require positioning the patient on the bed's edge. If the patient is lying on his side, better lateral views are obtained from below than above because of the shorter distance between the camera face and the left ventricle. Fig. 15-1 shows negative [^{99m}Tc] PYP myocardial scinti-

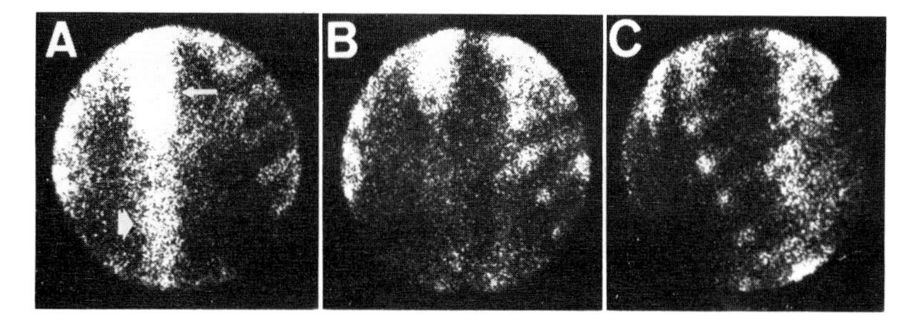

Fig. 15-1. Negative scintigrams of patient without acute myocardial infarction shown in, **A,** anterior, **B,** 45-degree LAO, and, **C,** left lateral views. Small arrow points to sternum with vertebral column indicated by large arrow. (From Parkey, B. W., Bonte, F. J., Meyer, S. L., et al.: Circulation **50:**540, 1974; by permission of the American Heart Association, Inc.)

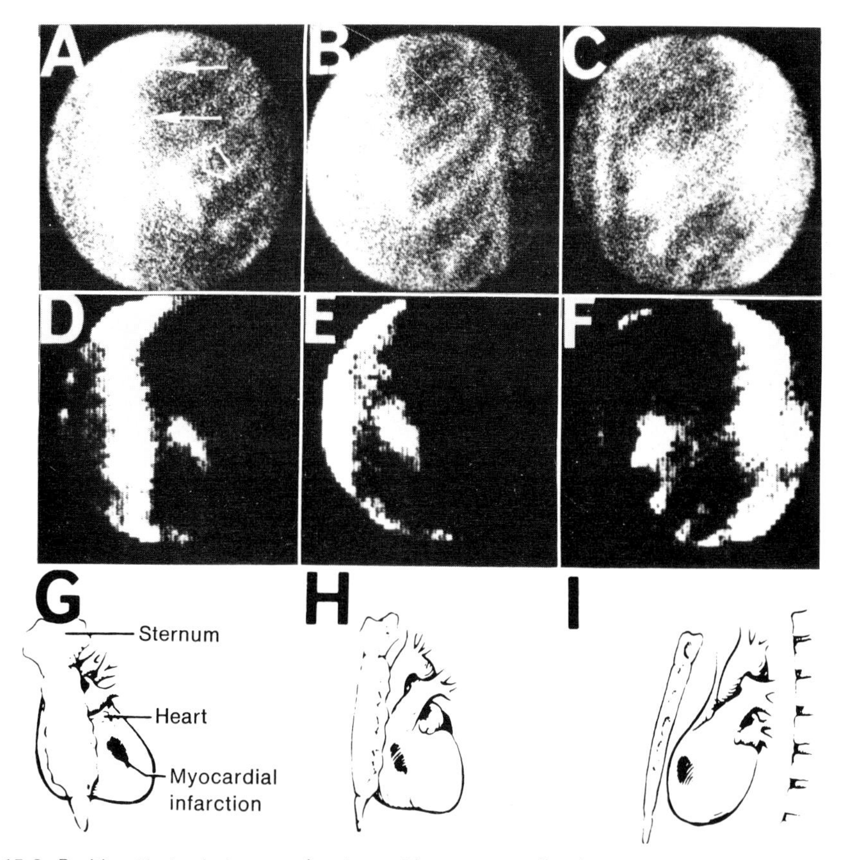

Fig. 15-2. Positive (4+) scintigrams of patient with anterior wall infarction. **A,** Anterior view. Sternum indicated by two smaller arrows, infarct by open arrow. **B,** Forty-five–degree LAO view. **C,** Left lateral view. **D** to **F,** Same as **A** to **C** after contrast enhancement. **G** to **I,** Drawings indicate how area of activity in myocardium rotates with anterior wall of heart. (From Parkey, B. W., Bonte, F. J., Meyer, S. L., et al.: Circulation **50:**540, 1974; by permission of the American Heart Association, Inc.)

grams in the three basic views. Fig. 15-2 demonstrates how these views are used to locate an acute anterior myocardial infarction.

About 400,000 counts are collected in each view. As in other imaging procedures, increased total counts improve resolution, but this has to be balanced against increased patient motion secondary to increased imaging times. From 300,000 to 500,000 total counts can be used, depending on the type of gamma camera and the patient's condition.

Centering each view is relatively easy when the sternum is well visualized. Anterior and LAO views are centered over the myocardium with the sternal image to one side. On these views, positioning too low visualizes the left kidney, which can be mistaken for a large myocardial infarct. Lateral views usually center on the myocardium between the sternum and the spine. The lateral view is the most difficult to position; sometimes the anterior myocardial wall is positioned out of the field of view.

Postinjection and postinfarction timing

Myocardial scintigrams are usually obtained 60 to 90 min after injection. Postinjection imaging time must fit in the "time window" between high blood concentration (early) and marked bone uptake (late). Fig. 15-3 shows sequential [^{99m}Tc] pyrophosphate myocardial scintigrams taken one day postinfarction in a dog in the lateral projection during the first hour after intravenous injection. Note that the infarct is obscured at 5 min by the cardiac blood pool but is visible as early as 12 min in this young dog with rapid blood clearance. Radiopharmaceuticals vary in their blood clearance and rate of bone uptake. More important is patient variation due to differences in renal excretion and bone metabolic rate. In general, delayed 90 min images are required more often in older patients. Again the most common cause of a false-positive interpretation is a persistent radionuclide blood pool. If the work load and time allow, anterior and lateral views within 5 min of injection identify the cardiac blood pool (Fig. 15-4). This information aids in localization

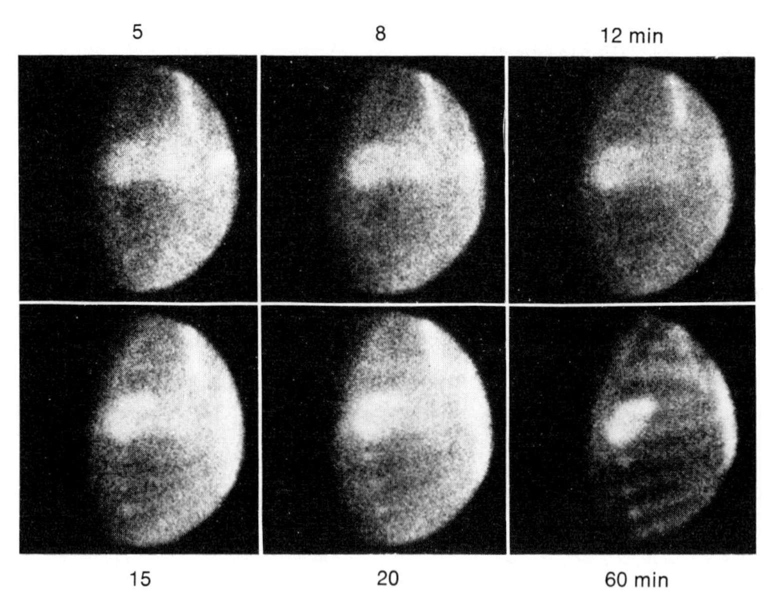

Fig. 15-3. Sequential [^{99m}Tc] PYP myocardial scintigrams obtained in lateral projection during first hour after intravenous injection of radionuclide in dog one day after it was subjected to proximal LAD occlusion. During first 8 min after injection, scintigrams show positive images of entire cardiac silhouette due to intense activity in cardiac blood pool. Scintigrams obtained at later intervals show progressive loss of activity from blood and progressive increase in visualization of radionuclide in skeletal structures and in region corresponding to site of anteroapicoseptal myocardial infarct. Selective concentration of [^{99m}Tc] PYP in region of infarct is questionably visible at 5 min and is readily apparent as early as 12 min after injection of radionuclide. (From Buja, L. M., Parkey, R. W., Dees, J. H., et al.: Circulation **52**:596, 1975; by permission of the American Heart Association, Inc.)

of abnormal uptake areas and is helpful in separating blood pool activity at 1 hr from diffusely positive uptake confined to one myocardial surface.

In experimental infarcts, high left anterior descending (LAD) occlusion, [^{99m}Tc] PYP images (Fig. 15-5) become visible 10 to 12 hr after coronary ligation. Activity is present in infarct tissue samples as early as 4 to 6 hr after infarction, but usually not in in vivo images. Localization increases and is at the maximum at 40 to 72 hr, fades after six to seven days, and ordinarily is absent by fourteen days. In humans, positive images have been seen at 10 to 12 hr (Fig. 15-6) with intensity increasing over the next 36 to 60 hr and usually fading after six to seven days. Unlike the animal model, patterns of uptake vary more in humans due, in part, to a difference in vascular microanatomy and the varying chronicity of underlying coronary disease. As will be discussed later, the extent of blood reflow modulates the level of [^{99m}Tc] PYP uptake. Some human infarcts show low levels of increased uptake for weeks (Fig. 15-7) and some, for months. Causes for continuing positive scintigrams are not known with certainty, but continued limited cell death, dystrophic calcification in the ventricular wall or pericardium, or developing aneurysms may play a role. These will be discussed on pp. 270 to 273. Serial imaging done at one and three days after onset of symptoms increases the diagnostic accuracy of the technique. If only one study can be performed, 48 to 72 hr postinfarction imaging time is recommended.

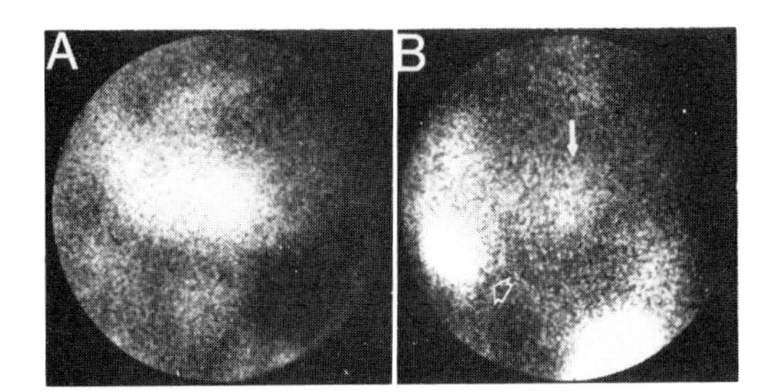

Fig. 15-4. Anterior scintigrams of patient with anterior myocardial infarction. **A,** Blood pool 5 min postinjection. **B,** One hour postinjection with infarct visible to left of sternum. Small arrow denotes infarct, open arrow, sternum.

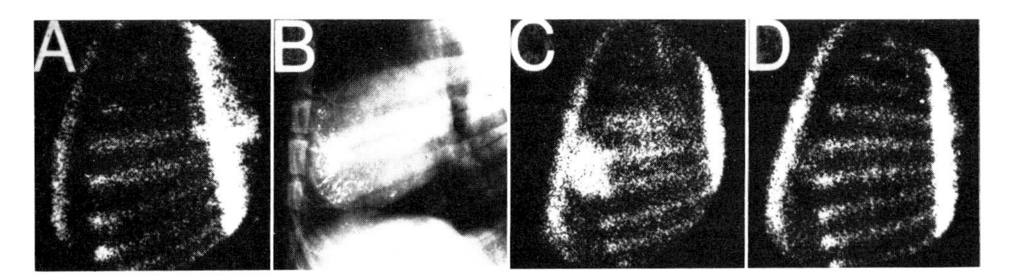

Fig. 15-5. A, Lateral scintiphotograph of chest of 20 kg dog made 1 hr after intravenous administration of 3.0 mCi [^{99m}Tc] PYP. Only normal skeletal distribution is seen. **B,** Left lateral chest roentgenogram of dog made immediately after instillation of 0.1 ml metallic mercury into branches of anterior descending artery. **C,** Left lateral scintiphotograph of dog made 24 hr after mercury embolization and 1 hr after intravenous administration of 3.0 mCi [^{99m}Tc] PYP. Note intense localization of radioactivity at site of myocardial infarct. **D,** Left lateral scintiphotograph of animal made 1 hr after tracer administration and eight days after infarction. Note that tracer localization has almost completely disappeared. Serial scintiphotographs had shown persistent localization only through the fifth day postinfarction. (From Bonte, F. J., Parkey, R. W., Graham, K. D., et al.: Radiology **110:**473, 1974.)

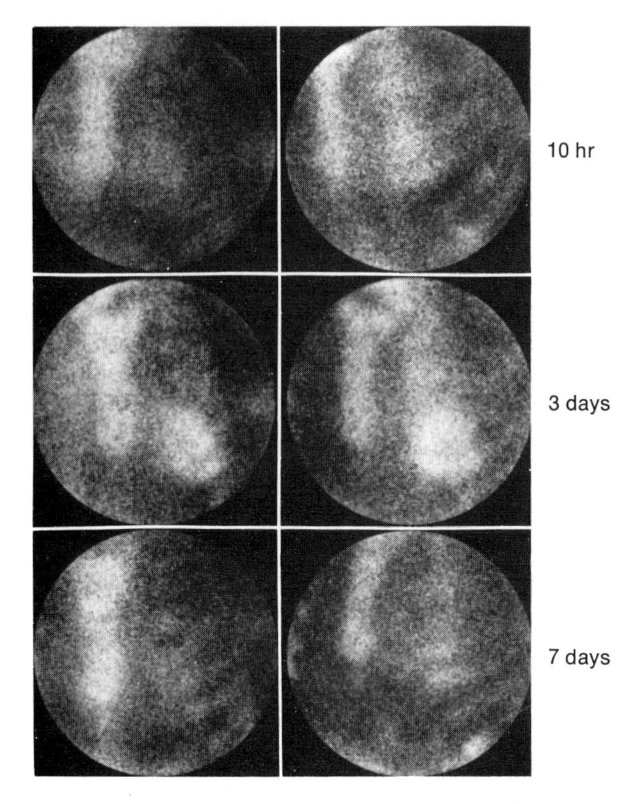

Fig. 15-6. Increase in intensity of [^{99m}Tc] PYP myocardial uptake that occurs with time in most patients with acute transmural myocardial infarctions. *Top,* Faintly positive [^{99m}Tc] PYP myocardial scintigrams approximately 10 hr after myocardial infarction. *Middle,* More intensely positive scintigrams obtained three days after infarction. *Bottom,* Marked reduction of [^{99m}Tc] PYP myocardial uptake seven days after infarction. (From Willerson, J. T., Parkey, R. W., Bonte, F. J., et al.: Circulation **51:**1046, 1975; by permission of the American Heart Association, Inc.)

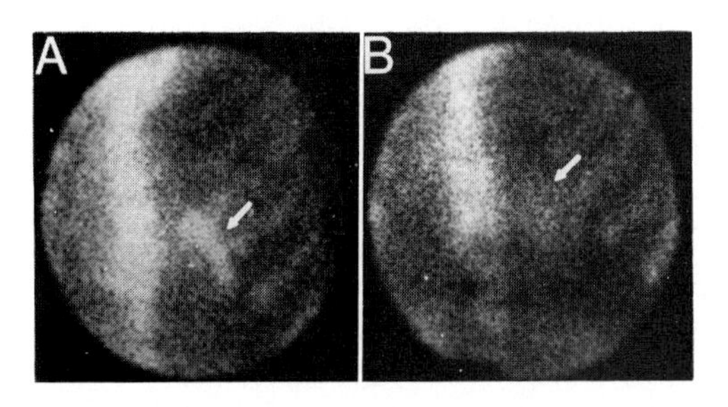

Fig. 15-7. Scintigrams of patient with anterior wall infarct (same patient as in Fig. 15-2). Anterior views **A,** four and **B,** eighteen days postinfarction, with arrows indicating area of damaged myocardium. Note decrease in radioactivity (4+ to 2+). (From Parkey, B. W., Bonte, F. J., Meyer, S. L., et al.: Circulation **50:**540, 1974; by permission of the American Heart Association, Inc.)

PATHOPHYSIOLOGY

A positive [^{99m}Tc] phosphate myocardial scintigram in acute myocardial infarction results from selective concentration of the agent in areas with partial to homogeneous myocardial necrosis, multifocal muscle cell calcification, and residual blood flow.*

The occurrence of significant levels of perfusion in the peripheral zones of transmural infarcts is explicable on the basis of collateral blood flow to these regions.[2,10,17,28] Residual blood flow in the periphery correlates with the presence of numerous irreversibly injured muscle cells with contraction bands and calcium deposits (Figs. 15-8 to 15-10). Experimental studies have indi-

*See references 5, 6, 9-12, 47, 48, 62a, 70, and 72.

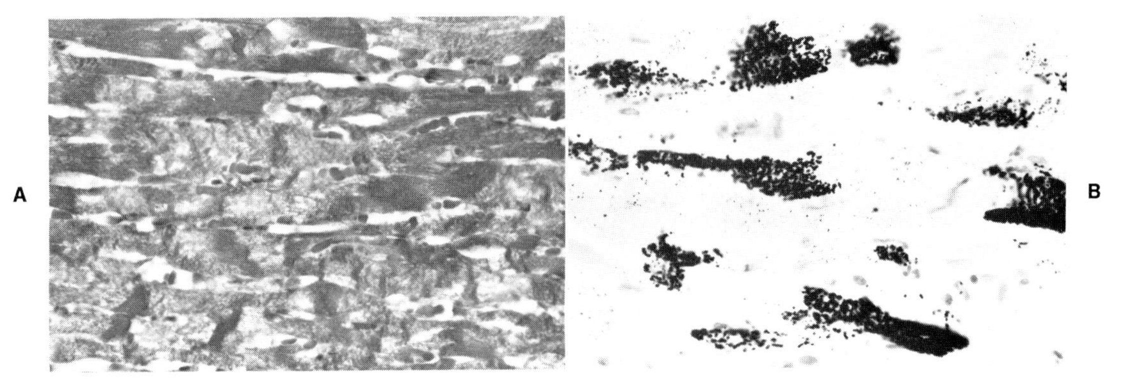

Fig. 15-8. Typical histopathologic features of infarct periphery. **A,** Necrotic muscle cells exhibit nuclear lysis and cytoplasmic disruption (× 435, hematoxylin and eosin stain). **B,** Several of these cells also show prominent calcification (× 435, Von Kossa stain for calcium salts). (From Parkey, R. W., Bonte, F. J., Buja, L. M., et al.: Semin. Nucl. Med. **7:**15-28, 1977; by permission.)

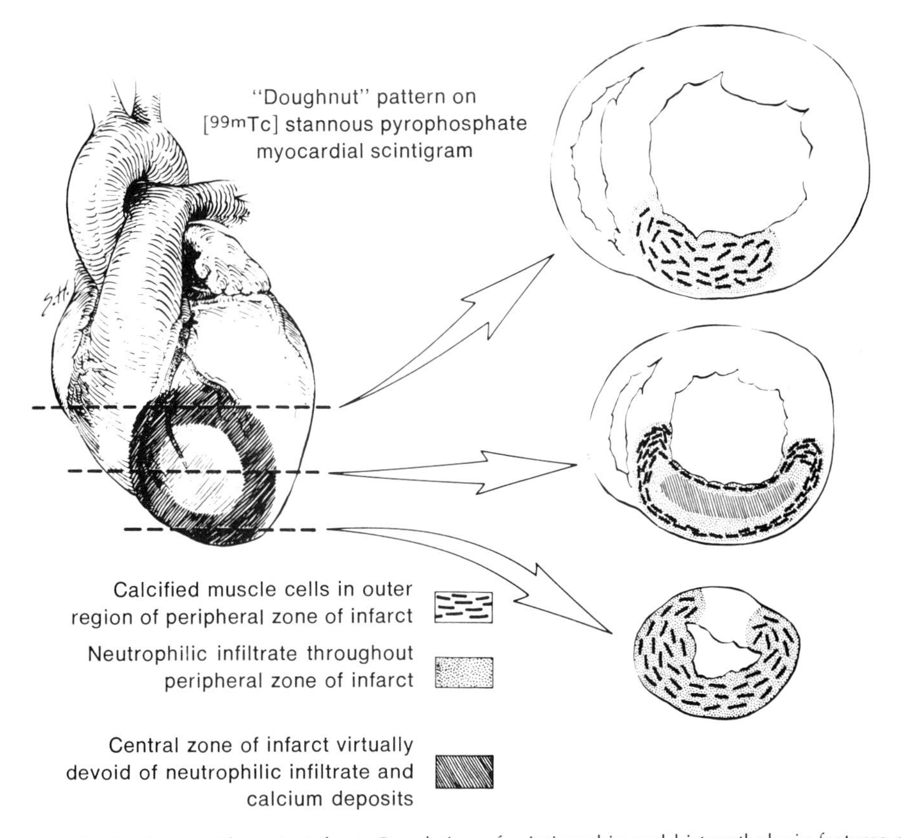

Fig. 15-9. Canine heart with acute infarct. Correlation of scintigraphic and histopathologic features of typical acute myocardial infarct produced in dogs by permanent occlusion of proximal LAD coronary artery. Histopathologic sections of transverse ventricular slices through infarct reveal large peripheral zone heavily infiltrated by neutrophils, surrounding subendocardial central zone devoid of neutrophils. Area of extensive muscle cell calcification is limited to outer region of peripheral zone of infarct. Doughnut pattern observed on [^{99m}Tc] PYP scintigram of heart is explicable on basis of selective concentration of [^{99m}Tc] PYP in this outer region of peripheral zone of infarct. (From Buja, L. M., Parkey, R. W., Dees, J. H., et al.: Circulation **52:**596, 1975; by permission of the American Heart Association, Inc.)

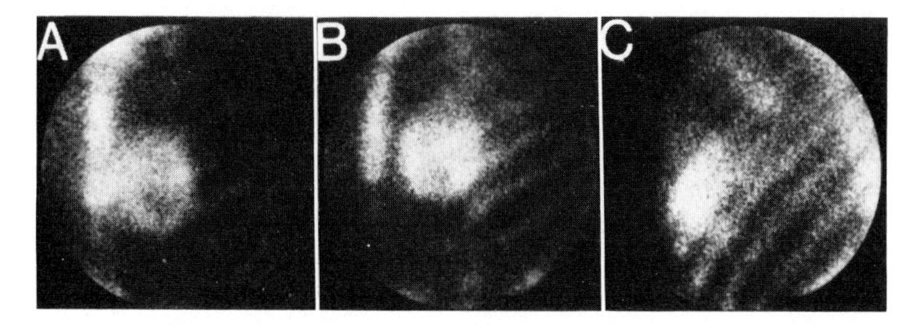

Fig. 15-10. Patient with acute anterior myocardial infarct imaged 48 hr after infarction. **A,** Anterior, **B,** LAO, and **C,** left lateral views. Note the doughnut pattern described in Fig. 15-9. (From Parkey, R. W., Bonte, F. J., Buja, L. M., et al.: Semin. Nucl. Med. **7**:15-28, 1977; by permission.)

cated that necrosis with contraction band formation and frequent calcification develop when myocardial perfusion persists throughout the period of injury or occurs at some time during the evolution of injury.[8,42] Ultrastructural studies have shown that calcium deposits in irreversibly injured muscle cells are localized primarily to the mitochondria and occur in the form of apatite-like spicules or deposits of subcrystalline, finely granular, amorphous calcium phosphate.[7,9,19,30,62] Central regions of the infarcts, which have severely reduced blood flow, contain necrotic muscle cells without contraction bands or calcium deposits. This pattern of myocardial necrosis is typical of areas rendered maximally ischemic by permanent coronary occlusion.[9,10,61,62] Such regions have also been shown to develop a severe impairment to myocardial reperfusion within 90 to 120 min after the onset of coronary occlusion.[42,69]

Recent autoradiographic studies in our laboratory have demonstrated that significant radionuclide labeling occurs over severely damaged border-zone muscle cells with marked lipid accumulation and, in some cases, early mitochondrial calcification, in addition to labeling of muscle cells with advanced necrosis. These findings show that the lack of observed correlation between extent of histologically demonstrable necrosis and level of [^{99m}Tc] phosphate uptake in a given infarct sample is due to labeling of severely damaged border-zone muscle cells in addition to greater blood flow and delivery of [^{99m}Tc] phosphate to the edges of the infarcts.* The autoradiographic data, however, also show that regions of partial necrosis with significant [^{99m}Tc] phosphate uptake involve only small, predominantly subepicardial areas at the edges of the infarcts[10] and that less severely injured border-zone muscle cells within 1 cm of the gross boundaries of the infarcts did not show increased labeling.[11,12] These observations are in accord with infarct sizing studies, showing that measurement of the area of increased [^{99m}Tc] phosphate uptake on myocardial scintigrams provides an accurate estimate of the size of transmural anterior infarcts.[5,63]

Further studies are needed to determine the ultimate fate of severely damaged border-zone muscle cells with increased [^{99m}Tc] phosphate uptake. Nevertheless, the demonstration of early mitochondrial calcification in some border-zone muscle cells shows they have progressed to an early stage of irreversible damage.† It appears unlikely that cellular injury of the type manifested by autoradiographically labeled border-zone myocardium occurs in the absence of associated cellular necrosis.

Considerable controversy has arisen regarding mechanisms of tissue concentration of ^{99m}Tc-labeled phosphate agents. Some workers have suggested a less important

*See references 5-7, 9-12, 18, 30, 47, 58, 63, and 70.
†See references 7, 11, 12, 30, 38, and 62.

role for complexing with calcium deposits than originally proposed.* In the case of acute myocardial infarction, Dewanjee has proposed that concentration of ^{99m}Tc-labeled phosphate agents results primarily from complexing with denatured proteins and other organic macromolecules.[21,23] It must be emphasized, however, that selective in vivo concentration of ^{99m}Tc-labeled phosphate compounds in altered soft tissues as well as bone is invariably associated with elevated tissue calcium content.† Our studies have also shown that cell fractionation procedures result in artefactually low retention of calcium and ^{99m}Tc-labeled phosphate compounds in the mitochondrial fraction; since ultrastructurally demonstrable calcific deposits are not retained in the mitochondrial fraction, ^{99m}Tc-labeled phosphate agents and calcium exhibit a directionally similar distribution in the various fractions.[11,12] Thus the available data strongly support the view that organic macromolecular complexing plays an essentially secondary role in tissue concentration of ^{99m}Tc-labeled phosphate compounds and that concentration of these agents results primarily from selective complexing of ^{99m}Tc-labeled phosphate agents with various forms of tissue calcium stores, including readily soluble and insoluble calcium deposits.‡

IMAGE INTERPRETATION

Interpretation of [^{99m}Tc] phosphate myocardial scintigrams is greatly simplified when the technical factors discussed earlier (camera-field uniformity, positioning, radiopharmaceutical stability, tagging efficiency, blood clearance, and postinjection and postinfarction timing) are well controlled. If these technical factors are not well controlled, the technique loses its sensitivity and ability to detect small amounts of myocardial necrosis.

Scintigrams are commonly graded 0 to 4+, depending on the activity over the myocardium. Grading refers to the visibility of the suspected lesion, not to the size—0 represents no activity; 1+ indicates minimal activity believed to be in blood pool or chest wall (negative scan, but with a lower confidence level), 2+, definite myocardial activity, 3+, activity equal to bone activity, and 4+, activity greater than bone activity. This system is arbitrary, but when used clinically has shown good correlation with electrocardiographic and enzymatic criteria for determining the presence of infarction. Fig. 15-11 shows examples of 0, 2+, and 4+ myocardial scintigrams. As discussed earlier, the amount of [^{99m}Tc] phosphate that localizes in an area of myocardial necrosis depends on the blood flow to the damaged area as well as the mass of injured tissue. In patients with acute transmural myocardial infarcts, approximately 80% have 3+ or 4+ [^{99m}Tc] PYP myocardial scintigrams with 20% being graded 2+. In patients with acute subendocardial infarcts, only 30% to 40% have 3+ or 4+ scintigrams, and about 60% to 70% are graded 2+. Fig. 15-12 demonstrates [^{99m}Tc] PYP myocardial scintigrams that are representative of the different types of transmural infarctions, all graded 3+ or 4+.

Localization of 3 or 4+ [^{99m}Tc] PYP infarcts is relatively easy when all three views are obtained. Fig. 15-2 demonstrates how an anterior infarct rotates with the sternum as the patient rotates. Fig. 15-13 shows how a posterior wall infarct rotates away from the sternum as the patient rotates. Myocardial damage characterized by 2+ [^{99m}Tc] PYP myocardial scintigrams is not always seen on all three views and is often harder to precisely locate. Inferior myocardial infarcts are visualized as platelike extensions of activity to the left of the sternum as seen on anterior views (Fig. 15-12, *2a* to *2c*). Acute subendocardial infarcts graded 3+ or 4+ (Fig. 15-14) localize easily, but those graded 2+ may be seen only on one or two views, and, as just noted, this makes exact localization more difficult.

Although computer processing of the

*See references 21, 23, 40, 41, 59, and 71.
†See references 11, 12, 25, 58, and 62a.
‡See references 3, 9-12, 58, and 62a.

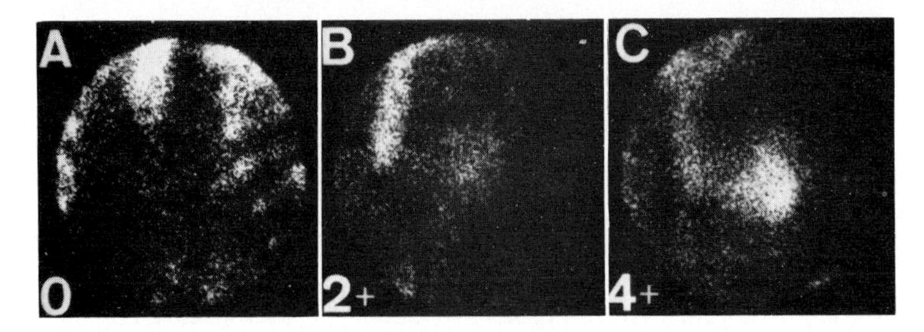

Fig. 15-11. Myocardial scintigrams obtained after [^{99m}Tc] PYP injection. **A,** Negative myocardial scintigram. **B,** 2+ activity. **C,** 4+ myocardial uptake of [^{99m}Tc] PYP. (From Willerson, J. T., Parkey, R. W., Bonte, F. J., et al.: Circulation **51:**1046, 1975; by permission of the American Heart Association, Inc.)

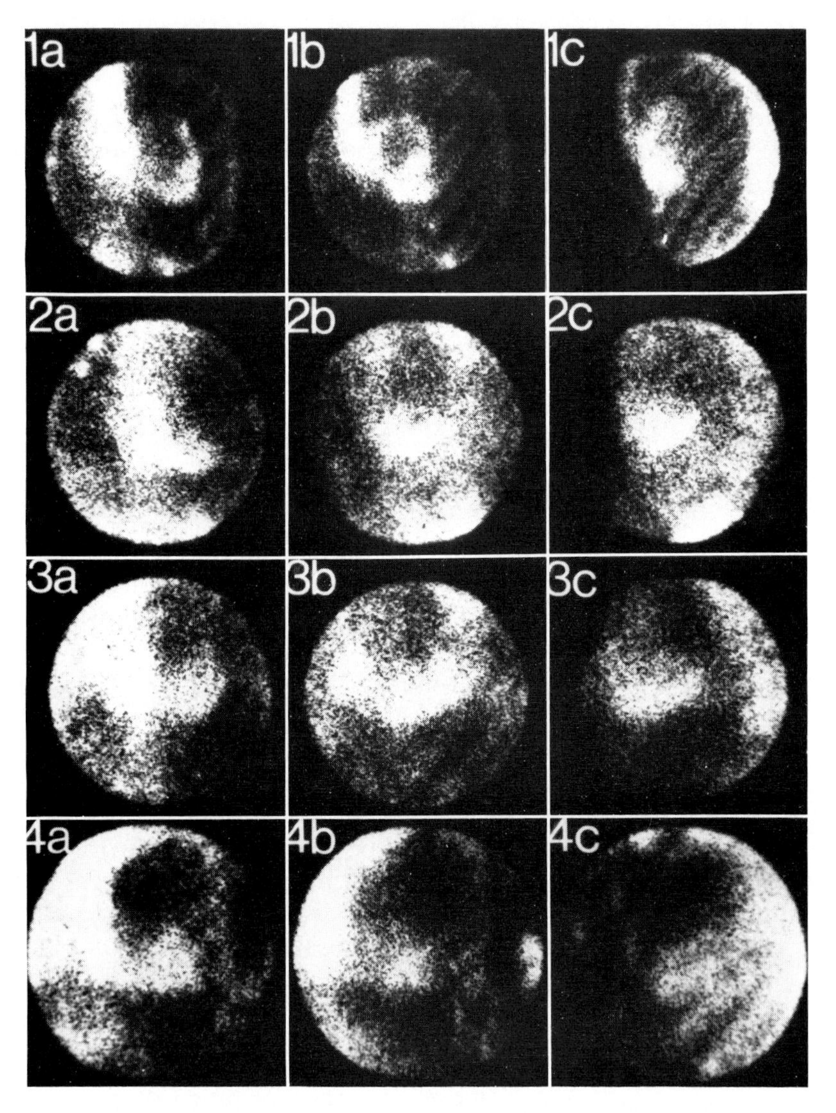

Fig. 15-12. Representative [^{99m}Tc] PYP myocardial scintigrams of different types of transmural myocardial infarction. In each horizontal row, left to right, anterior view, LAO view, and left lateral view. *1a* to *1c,* Anterior myocardial infarction; *2a* to *2c,* inferior myocardial infarction; *3a* to *3c,* anterolateral myocardial infarction; *4a* to *4c,* true posterior myocardial infarction. (From Willerson, J. T., Parkey, R. W., Bonte, F. J., et al.: Circulation **51:**1046, 1975; by permission of the American Heart Association, Inc.)

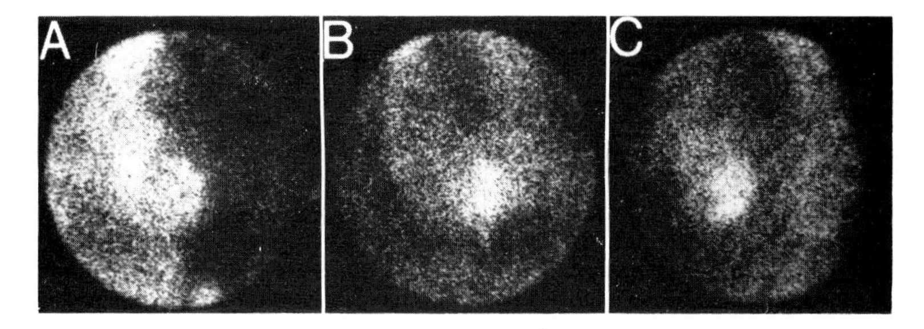

Fig. 15-13. Positive (4+) scintigram of patient with true posterior wall myocardial infarct. Note how activity rotates away from sternum. **A,** anterior, **B,** LAO, and **C,** left lateral views. (From Parkey, R. W., Bonte, F. J., Buja, L. M., et al.: Semin. Nucl. Med. **7**:15-28, 1977; by permission.)

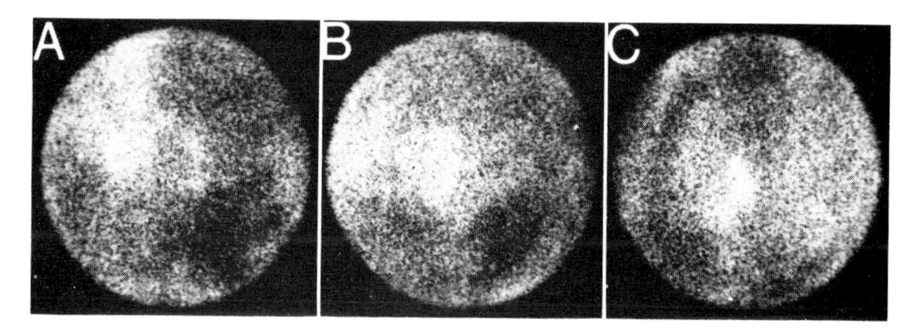

Fig. 15-14. Patient with acute posterior subendocardial myocardial infarct with positive (3+) [^{99m}Tc] PYP scintigrams. Note how infarct rotates away from sternum. **A,** Anterior, **B,** LAO, and **C,** left lateral views. (From Willerson, J. T., Parkey, R. W., Bonte, F. J., et al.: Circulation **51**:436, 1975; by permission of the American Heart Association, Inc.)

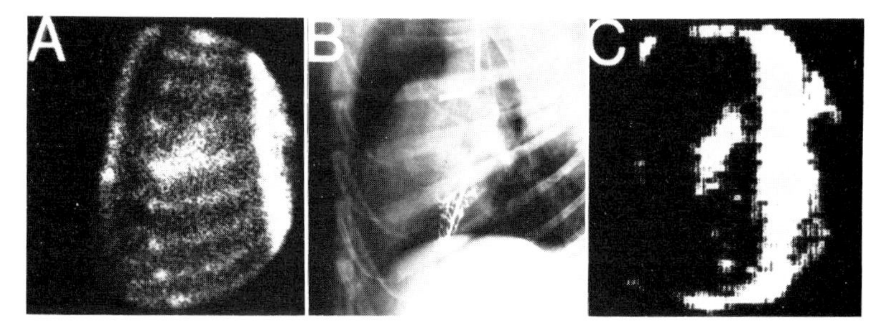

Fig. 15-15. Left lateral scintigrams of dog with posterior wall acute myocardial infarct. **A,** Unprocessed scintigram. Ribs overlying infarct. **B,** X-ray photograph. Location of infarction caused by mercury emboli. **C,** Image after band-reject filtering to improve visibility of infarct. (From Parkey, R. W., Bonte, F. J., Buja, L. M., et al.: Semin. Nucl. Med. **7**:15-28, 1977; by permission.)

images is useful in sizing the infarcts, it is not usually necessary to visualize them. Of the three views, the lateral view benefits most from processing because of rib activity. Simple background subtraction and contrast enhancement are all that is usually required, but rib structures can be removed from the images using a one-dimensional, recursive, band-reject digital filter (Fig. 15-15).

Increased activity can be seen in the chest wall if there is tissue damage present. By using the three views, the activity can usually be shown to be separate from the myocardium. Chest wall muscle damage from surgery (Fig. 15-16) or burns caused

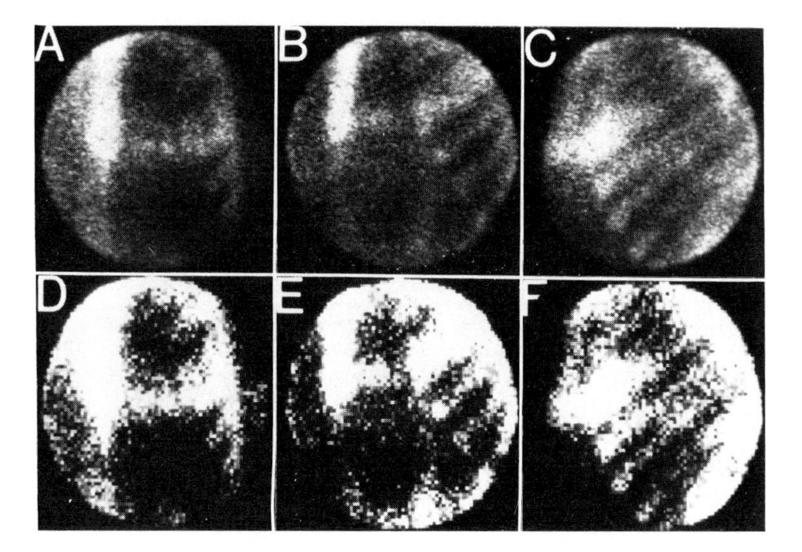

Fig. 15-16. Patient who had his chest opened in an emergency procedure because of stab wound. Note entire area of incision shows activity. **A,** Anterior, **B,** LAO, and **C,** left lateral views. Activity on anterior view extends from sternum to axilla, whereas on lateral view it is limited to anterior chest wall. **D** to **F** are same as **A** to **C** after processing. (From Parkey, R. W., Bonte, F. J., Buja, L. M., et al.: Semin. Nucl. Med. **7**:15-28, 1977; by permission.)

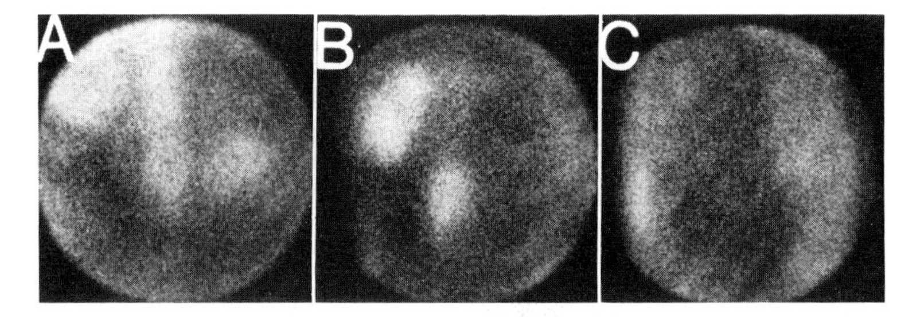

Fig. 15-17. A, Anterior, **B,** LAO, and **C,** left lateral views of patient cardioverted six times. Area of chest wall muscle damage in right upper chest is not a problem, but damage over myocardium could be mistaken for positive myocardial scintigram. Lateral view places activity in chest wall, not in myocardium. Chest wall muscle that has increased activity has been shown to have necrosis present. (From Pugh, B. R., Buja, L. M., Parkey, R. W., et al.: Circulation **54**:399, 1976; by permission of the American Heart Association, Inc.)

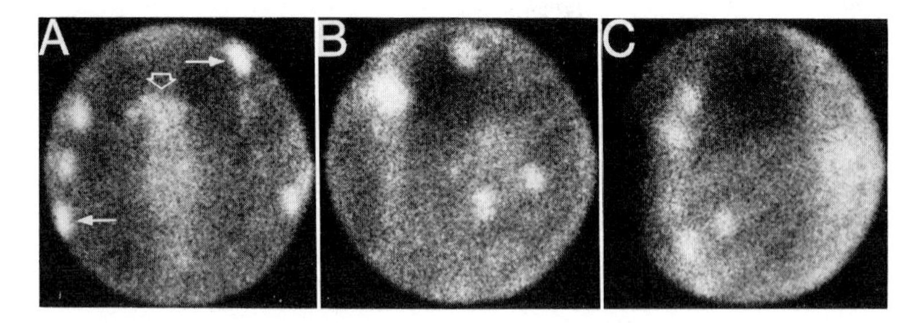

Fig. 15-18. A, Anterior, **B,** LAO, and **C,** left lateral scintigrams of patient who had vigorous closed chest massage. Note multiple broken ribs *(small arrows)* and fractured sternum *(large arrow)*. Areas of increased activity move with chest wall and not myocardium. (From Parkey, R. W., Bonte, F. J., Buja, L. M., et al.: Semin. Nucl. Med. **7**:15-28, 1977; by permission.)

by multiple cardioversions (Fig. 15-17) could lead to false-positive scintigrams. Abnormal activity in bone from trauma (Fig. 15-18) or tumors (Fig. 15-19) is ordinarily easier to separate from true myocardial injury.

What is the accuracy of this technique? We have found less than 4% false-negative scintigrams when serial [99mTc] PYP imaging is performed at the optimal postinjection time. Other groups have reported as high as 10% false-negative results, but this is often without the benefit of serial myocardial imaging. Serial imaging on all patients probably accounts for our low incidence of false-negative tests. Our false-positive results range from 8% to 12% if the ECG is used as the "gold standard" for recognizing acute myocardial infarcts. Half these patients have the syndrome of unstable angina pectoris and could have myocardial necrosis detectable only by scintigraphy. Histologic data on some patients appear to support this possibility,[56a] but more patient histologic data is needed before we can determine exactly how many patients with unstable angina have myocardial necrosis detectable only scintigraphically. Other groups have reported false-positive [^{99m}Tc] PYP myocardial scintigrams ranging as high as 20%.

If not recognized, disease processes in which false-positive images could be expected are functioning breast parenchyma in premenopausal females, breast inflammation or tumors, and chest wall disease of the muscle or ribs. One must also keep in mind that persistently positive [^{99m}Tc] PYP myocardial scintigrams obtained after an earlier myocardial infarction and myocardial injury related to metastatic carcinoma, cardioversion, and possibly, in some patients, coronary spasm can be additional reasons for positive [^{99m}Tc] PYP myocardial scintigrams without other clinical evidence of recent myocardial infarction. Lyons and associates[45] reported as high as 50% persistently positive myocardial scintigrams more than six weeks after acute infarction. Follow-up scintigrams were usually only 2+ positive. Our incidence of persistently positive scintigrams is as high as 40%.[56a] It is important to note that the follow-up scintigrams usually show decreased activity in the old area of infarction, and continued 3 or 4+ activity is a bad prognostic sign, suggesting significant ongoing necrosis. Follow-up scintigrams can identify reinfarction or necrosis involving a different myocardial surface.

CLINICAL USEFULNESS

How does [^{99m}Tc] phosphate imaging aid physicians in evaluating patients suspected of having myocardial infarcts? Does imaging add different information from that obtained from serial electrocardiograms and serum enzyme studies? First, it is important to understand that [^{99m}Tc] phosphate imaging has a different pathophysiologic basis from ECGs or serum enzyme tests. These tests should complement, not

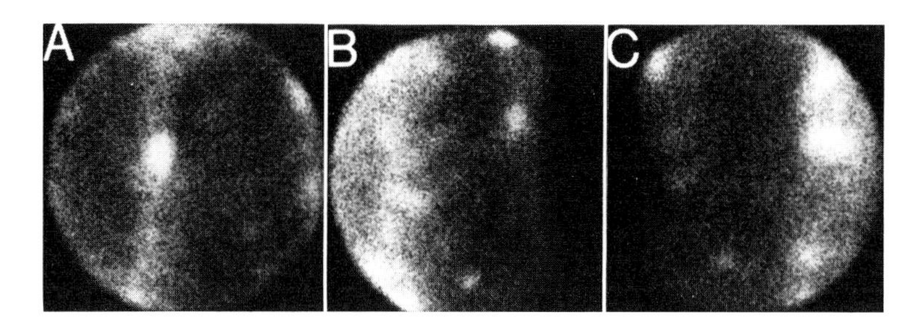

Fig. 15-19. A, Anterior, **B,** LAO, and **C,** left lateral scintigrams of 70-year-old patient with anterior chest pain. Scintigrams show increased bone activity in sternum, ribs, and spine secondary to carcinoma of prostate. (From Parkey, R. W., Bonte, F. J., Buja, L. M., et al.: Semin. Nucl. Med. **7**:15-28, 1977; by permission.)

compete with, one another. [^{99m}Tc] phosphate imaging is particularly valuable in detecting (1) small transmural infarcts (3 g and larger in size), (2) new acute transmural infarcts in or near regions of old infarction, (3) acute subendocardial infarcts (larger than 3 g in size), (4) acute infarction in patients with left bundle-branch block, and (5) perioperative myocardial infarction. Localization of inferior and posterior myocardial infarction is improved with imaging. Sizing of acute anterior and lateral infarcts has been done accurately in dogs[5,63] and should prove helpful in patients.

Extensive evaluation in both experimental animals and patients has shown [^{99m}Tc] phosphate myocardial imaging to be a useful clinical tool; it may be one of the most sensitive noninvasive methods presently available to identify acute myocardial necrosis.

ACKNOWLEDGMENTS

We gratefully acknowledge Dorothy Gutekunst and Katie Wolf for technical and photographic assistance and Gayle Blust for secretarial assistance.

REFERENCES

1. Berman, D. S., Amsterdam, E. A., Salel, A. F., et al.: Improved diagnostic assessment of acute myocardial infarction: sensitivity and specificity of Tc-99m pyrophosphate scintigraphy, abstracted, J. Nucl. Med. **17**:523, 1976.
2. Bloor, C. M.: Functional significance of the coronary collateral circulation. A review, Am. J. Pathol. **76**:562, 1974.
3. Bonte, F. J., Parkey, R. W., Graham, K. D., et al.: A new method for radionuclide imaging of myocardial infarcts, Radiology **110**:473, 1974.
4. Bonte, F. J., Parkey, R. W., Graham, D. K., et al.: Distribution of several agents useful in imaging myocardial infarcts, J. Nucl. Med. **16**:132, 1975.
5. Botvinick, E. H., Shames, D., Lappin, H., et al.: Noninvasive quantitation of myocardial infarction with technetium-99m pyrophosphate, Circulation **52**:909, 1975.
6. Bruno, F. P., Cobb, F. R., Rivas, F., et al.: Evaluation of ^{99m}technetium stannous pyrophosphate as an imaging agent in acute myocardial infarction, Circulation **54**:71, 1976.
7. Buja, L. M., Dees, J. H., Harling, D. F., et al.: Analytical electron microscopic study of mitochondrial inclusions in canine myocardial infarcts, J. Histochem. Cytochem. **24**:508, 1976.
8. Buja, L. M., Ferrans, V. J., and Graw, R. G., Jr.: Cardiac pathologic findings in patients treated with bone marrow transplantation, Hum. Pathol. **7**:17, 1976.
9. Buja, L. M., Parkey, R. W., Dees, J. H., et al.: Morphological correlates of ^{99m}technetium-stannous pyrophosphate of acute myocardial infarcts in dogs, Circulation **52**:596, 1975.
10. Buja, L. M., Parkey, R. W., Stokely, E. M., et al.: Pathophysiology of technetium-99m stannous pyrophosphate and thallium-201 scintigraphy of acute anterior myocardial infarcts in dogs, J. Clin. Invest. **57**:1508, 1976.
11. Buja, L. M., Tofe, A. J., Mukherjee, A., et al.: Role of elevated tissue calcium in myocardial infarct scintigraphy with technetium phosphorous radiopharmaceuticals, abstracted, Circulation **54** (suppl. II): II219, 1976.
12. Buja, L. M., Tofe, A. J., Mukherjee, A., et al.: Sites and mechanisms of localization of technetium-99m phosphorous radiopharmaceuticals in acute myocardial infarcts and other tissues J. Clin. Invest. **60**:724, 1977.
13. Campeau, R. J., Gottlieb, S., Chandarlapaty, S. C. K., et al.: Accuracy of technetium-99m labeled phosphates for detection of acute myocardial infarction, abstracted, J. Nucl. Med. **16**:518, 1975.
14. Carr, E. A., Jr., Beierwaltes, W. H., Patno, M. E., et al.: The detection of experimental myocardial infarcts by photoscanning, Am. Heart J. **64**:650, 1962.
15. Chandler, W. M., and Shuck, L. D.: Abnormal technetium-99m pertechnetate imaging following stannous pyrophosphate bone imaging, abstracted, J. Nucl. Med. **16**:518, 1975.
16. Coleman, R. E., Klein, M. S., Roberts, R., et al.: Improved detection of myocardial infarction with technetium-99m stannous pyrophosphate and serum MB creatine phosphokinase, Am. J. Cardiol. **37**:732, 1976.
17. Cox, J. L., Pass, H. I., Wechsler, A. S., et al.: Coronary collateral blood flow in acute myocardial infarction, J. Thorac. Cardiovasc. Surg. **69**:117, 1975.
18. D'Agostino, A. W.: An electron microscopic study of cardiac necrosis produced by a 9-α-fluorocortisol and sodium phosphate, Am. J. Pathol. **45**:633, 1964.
19. D'Agostino, A. W., and Chiga, M.: Mitochondrial mineralization in human myocardium, Am. J. Clin. Pathol. **53**:820, 1970.
20. Davis, M. A., and Jones, A. G.: Comparison of ^{99m}Tc-labeled phosphate and phosphonate agents for skeletal imaging, Semin. Nucl. Med. **6**:19, 1976.
21. Dewanjee, M. K.: Localization of skeletal-imaging ^{99m}Tc-chelates in dead cells in tissue culture: concise communication, J. Nucl. Med. **17**:993, 1976.
22. Dewanjee, M. K., and Kahn, P. C.: Myocardial mapping techniques and the evaluation of new

^{113m}In-labeled polymethylenephosphonates for imaging myocardial infarct, Radiology **117**:723, 1975.

23. Dewanjee, M. K., and Khan, P. C.: Mechanism of localization of ^{99m}Tc-labeled pyrophosphate and tetracycline in infarcted myocardium, J. Nucl. Med. **17**:639, 1976.
24. Donsky, M. S., Curry, G. C., Parkey, R. W., et al.: Unstable angina pectoris: clinical, angiographic, and myocardial scintigraphic observations, Br. Heart J. **38**:257, 1976.
25. Francis, M. D., Slough, C. L., Tofe, A. J., et al.: Factors affecting uptake and retention of Tc-99m-diphosphonate and Tc-99m-pertechnetate in osseous, connective and soft tissue, Calcif. Tissue Res. **20**:303, 1976.
26. Go, R. T., Doty, D. B., Chiu, C. L., et al.: A new method of diagnosing myocardial contusion in man by radionuclide imaging, Radiology **116**:107, 1975.
27. Gorten, R. J., Hardy, L. B., McCraw, B. H., et al.: The selective uptake of Hg-203 chlormerodrin in experimentally produced myocardial infarcts, Am. Heart J. **72**:71, 1966.
28. Gregg, D. E.: The natural history of coronary collateral development, Circ. Res. **35**:335, 1974.
29. Grossman, Z. D., Foster, A. B., Richardson, R., et al.: Uptake of six technetium-99m radiopharmaceuticals and strontium-85 in vasopressin-induced rabbit myocardial infarction, abstracted, J. Nucl. Med. **17**:534, 1976.
30. Hagler, H. K., Burton, K. P., Browne, R. H., et al.: Energy dispersive x-ray spectroscopic (EDS) analysis of small particulate inclusions in hypoxic and ischemic myocardium. Proceedings of the Illinois Institute of Technology Research Institute Scanning Electron Microscopy Symposium, Chicago, 1977, Chicago Press Corp., vol. 2, p. 145.
31. Harris, R. A., Parkey, R. W., Bonte, F. J., et al.: Sizing acute myocardial infarction in patients utilizing technetium-99m stannous pyrophosphate myocardial scintigrams, abstracted, Clin. Res. **23**: 1, 1975.
32. Henning, H., Schelbert, H., O'Rouke, R. A., et al.: Dual myocardial imaging with Tc-99m pyrophosphate and thallium-201 for diagnosing and sizing acute myocardial infarction, abstracted, J. Nucl. Med. **17**:524, 1976.
33. Holman, B. L., Dewanjee, M. K., Idoine, J., et al.: Detection and localization of experimental myocardial infarction with ^{99m}Tc-tetracycline, J. Nucl. Med. **14**:595, 1973.
34. Holman, B. L., Ehrie, M., and Lesch, M.: Correlation of acute myocardial infarct scintigraphy with postmortem studies, Am. J. Cardiol. **37**:311, 1976.
35. Holman, B. L., Tanaka, T. T., and Lesch, M.: Evaluation of radiopharmaceuticals for the detection of acute myocardial infarction, Radiology **121**:427, 1976.
36. Howe, W. R., Goodrich, J. K., Burno, F. P., et al.: Technetium-99m pyrophosphate detection of perioperative myocardial infarction in cardiac surgery patients, abstracted, J. Nucl. Med. **17**:524, 1976.
37. Hubner, P. J. B.: Radioisotopic detection of experimental myocardial infarction using mercury derivatives of fluorescein, Cardiovasc. Res. **4**:509, 1970.
38. Jennings, R. B., and Ganote, C. E.: Mitochondrial structure and function in acute myocardial ischemic injury, Circ. Res. **38**(suppl. I):I80, 1976.
39. Karunaratne, H. B., Walsh, W. F., Fill, H. R., et al.: Technetium-99m pyrophosphate myocardial scintigraphy in patients with chest pain—lack of diagnostic specificity, abstracted, J. Nucl. Med. **17**:523, 1976.
40. Kaye, M., Silverton, S., and Rosenthall, L.: ^{99m}Tc-pyrophosphate: studies in vivo and in vitro, J. Nucl. Med. **16**:40, 1975.
41. Klein, M. S., Coleman, R. E., Ahmed, S. A., et al.: ^{99m}Tc(Sn) pyrophosphate scintigraphy: sensitivity, specificity, and mechanisms, abstracted, Circulation **52**(suppl. II):II52, 1975.
42. Kloner, R. A., Ganote, C. E., and Jennings, R. B.: The "no-flow" phenomenon after temporary coronary occlusion in the dog, J. Clin. Invest. **54**:1496, 1974.
43. Kramer, R. J., Goldstein, R. E., Hirshfeld, J. W., et al.: Accumulation of gallium-67 in regions of acute myocardial infarction, Am. J. Cardiol. **33**: 861, 1974.
44. Logic, J. R., Mantel, J. A., Rogers, W. J., et al.: quantitation of serial myocardial scintigrams in patients with acute myocardial infarction, abstracted, J. Nucl. Med. **17**:524, 1976.
45. Lyons, D. K., Olson, H. G., Brown, W. T., et al.: Persistence of an abnormal pattern on ^{99m}Tc-pyrophosphate myocardial scintigraphy following acute myocardial infarction, Clin. Nucl. Med. **1**:253, 1977.
46. Malek, P., Vavrejn, B., Ratusky, J., et al.: Detection of myocardial infarction by *in vivo* scanning, Cardiologia **51**:22, 1967.
47. Marcus, M. L., Tomanek, R. J., Ehrardt, J. C., et al.: Relationships between myocardial perfusion, myocardial necrosis and ^{99m}technetium pyrophosphate uptake in dogs subjected to sudden coronary occlusion, Circulation **54**:647, 1976.
48. McLaugh, P., Coates, G., Wood, D., et al.: Detection of acute myocardial infarction by technetium-99m polyphosphate, Am. J. Cardiol. **35**:390, 1975.
49. McRae, J., Sugar, R. M., Shipley, B., et al.: Alternation in tissue distribution of ^{99m}Tc-pertechnetate in rats given stannous tin, J. Nucl. Med. **15**:151, 1974.
50. Parkey, R. W., Bonte, F. J., Buja, L. M., et al.: Myocardial infarct imaging with technetium-99m phosphates, Sem. Nucl. Med. **7**:15, 1977.
51. Parkey, R. W., Bonte, F. J., Meyer, S. L., et al.: A new method for radionuclide imaging of acute myocardial infarctions in humans, Circulation **50**:540, 1974.

52. Parkey, R. W., Bonte, F. J., Stokely, E. M., et al.: Acute myocardial infarction imaged with technetium-99m stannous pyrophosphate and thallium-201: a clinical evaluation, J. Nucl. Med. **17:**771, 1976.
53. Perez, L.: Clinical experience: technetium-99m labeled phosphates in myocardial imaging, Clin. Nucl. Med. **1:**2, 1976.
54. Perez, L. A., Hayt, D. B., and Freeman, L. M.: Localization of myocardial disorders other than infarction with ^{99m}Tc-labeled phosphate agents, J. Nucl. Med. **17:**241, 1976.
55. Platt, M. R., Mills, L., Parkey, R. W., et al.: Perioperative myocardial infarction diagnosed by technetium-99m stannous pyrophosphate myocardial scintigrams, Circulation **54**(suppl. III):III-24, 1976.
56. Platt, M. R., Parkey, R. W., Willerson, J. T., et al.: Technetium stannous pyrophosphate myocardial scintigrams in the recognition of myocardial infarction in patients undergoing coronary artery revascularization, Ann. Thorac. Surg. **21:**311, 1976.
56a. Polimer, L. R., Buja, L. M., Parkey, R. W., et al.: Clinicopathologic findings in 52 patients studied by technetium-99m stannous pyrophosphate myocardial scintigraphy, Circulation **59:**257, 1979.
57. Pugh, B. R., Buja, L. M., Parkey, R. W., et al.: Cardioversion and its potential role in the production of "false-positive" technetium-99m stannous pyrophosphate myocardial scintigrams, Circulation **54:**399, 1976.
58. Reimer, K. A., Martonffy, K., Schumacher, B. L., et al.: Cardiac localization of ^{99m}Tc-pyrophosphate after temporary or permanent coronary occlusion in dogs, Proc. Soc. Exp. Biol. Med. **156:**272, 1977.
59. Schelbert, H. R., Ingwall, J. S., Sybers, H. D., et al.: Uptake of infarct-imaging agents in reversibly and irreversibly injured myocardium in cultured fetal mouse heart, Circ. Res. **39:**860, 1976.
60. Schelbert, H., Ingwall, J., Sybers, H., et al.: Uptake of Tc-99m pyrophosphate and calcium in irreversibly damaged myocardium, abstracted, J. Nucl. Med. **17:**534, 1976.
61. Shen, A. C., and Jennings, R. B.: Kinetics of calcium accumulation in acute myocardial ischemic injury, Am. J. Pathol. **67:**441, 1972.
62. Shen, A. C., and Jennings, R. B.: Myocardial calcium and magnesium in acute ischemic injury, Am. J. Pathol. **67:**417, 1972.
62a. Siegel, B. A., Engel, W. K., and Derrer, E. C.: Localization of technetium-99m diphosphonate in acutely injured muscle. Relationship to muscle calcium deposition, Neurology **27:**230, 1977.
63. Stokely, E. M., Buja, L. M., Lewis, S. E., et al.: Measurement of acute myocardial infarcts in dogs with ^{99m}Tc-stannous pyrophosphate myocardial scintigrams, J. Nucl. Med. **17:**1, 1976.
64. Stokely, E. M., Parkey, R. W., Bonte, F. J., et al.: Gated blood pool imaging following technetium-99m phosphate scintigraphy, Radiology **120:**433, 1976.
65. Stokely, E. M., Parkey, R. W., Bonte, F. J., et al.: Radionuclide angiocardiography using in vivo labeling of erythrocytes with technetium-99m pertechnetate, abstracted, J. Nucl. Med. **17:**565, 1976.
66. Tofe, A. J., Buja, L. M., Parkey, R. W., et al.: Scintigraphy of canine myocardial infarcts: role of calcium and Tc-99m diphosphonate/pyrophosphate bone agents, abstracted, J. Nucl. Med. **17:** 534, 1976.
67. Willerson, J. T., Parkey, R. W., Bonte, F. J., et al.: Acute subendocardial myocardial infarction in patients: its detection by technetium-99m stannous pyrophosphate, Circulation **51:**436, 1975.
68. Willerson, J. T., Parkey, R. W., Bonte, F. J., et al.: Technetium stannous pyrophosphate myocardial scintigrams in patients with chest pain of varying etiology, Circulation **51:**1046, 1975.
69. Willerson, J. T., Watson, J. T., Hutton, I., et al.: Reduced myocardial reflow and increased coronary vascular resistance following prolonged myocardial ischemia in the dog, Circ. Res. **36:**771, 1975.
70. Zaret, B. L., DeCola, V. C., Donabedian, R. K., et al.: Dual radionuclide study of myocardial infarction: relationships between myocardial uptake of potassium-43, technetium-99m stannous pyrophosphate, regional myocardial blood flow and creatine phosphokinase depletion, Circulation **53:**422, 1976.
71. Zimmer, A. M., Isitman, A. T., and Holmes, R. A.: Enzymatic inhibition of diphosphonate: a proposed mechanism of tissue uptake, J. Nucl. Med. **16:**352, 1975.
72. Zweiman, F. G., Holman, B. L., O'Keefe, A., et al.: Selective uptake of ^{99m}Tc complexes and ^{67}Ga in acutely infarcted myocardium, J. Nucl. Med. **16:**975, 1975.

16 □ Dual radionuclide imaging of myocardial infarction

Barry L. Zaret
Harvey J. Berger

Several questions are relevant to the understanding, management, and investigation of acute myocardial infarction. Has an infarct occurred? If so, when did it occur? What is its location? How large is it? What is its functional significance? Furthermore, it is becoming clear that acute myocardial infarction is not a homogeneous pathologic event with uniform tissue damage, but rather involves a mixture of cell populations that are irreversibly and reversibly damaged as well as normal. In addition, a portion of myocardium destined for necrosis might be salvaged if the appropriate interventions were instituted at the appropriate time. Techniques must be developed to monitor myocardial viability and infarct size during critical periods in the early evolution of the infarct. One recent approach to the clinical and experimental assessment of myocardial infarction involves the use of radionuclide imaging for definition of zones of myocardial necrosis and/or ischemia. This method has the potential for answering many of the questions just posed.

Myocardial infarction can currently be evaluated by radionuclide imaging in two distinct ways (Fig. 16-1). The infarct can be defined as a zone of relatively reduced radioactive tracer accumulation adjacent to normal myocardium containing maximal activity. This technique displays the infarct as a so-called cold spot and will demonstrate *both* old and new infarction in a comparable manner. The cold spot radionuclide employed most widely at the present time is ^{201}Tl. Alternatively, the infarct zone can be defined as a region of increased radionuclide activity. This phenomenon occurs as a direct result of avid binding of the radioactive tracer to chemical constitutents of necrotic myocardial cells within the infarct zone and presumably demonstrates *only* zones of acute infarction. The Tc-labeled bone-seeking phosphates, of which [^{99m}Tc] PYP is the prototype, are the radiopharmaceuticals most widely employed for this particular technique. The myocardial uptake of each type of radionuclide is governed by independent and different pathophysiologic events. Dual imaging with both radionuclides may provide comprehensive and complementary data concerning myocardial viability, scar, and acute necrosis during the various phases of acute myocardial infarction. The dual nuclide approach is prototypic for what in the future will be a multitracer evaluation of patients with coronary disease based on imaging with a variety of newly developed intracellular radioactive markers and performance of radionuclide ventricular function studies.

PATHOPHYSIOLOGIC STUDIES IN ANIMAL MODELS

^{201}Tl is considered a potassium analog because its biologic behavior is in many ways similar to that of intracellular monovalent cations. ^{201}Tl follows a long line of

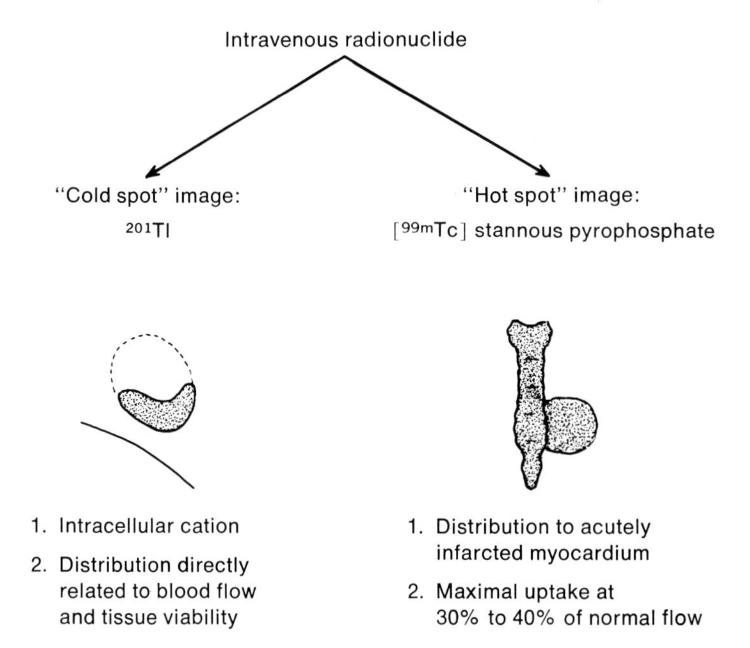

Fig. 16-1. Schematic representation of cold spot and hot spot images obtained following myocardial infarction.

previously employed radioactive potassium analogs, which included isotopes of cesium, rubidium, and potassium itself.[21,30,46] It is currently the cold spot radionuclide of choice because of its favorable physical properties, which include an energy spectrum more suitable for conventional scintillation camera imaging than the higher energy spectra of its predecessors and a physical half-life of approximately three days, allowing for a modest shelf life.[17]

After intravenous injection, ^{201}Tl is rapidly accumulated in the myocardium and distributed within the left ventricular muscle in proportion to the regional blood flow (Fig. 16-2) (Chapter 12). Several laboratories have noted excellent correlations between ^{201}Tl myocardial uptake and regional myocardial blood flow measured by the radioactive microsphere technique.[9,23,35] These correlations are present in preparations involving acute ischemia, as well as 24 hr old infarcts. However, in very low blood flow subendocardial infarct zones, ^{201}Tl appears to lead to overestimation of regional blood flow. This is probably due to enhanced tissue extraction in low blood flow, slow-transit myocardial regions. Furthermore, in extremely high blood flow situations, such as postocclusive reactive hyperemia or pharmacologically induced vasodilation, ^{201}Tl myocardial activity causes underestimation of actual blood flow.[13,35] Nevertheless, within the range of blood flows most commonly seen in experimental acute myocardial infarction, the ^{201}Tl myocardial distribution represents regional myocardial blood flow reasonably accurately.

The relationship of [^{99m}Tc] PYP myocardial uptake to regional blood flow is more complex. In acute myocardial infarction, [^{99m}Tc] PYP accumulation is governed by at least two specific pathophysiologic events: (1) the degree of myocardial necrosis and (2) the amount of residual myocardial blood flow present. Residual blood flow is necessary for the infarct-avid radionuclide to gain entry into the zone of myocardial necrosis.[5,19,44] Because of this blood flow relationship, [^{99m}Tc] PYP accumulation will be maximal in infarct zones that maintain regional myocardial blood flow at 30% to 40% of normal (Fig. 16-3). In infarct zones with

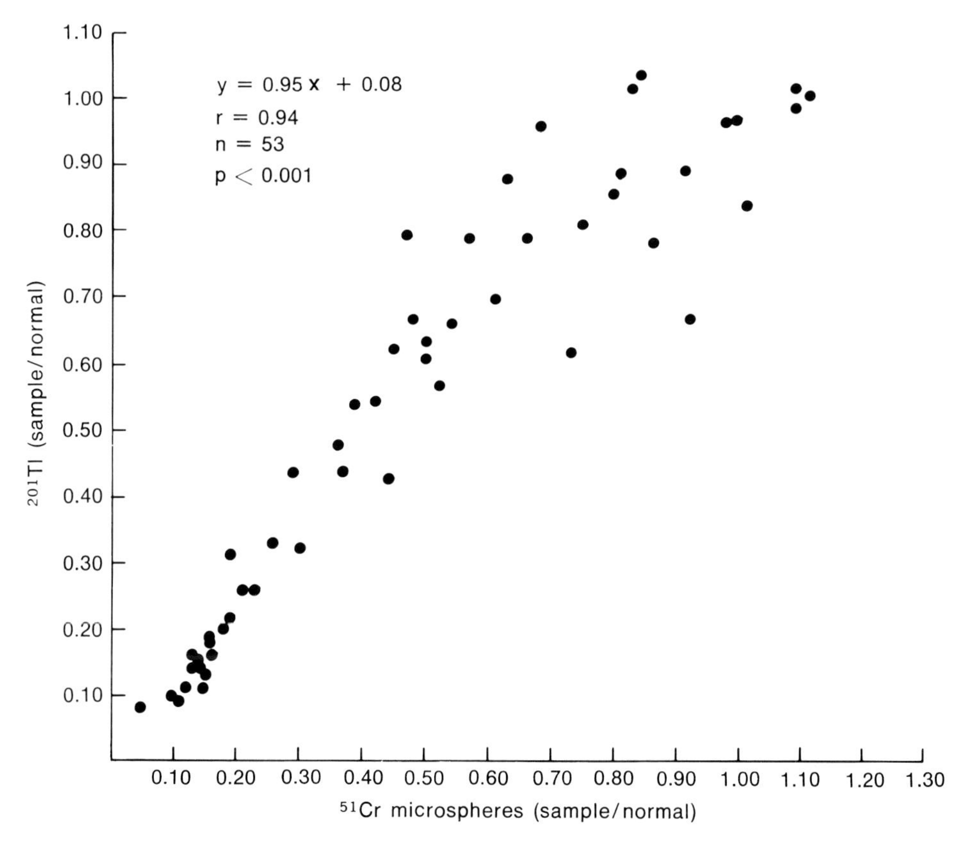

Fig. 16-2. Relationship of transmural ^{201}Tl uptake and microsphere estimates of myocardial blood flow in experimental myocardial infaction. (From DiCola, V. C., Downing, S. E., Donabedian, R. K., and Zaret, B. L.: Cardiovasc. Res. **11**:141, 1977.)

lower relative blood flow, [^{99m}Tc] PYP accumulation actually decreases. This decrease occurs in a linear fashion from zones with 30% to 40% of normal blood flow and maximal [^{99m}Tc] PYP uptake to zones with essentially no residual myocardial blood flow and minimal [^{99m}Tc] PYP uptake. As a corollary of this phenomenon, the same [^{99m}Tc] PYP tissue level can be seen in zones with marked necrosis and greatly reduced residual blood flow, as in zones with only moderate necrosis and significant residual blood flow (Chapters 14 and 15).

The relationship of myocardial uptake of both radionuclides to regional blood flow has a direct bearing on other pathophysiologic correlates of nuclide accumulation. In 24 hr infarct preparations, there is a direct relationship between ^{201}Tl activity and tissue viability as assessed by local serum enzyme activity[9] (Fig. 16-4). On the other hand, no such direct relationship exists between local serum enzyme depletion and tissue [^{99m}Tc] PYP activity.[44] Decreased ^{201}Tl uptake is a sensitive marker for histopathologic evidence of myocardial necrosis,[9] whereas a direct relationship is less evident with [^{99m}Tc] PYP.[5,19]

By contrast to acute infarction, in nonischemic necrosis (as seen in the canine countershock model), in which regional myocardial blood flow remains relatively intact, direct relationships do exist between [^{99m}Tc] PYP tissue activity, serum enzyme depletion, and histopathologic evidence of tissue necrosis.[10,13] Thus, under appropriate experimental conditions, [^{99m}Tc] PYP activity directly reflects the magnitude of tissue damage.

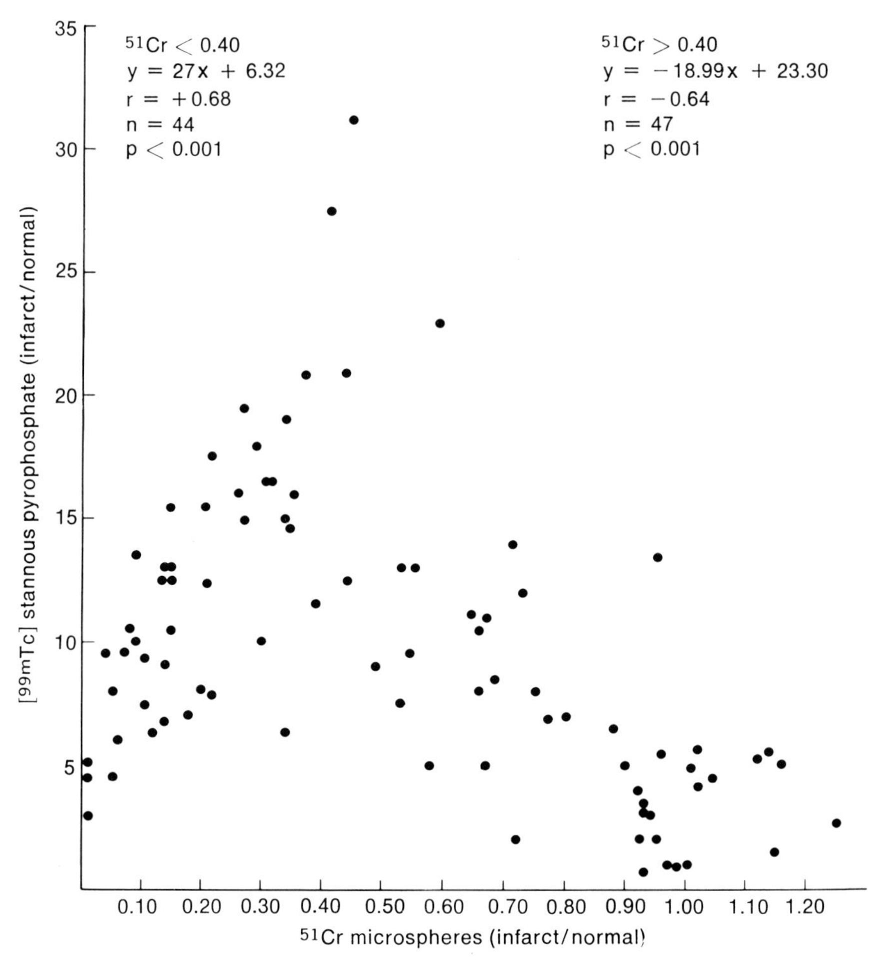

Fig. 16-3. Relationship between myocardial uptake of PYP and microsphere estimate of blood flow 24 hr following infarction. Note how uptake is maximal in range of 0.30 to 0.40 of normal flow. (From Zaret, B. L., DiCola, V. C., Donabedian, R. K., Puri, S., Wolfson, S., Freedman, G. S., and Cohan, L. S.: Circulation **53:**422, 1976, by permission of the American Heart Association, Inc.)

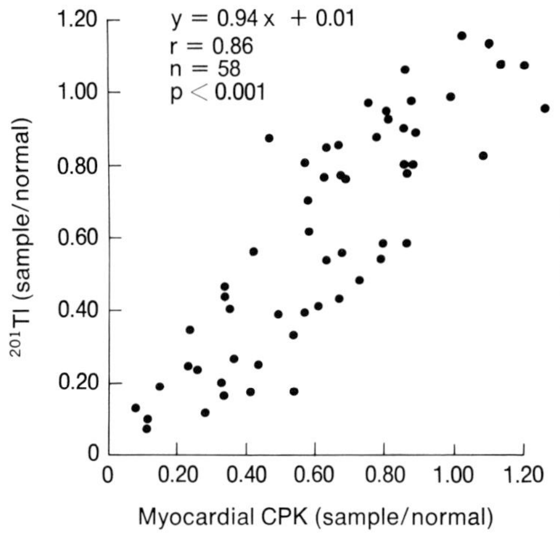

Fig. 16-4. Relationship between ^{201}Tl myocardial uptake and creatine phosphokinase (CPK) tissue levels. Note linear relationship. (From DiCola, V. C., Downing, S. E., Donabedian, R. K., and Zaret, B. L.: Cardiovasc. Res. **11:**141, 1977.)

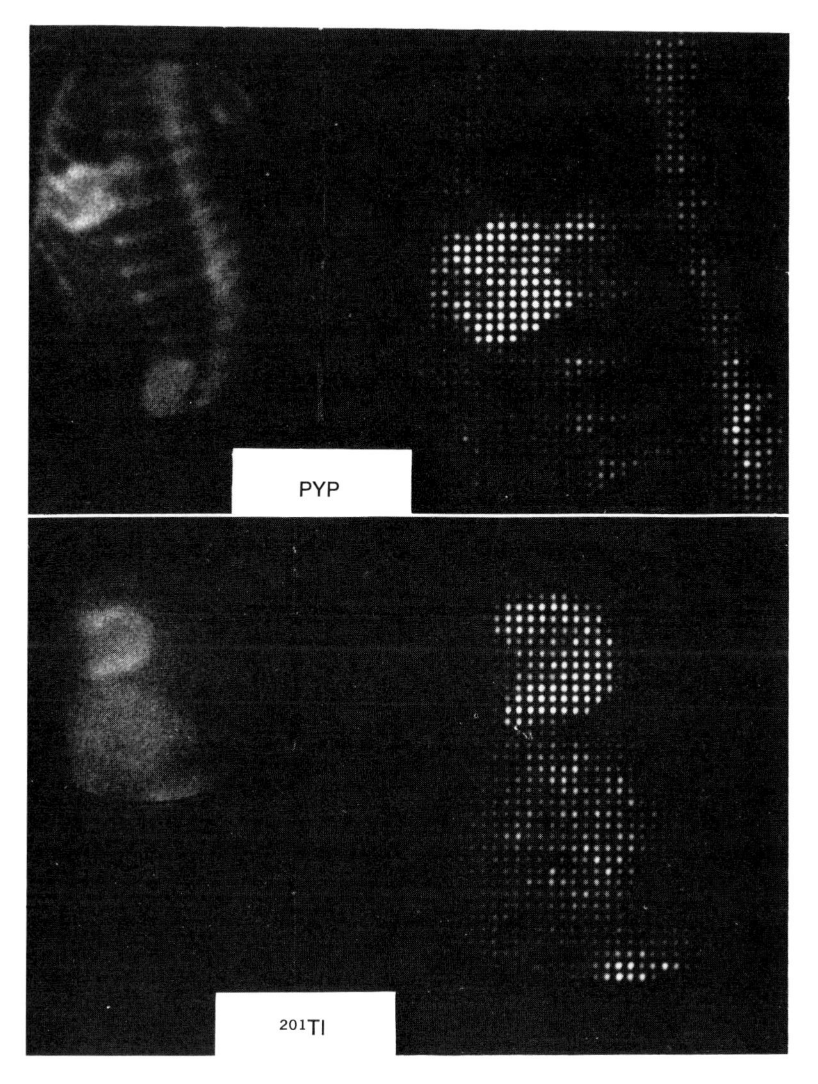

Fig. 16-5. Infarct images obtained in a dog 24 hr following closed chest infarction. *Top:* PYP. *Bottom:* ^{201}Tl. Analog images are shown on left, computer processed images on right. Note obvious size discordance between infarct size on ^{201}Tl and PYP images.

IMAGING AND SIZING OF INFARCTS IN EXPERIMENTAL ANIMALS

Both ^{201}Tl and [^{99m}Tc] PYP imaging can be used to quantify the size of experimental anterior wall myocardial infarction.[4,34,45] The use of computer techniques is more critical for ^{201}Tl infarct sizing than for [^{99m}Tc] PYP sizing. It has now been demonstrated that infarct size can be obtained as reliably from the planimetered areas of conventional [^{99m}Tc] PYP images as from computerized images. At least for [^{99m}Tc] PYP, results may be improved with application of three-dimensional reconstruction techniques.[18] In a 24 hr canine infarct model, in which animals were imaged initially with ^{201}Tl and several hours later with [^{99m}Tc] PYP, our laboratory noted good correlations between both ^{201}Tl and [^{99m}Tc] PYP infarct areas and pathologic infarct size.[45] Of interest was a significant increase in the size of the [^{99m}Tc] PYP infarct zone as compared to the ^{201}Tl infarct zone (Figs. 16-5 and 16-6). In only one instance was the area

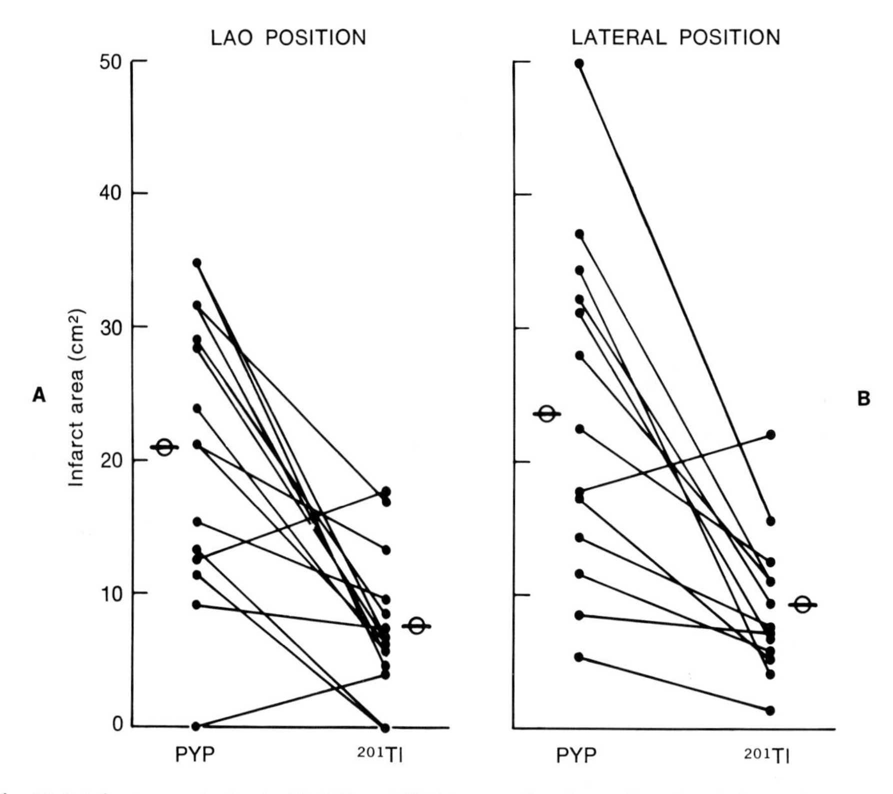

Fig. 16-6. Infarct area obtained with PYP and ^{201}Tl in animals subjected to closed chest infarction. Data are given for both LAO **(A)** and left lateral **(B)** positions. Data from each animal are connected by single line. Note how PYP area is significantly larger than that seen with ^{201}Tl.

of the decreased ^{201}Tl uptake infarct defect larger than the area of increased [^{99m}Tc] PYP uptake.

After the animals were imaged, their hearts were removed and sectioned in transverse slices. These slices were imaged, and the extent of infarction in the slices as measured by each radionuclide distribution was related to a similar histopathologic analysis. In this instance, the relationship of the ^{201}Tl infarct estimation to histopathologic assessment was excellent ($r = 0.93$), whereas that of [^{99m}Tc] PYP was less precise ($r = 0.70$). In selected instances, both ^{201}Tl and [^{99m}Tc] PYP appeared to lead to overestimation of the in vitro histopathologic assessment of the infarct. This trend was most evident in the slices with the largest degree of infarction. In the images of slices, [^{99m}Tc] PYP assessment also led to a greater infarct area size estimate than that achieved with ^{201}Tl imaging.

These studies indicate that at a specified point in time, infarction can be quantified with both tracer techniques. These techniques, however, will be limited by the physiologic characteristics of the tracers employed and problems associated with ventricular geometry and two-dimensional imaging of a three-dimensional process.

Further qualifying statements are necessary. When comparisons are made between studies, care must be taken to account for temporal variability intrinsic to both [^{99m}Tc] PYP and ^{201}Tl imaging. [^{99m}Tc] PYP images frequently are not positive for 12 to 24 hr and reach maximum abnormality at 2 to 3 days. Comparison of [^{99m}Tc] PYP image data at different points in time would obviously yield erroneous conclusions. Significant temporal variations are also seen with ^{201}Tl. In a recent study, dogs subjected to closed-chest anterior wall myocardial infarction were serially imaged 4 hr and again

24 hr after infarction. ^{201}Tl image defects measured by computer techniques decreased in size by an average of 30% within this 20 hr period.[16] The decrease in ^{201}Tl defect size was independent of final pathologic infarct size or initial radionuclide image infarct estimate. This temporal variability will hinder the use of any single technique for sequentially evaluating the effects of an intervention on infarct size. However, such temporal variation illustrates a potential utility of the dual radionuclide approach. When used together appropriately, each radionuclide provides complementary data, each optimized for study at different points in time after infarction.

Finally, assessment of infarct size in both animals and humans, for the most part, has been made in anterior wall myocardial infarction. This particular model has been utilized because it results in relatively large infarcts and provides a lesion that can best be imaged by the scintillation camera within the constraints imposed by ventricular geometry. No comparable data are available concerning inferior infarction, a lesion generally involving a smaller muscle mass, which, because of geometric considerations, is also not displayed in a manner distinguishing its full extent. Nontransmural infarcts, involving a limited extent of the total ventricular wall thickness, present great difficulties in terms of accurate sizing. For such infarcts, cross-sectional tomographic techniques are essential.

PHARMACOLOGIC MODIFICATION OF RADIONUCLIDE MYOCARDIAL UPTAKE

Recently our laboratory studied the effects of several drugs on ^{201}Tl myocardial uptake and kinetics in canine cardiac skeletal muscle. These phenomena must be defined prior to use of quantitative myocardial imaging in conjunction with pharmacologic administration in both clinical and protocol situations. These studies have demonstrated that propranolol and digitalis significantly reduce ^{201}Tl concentration in both types of muscle.[8] The most profound decrement in tracer uptake was noted with propranolol. In skeletal muscle, this effect is independent of blood flow and contraction strength. In contrast, administration of the beta-adrenergic agent isoproterenol results in enhanced uptake of ^{201}Tl. Comparable effects were seen with radioactive potassium in the same preparations. Similar pharmacologic modification of ^{201}Tl uptake has been noted in the fetal mouse heart organ culture preparation.[32]

The effects of the diuretic furosemide on ^{201}Tl uptake have also been studied, since earlier work indicated that diuretics may have intrinsic effects on myocardial cation flux. Neither acute nor chronic furosemide administration resulted in significant alteration of ^{201}Tl uptake. However, in the presence of induced hypokalemia, chronic furosemide therapy resulted in a significant decrease in myocardial ^{201}Tl accumulation. No comparable decrease in ^{201}Tl uptake was noted in the presence of hypokalemia alone.[12] This represents another clinical situation in which pharmacologic and physiologic variations may alter radionuclide uptake.

The clinical relevance of these observations remains to be defined. If normal-zone radioactivity is decreased by drugs without a comparable change in abnormal-zone radioactivity, this will result in less clear separation of normal and abnormal zones based on ^{201}Tl activity ratios. Such events could result in a decreased clarity of the infarct zone and poorer definition of its margins. Other pharmacologic interventions, such as bicarbonate or glucose and insulin, have been shown to increase ^{201}Tl myocardial accumulation, a phenomenon that would have different consequences for quantitative imaging.[15,31]

[^{99m}Tc] PYP activity may also be modified by pharmacologic intervention. Recently this was demonstrated in the canine countershock model after pretreatment with corticosteroids and use of [^{99m}Tc] PYP tissue levels as a means of assessing necrosis.[33] These studies showed that steroid administration resulted in a decrease in normal-zone myocardial [^{99m}Tc] PYP accumulation. This decrease resulted from a lowered level of circulating [^{99m}Tc] PYP, with which the normal myocardium was in equilibrium. The lowered blood level was a direct result of enhanced [^{99m}Tc] PYP renal excretion, a phenomenon in turn secondary to a steroid-induced increase in glomerular filtration rate. Thus pharmacologic agents may alter myocardial radionuclide distributions not only by direct blood flow–related or metabolic cardiac effects, but also by altering the total body distribution and metabolic clearance of the radionuclide.

Since drugs probably will affect each radionuclide in a different manner, situations involving pharmacologic intervention provide another circumstance under which the dual-imaging approach is of particular value. For example, after intravenous administration of propranolol, [^{99m}Tc] PYP imaging may prove more reliable than ^{201}Tl, whereas after administration of corticosteroids the reverse may be the case.

DUAL RADIONUCLIDE IMAGING OF INFARCTION IN HUMANS

The animal studies described earlier provide an experimental basis for understanding the pathophysiology of myocardial radionuclide accumulation under conditions of infarction, necrosis, and ischemia. Such an understanding is necessary if imaging techniques are to be applied meaningfully to clinical situations. The remainder of this chapter will deal with radionuclide imaging in humans.

Since ^{201}Tl and [^{99m}Tc] PYP imaging techniques each have advantages and limitations and the uptake of each radionuclide is governed by different pathophysiologic events, studies in our laboratory have been directed at evaluating the clinical utility of

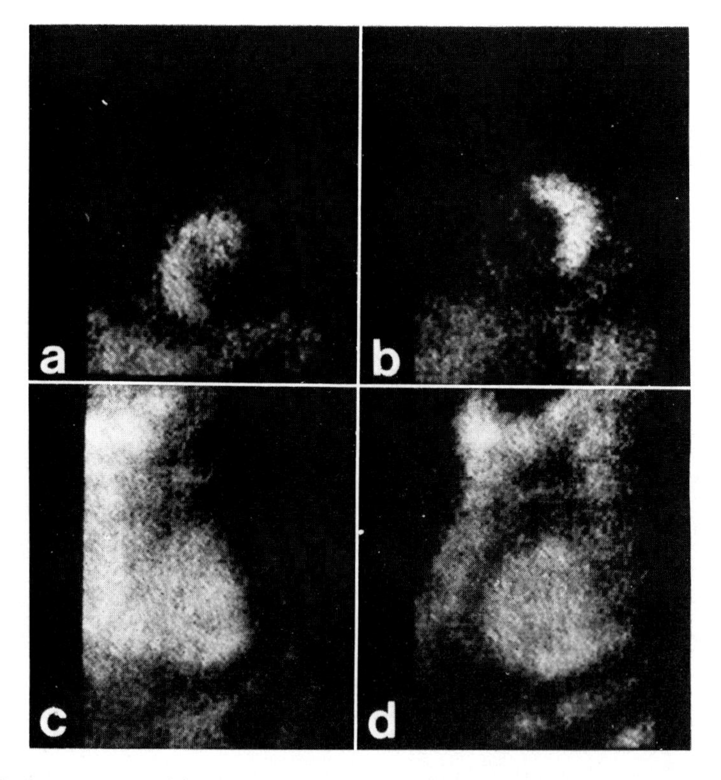

Fig. 16-7. ^{201}Tl images (*a* and *b*) and PYP images (*c* and *d*) obtained following acute anterior wall myocardial infarction. Anterior images on *left,* LAO images on *right.* Note significant anterior and septal ^{201}Tl defects and large area of PYP accumulation. Note as well significant difference in size of PYP area compared to ^{201}Tl defect.

dual radionuclide imaging of acute myocardial infarction. It was believed that from a diagnostic standpoint the sensitivity of each technique for infarct detection might be enhanced when both myocardial distributions were imaged in the same patient. Furthermore, differentiation of recent from remote infarction might be facilitated, and areas of detectable overlap between the imaged dual distributions might help define reversibly ischemic regions or sites of nontransmural necrosis.

In eighty patients with documented acute infarction undergoing ^{201}Tl and [^{99m}Tc] PYP imaging, one of the two imaging studies was positive in each patient.[3] Although single, individual nuclide studies occasionally were falsely negative, when the two radionuclides were employed together, all infarcts were detected (Figs. 16-7 to 16-11). Those patients in whom a single imaging study was negative generally had sustained either small inferior wall transmural myocardial infarcts or small nontransmural infarcts. In the case of ^{201}Tl, the presence of left ventricular hypertrophy (a situation in which radioactivity in a larger than usual noninfarcted myocardial mass is superimposed on the infarct zone) also decreased the sensitivity of infarct detection. The detection of nontransmural infarction with ^{201}Tl imaging was surprisingly good in this series. However, it must be understood that a significant number of patients within this group had sustained a previous myocardial infarction, and the ^{201}Tl defects noted might represent old as well as new infarcts.

The sensitivity for detection of infarction with ^{201}Tl in this study was enhanced significantly by the use of isocount computer smoothing and color displays (Figs. 16-12 to 16-14). This particular display was

Text continued on p. 296.

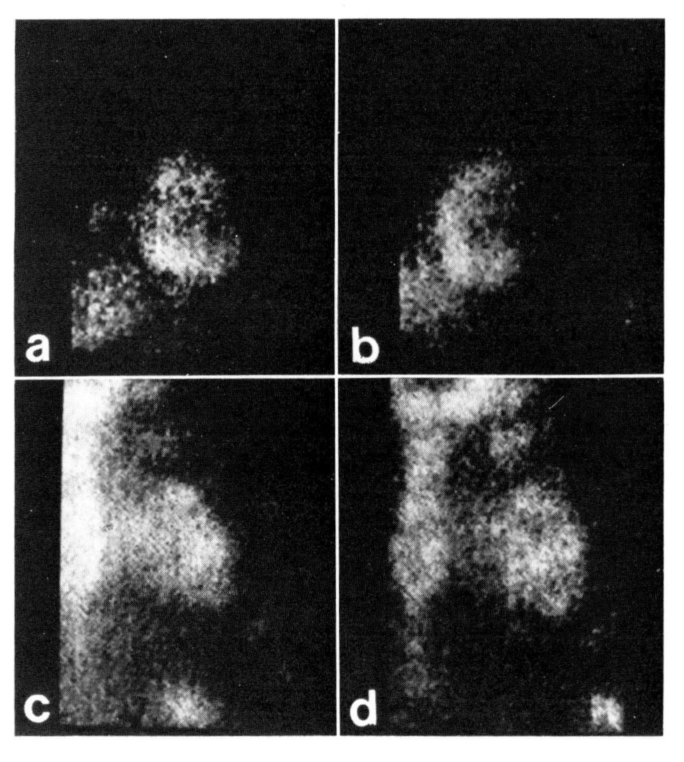

Fig. 16-8. ^{201}Tl and PYP images obtained following acute anterolateral myocardial infarction. Format same as Fig. 16-7. Again note size discordance, with PYP larger than ^{201}Tl. (From Berger, H. J., Gottschalk, A., and Zaret, B.: Ann. Int. Med. **88:**145, 1978.)

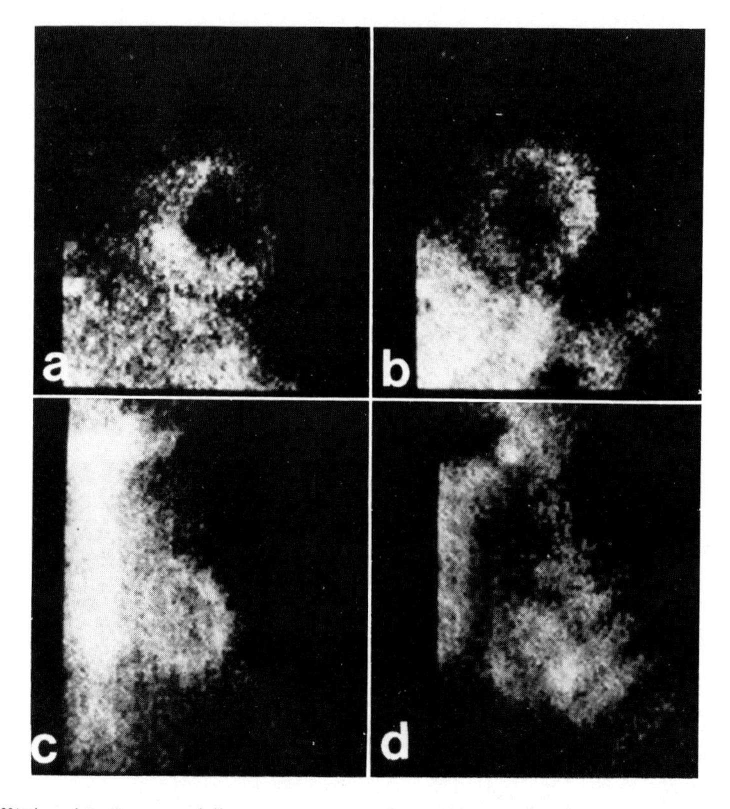

Fig. 16-9. ^{201}Tl and PYP images following anteroseptal myocardial infarction. Format same as Fig. 16-7. Note that in PYP images there is a zone of decreased tracer accumulation in center of infarct area. This is consequence of markedly decreased flow in center of infarct zone.

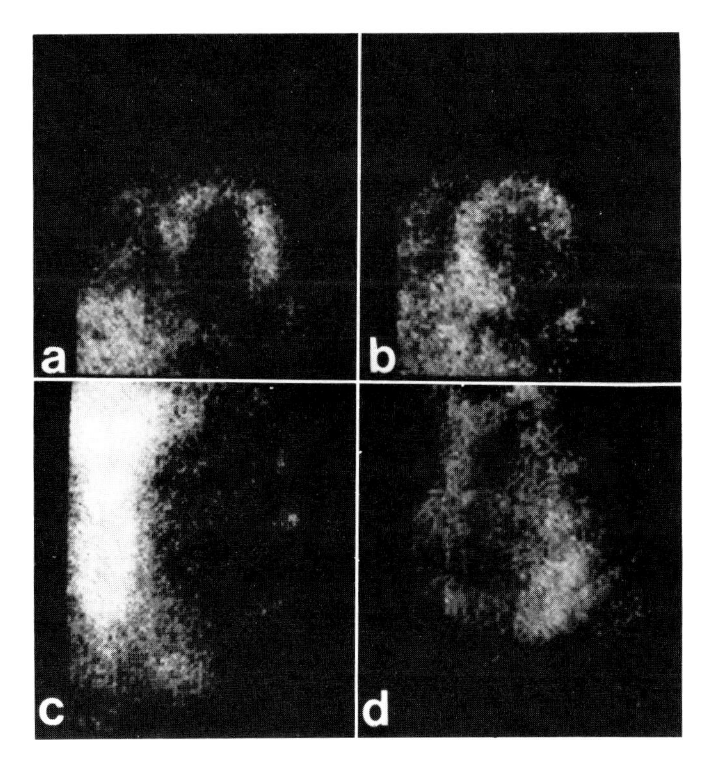

Fig. 16-10. ^{201}Tl and PYP images following acute inferior wall myocardial infarction. Format same as Fig. 16-7. Note abnormally imaged zones obtained with both radionuclides. (From Berger, H., J., Gottschalk, A., and Zaret, B. L.: Ann. Int. Med. **88:**145, 1978.)

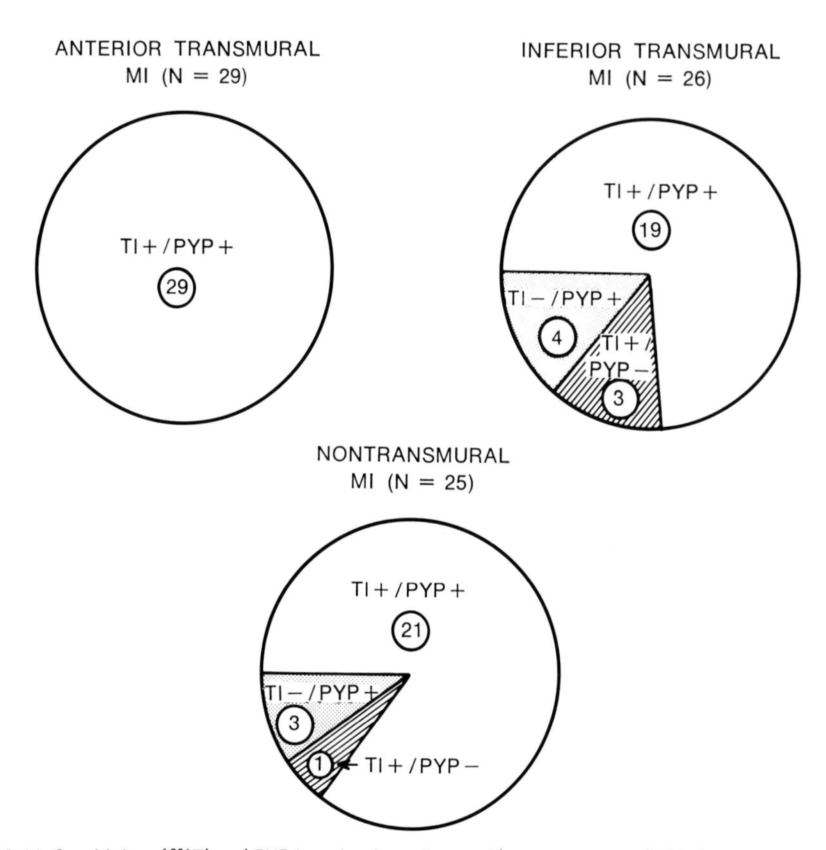

Fig. 16-11. Sensitivity of ^{201}Tl and PYP imaging in patients with acute myocardial infarction. Note that all patients were imaged correctly with at least one of the two radionuclides.

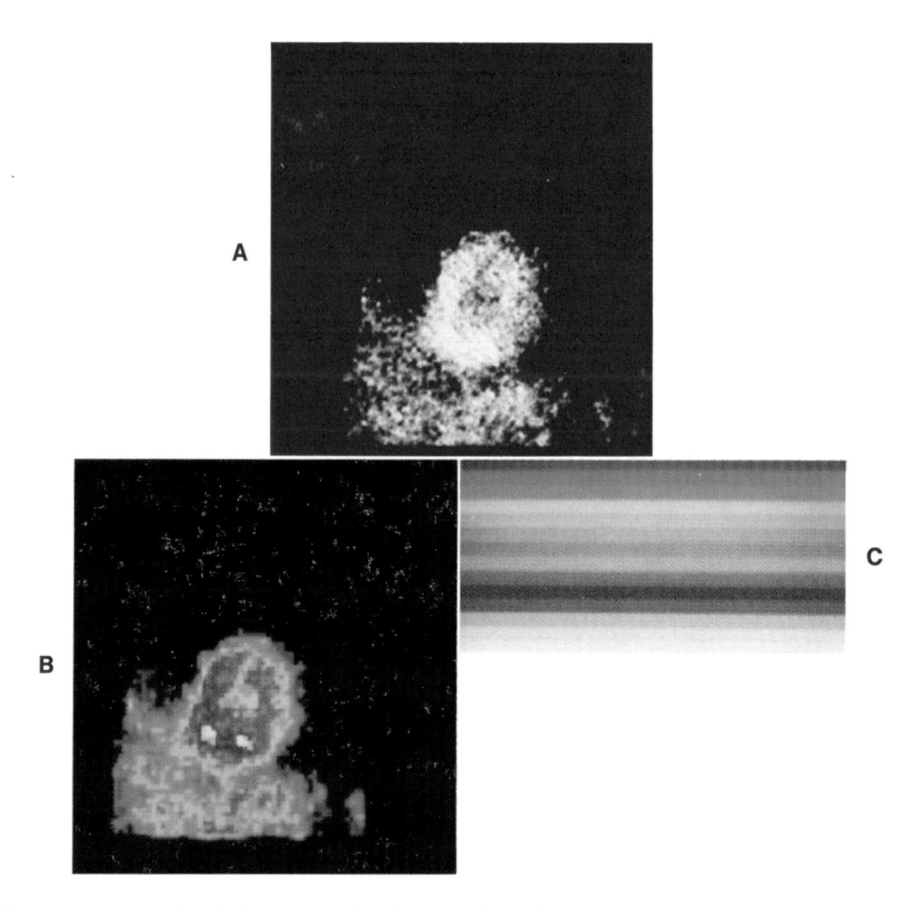

Fig. 16-12. Conventional black and white **(A)** and color-coded isocount contour **(B)** displays in LAO in patient with posterolateral myocardial infarction. **C,** Color spectrum. Maximal activity is white and light yellow, minimal activity is turquoise. Posterolateral defect is readily appreciated in color image; this is not the case in conventional black and white display. (From Berger, H. J., Gottschalk, A., and Zaret, B. L. Ann. Int. Med. **88:**145, 1978.)

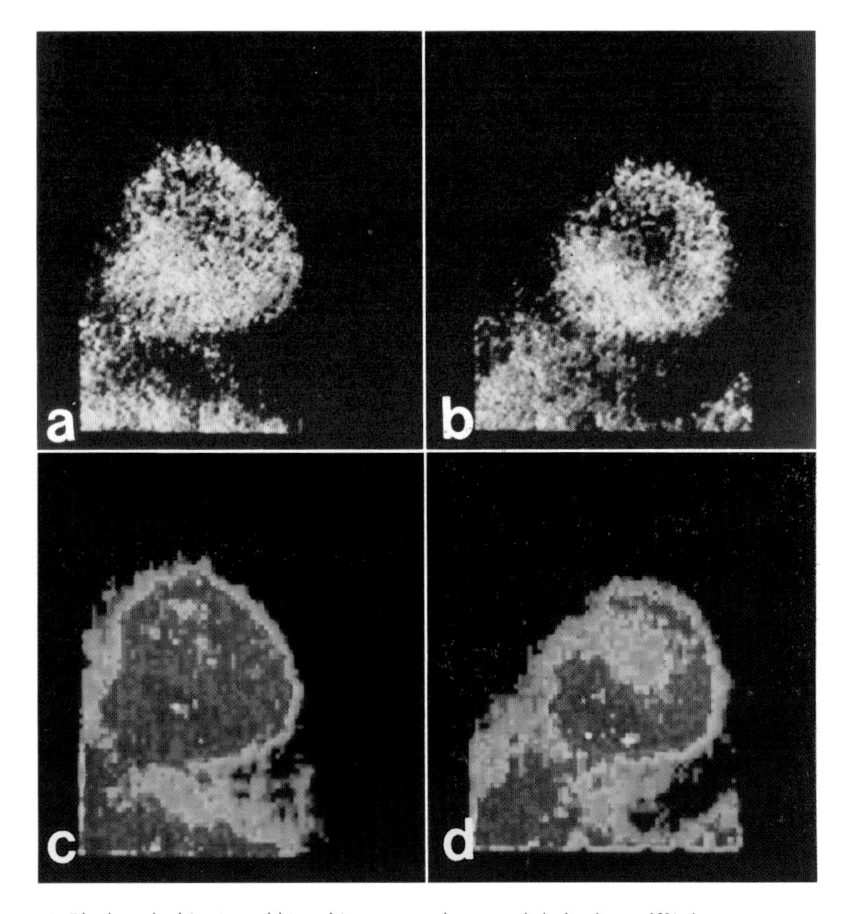

Fig. 16-13. Black and white (*a* and *b*) and isocount color (*c* and *d*) displays of ^{201}Tl images in patient with anterior wall myocardial infarction and left ventricular hypertrophy. Note enhanced ability to detect infarct in color-coded studies. Anterior images are shown on *left,* LAO images on *right.*

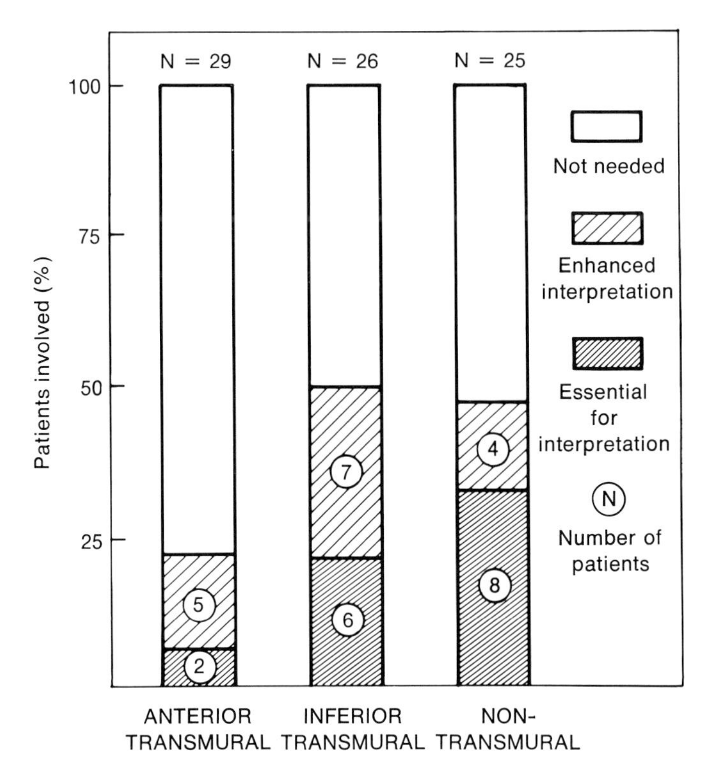

Fig. 16-14. Diagrammatic display of effect of computer smoothing and color display on ^{201}Tl image interpretation in eighty patients with myocardial infarction.

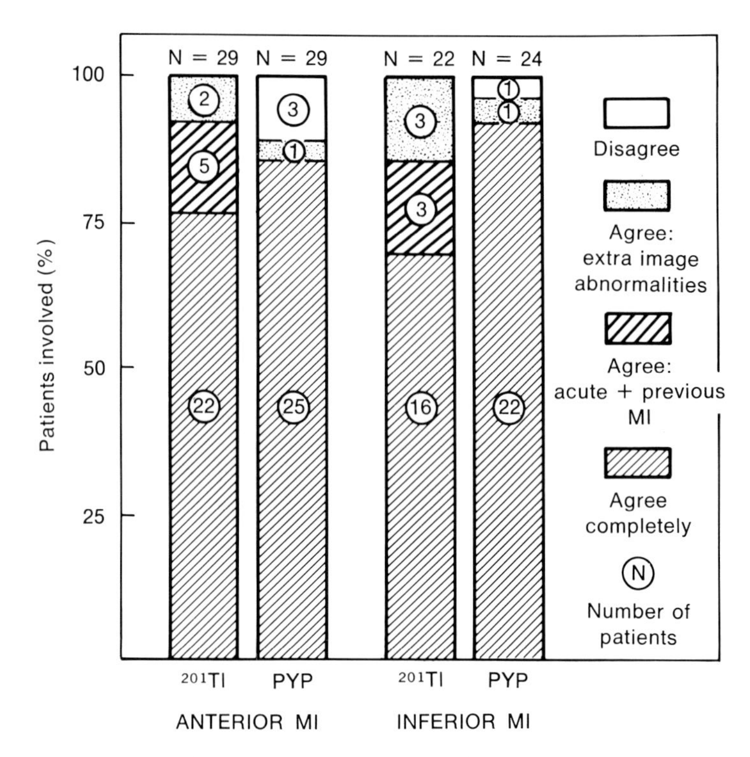

Fig. 16-15. Comparison of ^{201}Tl, PYP, and electrocardiographic localization of acute transmural myocardial infarction. Analysis only based on patients with transmural infarction in whom images were positive.

essential in defining perfusion abnormalities in sixteen of eighty patients. The computerized approach was of greatest value in patients with relatively small infarcts or left ventricular hypertrophy. The computerized color display did not result in lowered specificity in patients without infarcts (i.e., it did not lead to false-positive interpretation). On the other hand, computer processing was of no value in increasing the sensitivity of [^{99m}Tc] PYP imaging. This is not surprising, since it is clearly much easier to define a hot spot than a cold spot. With ^{201}Tl, problems of edge detection, overlap of normal and abnormal myocardium, and ventricular geometry are much greater.

As noted in Fig. 16-15, ^{201}Tl and [^{99m}Tc] PYP both accurately localized the site of infarction. In patients with transmural infarction, in whom accurate ECG localization was possible, ^{201}Tl correctly localized the site of infarction in 100% of the patients with positive images, whereas [^{99m}Tc] PYP localized the infarct site in 93%. In all eight patients with previous infarction, ^{201}Tl images correctly identified its presence and location, whereas [^{99m}Tc] PYP, appropriately, did not. The dual-imaging technique distinguished old from new infarction in the same patient. This will continue to be a major attribute of the dual-imaging approach. Both radionuclides identified individual abnormal regions not associated with comparable ECG changes. Learning whether this represents imaging of electrocardiographically silent infarction and/or ischemia or oversensitivity of the technique will require further pathologic study. In selected patients with transmural infarction there are good correlations between the ^{201}Tl scintigraphic infarct defect size and pathologic estimates obtained postmortem.[44]

There was a significant discordance between the visually estimated sizes of the

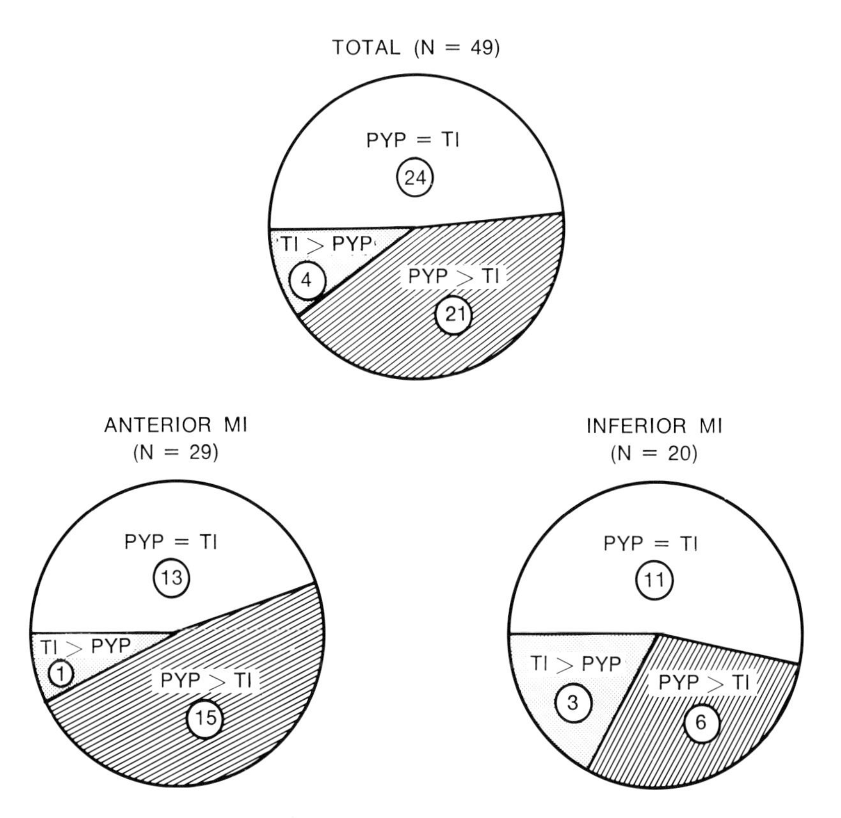

Fig. 16-16. Diagrammatic display of incidence of size discordance between infarct regions on ^{201}Tl and PYP images in patients with transmural myocardial infarction.

infarct zone on the ^{201}Tl and [^{99m}Tc] PYP images (Fig. 16-16). The predominant finding was that the area of increased [^{99m}Tc] PYP uptake was substantially larger than the area of decreased ^{201}Tl myocardial accumulation. This was noted in approximately half the patients with transmural infarction in whom both images were positive (Figs. 16-7, 16-8, and 16-13). In the case of nontransmural infarction, in which a diffuse pattern of [^{99m}Tc] PYP uptake is frequently noted, this comparison would be of limited value and was not employed. The significance of this size discordance and the mechanisms for its occurrence remain to be defined. One explanation might involve the pathologic nature of transmural infarction, which frequently extends peripherally from a central infarct zone of transmural necrosis to adjacent subendocardial regions. This nontransmural peripheral zone would be visualized readily by [^{99m}Tc] PYP imaging but not by ^{201}Tl (Fig. 16-17). Problems resulting from ventricular geometry may also play a significant role. Other factors to be considered include [^{99m}Tc] PYP uptake in ischemic noninfarcted zones and significant ^{201}Tl uptake in areas of necrosis such that the defect size is not demonstrated accurately. These findings of size discordance are similar to those noted previously in the animal model.[45] This size discordance may have prognostic importance in identifying patients at risk of either infarct extension into a peripheral zone with initially only subendocardial necrosis or reinfarction within a short period of time. Resolution of this question will require further observation and long-term follow-up.

The results of this dual nuclide study can be contrasted to previous individual tracer studies employing either ^{201}Tl or [^{99m}Tc] PYP. Wackers and associates observed that the sensitivity of ^{201}Tl imaging for infarct

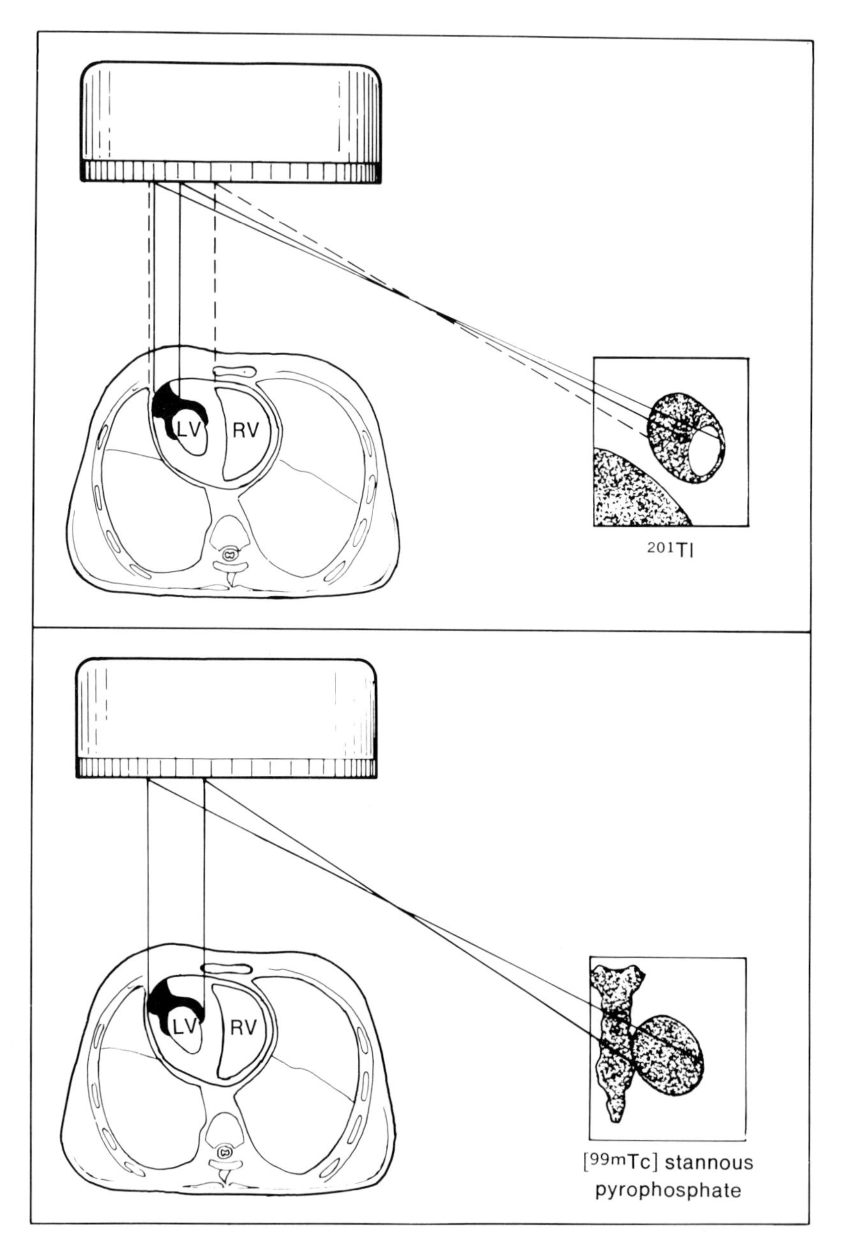

Fig. 16-17. Diagrammatic explanation of one reason for size discordance between PYP and ^{201}Tl infarct regions. Transverse section through chest at level of left ventricle is depicted with scintillation detector above and the images displayed to the right. Solid lines represent edges of infarct as detected by imaging; broken lines represent outer margins of normal left ventricle as seen with ^{201}Tl. Myocardial infarct is black zone. Although infarct may be classified as transmural based on standard electrocardiographic criteria, such an infarct may not be transmural from edge to edge, as illustrated. ^{201}Tl defect reflects only area of central transmural infarction, whereas PYP zone reflects edge-to-edge zone of infarct. Net result would be larger PYP infarct area than that noted with ^{201}Tl.

detection was greatest within the first 24 hr after infarction.[41] They noted positive images in all forty-four infarct patients studied within 6 hr of onset of symptoms. Furthermore, positive images were demonstrated in ninety of ninety-six patients studied within 24 hr after onset of symptoms and seventy-five of 104 studied later. In a number of patients, image defects became less evident or were no longer apparent on repeat scintigrams done in the early phase of infarction. These data have distinct correlates in the experimental studies discussed earlier.[16]

Because ^{201}Tl infarct imaging has its greatest sensitivity early in the course of infarction, it may provide a means for screening patients for admission to the coronary care unit. A pilot study has been reported dealing with 203 patients in whom the initial diagnosis of infarction was in question.[38] In this study, a positive ^{201}Tl scintiscan on admission appropriately selected those patients requiring intensive care hospitalization at a point when conventional clinical indices for diagnosing infarction were not helpful. In this instance [^{99m}Tc] PYP imaging would be of no value, since a longer time period is generally required for [^{99m}Tc] PYP infarct images to become reliably positive.

[^{99m}Tc] PYP imaging has been employed by a number of investigators with varying results in terms of both sensitivity and specificity. The clinical results described in this chapter are comparable to those reported by Parkey, Willerson, and colleagues.[24,42,43] In general, [^{99m}Tc] PYP appears to be a highly sensitive, somewhat less specific technique. These investigators, as well as others, have also studied a limited number of patients with the dual imaging approach.[14,25,41a]

In summary, it appears that dual infarct imaging with ^{201}Tl and [^{99m}Tc] PYP can be performed effectively within the coronary care unit and provides a high degree of sensitivity for detection of myocardial infarction. The two radionuclides would appear to be complementary in several instances. The combined imaging allows definition of old and new infarction in the same patient. Size discordance of the infarct as imaged with each radionuclide is common and will require further study to establish its clinical relevance. The sensitivity of each for infarct detection depends on the time at which the patient is evaluated after infarction. Early after infarction, ^{201}Tl is the nuclide of choice, whereas two to seven days after infarction, [^{99m}Tc] PYP will be more sensitive. The temporal imaging sequence is extremely important and must be emphasized continually if comparisons are to be made between patients in different series or between the same patient studied sequentially. Sizing of infarction has been demonstrated primarily with anterior wall infarcts. In the instance of nontransmural infarction, one might expect that [^{99m}Tc] PYP will provide a greater sensitivity, although this remains to be determined in larger series.

It is obvious from this discussion that false-negative studies often will be related to the time at which patients are studied after infarction. In addition, false-negative ^{201}Tl studies will be most noticeable in patients with relatively small infarcts involving inferior or nontransmural distributions or in patients with left ventricular hypertrophy. Small infarcts may be difficult to detect with either nuclide. In animal studies, the smallest infarct mass that can be reliably detected with [^{99m}Tc] PYP is 3 g[26] and with ^{201}Tl, 5 g.[23]

So-called false-positive results (that is, abnormal images in the absence of infarction) are noted with both radioactive tracers. The major causes of abnormal images (in conditions other than acute myocardial infarction) obtained with ^{201}Tl are previous infarction, unstable angina, left ventricular hypertrophy, idiopathic hypertrophic subaortic stenosis, and congestive or infiltrative cardiomyopathy. Unstable angina, left ventricular aneurysm, repeated countershock, blunt chest trauma, metastatic carcinoma to the heart, endocarditis, and persistent blood pool activity can give false-positive results when using [^{99m}Tc] PYP. However, the pathophysiologic relevance of these false-positive studies remains to be defined. It is quite possible that the imaging data provide pathophysiologic insights that are not available when only conventional clinical indices of infarction and necrosis are employed.

A major use of the dual tracer approach would be to distinguish old infarcts from new. [^{99m}Tc] PYP images are positive in instances of acute infarction, whereas ^{201}Tl images are positive in both old and new infarction. However, abnormal [^{99m}Tc] PYP uptake occurs in instances of ventricular aneurysm.[1] The mechanism for this uptake is not clear. It may be due to dystrophic calcification within the aneurysm wall or myocytolysis.[6] Additional false-positive [^{99m}Tc] PYP images have been noted in unstable angina, in which approximately 30% of the patients will have an abnormal myocardial [^{99m}Tc] PYP uptake.[11] This situation is not unique for [^{99m}Tc] PYP, since approximately 20% to 30% of patients with unstable angina will also have abnormal ^{201}Tl images.[39] Similarly, patients with variant angina will have markedly abnormal ^{201}Tl images if the nuclide is administered during pain.[22] Since ^{201}Tl uptake is known to be diminished in the presence of ischemia, the mechanism of abnormal distributions in this clinical circumstance can more readily be understood for this particular tracer. In the case of [^{99m}Tc] PYP, abnormal images may represent either uptake in chronically ischemic cells or, more likely, detection of subclinical myocardial necrosis. [^{99m}Tc] PYP uptake has been noted in situations of necrosis induced by conditions other than myocardial infarction, such as metastatic carcinoma involving the heart, blunt chest trauma, and external countershock.[10,27] Countershock presents specific problems in image analysis because of associated [^{99m}Tc] PYP uptake in damaged skeletal muscle overlying the cardiac region. Furthermore, true false-positive [^{99m}Tc] PYP images may result from imaging of persistent blood pool activity either due to altered tracer clearance, as in renal insufficiency, or to poor tracer preparation.

Abnormal ^{201}Tl myocardial distributions will be noted in conditions of idiopathic hypertrophic subaortic stenosis, left ventricular hypertrophy, and congestive or infiltrative cardiomyopathy with dilated left ventricular chambers. In these clinical instances, [^{99m}Tc] PYP uptake might be of particular value in assessing associated myocardial necrosis.

These data have been derived from the coronary care unit in a population of medical patients with myocardial infarction. Combined radionuclide tracer techniques also should be of value in evaluation of postoperative patients undergoing a variety of cardiovascular procedures, particularly coronary artery bypass surgery. Here the role of conventional indices for detection of myocardial infarction is limited because myocardial enzyme release is routine, and abnormal ECG changes, which generally are interpreted as "nonspecific," are noted in most patients. [^{99m}Tc] PYP uptake in this instance has already been shown to be of clinical value.[28] The combined imaging approach might be anticipated to provide additional information.

This discussion has concentrated primarily on a dual nuclide hot and cold spot imaging of infarction. These techniques will provide answers concerning the presence of infarction, its location, and its extent. They will provide no insight into another major question concerning myocardial infarction, that of the functional significance of the infarct. Additional radionuclide techniques must be employed to deal with this issue. These techniques involve assessment of global and regional left ventricular performance using either first-pass quantitative radionuclide angiocardiography or gated equilibrium cardiac blood pool techniques.[20,36] These techniques, combined with dual imaging, provide an accurate assessment of the relationship between perfusion, necrosis, and function in a given myocardial segment (Fig. 16-18). Such relationships have been demonstrated previ-

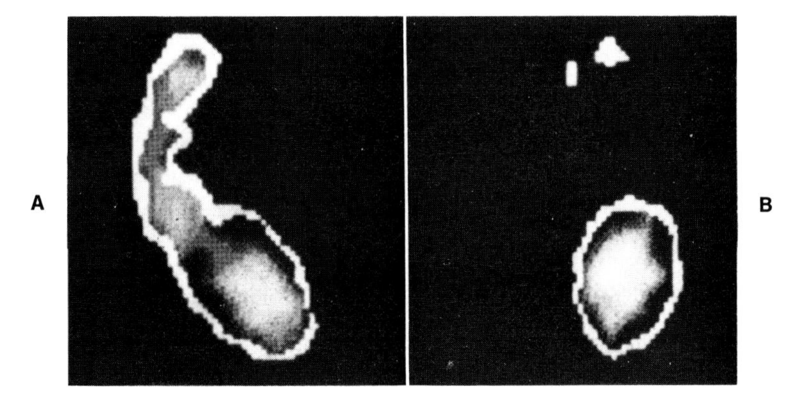

Fig. 16-18. Regional wall motion studies obtained following first-pass radionuclide angiocardiogram. End of diastole is depicted by outer margin, end of systole by image superimposed within this margin. **A,** Anterior position study. **B,** LAO position study. Note anteroapical zone of akinesis in anterior study and area of septal akinesis noted on LAO study.

ously in prototypic studies using radioactive K imaging[29,47] and more recently ^{201}Tl.[7] The functional studies have additional value for sequential repeated evaluation, particularly during the first 12 hr of infarction. Intracellular tracer distributions of ^{201}Tl and [^{99m}Tc] PYP are limited in the number of times they can be repeated. Functional studies, although providing an indirect index of what is happening at a cellular level, do supply relevant quantified physiologic data and, importantly, can be repeated on multiple occasions. The ability to repeat studies at regular intervals allows direct measure of the effects of interventions.

FUTURE

It is worthwhile to reemphasize that defining and assessing myocardial infarction with radionuclide techniques is in the very early stages of development. These techniques have only been introduced into clinical practice within the past three to four years and must constantly be updated in the light of new technologic advances and reassessed in view of broader clinical experience. Although ^{201}Tl and [^{99m}Tc] PYP are currently the radionuclide imaging agents of choice, they probably will not be five to ten years from now. Surely new radionuclides will be developed based on a variety of new physical and biologic principles. It is hoped that current concepts of the pathophysiology of ischemic heart disease can be applied to the development of new radiopharmaceuticals. Radiolabeled myosin-specific antibodies represent one such example.[2] Recent work by Thakur and associates[37] would indicate that radiolabeled polymorphonuclear leukocytes can be used to define and image myocardial infarction when the inflammatory response to tissue necrosis occurs. Techniques such as these should markedly enhance the specificity and sensitivity of infarct imaging and promote a physiologic approach to image pattern identification. It is reasonable to expect that in the future multiple tracers will be available for studying the ischemic heart. Each tracer, based on a specific physiologic principle, will provide unique information about a specific region of the heart. Various metabolic pathways as well as various states of necrosis or inflammation may be defined solely on the external imaging of specific radionuclide myocardial distributions.

Changes based on advances in instrumentation will also be forthcoming. Definition of transmural radionuclide distributions is essential. Currently evaluation of myocardial tracer distributions is limited by the geometry of the left ventricle and the uptake of radioactive tracers in associated noncardiac structures. Cross-sectional tomographic approaches (positron or gamma emitting) may overcome these geometric constraints and allow specific evaluation of the subendocardium, the area of greatest concern in the pathophysiologic process of ischemia and infarction.

REFERENCES

1. Ahmad, M., Dubiel, J. P., Verdon, T. A., and Martin, R. H.: Technetium-99m stannous pyrophosphate myocardial imaging in patients with and without left ventricular aneurysm, Circulation **53:** 833, 1976.
2. Beller, G. A., Khaw, B. A., Haber, E., and Smith, T. W.: Localization of radiolabeled cardiac myosin-specific antibody in myocardial infarcts, Circulation **55:**74, 1977.
3. Berger, H. J., Gottschalk, A., and Zaret, B. L.: Dual radionuclide study of acute myocardial infarction: comparison of thallium-201 and technetium-99m stannous pyrophosphate imaging in man, Ann. Int. Med. **88:**145, 1978.
4. Botvinick, E. H., Shames, D., Lappin H., Tyberg, J. V., Townsend, R., and Parmley, W. W.: Noninvasive quantitation of myocardial infarction with technetium-99m pyrophosphate, Circulation **52:** 909, 1975.
5. Buja, L. M., Parkey, R. W., Dees J. H., Stokely, E. M., Harris, R. A., Bonte, F. S., and Willerson, J. T.: Morphologic correlates of technetium-99m stannous pyrophosphate imaging of acute myocardial infarcts in dogs, Circulation **52:**596, 1975.
6. Buja, L. M., Poliner, L. R., Parkey, R. W., Pulido, J. I., Hutcheson, D., Platt, M. R., Mills, L. J., Bonte, F. J., and Willerson, J. T.: Clinicopathologic study of persistently positive technetium-99m stannous pyrophosphate myocardial scintigrams and myocytolytic degeneration after acute myocardial infarction, Circulation **56:**1016, 1977.
7. Bulkley, B. H., Hutchins, G. M., Bailey, I., Strauss, H. W., and Pitt, B.: Thallium-201 imaging and gated cardiac blood pool scans in patients with ischemic and idiopathic congestive cardiomyopathy, Circulation **55:**753, 1977.
8. Costin, J. C., and Zaret, B. L.: Effect of propranolol and digitalis upon radioactive thallium and potassium uptake in myocardial and skeletal muscle, abstracted, J. Nucl. Med. **17:**535, 1976.
9. DiCola, V. C., Downing, S. E., Donabedian, R. K., and Zaret, B. L.: Pathophysiologic correlates of thallium-201 myocardial uptake in experimental infarction, Cardiovasc. Res. **11:**141, 1977.
10. DiCola, V. C., Freedman, G. S., Downing, S. E., and Zaret, B. L.: Myocardial uptake of technetium-99m stannous pyrophosphate following direct current transthoracic countershock, Circulation **54:**980, 1976.
11. Donsky, M. S., Curry, G. C., Parkey, R. W., Meyer, S. L., Bonte, F. J., Platt, M. R., and Willerson, J. T.: Unstable angina pectoris: clinical, angiographic and scintigraphic observations, Br. Heart J. **38:**257, 1976.
12. Duchin, S., Costin, J. C., and Zaret, B. L.: Effect of furosemide and hypokalemia upon thallium-201 uptake in canine myocardium, medical school thesis, 1977, Yale University.
13. Gould, K. L.: Noninvasive assessment of coronary stenoses by myocardial perfusion imaging during pharmacologic coronary vasodilatation. I, Physiologic basis and experimental validation, Am. J. Cardiol. **41:**267, 1978.
14. Henning, H., Schelberg, H. R., Righetti, A., Ashburn, W. L., and O'Rourke, R. A.: Dual myocardial imaging with technetium-99m pyrophosphate and thallium-201 for detecting, localizing and sizing acute myocardial infarction, Am. J. Cardiol. **40:**147, 1977.
15. Hetzel, K. R., Westerman, B. R., Quinn, J. L., Meyers, S., and Barresi, V.: Myocardial uptake of thallium-201 augmented by bicarbonate, J. Nucl. Med. **18:**24, 1977.
16. Lange, R. C., Umbach, R., Lee, J. C., and Zaret, B. L.: Temporal changes in sequential quantitative thallium-201 imaging following myocardial infarction in dogs: comparison of four hour and twenty-four hour infarct images, Yale J. Biol. Med. (in press).
17. Lebowitz, E., Greene, M. W., Bradley-Moore, P., Atkins, H., Ansari, A., Richards, P., and Belgrave, E.: ^{201}Tl for medical use, J. Nucl. Med. **14:**421, 1973.
18. Lewis, M., Buja, L. M., Saffer, S., Mishelevich, D., Stokely, E., Lewis, S., Parkey, R., Bonte, F., and Willerson, J. T.: Experimental infarct sizing using computer processing and a three-dimensional model, Science **197:**167, 1977.
19. Marcus, M. L., Tomanek, R. J., Ehrhardt, J. C., Kerber, R. E., Brown, D. D., and Abboud, F. M.: Relationships between myocardial perfusion, myocardial necrosis and technetium-99m pyrophosphate uptake in dogs subjected to sudden coronary occlusion, Circulation **54:**647, 1976.
20. Marshall, R. C., Berger, H. J., Costin, J. C., Freedman, G. S., Wolberg, J., Cohen, L. S., Gottschalk, A., and Zaret, B. L.: Assessment of cardiac performance with quantitative radionuclide angiocardiography: sequential left ventricular ejection fraction, normalized left ventricular ejection rate, and regional wall motion, Circulation **56:**820, 1977.
21. Martin, M. D., Zaret, B. L., McGowan, R. L., Wells, H. P., and Flamm, M.D.: Rubidium-81: a new myocardial imaging agent, Radiology **111:**651, 1974.
22. Maseri, A., Parodi, O., Severi, S., and Pesola, A.: Transient transmural reduction of myocardial blood flow, demonstrated by thallium-201 scintigraphy as a cause of variant angina, Circulation **54:**280, 1976.
23. Mueller, T. M., Marcus, M. L., Ehrhardt, J. C.,

Chaudhuri, T., and Abboud, F. M.: Limitations of thallium-201 myocardial perfusion scintigrams, Circulation **54**:640, 1976.

24. Parkey, R. W., Bonte, F. J., Meyer, S. L., Atkins, J. M., Curry, G. L., Stokely, E. M., and Willerson, J. T.: A new method for radionuclide imaging of acute myocardial infarction in humans, Circulation **50**:540, 1974.
25. Parkey, R. W., Bonte, F. J., Stokely, E. M., Lewis, S. E., Graham, K. D., Buja, L. M., and Willerson, J. T.: Acute myocardial infarction imaged with ^{99m}Tc-stannous pyrophosphate and ^{201}Tl: a clinical evaluation, J. Nucl. Med. **17**:771, 1976.
26. Poliner, L. R., Buja, L. M., Parkey, R. W., Stokley, E. M., Stone, M. J., Harris, R., Saffer, S. W., Templeton, G. H., Bonte, F. J., and Willerson, J. T.: Comparison of different noninvasive methods of infarct sizing during experimental infarction, J. Nucl. Med. **18**:517, 1977.
27. Pugh, B. R., Buja, L. M., Parkey, R. W., Poliner, L. R., Stokely, E. M., Bonte, F. J., and Willerson, J. T.: Cardioversion and "false positive" technetium-99m stannous pyrophosphate myocardial scintigrams, Circulation **54**:309, 1976.
28. Righetti, A., Crawford, M. H., O'Rourke, R. A., Hardarson, T., Schelbert, H., Daily, P. O., DeLuca, M., Ashburn, W., and Ross, J.: Detection of perioperative myocardial damage after coronary artery bypass surgery, Circulation **55**:173, 1977.
29. Rigo, P., Strauss, H. W., and Pitt, B.: The combined use of gated cardiac blood pool scanning and myocardial imaging with potassium-43 in the evaluation of patients with myocardial infarction, Radiology **115**:387, 1975.
30. Romhilt, D. W., Adolph, R. J., Sodd, V. C., Levenson, N. I., August, L. S., Nishiyama, H., and Berke, R. A.: Cesium-129 myocardial scintigraphy to detect myocardial infarction, Circulation **48**:1242, 1973.
31. Schelbert, H. R., Ashburn, W. L., Chauncey, D. M., and Halpern, S.: Comparative myocardial uptake of intravenously administered radionuclides, J. Nucl. Med. **15**:1092, 1974.
32. Schelbert, H. R., Ingwall, J., Watson, R., and Ashburn, W.: Factors influencing the myocardial uptake of thallium-201, abstracted, J. Nucl. Med. **18**:598, 1977.
33. Schneider, R. M., Hayslett, J., Downing, S. E., Berger, H. J., Donabedian, R. K., and Zaret, B. L.: Effect of methylprednisolone upon technetium-99m pyrophosphate assessment of myocardial necrosis in the canine countershock model, Circulation **56**:1029, 1977.
34. Stokely, E. M., Buja, L. M., Lewis, S. E., Parkey, R. W., Bonte, F. J., Harris, R. A., and Willerson, J. T.: Measurement of acute myocardial infarct in dogs with ^{99m}Tc-stannous pyrophosphate scintigrams, J. Nucl. Med. **17**:1, 1976.
35. Strauss, H. W., Harrison, K., Langan, J. K., Lebowitz, E., and Pitt, B.: Thallium-201 for myocardial imaging: relation of thallium-201 to regional myocardial perfusion, Circulation **51**:641, 1975.
36. Strauss, H. W., Zaret, B. L., Hurley, P. J., Natarajan, T. K., and Pitt, B.: A scintiphotographic method for measuring left ventricular ejection fraction in man without cardiac catheterization, Am. J. Cardiol. **28**:575, 1971.
37. Thakur, M., Gottschalk, A., and Zaret, B. L.: Indium-labelled white blood cell imaging of acute myocardial infarction, Circulation (in press).
38. Wackers, F. J., Lie, I. K., Liem, K. L., Sokole, E. B., Samson, G., Schoot, J. B., and Durrer, D.: Potential value of thallium-201 scintigraphy as a means of selecting patients for the coronary care unit, Br. Heart J. **41**:111, 1979.
39. Wackers, F. J., Lie, K. I., Liem, L., Sokole, E. B., Samson, G., Schoot, J., Wellens, H., J. J., and Durrer, D.: Thallium-201 scintigraphy in unstable angina, Circulation **57**:738, 1978.
40. Wackers, J. T., Becker, A. E., Samson, G., Sokole, E. B., Schoot, J. B., Vet, A. J. T. M., Lie, K. I., Durrer, D., and Wellens, H.: Location and size of acute transmural myocardial infarction estimated from thallium-201 scintigrams, Circulation **56**:72, 1977.
41. Wackers, J. T., Sokole, E. B., Samson, G., Schoot, J. B., Lie, K. I., Liem, K. L., and Wellens, H. J. J.: Value and limitations of thallium-201 scintigraphy in the acute phase of myocardial infarction, N. Engl. J. Med. **295**:1, 1976.

41a. Walton, S., Kafetzakis, E., Shields, R. A., Testa, H. J., and Rowlands, D. J.: Use of ^{129}caesuim, ^{99m}Tc stannous pyrophosphate, and a combination of the two in the assessment of myocardial infarction, Br. Heart J. **40**:874, 1978.

42. Willerson, J. T., Parkey, R. W., Bonte, F. J., Meyer, S. L., Atkins, J. M., and Stokely, E. M.: Technetium stannous pryophosphate myocardial scintigrams in patients with chest pain of varying etiology, Circulation **51**:1046, 1975.
43. Willerson, J. T., Parkey, R. W., Bonte, F. J., Meyer, S. L., and Stokely, E. M.: Acute subendocardial myocardial infarction in patients: its detection by technetium-99m stannous pyrophosphate myocardial scintigrams, Circulation **51**:436, 1975.
44. Zaret, B. L., DiCola, V. C., Donabedian, R. K., Puri, S., Wolfson, S., Freedman, G. S., and Cohen, L. S.: Dual radionuclide study of myocardial infarction: relationships between myocardial uptake of potassium-43, technetium-99m stannous pyrophosphate, regional myocardial blood flow and creatine phosphokinase depletion, Circulation **53**:422, 1976.
45. Zaret, B. L., Lange, R. C., and Lee, J. C.: Comparative assessment of infarct size with quantitative thallium-201 and technetium-99m pyrophosphate dual myocardial imaging in the dog, abstracted, Am. J. Cardiol. **39**:309, 1977.

46. Zaret, B. L., Strauss, H. W., Martin, N. D., Wells, H. P., and Flamm, M. D.: Noninvasive regional myocardial perfusion with radioactive potassium: study of patients at rest, exercise and during angina pectoris, N. Engl. J. Med. **288:**809, 1973.

47. Zaret, B. L., Vlay, S. C., Freedman, G. S., Wolfson, S., and Cohen, L. S.: Quantitative relationships between potassium-43 imaging and left ventricular cineangiography following myocardial infarction in man, Circulation **52:**1076, 1975.

SECTION FOUR

CARDIOMYOPATHIES

17 □ Cardiomyopathy: morphologic features

William C. Roberts

This chapter describes certain morphologic changes observed *at autopsy* in patients with various noninflammatory (cardiomyopathy) and inflammatory (myocarditis) heart muscle diseases. The disorders will be presented in the following order (Fig. 17-1):

I. Noninflammatory heart muscle diseases (cardiomyopathy)
  A. Idiopathic disease
    1. Idiopathic dilated cardiomyopathy
    2. Idiopathic hypertrophic cardiomyopathy
      a. Asymmetric hypertrophy
        1. With subaortic stenosis
        2. Without subaortic stenosis
      b. Symmetric hypertrophy
        1. With subaortic stenosis
        2. Without subaortic stenosis
  B. Endomyocardial disease
    1. With eosinophilia (Löffler's fibroplastic parietal endocarditis)
    2. Without eosinophilia (endomyocardial fibrosis of Davies)
  C. Infiltrative disease
    1. Amyloid
    2. Iron (hemosiderosis)
    3. Glycogen
    4. Lipids
    5. Calcium
II. Inflammatory heart muscle disease (myocarditis)
  A. Associated with known infectious agent (bacterium, virus, parasite, rickettsia, spirochete, treponema, and fungus)
    1. Primary
    2. Secondary
  B. Associated with unknown agent (idiopathic)
  C. Associated with another known systemic condition
    1. Collagen disease
      a. Acute rheumatic fever
      b. Rheumatiod arthritis
      c. Systemic lupus erythematosus
      d. Other
    2. Sarcoidosis

NONINFLAMMATORY HEART MUSCLE DISEASES (CARDIOMYOPATHY)

Idiopathic disease

Idiopathic dilated cardiomyopathy (Fig. 17-2). Idiopathic cardiomyopathy may be defined as a disease of the myocardium not resulting from coronary, valvular, congenital, hypertensive, or pulmonary heart disease.[26] Idiopathic cardiomyopathy is of two types: the ventricular dilated type and the hypertrophic (nonventricular dilated) type. Morphologic differences between these two types are summarized in Table 17-1.

Idiopathic cardiac enlargement, heart disease of unknown origin, primary myocardial disease, and *congestive cardiomyopathy* are other terms that have been used to describe what I prefer to call idiopathic dilated cardiomyopathy (DC). The reason for the use of the latter term is that the major morphologic feature of this condition is dilation of both ventricular cavities. Although the degree of dilation of both cavities is usually similar, occasionally one ventricle is more dilated than the other. The

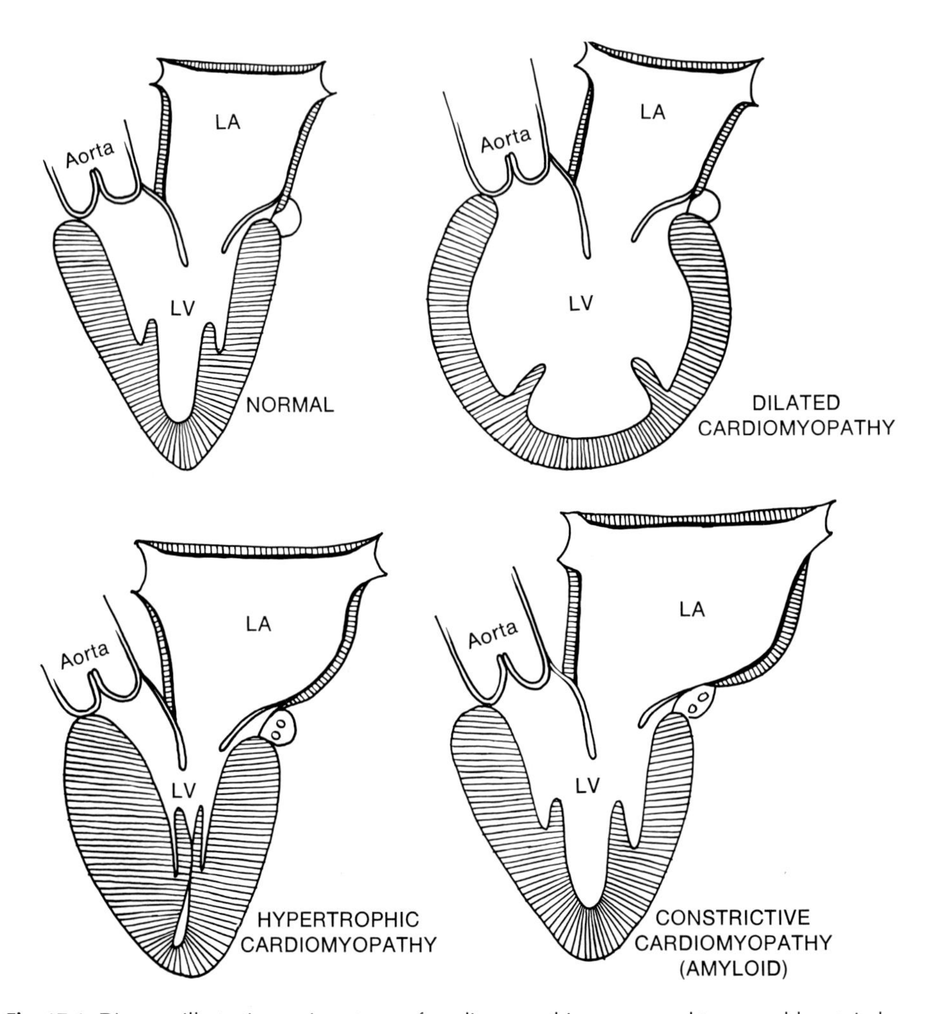

Fig. 17-1. Diagram illustrating various types of cardiomyopathies, compared to normal heart. In hypertrophic cardiomyopathy, left ventricular cavity is small, and in constrictive variety, illustrated by amyloidosis, left ventricular cavity is of normal size. In dilated type, largest circumference of left ventricle is not at its base but about midway between apex and base.

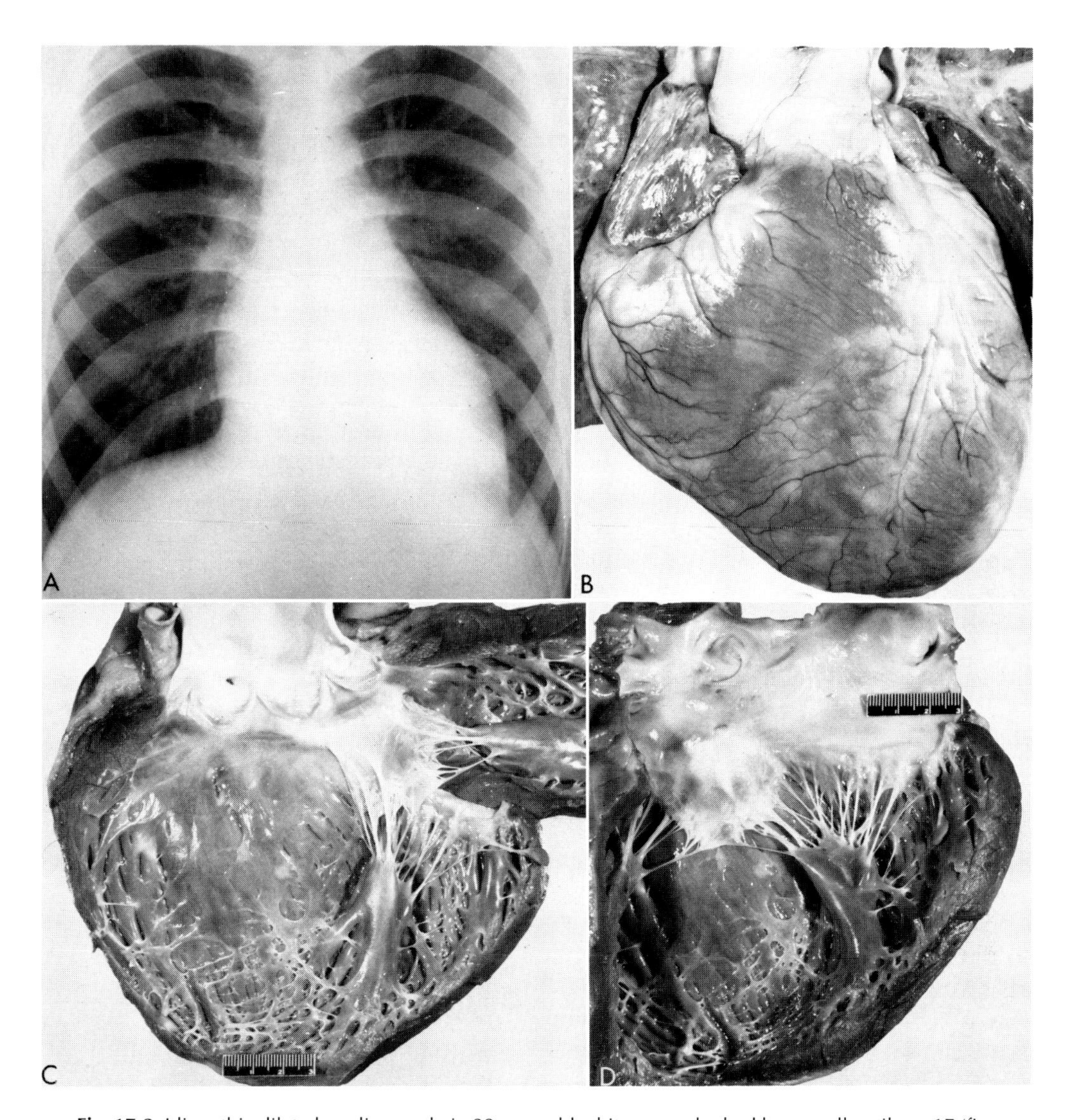

Fig. 17-2. Idiopathic dilated cardiomegaly in 22-year-old white man who had been well until age 17 (five years before death), when he experienced onset of pleuritic-type pain followed by congestive cardiac failure. Electrocardiogram showed nonspecific T-wave changes. After a four-month hospitalization, he was entirely well until two months before death, when an upper respiratory infection prompted overt congestive heart failure. Thereafter, heart failure progressively worsened and heart enlarged. Grade 2/6 apical systolic murmur appeared.

At necropsy, heart weighed 570 g and both ventricles were very dilated (**A** to **D**). Valves and coronary arteries were normal. Histologic sections of left ventricular wall showed only mild interstitial fibrosis. **A,** Chest roentgenogram during last week of life. **B,** Anterior aspect of heart. Pericardial sac contained 50 ml of serous fluid. **C,** Opened left ventricle, aortic valve, and aorta. Small focal endocardial thickenings are present. **D,** Opened left atrium, mitral valve, and left ventricle. Mitral anulus is not dilated.

Table 17-1. Morphologic differences between idiopathic dilated and hypertrophic cardiomyopathies

	Dilated type	Hyper-trophic type
Dilated ventricular cavities	+	0
Intracardiac thrombi	+	0
Asymmetric septal hypertrophy	0	+
Myocardial fiber disorientation	0	+
Endocardial plaque, left ventricular outflow tract (LVOT)	0	+
Thickened mitral valve	0	+
Abnormal intramural coronary arteries	0	+
Ventricular scarring	0 to +	+
Thickened ventricular walls	0 to +	+
Increased cardiac weight	+	+
Dilated atrial cavities	+	+

Present = +, absent = 0.

ventricular dilation is associated with poor ventricular contractions, which in turn produce low ejection fractions and high end-systolic volumes. The latter appear to limit atrial emptying, which then leads to high atrial end-diastolic volumes with resulting atrial dilation.

The large end-systolic ventricular volumes produce relative stasis of the blood in the apical portions of the ventricular cavities, and this frequently results in intracavitary thrombosis. Thrombi are also frequent in one or both atrial appendages, presumably the consequence of poor atrial emptying and relative stasis of the blood in the appendages, also. The most frequent locations of thrombi in patients with DC are (in order of frequency) the left ventricle, the right ventricle, the right atrial appendage, and the left atrial appendage. About 75% of these patients have left ventricular thrombi at necropsy. I have rarely observed thrombi in the atria or right ventricle without thrombi also being present in the left ventricle. In addition to fibrin thrombi, focal endocardial thickenings are often present, particularly near the apices of the ventricles, and these probably represent the end result of thrombi organization. The intracardiac thrombi often give rise to pulmonary and systemic emboli.

The hearts were always increased in weight; I found the average increase in a study of 100 necropsy patients to be just under 600 g. Despite the increased weight, the maximal thickness of the left ventricular free wall and ventricular septum is usually ≤ 1.5 cm.

Although interstitial myocardial fibrosis of varying degrees is usually detected by histologic examination, left ventricular free wall scarring or ventricular septal scarring or both may be observed by gross inspection in only about 25% of the patients. Even when scarring is observed grossly, it is usually limited to the papillary muscle and the subendocardium (inner half of wall). Thus the poor myocardial contractility rarely can be attributed to scarring of the left ventricular wall.

The leaflets of the four cardiac valves are usually normal, but occasionally the margins of the mitral and tricuspid valve leaflets are focally thickened by fibrous tissue. The latter is most frequent in patients with atrioventricular valvular regurgitation for long periods. These valvular thickenings appear to be secondary to the regurgitation, and do not themselves cause valvular incompetence. The cause of valvular regurgitation in these patients appears to be papillary muscle dysfunction because the tricuspid and mitral valvular anuli are usually only mildly dilated (25% above normal).[5] The aorta and major pulmonary arteries are of normal size.

Histologic study of the myocardial walls in these patients discloses nonspecific changes. Many myocardial cells appear hypertrophied, others, atrophied. The amount of fibrous tissue between the myocardial cells is usually increased. Inflammatory cells are absent. The intramural coronary arteries are normal. Electron microscopy has confirmed these histologic observations and demonstrated other nonspecific changes, including cellular edema; increased lipid droplets, lysosomes, and lipofuscin granules; dilation of the tubules of the sarcoplasmic reticulum and the T system; and mild myofibrillar damage and various mitochondrial alterations.[8] The mitochondria vary in size; many are smaller than normal. No virus particles have been observed in the myocardium of patients with fatal cardiomyopathy. The changes in myocardium in patients with histories of habitual alcoholism are indistinguishable from those in patients without such histories.[8] In general, the extent of the degenerative changes in the myocardial cells correlates with the duration and severity of cardiac dysfunction.

Although it is well recognized that DC may be confused clinically with coronary heart disease, separation of the two at necropsy may present problems if one adheres rigidly to a definition that includes "absence of coronary arterial disease." In other words, how much coronary arterial narrowing is permissible to still allow the diagnosis of DC to be made? Initially, I included in a definition of this entity "less than 50% cross-sectional area luminal narrowing" of a major epicardial coronary artery. Because fatal coronary (atherosclerotic) heart disease is always associated with cross-sectional area luminal narrowing of greater than 75%, a 75% cutoff point appears more reasonable.[24,25] Furthermore,

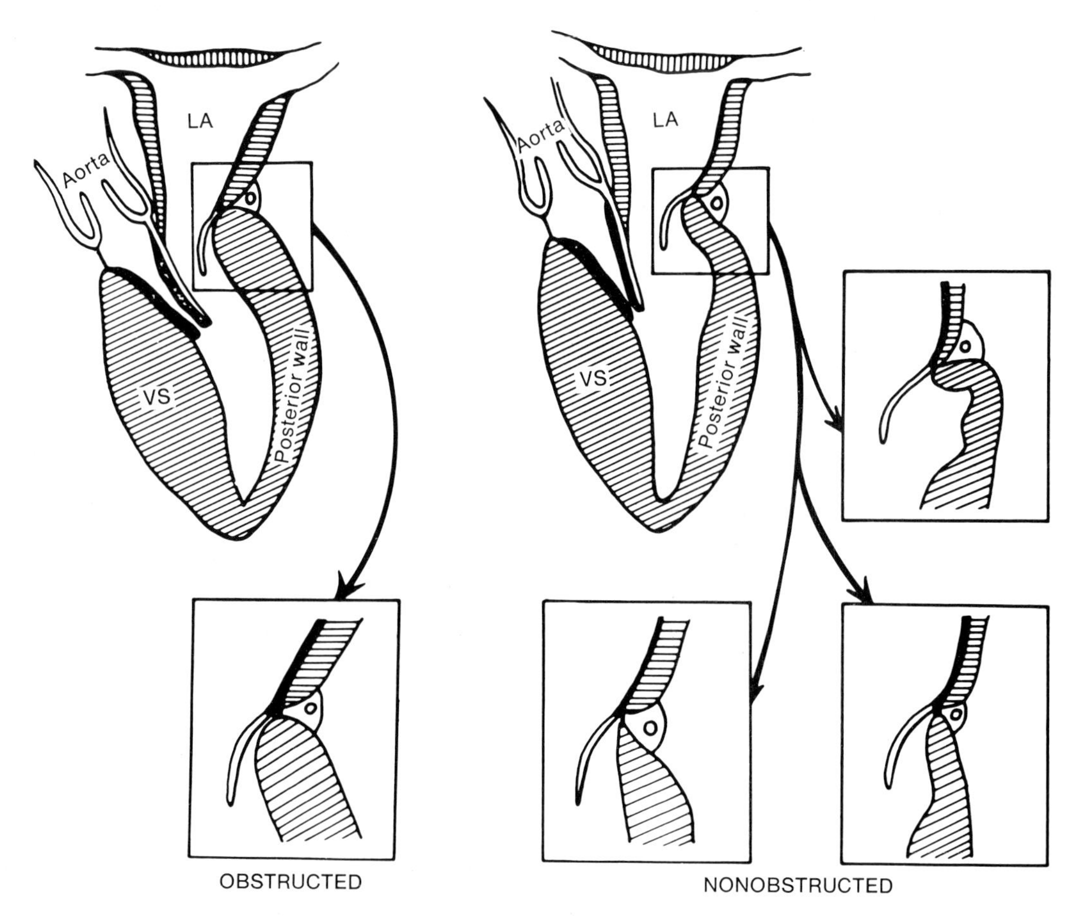

Fig. 17-3. Diagram illustrating major differences between obstructive and nonobstructive types of hypertrophic cardiomyopathy. In obstructive type, most basal portion of posterior and lateral left ventricular free walls is rounded and thick. In nonobstructive type, left ventricular free wall beneath posterior mitral leaflet is thinned and pointed.

there is apparently no reduction in blood flow through a tube (coronary artery) until more than 75% of the cross-sectional area of the lumen is obliterated. Obviously, the associated coronary atherosclerosis rarely presents a differential problem in the young patient, but it may in the older patient.

Idiopathic hypertrophic cardiomyopathy (Figs. 17-3 and 17-4). Hypertrophic cardiomyopathy (HC) is strikingly different both functionally and morphologically from DC (Table 17-1).[7,26,31,32] The major problem is a *hyper*contracting left ventricle rather than a *hypo*contracting ventricle as in DC. The morphologic counterpart of this hypercontracting ventricle is a massive increase in ventricular muscle mass with resulting small ventricular cavities. HC is the classic example of the "muscle-bound heart"; DC, in contrast, is the classic example of the "flabby heart," mostly cavity with comparatively little myocardial mass.

ASYMMETRIC HYPERTROPHY. In his original description, Teare[32] applied the term *asymmetrical hypertrophy of the heart* to this condition. Subsequently, other terms including *functional subaortic stenosis, diffuse muscular subaortic stenosis, idiopathic hypertrophic subaortic stenosis (IHSS), hypertrophic obstructive cardiomyopathy,* and *asymmetric septal hypertrophy (ASH)* have been applied. The terms that include the word "stenosis" or "obstruction" appear to be inappropriate because obstruction is more often absent than present.[7] The terms that stress the asymmetry of the walls bordering the left ventricular cavity have served a useful purpose in emphasizing the characteristic morphologic feature of this condition, but it is now apparent that the ventricular septum may be thicker than the left ventricular free wall in some conditions other than HC,[12-14] that a few (about 5%) patients with HC do not have ASH (i.e., the septum thicker than free wall),[16,26] and that on rare occasions the free wall in HC is not thickened.[26] Following are causes of ASH:

A. Hypertrophic cardiomyopathy
    1. With obstruction (IHSS)
    2. Without obstruction
B. Aortic stenosis
    1. Valvular
    2. Discrete subvalvular
    3. Supravalvular
    4. Tunnel
C. Systemic hypertension
D. Right ventricular systolic hypertension
E. Myocardial infarction
F. Parachute mitral valve syndrome
G. Cardiac neoplasm

Study at necropsy of patients with HC has demonstrated the following morphologic features of this condition[26]: (1) greater thickening of the ventricular septum than the left ventricular free wall (95%), (2) small or normal-sized left and right ventricular cavities (95%), (3) mural endocardial plaque, LVOT (75%—adults only), (4) thickened mitral valve (75%—adults only), (5) dilated atria (100%—adults only), (6) abnormal intramural coronary arteries (50%), and (7) disorganization of myocardial fibers in the ventricular septum (95%).

The chief morphologic abnormality in HC resides in the *ventricular septum.* As pointed out by Teare[32] in 1958, the ventricular septum is nearly always thicker than the left ventricular free wall. Among adult necropsy patients at our laboratory, the maximal thickness of the ventricular septum averaged 3.0 cm and the maximal thickness of the left ventricular free wall, 1.8 cm. The thickest portion of the septum is that part located about midway between the aortic valve and left ventricular apex. This level corresponds approximately to the apex of the right ventricle. Although in all the necropsy patients at our laboratory the ventricular septum was thicker than normal, the left ventricular free wall was usually, but not always, thicker than normal.

SYMMETRIC HYPERTROPHY. Although *ventricular asymmetry,* that is, greater maximal thickness of the ventricular septum than of the left ventricular free wall, is the characteristic morphologic feature of HC, about 5% of our patients at necropsy with this condition had *ventricular symmetry,* that is, equal maximal thicknesses of the ventricular septum and left ventricular free wall. Among siblings with HC, I have observed

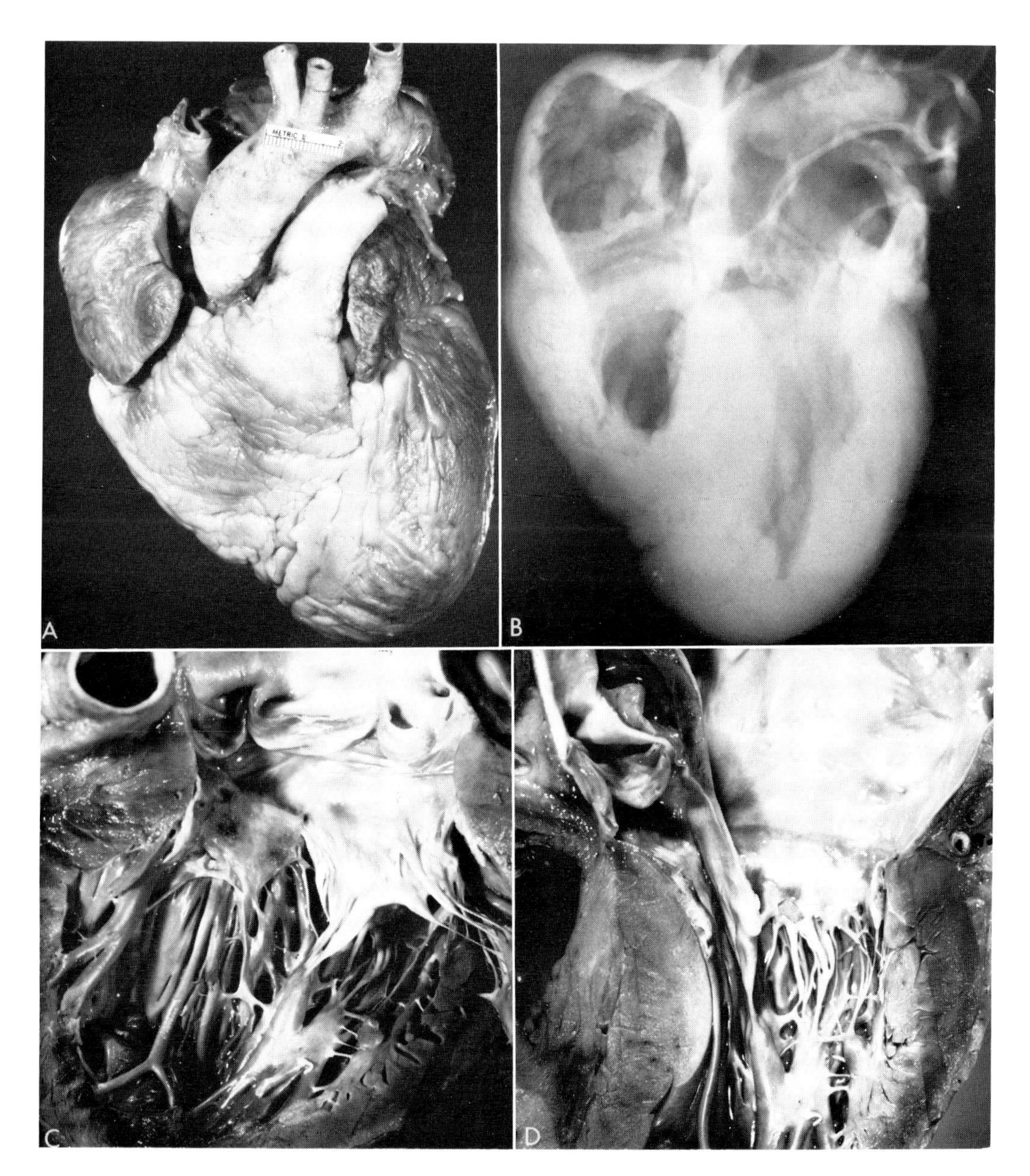

Fig. 17-4. Hypertrophic cardiomyopathy. This 33-year-old man had no systolic pressure gradient at rest between left ventricle and brachial artery, but 33 mm Hg gradient was provoked by isoproterenol. He died of sudden unilateral hemiparesis, which appeared six days after cardiac catheterization. He had been asymptomatic when examined at age 26; grade 2 to 3/6 precordial systolic murmur, loudest at apex, was heard. Catheterization at that time showed a 30 mm Hg peak systolic gradient at rest between left ventricle and brachial artery, and it rose to 85 mm Hg with ouabain. He became symptomatic (exertional and nocturnal dyspnea) six months before death, when atrial fibrillation appeared. Precordial murmur was then only intermittently audible.

At necropsy, heart weighed 450 g. **A,** Both atria were dilated but neither ventricle was dilated. **B,** Ventricular septum was thicker than left ventricular free wall. **C** and **D,** Mural endocardial plaque was present in left ventricular outflow tract and corresponding anterior mitral leaflet was thickened. Longitudinal cut of heart in **D** clearly shows that ventricular septum is thicker than left ventricular free wall.

one family in whom one member had ventricular asymmetry and another had ventricular symmetry. Thus ventricular asymmetry is the usual, but it is not a requisite for diagnosis of HC.

Among patients with HC, the thickness of the ventricular septum does not depend on the presence of a LVOT obstruction at rest, and, indeed, the thickness of the septum is similar in patients both with and without outflow obstruction at rest.[26] Although examination of the septum in patients with HC is not helpful in separating patients *with* from those *without* LVOT obstruction at rest, examination of the basal portion of the left ventricular free wall (posterior or posterolateral portion) does permit delineation between these two groups of patients. In the patients with the obstructive type of HC, this posterobasal portion of left ventricular free wall represents the thickest portion of free wall, whereas in the patients with the nonobstructive type of HC, this posterobasal portion is thinner than normal and often pointed like a bird's bill (Fig. 17-4).[11] In patients with obstruction at rest, the thickest portion of left ventricular free wall is about midway between the base of posterior mitral leaflet and apex of left ventricle.

Another abnormality in the ventricular septum in patients with HC is focal myocardial fiber disarray or disorganization.[7,9,26,32] This abnormality, also described by Teare,[32] occurs in 90% to 100% of these patients, depending on how "septal disorganization" is defined.[19] Small foci of nonparallel myocardial fiber groups in the ventricular septum are common in normal developing and adult hearts.[19] When, however, the definition of septal disorganization is rigid and quantitation of the amount of disorganization is introduced, about 90% of the patients with HC can be clearly delineated from persons with normal hearts or from patients with cardiac conditions other than HC.[19] Studies by Maron and associates[19] have shown that among patients with HC nearly 90% showed myofiber disorganization in 5% or more of the ventricular septum. In contrast, among persons with normal hearts or heart conditions other than HC, less than 5% of the ventricular septum showed myofiber disorganization.

Although septal disorganization of greater than 5% of the ventricular septum occurs in about 90% of the patients with HC, about 10% of these patients have less than 5% of the septum showing disarray and of them a few patients have absolutely no disarray at all. Thus, like ventricular asymmetry, septal disorganization of greater than 5% of the septum occurs in most patients with HC, but this abnormality may involve a smaller area of the septum or, rarely, may be entirely absent.

Not only does disorganization of *myocardial fibers* occur, but, in addition, ultrastructural studies have shown disarray of *myofibrils* and *myofilaments* within individual cells.[9,15] Again, the abnormalities of myofibrils and myofilaments are not absolutely specific for HC, but when disorganization of these subcellular components is found in other cardiac conditions, the abnormal cells are in small numbers. The individual septal myocardial cells in HC also frequently show increased amounts of Z-band material and nonspecific changes of cellular hypertrophy and degeneration.[9]

Myofiber disarray in the left and right ventricular free walls is more difficult to determine by light microscopy than that in the ventricular septum. Ultrastructural examination of the left and right ventricular free walls, however, has disclosed many bizarrely shaped, disorganized cells in the ventricular free walls of patients *without* obstruction, whereas the bizarrely shaped and disorganized myocardial fibers are virtually entirely absent from the ventricular free walls in the patients *with* obstruction.[15] Thus the disorientation of groups of myocardial cells and the disarray of myofibrils and myofilaments within individual myocardial cells is not the result of LVOT obstruction or high intraventricular systolic pressures.

The functional or clinical significance of septal disorganization among patients with HC is not entirely clear. Among the patients studied at necropsy by Maron and associates,[15] those with large areas of the septum showing disorganization were younger, had a higher frequency of premature deaths involving more than one member of the same family, and thicker ventricular septa than those with smaller areas of the septum involved. Comparison of the presence or degree of LVOT obstruction, length of symptoms of cardiac dysfunction, mode of death, arrhythmias, etc., were not significantly different in the patients with high percentages of the septal area involved by disorganization, compared to those with low percentages.

Another abnormality of the ventricular septum in patients with HC is the presence of abnormal intramural coronary arteries in about half of the patients.[20] The abnormalities observed in the intramural coronary arteries in these patients are striking and consist of increased number and sizes of arteries with thickened walls and narrow lumens. The thickening results from proliferation of smooth muscle cells and collagen in both the media and the intima. Also, mucoid deposits (acid mucopolysaccharide material) are increased. Like disproportional septal thickening (causing ventricular asymmetry) and septal disorganization, however, abnormality of the intramural coronary arteries is not completely specific for HC, but it is certainly more striking in HC than in most other conditions. Similar abnormality of the intramural coronary arteries may be observed in patients (also Newfoundland dogs) with discrete subaortic stenosis,[22] in patients with so-called tunnel subaortic stenosis,[19] and in newborns with aortic valve atresia.[29]

The significance of the changes in the intramural coronary arteries in patients with HC is unknown. The presence or absence of the changes does not correlate with any analyzed clinical parameter, encompassing the presence or degree of LVOT obstruction; age; sex; presence of cardiac dysfunction, including chest pain; or length of symptoms of cardiac dysfunction.

The cause of the striking abnormalities of the intramural coronary arteries is also uncertain. In each of the four conditions in which this degree of abnormality has been observed, namely, HC, discrete and tunnel subaortic stenosis, and aortic valve atresia, the septum is extremely thick, and movement of the septum during life as shown by echocardiogram is extremely limited. It is possible that the severe septal thickening prevents adequate expansion of these vessels during ventricular diastole so that flow through them during this phase of the cardiac cycle is greatly reduced. The fibrous smooth muscle cell response, in turn, may serve to obliterate the unused lumen. This explanation, however, does not explain the large size of these arteries. It is possible that the proliferation of the smooth muscle cells in the walls of these arteries represents a response to the same stimulus that is causing severe hypertrophy and abnormal configuration of many striated myocardial cells.

A fourth abnormality of the ventricular septum in adults with HC is the occurrence of a fibrous plaque on the mural edocardium of the outflow portion of the septum. This fibrous thickening is located in direct apposition to the ventricular aspect of the anterior mitral leaflet and is the result of contact between the valvular and mural endocardium. The mural plaque, in other words, is the anatomic equivalent of the *systolic anterior motion (SAM)* of the anterior mitral leaflet observed on echocardiogram in HC and occasionally in other conditions.[16] The obstruction in HC when present almost certainly begins at a level corresponding to the distal margin of the anterior mitral leaflet and with the caudal margin of the septal endocardial plaque. The mural endocardial septal plaque is present more frequently and is thicker in patients with obstruction at rest, compared to those without this obstruction.[26]

It appears that HC is primarily a disease of the ventricular septum and that the morphologic abnormalities in it are apparent at the gross, microscopic, and ultrastructural levels. Other portions of the heart, however, are affected in HC, but it seems likely that these other abnormalities are secondary to the abnormalities of the ventricular septum. These include decreased ventricular cavity size or at least lack of ventricular cavity dilation, increased atrial cavity size, varying degrees of myocardial fibrosis, thickening of posterior as well as anterior mitral leaflet, and, in elderly patients, calcification of the mitral anulus.

The cause of the relatively small sizes of the ventricular cavities in HC is unclear. The septum, being on the average about twice its normal thickness, surely occupies space normally represented by cavity. Furthermore, the ventricular free walls, on the average about a third thicker than normal, probably also slightly compromise the intracavitary space. Additionally, the greater thickness of the septum makes this structure relatively noncompliant (as demonstrated by its absent or minimal movement on echocardiogram). Although the left ventricular free wall in HC (detected by echocardiogram) usually contracts vigorously (in contrast to the septum), it appears to have little capacity to distend outwardly. Whatever the explanation, ventricular cavity size is rarely enlarged in these patients. Those with dilated ventricular cavities (seven of sixty-five necropsy patients that I personally studied) have either nonscarred left and/or right ventricular free walls of normal thickness or scarred ventricular free walls of less than normal thickness.[26] Dilation may also occur after operative septotomy-septectomy in HC.[21]

The impression of ventricular cavity size as observed at necropsy in patients with HC may be somewhat misleading because the cavity represents the size at the end of ventricular systole.[17] This information was derived by comparing the left ventricular free wall thickness at necropsy to that measured by echocardiogram during life.[17] The thickness at necropsy corresponds to the thickness during ventricular systole, not diastole as observed by echocardiogram.

In contrast to the ventricular cavities, the atrial cavities in HC, at least in adults, are always dilated. The dilation appears to be a response to difficulty in filling the relatively small ventricular cavities.

Not only is the anterior mitral leaflet thickened, but often (more frequently in the patients with obstruction) the posterior mitral leaflet is thickened as well. The thickening of both leaflets is probably the consequence of the small left ventricular cavity. It is important to recall that the mitral valve resides in the left ventricle, which, being small, causes abnormal contact of the leaflets with themselves and also causes the anterior one to contact the septum. In the patients without obstruction, the basal portion of the left ventricular cavity is larger than in the patients with obstruction because of the thinning of the basal portion of free wall posteriorly and laterally. The focal fibrous thickening of the posterior mitral leaflet probably results from abnormal contact during ventricular systole because there is not enough room in the ventricular cavity to easily accommodate the leaflets and chordae. The ventricle and mitral leaflets in this respect may be regarded as being similar to an accordion. The mitral leaflet thickening in HC is somewhat analogous to that which occurs normally as a consequence of aging. With aging, the left ventricular cavity becomes smaller, presumably in response to the lowered cardiac output; the mitral leaflets become thicker, and the left ventricular muscle, less compliant. The left atrium dilates in response to the increased work required to fill the small left ventricle. Similar mechanisms appear to be accelerated in HC.

Most patients with HC have some degree of mitral regurgitation. It is not related to anular dilation because the mitral anulus in this condition is smaller than normal.[26] Furthermore, the leaflet thickening, in and of itself, is not extensive enough to cause valvular regurgitation. The most likely explanation appears to be abnormal bending of the papillary muscles, particularly the anterolateral.[7] These structures may be bent abnormally by the bulging ventricular septum with resulting excessive tension on the chordae tendineae, preventing closure of the mitral orifice. Focal scars are usually present in the papillary muscles in HC, but they cannot account for the regurgitation.

Although mitral regurgitation is usually mild in HC, it may be the predominant clinical feature of this condition. Mitral valve replacement in HC is hazardous because the small left ventricular cavity is rarely capable of freely accommodating a prosthesis, and the mitral regurgitation in patients with HC nearly always either disappears or is strikingly lessened by adequate septotomy-septectomy.[23]

Endomyocardial disease (with and without eosinophilia)

Löffler in 1936 described two patients with "fibroplastic parietal endocarditis with blood eosinophilia."[26] In 1957, Bousser summarized reports of thirteen patients with "eosinophilic leukemia," and a number of reports have appeared describing each of these two conditions.[26] In each, there is endocardial fibrosis of one or both ventricles and severe eosinophilia. Death is usually the result of congestive cardiac failure. From examination of previous reports and of eight necropsy patients at my laboratory,[26] it appears that the clinical and morphologic features of eosinophilic leukemia and Löffler's endocarditis are similar and that these two conditions are indeed the same disease. Certain similarities also exist between African endomyocardial fibrosis (EMF) and EMF with eosinophilia, described by others as Löffler's endocarditis or eosinophilic leukemia. The mural endocardial thickening in the ventricles is seen in both conditions, and blood eosinophilia may occur in EMF.

The possibility that Löffler's parietal fibroplastic endocarditis, eosinophilic leukemia (without abnormal myelopoiesis), and African EMF are the same disease at different stages of development is an attractive hypothesis (Fig. 17-5). Patients with eosinophilia associated with a transient, benign febrile illness, such as tropical eosinophilia or Löffler's pneumonia, may represent one end of the spectrum. Patients with severe EMF and no blood eosinophilia may represent the other end. Between these two extremes there may be patients who have eosinophilia but less extensive endocardial scarring. The latter process is still active, and thus eosinophilia is still present, but as the inflammatory process dies down, cardiac scar tissue remains and eosinophilia disappears.

Eosinophilia from any cause, as long as the quantity of eosinophils is sufficient, appears capable of damaging ventricular mural endocardium. Recently, I studied a patient with severe eosinophilia caused by trichinosis; at necropsy the mural endocardium of both ventricles was extensively damaged and covered by fibrin.[1] Similarly damaged mural endocardium was observed by Fishbein[11a] in a patient with severe blood eosinophilia associated with fatal carcinoma of the lung.

Infiltrative diseases

The infiltrative cardiomyopathies include various infiltrates into the walls of the cardiac chambers. The infiltrates may be localized to *myocardial interstitium* (amyloid) or within *myocardial cells* (iron, glycogen, lipids, calcium).

Amyloid. Amyloid may be deposited in any portion of the heart, and the deposits may be large (grossly visible) or small (microscopic).[3] When cardiac dysfunction results from cardiac amyloidosis, the amyloid deposits are grossly visible and diffusely distributed throughout the ventricular walls, which have a firm, rubbery consistency (Fig. 17-6). Histologically, most ventricular myocardial cells are surrounded

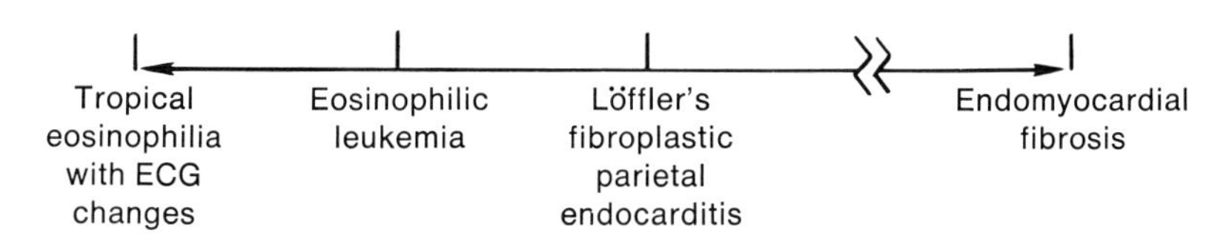

Fig. 17-5. Diagram illustrating spectrum of endomyocardial disease with and without eosinophilia.

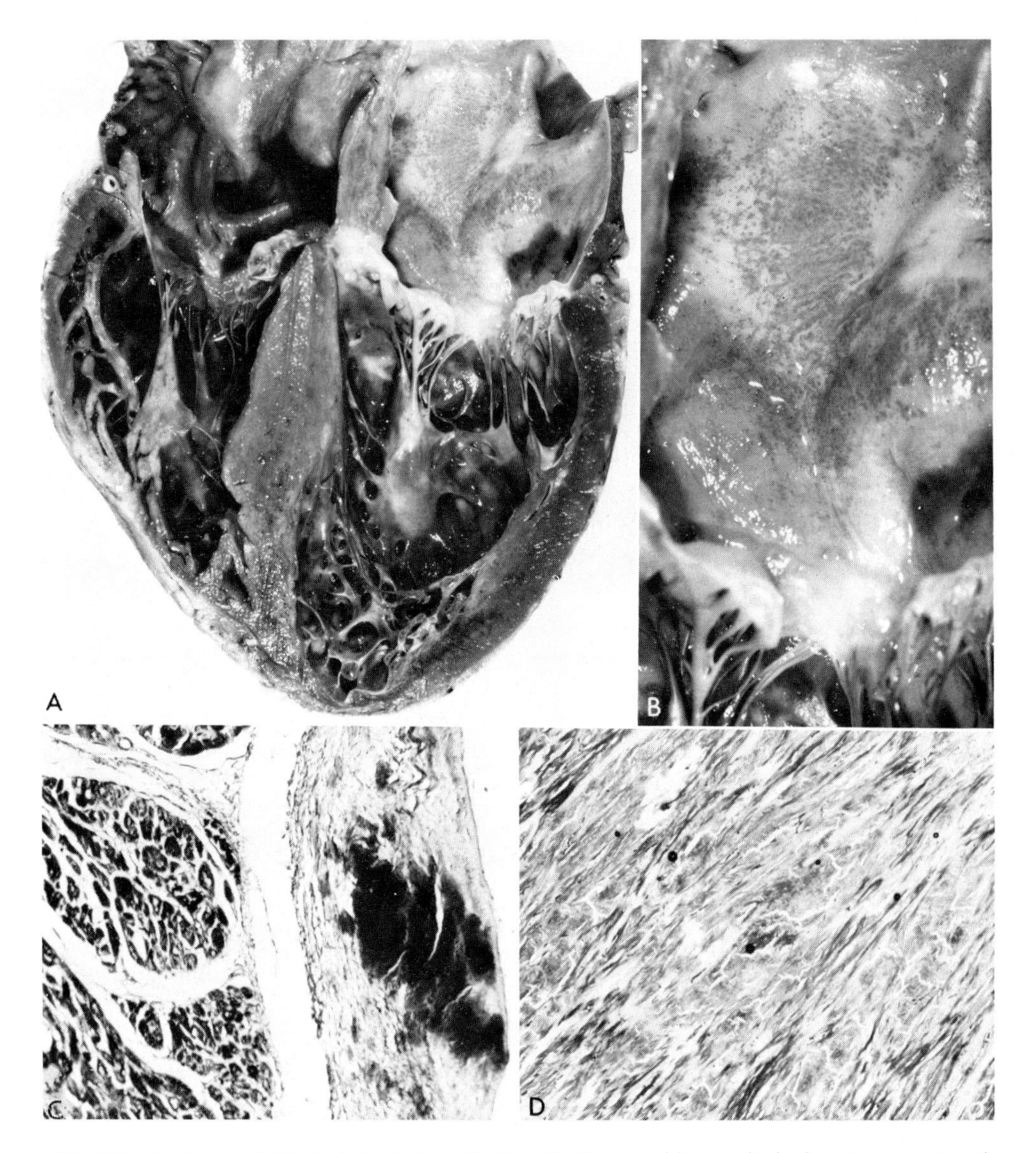

Fig. 17-6. Cardiac amyloidosis. **A,** Posterior half of heart in 86-year-old man who had angina pectoris and healed myocardial infarction as well as large amyloid deposits. **B,** Close-up of left atrial endocardium. Wavy lesion on this endocardium represents amyloid deposits and is indicative of extensive ventricular amyloid deposits. **C,** Photomicrograph of left atrial endocardial amyloid deposit (×107, crystal violet stain). **D,** Amyloid has extensively infiltrated left ventricular papillary muscle (×32, crystal violet stain).

by amyloid fibrils. Thus, not only is ventricular motion restricted, but also most myofibers are constricted by the amyloid deposits. In addition to being in the myocardial interstitium, amyloid is deposited in the walls and lumina of the intramural coronary arteries, in the mural endocardium (particularly in the atria), in the valvular endocardium, in the epicardium, and in both conduction tissue and cardiac nerves. The endocardial deposits are usually not large enough to produce valvular dysfunction. The deposits in the lumina of the small coronary arteries, however, may lead to focal myocardial ischemia with resulting necrosis and fibrosis. The amount of amyloid deposition in the conducting myocardium is minimal compared to that in the contracting myocardium. Nevertheless, conduction and rhythm disturbances are more common in these patients than in those of similar age and sex without cardiac amyloidosis. Amyloid may be viewed as an "acellular cancer," in that the deposits infiltrate within and between normal tissues, but no cells are present in the proliferating material.

Although the ventricular walls are made rigid by heavy amyloid deposits, the ventricular cavities are not dilated unless another cardiac condition is also present. Thus cardiac amyloidosis joins hypertrophic cardiomyopathy and constrictive pericarditis as a cause of congestive cardiac failure in the absence of ventricular cavity dilation. The atria, in contrast, are usually dilated. The enlargement of the cardiac silhouette on roentgenograms in this condition may well be related to the atrial dilation. Most patients with symptomatic cardiac amyloidosis have electrocardiographic low voltage, conduction or rhythm disturbances, and congestive cardiac failure unresponsive to digitalis therapy. Diagnosis of clinically significant cardiac amyloidosis should be questioned in the absence of congestive cardiac failure and electrocardiographic low-voltage disturbances.[3] Not only is the congestive cardiac failure unresponsive to digitalis, but also evidence of digitalis toxicity is extremely common in patients with diffuse cardiac amyloidosis, sometimes even when only very small doses of digitalis are administered.

Iron (hemosiderosis). For years it was debated whether or not iron deposition in the heart could cause cardiac dysfunction. It is now clear, just as with amyloid, that cardiac dysfunction will occur if the cardiac iron deposits are large enough,[4] that cardiac dysfunction rarely occurs in patients with only microscopically visible iron deposits, and that cardiac dysfunction usually does occur when the deposits are grossly visible. In contrast to amyloid deposits, which are between myocardial cells, the iron deposits are within the myofibers (Fig. 17-7). The iron deposits are most extensive in ventricular myocardium, less in atrial myocardium, and the least in conducting, as contrasted to contracting, myocardium. Most patients with large cardiac iron deposits have abnormal electrocardiograms, mainly from arrhythmias and conduction disturbances, and congestive cardiac failure. In contrast to amyloid, the ventricular cavities are typically dilated. Thus the "iron heart" is a weak heart, not a strong one.

Glycogen. A deficiency of one or more of the enzymes involved in the biosynthesis and degradation of glycogen produces glycogen storage disease, of which at least eight types have been described.[26] Cardiac involvement occurs in types II, III, and IV. Most patients with glycogen storage disease causing cardiomegaly have type II (Pompe's disease), caused by a deficiency of α-1,4-glucosidase, a lysosomal enzyme that hydrolyzes glycogen to glucose. As a result the heart enlarges, sometimes to a marked degree, and congestive cardiac failure supervenes. Survival rarely extends beyond infancy or early childhood. Microscopically, the muscle cells show a characteristic lacework pattern, with large, clear central spaces (which represent the sites of glycogen deposition) and a peripheral rim of compressed cytoplasm. Endocardial fibroelastosis of at least the left ventricle is usually also present. Although the deposits of glycogen generally involve all myocardial cells relatively uniformly, on occasion the fibers in the ventricular septum may contain disproportionately excessive quantities of glycogen and lead to subaortic stenosis. Myocardial involvement occurs to a variable extent in type III glycogenosis (glycogen debranching enzyme deficiency). In type IV glycogenosis (branching enzyme deficiency) deposits of abnormal glycogen are found free in the cytoplasm of cardiac muscle cells, hepatocytes, and a variety of other cell types. These glycogen deposits are basophilic and are composed of nonbranching fibrils that measure 40 to 80 Å in diameter and resemble those in basophilic degeneration of the heart.

Lipids. Infiltration of the cytoplasm of car-

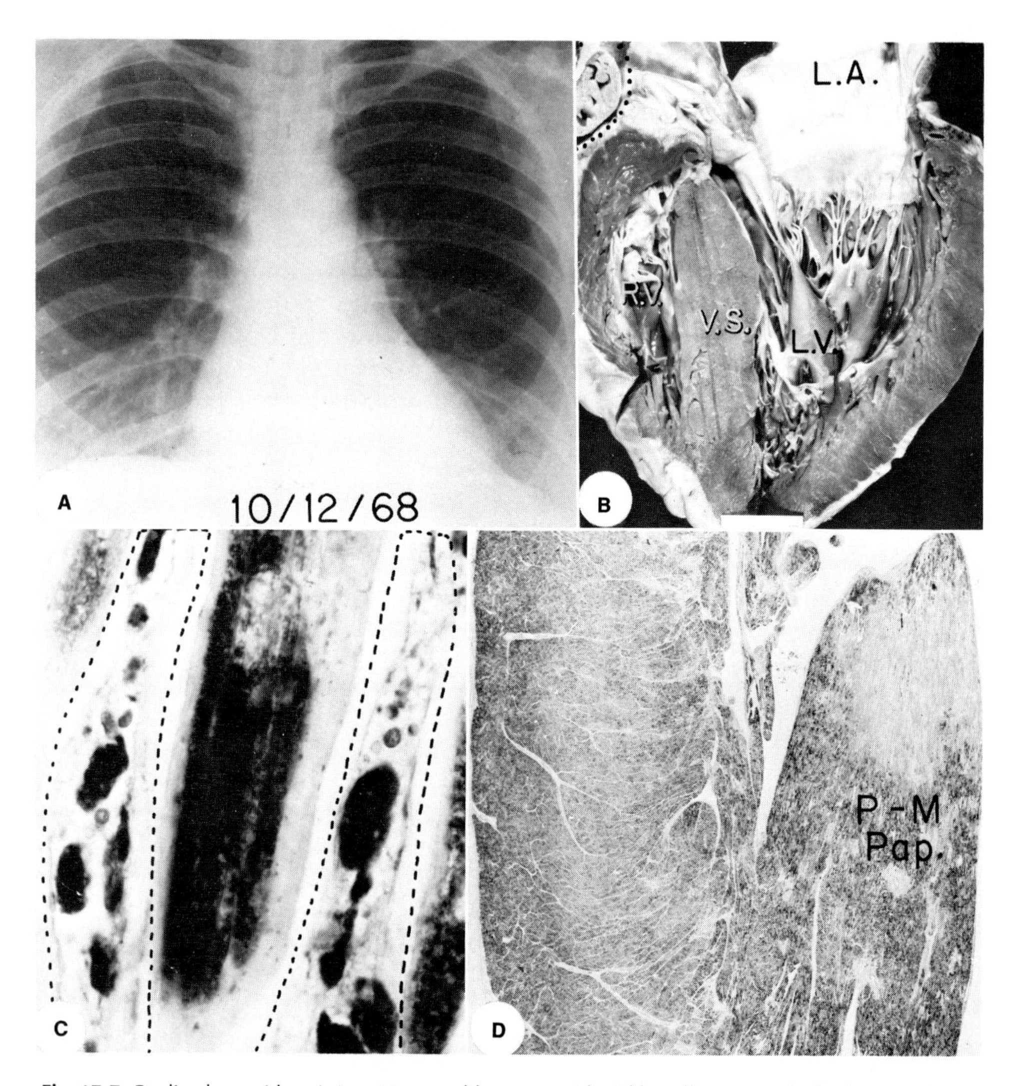

Fig. 17-7. Cardiac hemosiderosis in a 42-year-old woman with sickle cell anemia. She had received 260 units of blood when congestive cardiac failure developed six years before death. **A,** Chest roentgenogram showed cardiomegaly. By time of death she had received 359 units of blood (90 g of iron).

B, At necropsy, walls of right and left ventricles and left atrium were rusty brown due to extensive iron deposits. Right atrial wall was, in contrast, tan and only minute particles of iron were found in it by histologic examination. **C,** Photomicrograph of several myocardial cells, showing huge deposits of iron in them. Despite large deposits of iron in working myocardium, no iron deposits were observed in conducting myocardium (×628, Prussian–blue iron stain). **D,** Longitudinal section of left ventricular wall, including posteromedial (P-M) papillary muscle (×3, Prussian–blue iron stain). Foci of necrosis and fibrosis are present. Necrotic and fibrotic areas probably are anatomic indicators of chronic myocardial hypoxia, result of chronic anemia.

diac muscle cells by lipid droplets (fatty degeneration) is a nonspecific finding in many disorders, including hypoxia. *Fabry's disease* (angiokeratoma corporis diffusum universale) may be singled out among the lipid storage diseases as one in which extensive and clinically significant (cardiomegaly and congestive cardiac failure) accumulation occurs in cardiac muscle cells. In this condition, glycolipid deposits are also present in endothelium and vascular smooth muscle and in most other body tissues. The disease results from a deficiency of ceramide trihexosidase, an enzyme that hydrolyzes ceramide trihexoside. In ordinary histologic preparations, the deposits are dissolved, and the cardiac muscle cells resemble those in type II glycogenosis.

Calcium. Calcium is most commonly observed in the heart in the coronary arteries, mitral anulus, aortic and scarred mitral valve cusps, large myocardial scars, and the pericardium. In these areas, the calcific deposits are extracellular. Calcific deposits, however, may also occur within individual myocardial cells. This occurrence is observed most commonly in patients with fatal prolonged shock or renal failure. Its cause is uncertain, although severe hypoxia is the most likely explanation, since the lesion may be produced experimentally by this means. High serum levels of calcium may also be a factor. The myofibers infiltrated by calcium are always necrotic. Massive myocardial calcification may occur and produce cardiac dysfunction.

INFLAMMATORY HEART MUSCLE DISEASES (MYOCARDITIS)

Heart muscle disease associated with known infectious agent

In contrast to noninflammatory heart diseases, inflammatory diseases are infrequently observed today at necropsy. Acute

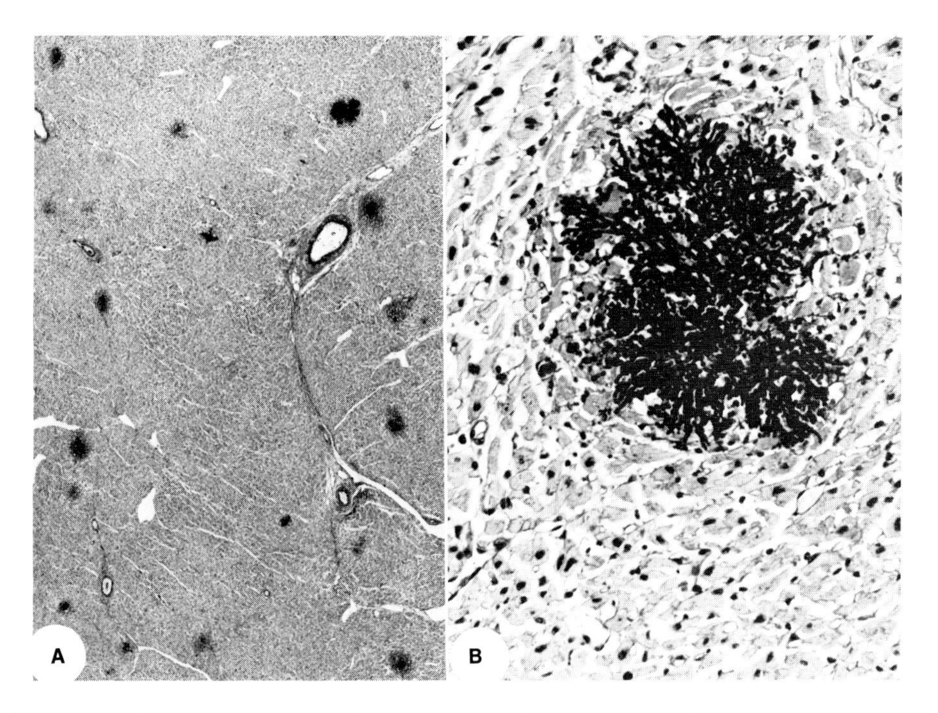

Fig. 17-8. Cardiac candidiasis. This 14-year-old boy had lymphosarcoma for seventeen months with terminal acute leukemic phase. He received a variety of chemotherapeutic agents, including prednisone, and terminally he developed *Candida albicans* septicemia. At necropsy, he had candida abscesses in most body organs, including heart. **A,** Low-power photomicrograph of left ventricular wall, showing multiple candida abscesses (×3, periodic acid–Schiff stain). **B,** Close-up of myocardial abscess (×182, periodic acid–Schiff stain). He never showed evidence of cardiac dysfunction.

myocarditis is now most commonly seen at necropsy only in patients with "overwhelming" systemic acute infections (Fig. 17-8), in those receiving immunosuppressive therapy, and those with active infective endocarditis.[2]

Virtually every known bacterium, virus, parasite, rickettsia, spirochete, treponema, and fungus has been shown to produce myocarditis. The myocardial involvement is rarely primary but is associated with involvement of other body organs and tissues.

Heart muscle disease associated with unknown agent (idiopathic)

Occasionally at necropsy a patient is observed to have foci of polymorphonuclear or mononuclear cell myocardial inflammation unassociated with stainable organisms or positive cultures. Most of these patients have received antibiotic therapy so that possibly the organisms, although not the inflammatory response, were eradicated.

Heart muscle disease associated with another known systemic condition

Collagen disease. Focal specific inflammatory infiltrates, namely, Aschoff bodies[30,33] and rheumatoid nodules,[27] are recognized lesions in the heart in acute rheumatic fever and rheumatoid arthritis, respectively. Fortunately, fatal acute rheumatic fever is infrequent today; when it does occur, the cardiac inflammatory response often has been altered by corticosteroid therapy. Rheumatoid nodules in the myocardium may also be affected by corticosteroid therapy. The latter occur in about 3% of patients with fatal rheumatoid arthritis. Aschoff bodies are found in most patients dying during the acute episode of rheumatic fever and at necropsy in about 5% of patients with mitral stenosis.[30]

At one time, focal myocarditis was believed to be fairly common in systemic lupus erythematosus. In the corticosteroid era, however, myocarditis is extremely rare.[6]

Sarcoidosis (Figs. 17-9 and 17-10). A neglected cause of cardiac dysfunction is cardiac sarcoidosis.[28] Hard granulomas are found in the heart in about 25% of the patients with sarcoidosis at necropsy. Recently, Ferrans, McAllister, and I[28] described certain clinical and morphologic observations in thirty-five patients with cardiac sarcoidosis (groups IA and IB) and summarized similar observations in seventy-eight previously described necropsy patients with cardiac sarcoidosis (groups

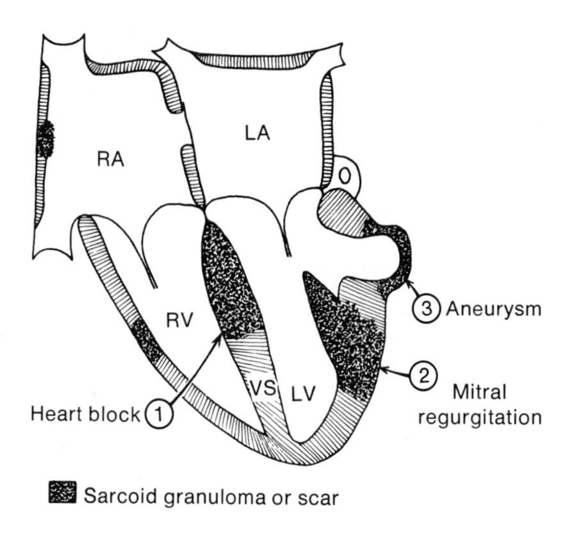

Fig. 17-9. Cardiac sarcoidosis. Diagram showing more frequent locations of sarcoid granulomas in heart and their common functional consequences.

IIA and IIB). All patients had nonnecrotic ("hard") granulomas in lymph nodes and either sarcoid granulomas in the heart (108 patients) or transmural myocardial scarring (five patients) unassociated with coronary arterial narrowing, that is, healed granulomas. The 113 patients (groups I and II) whose data were analyzed were also divided into those in whom cardiac dysfunction clearly was the result of sarcoid granulomatous infiltration of the heart (eighty-nine patients) (groups IA and IIA) and those in whom cardiac dysfunction, if present, was clearly not the result of sarcoid involvement of the heart (twenty-four patients) (groups IB and IIB).

Analysis of the eighty-nine patients in groups IA and IIA disclosed that death was sudden (arrhythmia) in sixty (67%), secondary to progressive congestive cardiac failure in twenty (23%), from recurring pericardial effusion in three (3%), and from other or unknown causes in six (7%). Sudden death was the initial manifestation of sarcoidosis

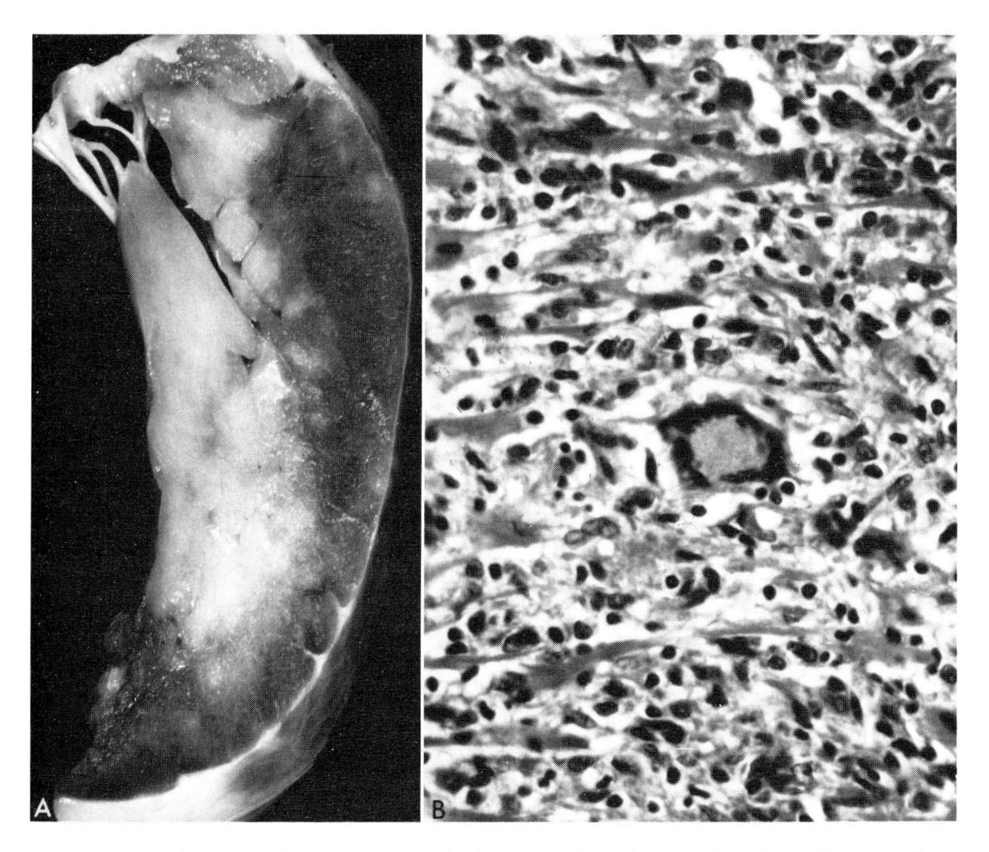

Fig. 17-10. Cardiac sarcoidosis. **A,** Longitudinal section through anterolateral papillary muscle in 26-year-old woman who had been asymptomatic until ten days before death, when dyspnea appeared. Dyspnea rapidly worsened, and when hospitalized on day of death, she was in acute pulmonary edema. Blood pressure was 80/70 mm Hg, heart rate 160 beats/min, and grade 3 to 4/6 pansystolic blowing murmur, which radiated to axilla, was audible. Chest roentgenogram showed congested lungs, cardiomegaly, and prominent hilar adenopathy. Electrocardiogram showed nonspecific ST-T wave changes and atrial hypertrophy. She developed complete heart block and died shortly thereafter.

At necropsy, large, firm, white deposits were present in walls of all four cardiac chambers and completely replaced both left ventricular papillary muscles (**A**). On histologic section, firm white areas represented hard granulomas typical of sarcoidosis (**B**) (×400, hematoxylin–eosin stain). Similar hard granulomas were present in lymph nodes, liver, spleen, and lung. Stains for acid-fast organisms, other bacteria, and fungi were negative.

in ten (17%) of the sixty patients who died suddenly. Other than premature ventricular beats, ventricular tachycardia was the most common arrhythmia (seventeen patients), and complete heart block was the most common conduction disturbance (twenty-five patients). Complete bundle-branch block occurred in twenty-one patients. Ventricular aneurysm (eight patients) and papillary muscle dysfunction (possibly sixteen patients) were other observed cardiac disturbances.

Although corticosteroid therapy tends to cause fibrous replacement of myocardial sarcoid granulomas, a possible consequence of this medication is the development of ventricular aneurysm.

Most patients with cardiac sarcoidosis causing dysfunction (groups IA and IIA) initially presented with manifestations related to the heart. Furthermore, most patients with cardiac sarcoidosis had little or no clinical evidence of dysfunction of an organ system other than the heart. Usually the course in patients with extensive cardiac sarcoidosis is not prolonged.

REFERENCES

1. Andy, J. J., O'Connell, J. P., Daddario, R. C., and Roberts, W. C.: Trichinosis causing extensive ventricular mural endocarditis with superimposed thrombosis. Evidence that severe eosinophilia damages endocardium, Am. J. Med. **63**:825, 1977.
2. Arnett, E. N., and Roberts, W. C.: Active infective endocarditis: a clinicopathologic analysis of 137 necropsy patients, Curr. Prob. Cardiol. **1**(7):1, 1976.
3. Buja, L. M., Khoi, N. B., and Roberts, W. C.: Clinically significant cardiac amyloidosis. Clinicopathologic findings in 15 patients, Am. J. Cardiol. **26**:394, 1970.
4. Buja, L. M., and Roberts, W. C.: Iron in the heart. Etiology and clinical significance, Am. J. Med. **51**:209, 1971.
5. Bulkley, B. H., and Roberts, W. C.: Dilatation of the mitral anulus. A rare cause of mitral regurgitation, Am. J. Med. **59**:457, 1975.
6. Bulkley, B. H., and Roberts, W. C.: The heart in systemic lupus erythematosus and the changes induced in it by corticosteroid therapy. A study of 36 necropsy patients, Am. J. Med. **58**:243, 1975.
7. Epstein, S. E., Henry, W. L., Clark, C. E., Roberts, W. C., Maron, P. I., Ferrans, V. I., Redwood, D. R., and Morrow, A. G.: Asymmetric septal hypertrophy, Ann. Intern. Med. **81**:650, 1974.
8. Ferrans, V. J., Massumi, R. A., Shugoll, G. I., Aki, N., and Roberts, W. C.: Ultrastructural studies of myocardial biopsies in 45 patients with obstructive or congestive cardiomyopathy. In Bajusz, E., Rona, G., with Brink, A. J., and Lochner, A., editors: Recent advances in studies on cardiac structure and metabolism. Vol. 2, cardiomyopathies, Baltimore, 1973, University Park Press, pp. 231-272.
9. Ferrans, V. J., Morrow, A. G., and Roberts, W. C.: Myocardial ultrastructure in idiopathic hypertrophic subaortic stenosis. A study of operatively excised left ventricular outflow tract muscle in 14 patients, Circulation **45**:769, 1972.
10. Henry, W. L., Clark, C. E., Roberts, W. C., Morrow, A. G., and Epstein, S. E.: Differences in distribution of myocardial abnormalities in patients with obstructive and non-obstructive asymmetric septal hypertrophy (ASH). Echocardiographic and gross anatomic findings, Circulation **50**:447, 1974.
11. Jaski, B. E., Goetzl, E. J., Said, J. W., and Fishbein, M. C.: Endomyocardial disease and eosinophilia. Report of a case, Circulation **57**:824, 1978.
12. Maron, B. J., Clark, C. E., Henry, W. L., Fukuda, T., Edwards, J. E., Mathews, E. C., Redwood, D. R., and Epstein, S. E.: Prevalence and characteristics of disproportionate ventricular septal thickening in patients with acquired or congenital heart disease: echocardiographic and morphologic findings, Circulation **55**:489, 1977.
13. Maron, B. J., Edwards, J. E., and Epstein, S. E.: Disproportionate ventricular septal thickening in patients with systemic hypertension, Chest **73**:466, 1978.
14. Maron, B. J., Edwards, J. E., Ferrans, V. J., Clark, C. E., Lebowitz, E. A., Henry, W. L., and Epstein, S. E.: Congenital heart malformations associated with disproportionate ventricular septal thickening, Circulation **52**:926, 1975.
15. Maron, B. J., Ferrans, V. J., Henry, W. L., Clark, C. E., Redwood, D. R., Roberts, W. C., Morrow, A. G., and Epstein, S. E.: Differences in distribution of myocardial abnormalities in patients with obstructive and nonobstructive asymmetric septal hypertrophy (ASH). Light and electron microscopic findings, Circulation **50**:436, 1974.
16. Maron, B. J., Gottdiener, J. S., Roberts, W. C., Henry, W. L., Savage, D. D., and Epstein, S. E.: Left ventricular outflow tract obstruction due to systolic anterior motion of the anterior mitral leaflet in patients with concentric left ventricular hypertrophy, Circulation **57**:527, 1978.
17. Maron, B. J., Henry, W. L., Roberts, W. C., and Epstein, S. E.: Comparison of echocardiographic and necropsy measurements of ventricular wall thicknesses in patients with and without dispro-

portionate septal thickening, Circulation **55**:341, 1977.

18. Maron, B. J., Redwood, D. R., Roberts, W. C., Henry, W. L., Morrow, A. G., and Epstein, S. E.: Tunnel subaortic stenosis. Left ventricular outflow tract obstruction produced by fibromuscular tubular narrowing, Circulation **54**:404, 1976.
19. Maron, B. J., and Roberts, W. C.: Quantitative analysis of cardiac muscle cell disorganization in the ventricular septum of patients with hypertrophic cardiomyopathy, Circulation **59**:689, 1979.
20. McReynolds, R. A., and Roberts, W. C.: The intramural coronary arteries in hypertrophic cardiomyopathy, abstracted, Am. J. Cardiol. **35**:120, 1975.
21. Merrill, W. H., Henry, W. L., Maron, B. J., Epstein, S. E., and Morrow, A. G.: Echocardiographic assessment of the left ventricle following myotomy and myectomy for idiopathic hypertrophic subaortic stenosis (IHSS), abstracted, Am. J. Cardiol. **41**:435, 1978.
22. Muna, W. F. T., Ferrans, V. J., Pierce, J. E., and Roberts, W. C.: Discrete subaortic stenosis in Newfoundland dogs: association of infective endocarditis, Am. J. Cardiol. **41**:746, 1978.
23. Roberts, W. C.: Operative treatment of hypertrophic obstructive cardiomyopathy. The case against mitral valve replacement, Am. J. Cardiol. **32**:377, 1973.
24. Roberts, W. C.: Coronary heart disease. A review of abnormalities observed in the coronary arteries, Cardiovasc. Med. **2**:29, 1977.
25. Roberts, W. C., and Buja, L. M.: The frequency and significance of coronary arterial thrombi and other observations in fatal acute myocardial infarction. A study of 107 necropsy patients, Am. J. Med. **52**:425, 1972.
26. Roberts, W. C., and Ferrans, V. J.: Pathologic anatomy of the cardiomyopathies. Idiopathic dilated and hypertrophic types, infiltrative types, and endomyocardial disease with and without eosinophilia, Hum. Pathol. **6**:287, 1975.
27. Roberts, W. C., Kehoe, J. A., Carpenter, D. F., and Golden, A.: Cardiac valvular lesions in rheumatoid arthritis, Arch. Intern. Med. **122**:141, 1968.
28. Roberts, W. C., McAllister, H. A., and Ferrans, V. J.: Sarcoidosis of the heart. A clinicopathologic study of 35 necropsy patients (group I) and review of 78 previously described necropsy patients (group II), Am. J. Med. **63**:86, 1977.
29. Roberts, W. C., Perry, L. W., Chandra, R. S., Myers, G. E., Shapiro, S. R., and Scott, L. P.: Aortic valve atresia: a new classification based on necropsy study of 73 cases, Am. J. Cardiol. **37**:753, 1976.
30. Roberts, W. C., and Virmani, R.: Aschoff bodies at necropsy in valvular heart disease. Evidence from an analysis of 543 patients over 14 years of age that rheumatic heart disease, at least anatomically, is a disease of the mitral valve, Circulation **57**:803, 1978.
31. Spray, T. L., Maron, B. J., Morrow, A. G., Epstein, S. E., and Roberts, W. C.: A discussion on hypertrophic cardiomyopathy, Am. Heart J. **95**:511, 1978.
32. Teare, D.: Asymmetrical hypertrophy of the heart in young adults, Br. Heart J. **20**:1, 1958.
33. Virmani, R., and Roberts, W. C.: Aschoff bodies in operatively excised atrial appendages and in papillary muscles. Frequency and clinical significance, Circulation **55**:559, 1977.

18 □ Radionuclide techniques in cardiomyopathy

Gerald M. Pohost
John T. Fallon
H. William Strauss

Myocardial disease may be classified according to whether its cause is *secondary* to a systemic disease process (e.g., sarcoidosis) or to extramyocardial cardiac disease (e.g., coronary artery disease) or *primary,* in which case direct involvement of the myocardium by a pathologic process leads to dysfunction (e.g., alcoholic cardiomyopathy)[27] (Table 18-1). Radionuclide methods are helpful in determining both the physiology and etiology of a variety of disorders affecting the myocardium.

EVALUATION OF CARDIOMYOPATHIES

Radionuclide techniques of value in the diagnostic evaluation of patients with cardiomyopathy include the gated cardiac blood pool scan, the first-pass radionuclide angiogram, and myocardial imaging with ^{201}Tl. The gated or multiple-gated cardiac blood pool scan is performed after the intravenous injection of 15 to 20 mCi of ^{99m}Tc-labeled human serum albumin or red blood cells to evaluate both chamber size and function. Multiple images are usually recorded in synchrony with the patient's heartbeat during the entire cardiac cycle. From these images, left ventricular ejection fraction and regional wall motion can be reliably determined.[49]

After intravenous administration of 1.5 to 2.0 mCi, the initial distribution of ^{201}Tl in the heart is related to the blood flow in the myocardium.[48] Frequently, the ^{201}Tl scan reveals a homogeneous distribution in patients with myocardial disease, suggesting diffuse rather than focal involvement. Focal regions of scar tissue or infiltrate appear as discrete zones of decreased tracer concentration, that is, cold spots on gamma camera images.[54] For example, myocardial involvement with sarcoid may produce multifocal regions of reduced ^{201}Tl activity corresponding to regions of granuloma or scar tissue.[11]

By depicting the cardiac chamber size and function and the configuration and thickness of the interventricular septum, the gated blood pool scan or the radionuclide angiogram and ^{201}Tl myocardial scan suggest the functional classification of the patient's heart muscle disease (Fig. 18-1). After the functional class is suggested, a number of etiologic possibilities should be considered (Table 18-1).

Congestive, or dilated, cardiomyopathy

Clinical findings of right and left heart failure are characteristic of congestive cardiomyopathy (CCM).[41] In general, the pathologic findings are nonspecific; heart weight is increased, although ventricular dilation in excess of myocardial hypertrophy is often apparent; apical thrombi are common in both ventricles, and the coronary arteries demonstrate no significant disease. Microscopic examination usually reveals diffuse interstitial fibrosis as well as foci of replace-

Table 18-1. Classification of most common disorders affecting heart muscle

Myocardial involvement (cause)	Functional class			
	Congestive or dilated cardiomyopathy (CCM)	Restrictive cardiomyopathy (RCM)		Hypertrophic cardiomyopathy (HCM)
		Infiltrative	Obliterative	
Primary				
Idiopathic	Idiopathic CCM Postpartum cardiomyopathy Endocardial fibroelastosis (dilated type)		Endocardial fibroelastosis Endomyocardial fibrosis Davies' disease Löffler's disease	Obstructive HCM Nonobstructive HCM
Toxic	Alcohol Heavy metals (e.g., arsenic, cobalt) Bleomycin Diphtheria			
Inflammatory	Chagas' disease Rheumatic myocarditis			
Secondary (associated with systemic disease)				
Inflammatory	Infective Viral Bacterial Mycoplasmal Rickettsial (e.g., Q fever)			
	Noninfective			
	Lupus erythematosus	Scleroderma		
	Sarcoidosis Polyarteritis nodosa			
Metabolic	Thiamin deficiency Acromegaly	Amyloidosis Hemochromatosis Glycogen storage disease		
Neuromuscular	Friedreich's ataxia Muscular dystrophy			
Secondary (associated with cardiac disease)				
Valvular Congenital Hypertensive	Chronic volume and/or pressure overload		Endocardial fibroelastosis	
Coronary	Ischemic cardiomyopathy			
Pericardial		Constrictive RCM		

ment fibrosis. Alcoholic cardiomyopathy[53] and postpartum cardiomyopathy[13] are examples of CCM. Other examples of CCM are given in Table 18-1.

An example of a gated cardiac blood pool scan from a patient with CCM is illustrated in Fig. 18-2. Patients with CCM demonstrate biventricular dilation and dysfunction. The left ventricle is usually more severely dilated than the right, and the left ventricular ejection fraction is severely reduced (typically to less than 30%). Right ventricular dysfunction is related to both myocardial involvement and increased pulmonary artery pressure due to left ventricular failure. Wall motion is concentrically reduced, although the apical segment may be akinetic or dyskinetic. The anterobasal segment and the basal septum demonstrate reduced wall motion in CCM, whereas these segments frequently move normally in ischemic heart disease. Bulkley and asso-

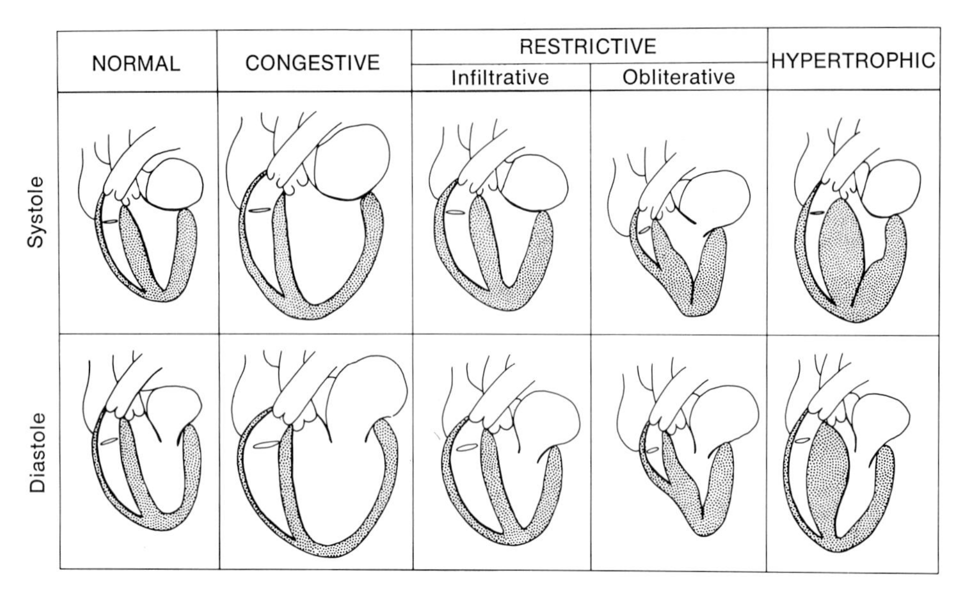

Fig. 18-1. Schematic illustration of functional classifications of cardiomyopathies in view equivalent to 40- to 50-degree LAO radionuclide images in end of systole and end of diastole. Size and contractile function of ventricles and configuration of chambers and interventricular septum provide characteristics that help define functional classes. (From Pohost, G. M., Vignola, P. A., McKusick, K. A., Block, P. C., Meyers, G. S., Walker, H. J., Copen, D. L., and Dinsmore, R. E.: Circulation **55**:92, 1977.)

ciates[9] have suggested that the percentage of the circumference involved by akinesis or dyskinesis is of diagnostic value in differentiating between CCM and ischemic heart disease. Greater than 40% circumferential involvement in any one oblique view was identified in 84% of the patients with ischemic cardiomyopathy, whereas it was found in only 25% of the patients with CCM. Occasionally, akinesis with flattening of the left ventricular apical segment can be demonstrated and is suggestive of a mural thrombus.

The left atrium is usually difficult to define on a normal gated blood pool study unless the left posterior oblique view is employed. However, biatrial dilation in CCM frequently permits ready visualization of the left atrial appendage in the 40- and 50-degree LAO projections and the right atrium in the anterior projection.

Myocardial imaging with ^{201}Tl usually demonstrates left ventricular dilation and either homogeneous or diffusely inhomogeneous uptake (Fig. 18-2). Severe left ventricular dysfunction resulting from coronary artery disease and multiple myocardial infarctions but without obvious ventricular aneurysm has been called "ischemic cardiomyopathy" and may have a clinical picture indistinguishable from CCM. Bulkley and co-workers have reported the utility of ^{201}Tl myocardial imaging and gated cardiac blood pool scanning in distinguishing between ischemic cardiomyopathy and CCM.[9] All patients with angiographically and/or autopsy-documented ischemic cardiomyopathy had a defect on the ^{201}Tl scan occupying over 40% of the circumference of the left ventricular image in any one projection. On the other hand, all but one patient with idiopathic CCM demonstrated a ^{201}Tl defect of under 20% of the circumference in any one projection. ^{201}Tl images from a patient with CCM are illustrated in Fig. 18-3 and those from a patient with ischemic cardiomyopathy in Fig. 18-4.

Serial evaluation of global ventricular volumes and ejection fraction using radionuclide angiography or gated blood pool scanning provide objective methods of evaluating changes in ventricular function

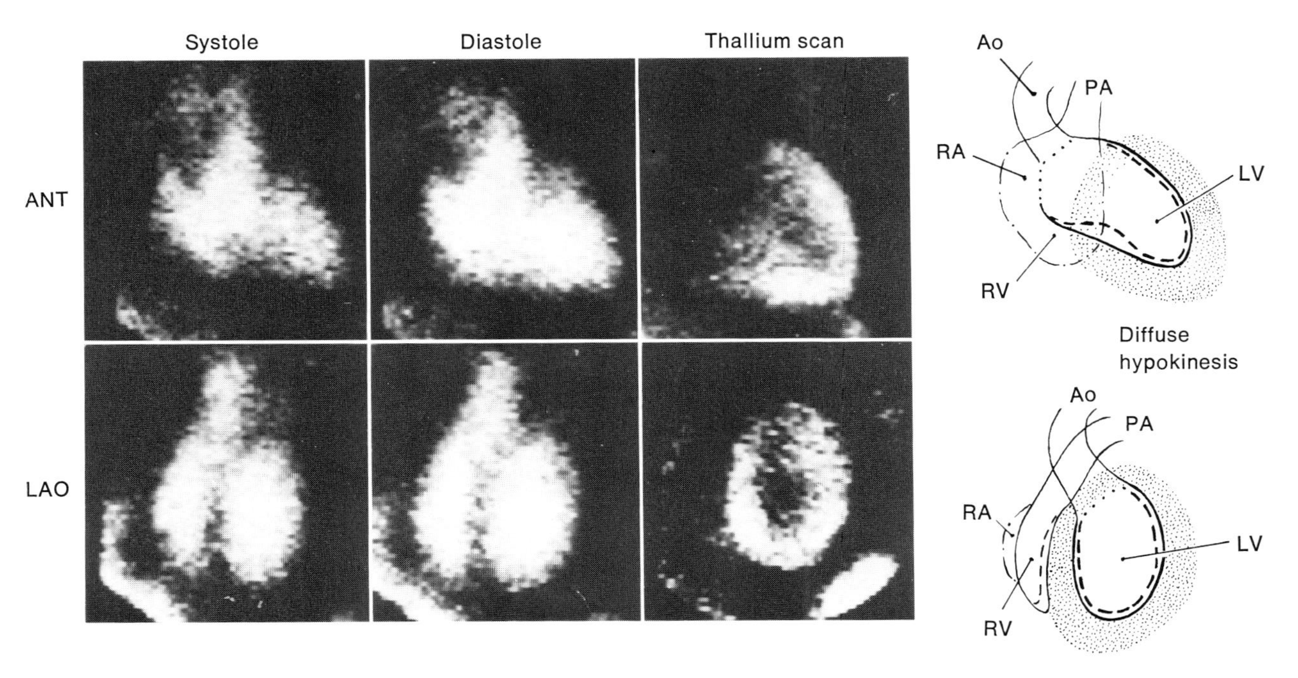

Fig. 18-2. ^{201}Tl scan reveals prominent left ventricular cavity. Tracer concentration in myocardium is uniform, and right ventricular myocardium is not well defined. Gated scan reveals increased end-diastolic volume and diffuse left ventricular hypokinesis. This combination of findings is suggestive of primary congestive cardiomyopathy. (From Strauss, W. H., Pitt, B., Rouleau, J., Bailey, I. K., and Wagner, H. N., editors: Atlas of cardiovascular nuclear medicine, St. Louis, 1977, The C. V. Mosby Co.)

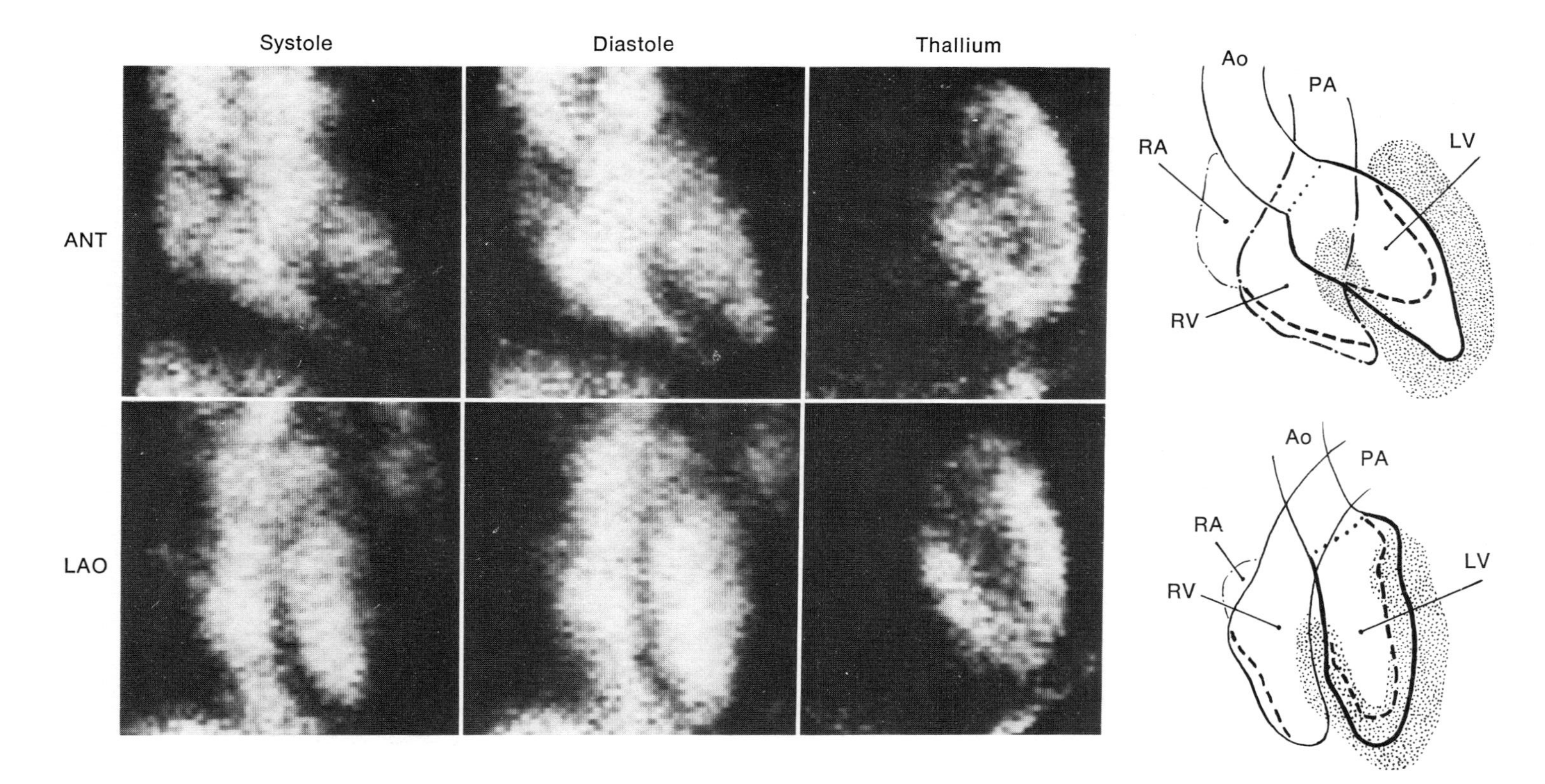

Fig. 18-3. Idiopathic congestive cardiomyopathy. Uniform ^{201}Tl uptake suggests that this patient's left ventricular dysfunction is not likely to be caused by previous myocardial .nfarction but rather by CCM. To account for left ventricular dysfunction, ^{201}Tl images in patients with infarction should show a large perfusion deficit. This patient's symptoms were attributed to idiopathic congestive cardiomyopathy. (From Strauss, W. H., Pitt, B., Rouleau, J., Bailey, I. K., and Wagner, H. N., editors: Atlas of cardiovascular nuclear medicine, St. Louis, 1977, The C. V. Mosby Co.)

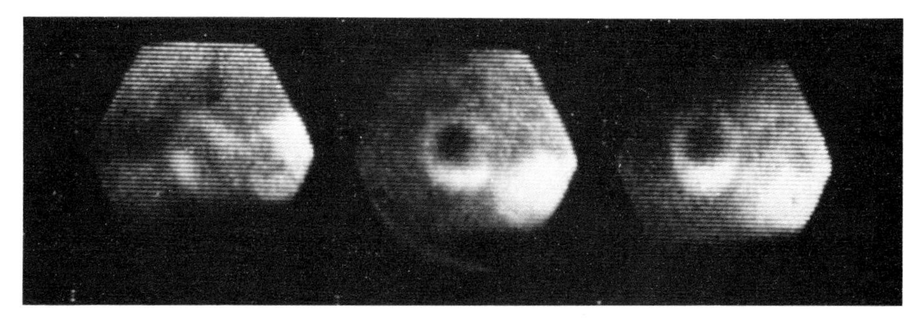

Fig. 18-4. Anterior position *(left)*, 45-degree LAO position *(middle)*, and 70-degree LAO *(right)*. Scan reveals marked decrease of ^{201}Tl concentration in distal anterior wall, apex, and septum. In addition, there is increased ^{201}Tl concentration in patient's lungs, suggesting congestive cardiac failure.

brought about by therapy. For example, initial therapy in a patient with alcoholic cardiomyopathy would typically consist of withdrawal from alcohol intake and administration of digitalis and diuretics. If there is no significant improvement in the clinical state, radionuclide ejection fraction, or ventricular size, additional therapy such as afterload reduction and/or long-term bed rest would be considered. We have observed a decrease in the size and an improvement in the ejection fraction of both ventricles in several patients with alcoholic CCM after two or more months of abstinence from alcohol.

Restrictive cardiomyopathy

Restrictive cardiomyopathy (RCM), the least common class of cardiomyopathy outside of the tropics, shows findings that simulate constrictive pericardial disease and normal-sized or small ventricles. This form of myocardial disease results in diminished ventricular compliance either due to an infiltrative process or to fibrosis. Physical findings include pulsus paradoxus, elevated jugular venous pressure with deep X and Y descents, and an increase in the jugular venous pressure with inspiration (Kussmaul's sign). Although congestive or dilated cardiomyopathy may have stigmata resembling constrictive disease, to avoid confusion we will also include in our definition of RCM that the ventricular end-diastolic cavity dimension or size, as it can be defined by gated blood pool scan or echocardiogram, is normal or reduced. RCM, so defined, may result from a variety of pathologic disorders, which can be subdivided into two basic types: (1) infiltrative RCM, in which the left or both ventricular cavities are normal to decreased in size and the walls are normal to increased in thickness, and (2) obliterative RCM, in which the involved ventricle(s) is (are) reduced in size and there is partial obliteration of one or both of the ventricular cavities. Fibrosis of the myocardium and endocardium usually leads to increased wall thickness in these disorders as well. In some functional classification schemes, obliterative cardiomyopathy has been employed as a separate category,[20] whereas other schemes have included both infiltrative and obliterative diseases under the single heading of restrictive cardiomyopathy.[36] Amyloid infiltration, which is the most common cause of infiltrative RCM in nontropical regions, leads to a waxy, stiff, and rubbery-appearing myocardium.[8] Other causes of myocardial infiltration, which may lead to RCM, include metastatic carcinoma,[19,24] lymphoma,[42] or invasion with leukemic cells.[40] Generally, infiltration involves both right and left ventricles to varying degrees, and thus patients with infiltrative RCM usually have clinical evidence of biventricular failure, although typically signs of right ventricular failure predominate.

Progressive fibrosis that involves the endocardium and myocardium, leading to compromise of the size of one or both ventricles, is characteristic of the obliterative form of RCM. In tropical zones, including eastern and western Africa, Indonesia, Ceylon, and Brazil, this process is known as endomyocardial fibrosis (EMF).[37] Obliterative cardiomyopathy occuring in temperate regions is rare and is known as Löffler's fibroplastic eosinophilic endocarditis.[29] The clinical picture in obliterative RCM is similar to that of infiltrative RCM.

Differentiation between constrictive pericarditis and RCM is frequently difficult on clinical grounds. Radionuclide imaging may be of some value. As illustrated in Fig. 18-5, in constrictive pericarditis, both ventricles usually demonstrate normal systolic function, whereas in RCM, biventricular function is usually reduced, although typically not as severely as in CCM (i.e., the ejection fraction generally exceeds 30%). In obliterative disease, the unusual configuration of the ventricles suggests the underlying pathologic process. Myocardial imaging with ^{201}Tl may be helpful in depicting increased mural thickness, as with amyloid disease, or focal myocardial involvement, as with metastatic disease.

However, constrictive pericardial disease can involve the myocardium and lead to depressed ventricular function, and conversely, early RCM with stigmata of constrictive disease diagnosed by physical examination may display normal or only mildly reduced ventricular function. If the clinical, noninvasive, and catheterization data cannot clearly distinguish between constrictive pericardial disease (potentially surgically curable) and RCM, myocardial biopsy or pericardiectomy may be necessary to ultimately resolve the issue.

The gated blood pool scan is useful in distinguishing between congestive, or dilated, cardiomyopathy and RCM by depicting the sizes of the cardiac chambers. As noted earlier, in CCM all four chambers are dilated, whereas in RCM the ventricles are normal to reduced in size while the atria are usually considerably dilated. In addition, the gated cardiac blood pool scan may be helpful in differentiating between the infiltrative and obliterative forms of RCM. In the infiltrative type, for example, amyloidosis, both ventricles are usually normal or reduced in size. By way of contrast, in the

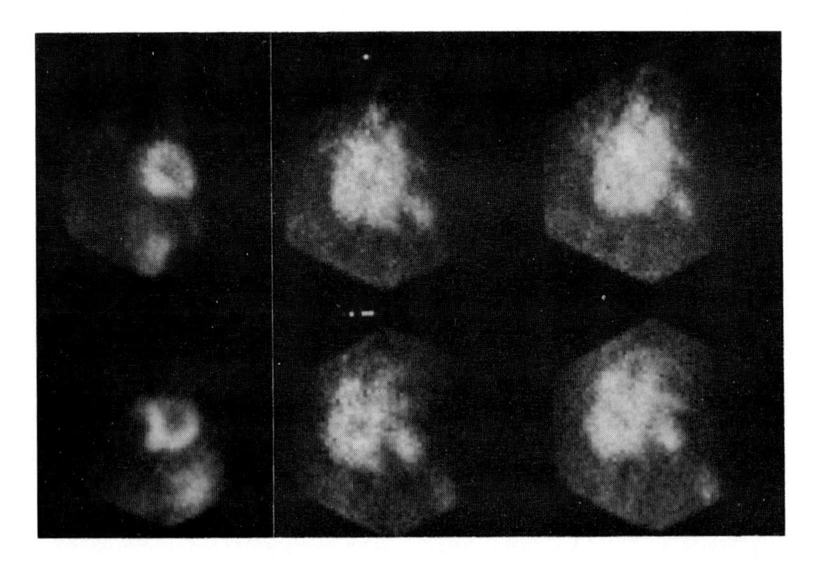

Fig. 18-5. ^{201}Tl scan *(left)* and gated blood pool scan at end of diastole *(center)* and end of systole *(right)* recorded in anterior position *(top)* and 45-degree LAO *(bottom)* from patient with infiltrative myopathy associated with marked eosinophilia. ^{201}Tl scan reveals marked increase in width of septum and of inferior wall, especially at base of left ventricle. Blood pool images reveal marked increase in myocardial wall thickness in these areas. Right atrium is enlarged and noncontractile—raising the likelihood of increased venous pressure (manifested clinically by distended neck veins).

obliterative type, either or both ventricles are generally smaller than normal. The right ventricle is frequently diminutive in EMF. ^{201}Tl imaging is of value when neoplastic involvement of the myocardium with solid tumors is a possibility, since the ^{201}Tl image may reveal multiple large defects that are present at rest or with exercise. In contrast, leukemic infiltration is usually homogeneous, and the myocardial images display uniform ^{201}Tl activity. The treatment of malignant disease with chemotherapeutic agents such as doxorubicin (Adriamycin) may result in heart failure due to myocardial toxicity. The radionuclide angiogram or gated blood pool scan shows this toxic cardiomyopathy to be of the congestive or dilated type, and ^{201}Tl distribution is homogeneous, helping to differentiate it from the RCM due to direct neoplastic involvement with solid tumors. Repetitive measurements of ventricular function, started before therapy and repeated at intervals during therapy, may permit the detection of incipient cardiac failure at a time when drug withdrawal permits a recovery of ventricular function.

Hypertrophic cardiomyopathy

Hypertrophic cardiomyopathy (HCM) is characterized by left ventricular hypertrophy of an unknown cause. Hypertrophy is generally asymmetric, with the interventricular septum disproportionately thickened in comparison to the remainder of the ventricle.[1,17,25,52] HCM may be obstructive or nonobstructive. Left ventricular outflow obstruction is caused by hypertrophy of the anterior superior aspect of the interventricular septum and abnormal motion of the anterior leaflet of the mitral valve.[5,26,55] Before echocardiography was developed, the detection of HCM largely depended on the presence of an obstruction with the resulting physical findings (e.g., bisferiens pulse and systolic ejection murmur, which increased with Valsalva's maneuver and decreased with squatting).[5,35] Early names for the disease mentioned the obstructive component. Braunwald and colleagues[6] coined the term idiopathic hypertrophic subaortic stenosis (IHSS), whereas Brent and coworkers[7] used the term muscular subaortic stenosis, both emphasizing obstruction to left ventricular outflow. On the other hand, Goodwin and asssociates used the term hypertrophic obstructive cardiomyopathy (HOCM), emphasizing the myocardial disease.[22]

Noninvasive imaging techniques now permit the diagnosis of HCM with or without obstruction. Asymmetric septal hypertrophy (ASH) is easily documented by echocardiography and has been employed as a marker of HCM[25]; however, it is not specific. For example, right ventricular hypertrophy may also lead to ASH.[51] Abnormal systolic anterior motion (SAM) of the anterior leaflet of the mitral valve, which can be documented echocardiographically and angiographically, has been thought to be highly specific for HCM with obstruction to left ventricular outflow,[46] although recent reports suggest that SAM is not as specific as it was once considered.[31,34]

Both the gated cardiac blood pool scan (Fig. 18-6) and myocardial imaging with ^{201}Tl (Fig. 18-7) are helpful alternatives to the echocardiogram in the detection of HCM.[10,38] The end-diastolic image of the gated blood pool scan in the LAO view allows evaluation of the interventricular septal configuration. The appearance of the interventricular septum by gated scan not only depends on septal configuration but is also affected by the angle of view. The normal interventricular septum is curvilinear, with its concavity facing the left ventricle. In a shallow LAO view, the apical septum appears thickest, whereas in a high oblique view, the upper septum appears thickest.[38] This phenomenon occurs because of the spiral geometry of the interventricular septum. The view in which the septum appears most uniform from top to bottom (usually the 40- to 50-degree LAO) is selected for evaluation of configuration.

In HCM, the interventricular septum has an abnormal configuration, which can be demonstrated by gated scan when evaluated as just described. In our experience,

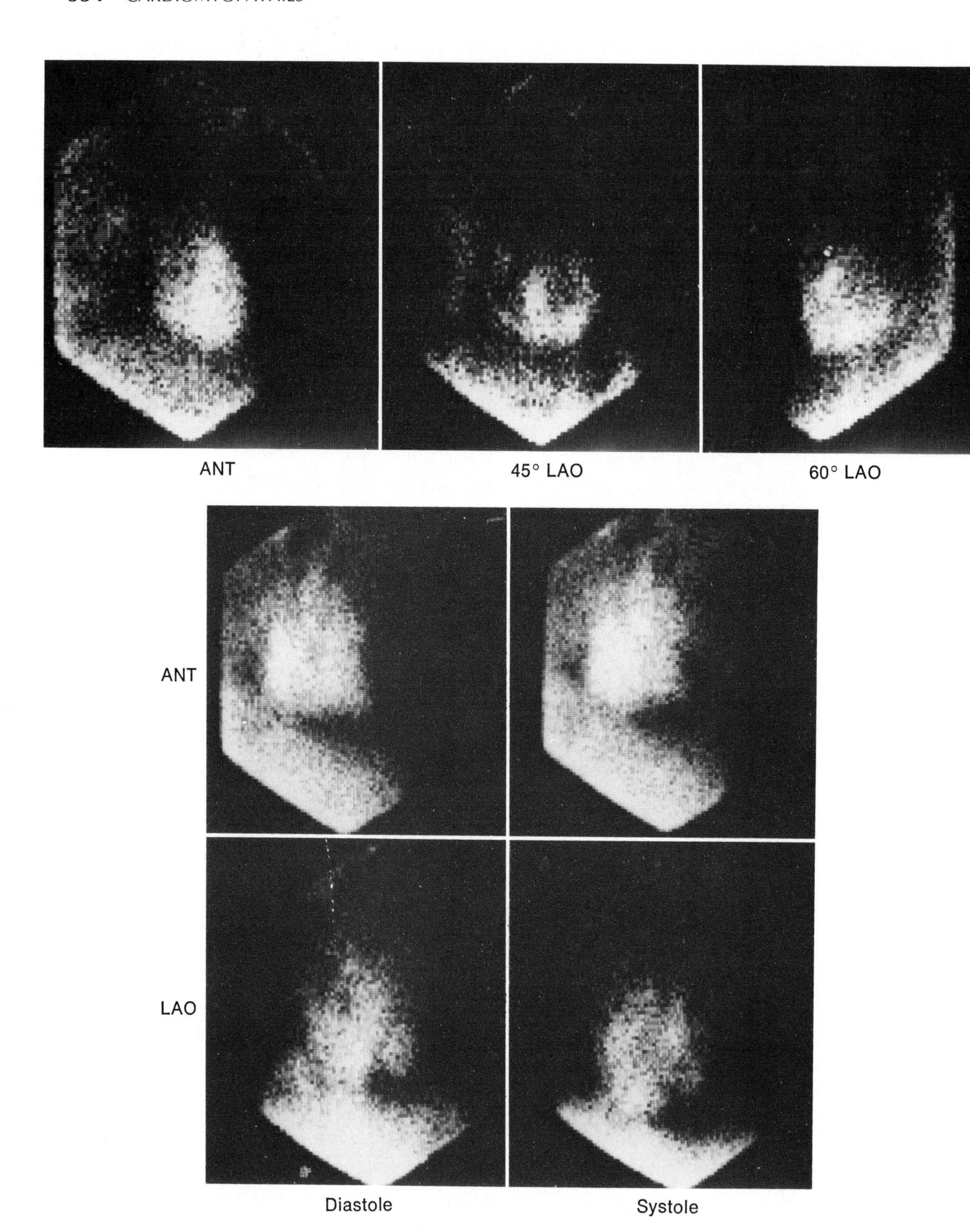

Fig. 18-6. Hypertrophic cardiomyopathy. In patients with hypertrophic cardiomyopathies such as IHSS **(A)** septum appears quite thickened. This may cause anterior view to appear very unusual, with septum appearing as stripe that covers entire left ventricular cavity. LAO position usually reveals septum to have marked increase in its basal dimension compared to that in upper portion. Gated scans on these patients have very characteristic appearance with marked separation between right and left ventricles best defined in LAO view **(B).** (From Strauss, W. H., Pitt, B., Rouleau, J., Bailey, I. K., and Wagner, H. N.: Atlas of cardiovascular nuclear medicine, St. Louis, 1977, The C. V. Mosby Co.)

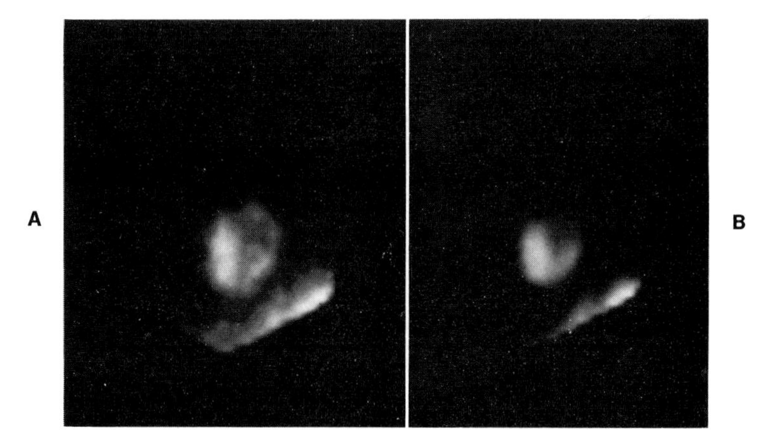

Fig. 18-7. ^{201}Tl scans performed in the LAO position, gated to depict end of diastole **(A)** and ungated **(B)**. On gated scan, septum appears thicker than posterior wall and straight in configuration. On ungated scan, septum appears more triangular in shape. Gating of ^{201}Tl images may be helpful in obtaining better definition of ASH and configuration of interventricular septum.

50% of the patients with HCM demonstrate disproportionate upper septal thickening (DUST) by gated blood pool scan.[38] The middle or upper septum appears thicker than the lower septum in any LAO projection. In addition, the majority of patients with HCM (73%) demonstrate loss of the normal concavity on the left ventricular aspect of the interventricular septum. Although this flattening of the septum is a frequent finding in HCM, it is less specific than DUST and can be observed in some patients with aortic stenosis as well as in some normal patients. The gated scan may also demonstrate left ventricular systolic cavity or apex obliteration and a circular defect in the left ventricular outflow, which is most likely related to the displacement of blood pool activity by the thickened upper septum. These abnormalities are best seen in the anterior or 30-degree RAO projection.

Bulkley and co-workers employed ^{201}Tl myocardial imaging to evaluate the ratio of septal to posterior wall thicknesses in patients with HCM.[10] As anticipated from echocardiographic studies, ASH was noted on the ^{201}Tl scan in all ten patients with HCM and confirmed in nine by cardiac catheterization. In addition, it was noted that the basal and midposterior walls were of equal thickness in patients with obstructive HCM, but that the basal was thinner than the midposterior wall in patients with nonobstructive HCM. Three patients with chronic pulmonary hypertension had ASH by ^{201}Tl scan, but in these, the right ventricular free wall thickness was equal to the septal thickness, which is consistent with right ventricular hypertrophy. The configuration of the interventricular septum on the ^{201}Tl scan was noted to be triangular with the maximal thickness inferoapically in contrast to DUST noted at end of diastole on gated blood pool studies. Since electrocardiographic gating was not employed with ^{201}Tl imaging, the images obtained represent a composite view of the beating heart over multiple cardiac cycles. As a result, these images include end of systole, when the muscle is thicker and the count density higher. At end of systole, obliteration of the left ventricular apex may give the erroneous impression that the apical septum is thickest (Fig. 18-7). Gated ^{201}Tl scanning demonstrates that the end-diastolic septal configuration in HCM is similar to that noted on gated blood pool studies, that is, the septum is flattened and occasionally shows DUST (Fig. 18-7). The inferoapical septal thickening, present at end of systole, disappears at end of diastole. ASH is clearly defined on the end-diastolic image.

Myocardial disease secondary to chronic volume or pressure overload

Myocardial dysfunction may result from valvular or congenital heart disease as a consequence of chronic volume or pressure overload.[44] Although the clinical presentation in patients with heart muscle dysfunction secondary to volume or pressure overload is usually dominated by the primary process, the myocardial pathology in the late stage may closely resemble that of CCM.[30]

Volume overload. In the majority of patients with chronic left ventricular volume overload, such as that due to aortic or mitral regurgitation, end-diastolic volume is increased in proportion to stroke volume with maintenance of a normal ejection fraction[14-16,39] (Fig. 18-2). However, chronic increase in stroke volume and concomitant dilation may lead to myofibrillar damage.[44] Similarly, right ventricular dysfunction may result from prolonged right ventricular volume overload, as most commonly observed in the presence of atrial septal defects. Left ventricular volume overload states may lead to dilation and reduction in ejection fraction,[18,33,47] which may be difficult to distinguish from CCM. Preoperative ventricular function is an important determinant of potential surgical benefit.[2,33,45,47] Radionuclide techniques provide a means of determining the ventricular function by quantification of ejection fraction and ejection velocity.[12,50] In addition, these methods allow repeated measurement of ventricular size and function and thus a means of following patients with volume overload, which might be helpful in deciding when to operate, for example, if ejection fraction begins to fall. Hammermeister and colleagues[23] recently observed that survival in chronic mitral valve disease is related to end-diastolic volume and ejection fraction. Furthermore, patients with a moderately reduced ejection fraction demonstrated the most significant benefit from surgery, compared to their medically treated cohorts. These data suggest that serial noninvasive evaluation of ejection fraction and end-diastolic volume by radionuclide techniques may be useful in determining when surgery is indicated in patients with mitral valve disease. Multiple-gated acquisition scanning affords the opportunity to determine the ejection velocity,[12] which has been suggested to be a sensitive marker of myocardial dysfunction in volume overload states.[16]

Borer and associates have recently reported their experience with multiple-gated blood pool imaging at rest and with supine exercise in a series of patients with aortic regurgitation.[3] These investigators found that many asymptomatic patients with aortic regurgitation develop a fall in ejection fraction with exercise, whereas normal subjects uniformly increase their ejection fraction. Although exercise continued to induce a fall in ejection fraction postoperatively, this fall was significantly less than that documented preoperatively. The magnitude of the fall in ejection fraction occurring with exercise may provide a mechanism of identifying impending left ventricular failure and thus allow a more sensitive means of determining when surgery should be employed than studies performed only in the resting state.

Pressure overload. Pressure overload due to chronic increase in ventricular systolic pressure usually leads to secondary hypertrophy with maintenance of a normal ejection fraction and end-diastolic volume.[28] Myocardial failure is associated with decreasing stroke volume and ejection fraction and some dilation. Left ventricular failure may complicate the clinical picture in advanced aortic stenosis.[15] Pressure overload in the left ventricle due to systemic hypertension and in the right ventricle due to pulmonary hypertension (primary or secondary) or pulmonic stenosis may similarly result in failure of the respective ventricle.

Concentric hypertrophy and normal ventricular ejection function are characteristic of uncomplicated pressure overload states. In an appropriate LAO projection (40 to 60 degrees), the interventricular septum can be evaluated, and if hypertrophy is moderate to marked, septal thickening may be suggested by the end-diastolic gated blood pool or the ^{201}Tl image. As mentioned earlier, the interventricular septal configuration can be evaluated by gated blood pool scan; in concentric hypertrophy due to chronically increased afterload, the septum is concave, with its concavity facing the left ventricle. Radionuclide scanning will also demonstrate ventricular cavity size and function. In chronic severe aortic stenosis, ventricular dilation and dysfunction may occur[28] and are associated with a poorer prognosis. Late in the course of aortic stenosis, this dilation and reduced ejection fraction may resemble congestive cardiomyopathy on a gated scan. At rest, patients with compensated aortic stenosis usually have a supernormal ejection fraction.[4] In a group of patients with normal resting ventricular function, Borer found a fall in ejection fraction[4] with exercise-gated scanning in the majority. The magnitude of the fall in ejection fraction with exercise was speculated to be related to the surgical prognosis. This concept is provocative, but more data are needed before the utility of stress-gated blood pool imaging for prognostication in valvular heart disease can be defined.

Ischemic heart disease

The clinical presentation of patients with ventricular failure related to coronary artery disease generally includes a history of angina pectoris and/or myocardial infarction. Myocardial damage occurs as a consequence of reduced coronary blood flow, leading to ischemia and infarction. As a result, large regions of myocardium are replaced by scar tissue. Chronic myocardial failure resulting from coronary artery disease is at times clinically indistinguishable from CCM. As has been discussed, both gated blood pool and myocardial imaging (Fig. 18-4) are helpful in differentiating between ischemic cardiomyopathy and CCM.[9]

SUMMARY

Radionuclide techniques provide information helpful in determining the diagnosis, prognosis, and therapeutic response in patients with myocardial disease. Gated blood pool scanning and radionuclide angiography permit evaluation of ejection fraction, ventricular size and wall motion, and interventricular septal thickness and configuration. With this information, the functional class of the cardiomyopathy can be established (congestive, restrictive [infiltrative or obliterative], or hypertrophic) and a number of etiologic possibilities suggested (Table 18-1). Nonmyocardial forms of heart disease, which secondarily involve the myocardium, can simulate a primary cardiomyopathy. The diagnosis can frequently be clarified by radionuclide techniques. In valvular heart disease, exercise-gated blood pool imaging is a provocative new modality that may be of value in detecting subclinical myocardial dysfunction and in determining prognosis. Myocardial imaging with ^{201}Tl is useful in detecting disease that replaces (e.g., fibrosis) or displaces (e.g., metastatic neoplasms) myocardium. It is of particular value in differentiating myocardial dysfunction resulting from coronary artery disease from that caused by congestive cardiomyopathy and in depicting septal abnormalities in hypertrophic cardiomyopathy.

Finally, although these radionuclide techniques can be quite helpful in the evaluation of patients with myocardial disease, application of other diagnostic modalities should be considered initially. These include electrocardiography, chest roentgenography, fluoroscopy, phonocardiography and other graphics techniques, and echocardiography. In the final analysis, the information provided by the appropriate noninvasive technique might establish the diagnosis and obviate the need for cardiac catheterization, angiography, and myocardial biopsy. If invasive studies are required to establish the diagnosis of a potentially treatable disease, serial imaging studies may be useful in following the therapeutic effect and thereby helping to delineate the appropriate therapeutic program.

ACKNOWLEDGMENTS

We wish to express our appreciation to Dr. Robert E. Dinsmore, Dr. Allen B. Nichols, Dr.

Andrew Hoffman, Dr. Charles Homcy, and Dr. Henry Gewirtz for their supply of clinical material, critical review, and constructive recommendations and to Ms. Eileen Fitzgerald for valuable help in the preparation of the manuscript.

REFERENCES

1. Abbasi, A. S., MacAlpin, R. N., Eber, L. M., and Pearce, M. L.: Endocardiographic diagnosis of idiopathic hypertrophic cardiomyopathy without obstruction. Circulation **46**:897, 1972.
2. Bonow, R. O., Henry, W. L., Kent, K. M., Borer, J. S., Redwood, D. R., Conkle D. M., McIntosh, C. L., Morrow, A. G., and Epstein, S. E.: Predictors of late deaths due to congestive heart failure following operation for aortic regurgitation, abstracted, Am. J. Cardiol. **41**:382, 1978.
3. Borer, J. S., Bacharach, S. L., Green, M. V., Kent, K. M., Epstein, S. E., and Johnston, G. S.: Exercise induced left ventricular dysfunction in symptomatic and asymptomatic patients with severe aortic regurgitation assessed by radionuclide cineangiography, abstracted, Am. J. Cardiol. **42**:351-357, 1978.
4. Borer, J. S., Bacharach, S. L., Green, M. V., Rosing, D. R., Seides, S. F., McIntosh, C. L., Conkle, D., Morrow, A. G., and Epstein, S. E.: Left ventricular function in aortic stenosis: response to exercise and effects of operation, Am. J. Cardiol. **41**:382, 1978.
5. Braunwald, E., Lambrew, C. T., Rockoff, S. D., Ross, J., Jr., and Morrow, A. G.: Idiopathic hypertrophic subaortic stenosis. I, a description of the disease based upon an analysis of 64 patients, Circulation **30**(suppl. IV):IV-3, 1964.
6. Braunwald, E., Morrow, A. G., Cornell, W. P., Aygen, M. M., Hilbish, T. F.: Idiopathic hypertrophic subaortic stenosis: clinical hemodynamic, and angiographic manifestations, Am. J. Med. **29**: 924, 1960.
7. Brent, L. B., Aburano, A., Fisher, D. L., Morgan, R. J., Myers, J. D., and Taylor, W. J.: Familial muscular subaortic stenosis: an unrecognized form of "idiopathic heart disease" with clinical and autopsy observations, circulation **21**:107, 1960.
8. Buja, L. M., Khoi, N. B., and Roberts, W. C.: Clinically significant cardiac amyloidosis. Clinicopathologic findings in 15 patients, Am. J. Cardiol. **26**:394, 1970.
9. Bulkley, B. H., Hutchings, G. M., Bailey, I., Strauss, H. W., and Pitt, B.: Thallium-201 imaging and gated cardiac blood pool scans in patients with ischemic and congestive cardiomyopathy: a clinical and pathologic study circulation **55**:753, 1977.
10. Bulkley, B. H., Rouleau, J., Strauss, H. W., and Pitt, B.: Idiopathic hypertrophic subaortic stenosis: detection by thallium-201 myocardial perfusion imaging, N. Engl. J. Med. **50**:421, 1971.
11. Bulkley, B. H., Rouleau, J., Strauss, H. W., and Pitt, B.: Sarcoid heart disease: diagnosis by thallium-201 myocardial perfusion imaging, abstracted, Am. J. Cardiol. **37**:125, 1976.
12. Burow, R. D., Strauss, H. W., Singleton, R., Pond, M., Rehn, T., Bailey, I., Griffith, L. C., Nickoloff, E. and Pitt, B.: Analysis of left ventricular function from multiple gated (MUGA) cardiac blood pool imaging: comparison to contrast angiography, circulation **56**:1024, 1977.
13. Demakis, J. G., and Rahimtoola, S. H.: Peripartum cardiomyopathy, Circulation **55**:753, 1977.
14. Dodge, H. T., and Baxley, W. A.: Hemodynamic aspects of heart failure, Am. J. Cardiol. **22**:24, 1968.
15. Dodge, H. T., and Baxley, W. A.: Left ventricular volume and mass and their significance in heart disease. Am. J. Cardiol. **23**:528, 1969.
16. Eckberg, D. L., Gault, J. H., Bouchard, R. L., Karliner, J. S., and Ross, J., Jr.: Mechanics of left ventricular contraction in chronic severe mitral regurgitation, Circulation **47**:1252, 1973.
17. Epstein, S. E., Henry, W. L., Clark, C. E., Roberts, W. C., Maran, B. J., Ferrand, V. J., Redwood, D. R., and Morrow, A. G.: Asymmetric septal hypertrophy, Ann. Intern. Med. **81**:650, 1974.
18. Gault, J. H., Covell, J. W., Braunwald, E., Ross, J., Jr.: Left ventricular performance following correction of free aortic regurgitation, Circulation **42**: 773, 1970.
19. Glancy, D. L., and Roberts, W. C.: The heart in malignant melanoma. A study of 70 autopsy cases, Am. J. Cardiol. **21**:555, 1968.
20. Goodwin, J. F.: Prospects and predications for the cardiomyopathies, Circulation **50**:210, 1974.
21. Goodwin, J. F., Gordon, H., Hollman, A., and Bishop, M. B.: Clinical aspects of cardiomyopathy, Br. Med. J. **1**:69, 1961.
22. Goodwin, J. F., Hollman, A., Cleland, W. P., and Teare, R. D.: Obstructive cardiomyopathy simulating aortic stenosis, Br. Heart J. **22**:169, 1960.
23. Hammermeister, K. E., Fisher, L., Kennedy, J. W., Samuels, S., and Dodge, H. T.: Prediction of late survival in patients with mitral valve disease from clinical, hemodynamic, and quantitative angiographic variables, Circulation **57**:341, 1978.
24. Hanfling, S. M.: Metastatic cancer to the heart: review of the literature and report of 12 cases, Circulation **22**:474, 1960.
25. Henry, W. L., Clark, C. E., and Epstein, S. E.: Asymmetric septal hypertrophy (ASH): echocardiographic identification of the pathognomonic anatomic abnormality of IHSS, Circulation **47**:225, 1973.
26. Henry, W. L., Clark, C. E., Griffith, J. M., and Epstein, S. E.: Mechanism of left ventricular outflow obstruction in patients with obstructive asymmetric septal hypertrophy, Am. J. Cardiol. **35**:337, 1975.

27. Hudson, R. E. B.: The cardiomyopathies: order from chaos, Am. J. Cardiol. **25**:70, 1970.
28. Kennedy, J. W., Twiss, R. D., Blackman, J. R., and Dodge, H. T.: Quantitative angiocardiography: III, relationship of left ventricular pressure, volume, and mass in aortic valve disease, Circulation **38**:838, 1968.
29. Löffler, W.: Endocarditis parietalis fibroplastica mit Bluteosinophile, Schweiz. Med. Wochenschr. **66**:817, 1936.
30. Maron, B. J., Ferrans, V. J., and Roberts, W. C.: Myocardial ultrastructure in patients with chronic aortic valve disease, Am. J. Cardiol. **35**:725, 1975.
31. Maron, B. J., Gotdiener, J. S., Roberts, W. C., Henry, W. L., Sarage, D. D., and Epstein, S. E.: Left ventricular outflow tract obstruction due to systolic anterior motion of the anterior mitral valve leaflet in patients with concentric left ventricular hypertrophy, Circulation **57**:527, 1978.
32. Meaney, E., Shabetai, R., Bhargava, V., Shearer, M., Weidner, C., Mangiardi, L. M., Smalling, R., and Paterson, K.: Cardiac amyloidosis, constrictive pericarditis and restrictive cardiomyopathy, Am. J. Cardiol. **38**:547, 1976.
33. Miller, G. A. H., Kirklin, J. W., and Swan, H. J. C.: Myocardial function and left ventricular volumes in acquired valvular insufficiency, Circulation **31**:374, 1965.
34. Mintz, G. S., Kotler, M. N., Segal, B. L., and Parry, W. R.: Systolic anterior motion in the absence of asymmetric septal hypertrophy, Circulation **57**:256, 1978.
35. Nellen, M., Beck, W., Vogelpoel, L., and Schrire, V.: Auscultatory phenomena in hypertrophic obstructive cardiomyopathy. In Wolstenholme, G. E. W., O'Connor, M., London, J., and Churchill, A. editors: Hypertrophic obstructive cardiomyopathy, Ciba Foundation Study Group No. 37, Summit, N.J., 1971, Ciba Foundation, p. 77.
36. Oakley, C. M.: Clinical recognition of the cardiomyopathies, Circ. Res. **35**(suppl. II):152, 1974.
37. Parry E. H. O.: Endomyocardial fibrosis. In Wolstenholme, G. E. W., O'Connor, M., London, J., and Churchill, A., editors: Cardiomyopathies: a Ciba Foundation Symposium, Summit, N.J., 1964, Ciba Foundation, p. 322.
38. Pohost, G. M., Vignola, P. A., McKusick, K. A., Block, P. C., Myers, G. S., Walker, H. J., Copen, D. L., and Dinsmore, R. E.: Hypertrophic cardiomyopathy: evaluation by gated cardiac blood pool scanning, Circulation **55**:92, 1977.
39. Rackley, C. E., and Hood, W. P., Jr.: Quantitative angiographic evaluation and pathophysiologic mechanisms in valvular heart disease, Prog. Cardiovasc. Dis. **15**:427, 1973.
40. Roberts, W. C., Bodey, G. P., and Wertlake, P. I.: The heart in acute leukemia. A study of 420 autopsy cases, Am. J. Cardiol. **21**:388, 1968.
41. Roberts, W. C., and Ferrans, V. J.: Pathologic anatomy of the cardiomyopathies. Idiopathic dilated and hypertrophic types, infiltrative types, and endomyocardial disease with and without eosinophilia, Human. Pathol. **6**:287, 1975.
42. Roberts, W. C., Glancy, D. L., and Devita, V. T., Jr.: Heart in malignant lymphoma (Hodgkin's disease, lymphoma, reticulum cell sarcoma and mycosis fungoides). A study of 196 autopsy cases, Am. J. Cardiol. **22**:85, 1968.
43. Schelbert, H. R., Verba, J. W., Johnson, A. D., Brock, G. W., Alazraki, N. P., Rose, F. J., and Ashburn, W. L.: Non-traumatic determination of left ventricular ejection fraction by radionuclide angiocardiography, Circulation **51**:902, 1975.
44. Schlant, R. C., and Nutter, D. O.: Heart failure in valvular heart disease, Medicine **50**:421, 1971.
45. Schuler, G., Ross, J., Jr., Johnson, A., Dennish, G., Schelbert, H., Francis, G., and Peterson, K.: Factors affecting recovery of left ventricular function after mitral valve surgery, abstracted, Am. J. Cardiol. **41**:383, 1978.
46. Shah, P. M., Gramiak, R., and Kramer, D. H.: Ultrasound localization of left ventricular obstruction in hypertrophic obstructive cardiomyopathy, Circulation **40**:3, 1969.
47. Shea, W. H., Boucher, C. A., Curfman, G. D., Dinsmore, R. E., and Pohost, G. M.: Early reversibility of left ventricular dilatation following relief of chronic valvular regurgitation, Circulation **53** and **54**(suppl.II):II-215, 1976.
48. Strauss, H. W., Harrison, K., Langan, J. K., Lebowitz, E., and Pitt, B.: Thallium-201 for myocardial imaging. Relation of thallium-201 to regional myocardial perfusion, Circulation, **51**:641, 1975.
49. Strauss, H. W., and Pitt, B.: Gated cardiac blood pool scan: use in patients with coronary artery disease, Prog. Cardiovasc. Dis. **20**:207, 1977.
50. Strauss, H. W., Zaret, B. L., Hurley, P. J., Natarajan, T. K., and Pitt, B.: A scintiphotographic method for measuring left ventricular ejection fraction in man without cardiac catheterization, Am. J. Cardiol. **28**:275, 1971.
51. Tarter, W. E., Allen, H. D., Sahn, D. J., and Goldberg, S. J.: The asymmetrically hypertrophied septum: further differentiation of its causes, Circulation **53**:19, 1976.
52. Teare, D.: Asymmetrical hypertrophy of the heart in young patients, Br. Heart J. **20**:1, 1958.
53. Tobin, J. R., Jr., Driscoll, J. F., Lim, M. T., Sutton, G. C., Szanto, P. B., and Gunnar, R. M.: Primary myocardial disease and alcoholism: clinical manifestations and course of the disease in a selected population of patients observed for three or more years, Circulation **35**:754, 1967.
54. Wackers, F. J., Becker, A. E., Samson, G., Sokole, E. B., van der Schoot, J. B., Vet, A. J., Lie, K. I., Durrer, D., and Wellens, H. J.: Location and size of acute transmural myocardial infarction estimated from thallium-201 scintiscams:

a clinicopathologic study, Am. J. Cardiol. **56**:72, 1977.

55. Wigle, E. D., Adelman, A. G., and Silver, M. D.: Pathophysiological consideration in muscular subaortic stenosis. In Wolstenholme, G. E. W., O'Connor, M., London, J., and Churchill, A., editors: Hypertrophic obstructive cardiomyopathy, Ciba Foundation Study Group No. 37, Summit, N.J., 1971, Ciba Foundation, p. 63.

SECTION FIVE

PERIPHERAL VASCULAR DISEASE

19 □ Detection of thrombi

Sol Sherry

Prevention and management of thromboembolism is one of the more common problems in the medical treatment of adults. The advent of angiography and venography has helped immeasurably in the confirmation of the diagnosis of certain thromboembolic events and in management planning, but for a number of reasons these procedures are still limited in usefulness, particularly as screening procedures for the detection of thrombi. In addition, recent evidence has demonstrated that contrast venography may *cause* thrombi to form in *50%* of the patients studied (probably due to the effect of the contrast material on the venous endothelium). Thus there is a need for new screening procedures to detect thrombi.

^{125}I-LABELED FIBRINOGEN SCAN

An important advance has occurred through the introduction of the ^{125}I-labeled fibrinogen test as a diagnostic tool for demonstrating incipient venous thrombosis and thrombus growth in the lower limbs of patients at high risk.[8] This test has reemphasized the frequency of deep-vein thrombosis as a complicating event in a variety of medical and surgical problems and is providing the best epidemiologic data ever obtained in the living patient. The basis for this examination is simple: ^{125}I-labeled fibrinogen is injected intravenously into patients at high risk and is deposited as ^{125}I-labeled fibrin if thrombosis occurs. The appearance of a thrombus is determined by daily measurements with a simple monitoring device over selected areas of the lower thigh and leg bilaterally. The test has been well correlated with venographic studies and enjoys its greatest usefulness in determining when and where thrombi form in the leg veins and whether they are stable, extending, or resolving. Since the test depends on new fibrin deposition, it is less useful for diagnosing previous thrombotic events unless a thrombus continues to grow rapidly. Recent observations with this procedure have indicated that among older patients venous thrombi occur in 30% after major abdominal surgery, in 60% following hip or extensive bone surgery, in 25% after relatively minor surgical procedures such as herniorrhaphy, and in 30% following acute myocardial infarction.

It has been claimed that this procedure shows that leg vein thrombi occur initially in the soleal arcade of the calf muscles, where they are usually unassociated with clinical manifestations unless the thrombus progresses and extends into the tibial, peroneal, popliteal, and femoral veins. Furthermore, only when such extension occurs is there evidence of deep-vein thrombophlebitis, such as clinical manifestations, alterations in the augmented Doppler ultrasound or the electrical impedance plethysmogram, or any risk of pulmonary embolism. On the basis of his observations, Kakkar estimated that of patients undergoing major surgery, 30% develop deep-vein thrombi in the soleal arcade; of these thrombi, one fifth (overall incidence of 6%) extend to the popliteal vein or above, and of

□ This work was supported by the National Heart & Lung Institute, Bethesda, Md., grant no. HL14217.

these extending thrombi, half (overall incidence of 3%) then embolize and are swept to the pulmonary artery or its branches,[8] where perhaps 10% of the latter are immediately fatal (overall incidence of 0.3%). However, after hip surgery, the incidence of thrombosis is higher; femoral or ileofemoral vein thrombosis may occur independent of calf vein thrombosis, and episodes of pulmonary embolism are more frequent.

The ^{125}I-fibrinogen test, besides providing a great opportunity for studying the risk factors and epidemiologic aspects of leg vein thrombosis, also has made available a most useful and sensitive technique by which preventive measures can be evaluated. For example, Wray and associates[20] studied the incidence of leg vein thrombi in patients after acute myocardial infarction. After documenting a high incidence of demonstrable thrombi, albeit mostly subclinical, a controlled trial of the efficacy of conventional anticoagulant therapy in preventing postinfarction leg vein thrombosis was undertaken. In a study of ninety-two patients, they found the incidence of deep-vein thrombosis in the control group to be 22% and the incidence in the anticoagulant group to be 6.5%. These differences were significant ($p < 0.05$) and readily confirmed what has been known for some time: conventional anticoagulation is an effective form of prophylaxis for deep-vein thrombosis.

More importantly, the ^{125}I-fibrinogen test has also served well for evaluating a new prophylactic regimen that has excited considerable interest,[9] namely, the use of heparin in amounts that do not increase the risk of bleeding but may be effective in the prevention of venous thrombosis. Two or three times daily heparin is injected subcutaneously in 5000-unit doses; the Lee-White clotting time, thrombin time, and activated partial thromboplastin time are not frequently prolonged, and the low blood levels of heparin that are achieved usually can only be detected by more sensitive techniques than are currently used in the clinical laboratory. Such levels do not appear to enhance bleeding in patients even when the therapy is given prior to, during, and after surgical procedures; yet they are capable of mediating the inhibition of the earlier stages of blood coagulation.

The efficacy of low-dose heparin therapy in the prevention of deep-vein thrombosis measured by the ^{125}I-fibrinogen test has now been confirmed in several independent studies,[7,10] and the final proof of its utility as a prophylaxis against fatal pulmonary embolism in patients undergoing major abdominothoracic surgery has been reported.[9]

The availability of this scanning technique has also made possible the critical evaluation of other new forms of prophylaxis: electrical calf muscle stimulation, intermittent pulsating compression devices, platelet function inhibitors, and agents capable of enhancing venous blood flow. Also, it has allowed comparisons between the different modes of prophylaxis. Considering the magnitude of the problem of venous thromboembolism and knowing that prevention rather than treatment of pulmonary embolism represents the only way in which the mortality from this potentially lethal complication can be reduced significantly, these newer developments in prophylaxis have been a most welcome addition. We now know what can be achieved by most of the various forms of prophylaxis, and the major problem that faces us is the education of physicians in the widespread use of acceptable yet more effective preventive measures.

The ^{125}I-fibrinogen test should be a powerful tool for the study of other therapeutic agents. For example, it could be used for the study of resolution rates as affected by different regimens of thrombolytic agents; pharmacologic data crucial to the development of safer and more effective dosage regimens could be obtained.

Finally, this examination may prove to be most helpful in answering certain fundamental questions, such as those concerning the importance and significance of thrombosis in the pathogenesis of acute myocardial infarction and strokes. In this respect, a recent study on the incorporation of ^{125}I-fibrinogen into coronary arterial thrombi in acute myocardial infarction is important.[5]

Although the ^{125}I-fibrinogen technique has opened up the field of thrombus detection by radionuclide scanning methods, it has a number of serious limitations. ^{125}I has a low emission energy; this limits its usefulness as a tracer to relatively superficial areas such as the extremities, and even in the legs, it is useful only below the midthigh. Therefore it is not capable of detecting extension of a thrombus above this area or of providing information about whether thrombi form independently in the proximal part of the femoral, in the iliac, or in the pelvic veins. This information is necessary for answering the question of whether pulmonary emboli arise primarily from thrombi that propagate from the soleal arcade or are more likely to occur from thrombi that arise independently and perhaps without association to calf vein thrombosis. This is a most important question, for the answer to it will determine the true clinical significance of the ^{125}I-fibrinogen test. Another limitation of this test is that it depends on new fibrin formation; thus it is relatively insensitive in detecting a thrombus that has formed earlier and that is in a static state as far as new growth is concerned. Consequently, further developments in radionuclide scans are needed for detecting thrombi forming in sites other than the distal parts of the extremities and for detecting preformed thrombi. Some of the investigative work relating to such developments are described in the following sections.

^{131}I-LABELED FIBRINOGEN SCAN

To circumvent some of the deficiencies inherent in the ^{125}I-labeled fibrinogen scan for detecting newly forming thrombi, investigators[3,4] have been developing a ^{131}I-labeled fibrinogen scan. Potentially, a technique in which this radionuclide is used could provide for total-body scanning, yield a permanent pictorial display, and be more readily adaptable to quantitation. The initial observations have been quite promising: the technique has worked well when used in experimental animals not only in detecting a forming thrombus but also in demonstrating a pulmonary embolus released from such a tagged thrombus. Preliminary studies in humans have also been promising: thrombi have been demonstrated in the ileofemoral area, and the practicality of total-body scanning has been established. At present this group is directing its efforts toward standardizing the procedure and evaluating its accuracy with appropriate venographic studies.

SCANS FOR DETECTION OF PREFORMED THROMBI

Unless a preformed thrombus is growing rapidly, new fibrin deposition in it is proceeding at a relatively slow rate. Thus labeled fibrinogen is unlikely to serve as a good scanning material for detecting a preformed thrombus. For this purpose, there is need for a labeled material that will be attracted to and penetrate a thrombus or specifically adhere to it or be absorbed by it. Accordingly, most of the investigative interest to date has focused on the use of labeled leukocytes and plasminogen activators, the former because a thrombus appears to be very leukotactic and the latter because of its presumed ability to bind to fibrin.

Kwaan and Grumet[11] demonstrated the feasibility of using a radionuclide scan with ^{51}Cr-labeled leukocytes for the detection of experimental thrombi in animals, and Charkes (personal communication) has confirmed this with a ^{99m}Tc label. Despite the promise inherent in this type of scan, there are many practical problems, not the least of which is the need to harvest and label autologous leukocytes.

As for tagged plasminogen activators, most of the earlier studies were performed with ^{131}I-labeled streptokinase. This material has worked well in animals with experimental thrombi when the tagged streptokinase has been injected into a vein whose flow is directed to the site of the thrombus. However, when the labeled streptokinase has been injected into a distant and totally unrelated vein, relatively little tagging of the thrombus has taken place. Whether this is due to dilution, inactivation of streptokinase in the general circulation, or other factors still remains to be determined. Furthermore, observations made with streptokinase in animals may not bear much relation to the human situation because of marked variations in species sensitivity to this substance.

More recently, encouraging studies[2,13] have been reported on the use of ^{99}Tm-labeled urokinase for thrombus detection, but other researchers have not been able to confirm its usefulness.[19]

Observations have also been made on the use of labeled platelets; although this is an attractive approach, the number of platelets required will probably be too great to allow this type of scan to be practical unless the tagging can be carried out readily in very high concentration and without damage to the platelet. Should this prove to be possible, the field of thrombosis detection by radionuclides would be considerably extended, for example, to include the detection of arterial and platelet thromboembolism as well.

Finally, it should be noted that radionuclide venography[6] can also be used as an alternative to contrast venography for the detection of preformed thrombi. Although the radionuclides used in this procedure may not have the specificity of those previously discussed, the value of the procedure as a diagnostic process may prove to be entirely adequate.

BIOCHEMICAL TECHNIQUES

Over the past several years, a number of biochemical techniques have been introduced that could have application to the detection of thrombosis in vivo. Originally, these centered on the demonstration, by a variety of procedures, of the presence of soluble fibrin complexes in the circulation or of breakdown products arising from recent fibrin deposits. For one reason or another, such as complexity of the assay or lack of specificity or sensitivity, none has as yet proved to be useful as a screening procedure for the detection of thrombi.

Recently, attention has been directed toward the development of a radioimmunoassay for the highly specific fibrinopeptides A and B, released from fibrinogen by the action of thrombin. It is this action on fibrinogen that is unique for thrombin; the resulting altered fibrinogen (fibrin monomer) then spontaneously polymerizes to form fibrin. Thus the appearance of these peptides in significant concentration in the circulation is strong evidence of fibrinogen-fibrin conversion. The feasibility of developing such an assay for fibrinopeptide A has now been established,[1,18] and the preliminary observations reported by Nossel and associates are sufficiently encouraging to make this one of the most promising new developments in thrombosis research.[17]

In addition, under current study for thrombus detection are radioimmunoassays for some of the specific proteins released from platelets during thrombosis, for example, platelet factor 4[16] and β-thromboglobulin.[14] Also being developed is a radioimmunoassay for a specific fibrin fragment, D-dimer,[12] which is distinct from any of the polypeptide fragments that can be derived from fibrinogen.

SUMMARY

The detection of thrombi through the use of radionuclide scans has become a reality with the introduction of the ^{125}I-fibrinogen radionuclide scanning technique. Although it has proved to be very useful for documenting the formation of thrombi in the lower extremities of patients at high risk and for evaluating prophylactic therapeutic procedures, the technique still has serious limitations as a diagnostic procedure. It is useful only over selected areas of the extremities and for the detection of thrombi that are either forming or growing rapidly.

For expansion of the usefulness of such radionuclide studies, evaluations are under way with a variety of materials and radionuclides that may overcome the inadequacies of the ^{125}I-fibrinogen test and extend the usefulness of scanning techniques for thrombus detection. The most promising of these involve the use of ^{131}I-fibrinogen, labeled leukocytes, and labeled plasminogen activators.

The detection of thrombi through biochemical techniques now centers on the evaluation of radioimmunoassays for specific peptides or polypeptides released into the circulation during or after thrombosis; at present, several are under active study.

REFERENCES

1. Budzynski, A. Z., and Marder, V. J.: Solid phase synthesis of human fibrinopeptides A and B. In Protides of the biological fluids, vol. 20, Elmsford, N. Y., 1972, Pergamon Press, Inc.
2. Cox, P. H., van der Pompe, W. B., and den Ottolander, G. J. H.: Visualization of thrombi with technetium-99m urokinase, Lancet **2**:572, 1976.
3. DeNardo, G., and DeNardo, S.: ^{131}I-fibrinogen scintigraphy, Semin. Nucl. Med. **7**:245, 1977.
4. Dugan, M. A., Kozar, J. J., III, Charkes, N. D., Maier, W., and Budzynski, A.: The use of iodinated fibrinogen for localization of deep venous thrombi by scintiscanning, Radiology **106**:445, 1973.
5. Erhardt, L. R., Lundman, T., and Mellstedt, H.: Incorporation of ^{125}I-labeled fibrinogen in coronary arterial thrombi in acute myocardial infarction in man, Lancet **1**:387, 1973.
6. Henkin, R. E., Yao, J. S. T., and Quinn, J. L., III, et al.: Radionuclide venography in lower extremity disease, J. Nucl. Med. **15**:171, 1973.
7. Hirsh, J., Genton, E., and Gent, M.: Low dose heparin prophylaxis for venous thromboembolism. In Fratantoni, J., and Wessler, S., editors: Prophylactic therapy of deep vein thrombosis and pulmonary embolism, DHEW publication no. (NIH) 76-866, Washington, D.C., 1975, U. S. Department of Health, Education and Welfare, p. 183.
8. Kakkar, V. V.: The ^{125}I-labeled fibrinogen test and phlebography in the diagnosis of deep-vein thrombosis. In Foster, C. S., Genton, E., Henderson, M., Sherry, S., and Wessler, S., editors: The epidemiology of venous thrombosis, Milbank Mem. Fund. Q. **50**(suppl. 2):206, 1972.
9. Kakkar, V. V., Corrigan, T. P., and Fossard, D. P.: Prevention of fatal postoperative pulmonary embolism by low doses of heparin: an international multicentre trial, Lancet **2**:45, 1975.
10. Kakkar, V. V., Nicolaides, A. N., Fields, E. S., Flute, P. T., Wessler, S., and Yin, E. T.: Low doses of heparin in the prevention of deep-vein thrombosis, Lancet **2**:669, 1971.
11. Kwaan, H. C., and Grumet, G.: The use of ^{51}Cr-labeled leukocytes in the detection of venous thrombosis, Clin. Res. **20**:788, 1972.
12. Marder, V. J., Budzynski, A. Z., and Barlow, G. T.: Comparison of the physicochemical properties of fragment D derivatives of fibrinogen and fragment D-D of cros linked fibrin, Biochim. Biophys. Acta **427**:1, 1976.
13. Millar, W. T., and Smith, J. F. B.: Localization of deep-venous thrombosis using technetium-99m-labeled urokinase. Lancet **ii**:695, 1974.
14. Moore, S., Peppar, D. S., and Cash, J. D.: The isolation and characterization of a platelet specific betaglobulin (β-thromboglobulin) and the detection of anti-urokinase and antiplasmin release from thrombin aggregated washed human platelets, Biochim. Biophys. Acta **379**:370, 1975.
15. Moschos, C. B., Oldewurtel, H. A., Haidler, B., and Regan, T. J.: Effect of coronary thrombus age on fibrinogen uptake, Circulation **54**:653, 1976.
16. Niewiarowski, S., Lowery, C. T., Hawiger, J., et al.: Immunoassay of human platelet factor 4 (PF4, antiheparin factor) by radial immunodiffusion, J. Lab. Clin. Med. **87**:720, 1976.
17. Nossel, H. L.: Radioimmunoassay of fibrinopeptides in relation to intravascular coagulation and thrombosis, N. Engl. J. Med. **295**:428, 1976.
18. Nossel, H. L., Younger, L. R., Wilner, G. D., Procupez, T., Canfield, R. E., and Butler, V. P., Jr.: Radioimmunoassay of human fibrinopeptide A, Proc. Natl. Acad. Sci. U.S.A. **68**:2350, 1971.
19. Weir, G. J. Jr., Wenzel, F. J., Roberts, R. C., and Sautter, R. D.: Visualization of thrombi with technetium-99m urokinase. Lancet **2**:341, 1976.
20. Wray, R., Maurer, D., and Shillingford, J.: Prophylactic anticoagulant therapy in the prevention of calf-vein thrombosis after myocardial infarction, N. Engl. J. Med. **288**:815, 1973.

20 □ Use of radioactive tracers in the evaluation of peripheral arterial disease

Michael E. Siegel

There is general acceptance that peripheral vascular disease is not only a major medical problem throughout the world but also, because of its associated incapacitation and disability, is an important socioeconomic problem as well. The study of peripheral vascular disease is conveniently classified etiologically into that which is arterial in origin and that which is venous in origin.

Arterial disease

The chronic, occlusive arterial diseases represent the most common forms of arterial involvement. These chronic diseases can be further categorized etiologically into that which is degenerative and that which is inflammatory (Buerger's disease). Most clinical disorders result from atherosclerosis, which is a degenerative, noninfectious, noninflammatory variety and only one of many types of arteriosclerosis recognized by pathologists. This disease is found in a shockingly large cross section of the population, as noted in a large necropsy study, which revealed that 50% of American soldiers (average age 22 years), having died of battle trauma, had gross evidence of atherosclerosis. The high incidence of the disease is further emphasized by the fact that by age 50 it rises to approximately 90% of the general population.

PHYSICAL EXAMINATION

The history and physical examination can often yield valuable clues as to the extent and location of intravascular disease.

The presence and intensity of the peripheral pulses should be noted and compared with the contralateral side because the absence of a particular pulse often correlates with the location of a proximal arterial occlusion. Exceptions to this finding are, however, quite common. In the region of the ankles, the posterior tibial pulse is the more important of the pulses to note because of the 10% incidence of congenital absence of the dorsalis pedis artery. Auscultation along the major arteries should also be performed, since the most common cause of a harsh bruit is a stenotic lesion of the vessel proximal to that level.

Because of variations in vasomotor control, gross evaluations of skin temperature or color are somewhat unreliable. However, asymmetry in these findings is valuable in grossly localizing arterial disease.

SYMPTOMATOLOGY

Chronic deprivation of the arterial supply to an extremity produces symptoms of essentially the same type, independent of the underlying pathology of the occluding lesion. Of these symptoms, one of the earliest is severe cramping pain in those muscle groups inadequately supplied with blood. The pain develops during exercise, is relieved by rest, and occurs when the arterial supply to the part is adequate at rest but cannot meet the hyperemic needs of exercise. This symptom is therefore called intermittent claudication. As the degree of arterial occlusion increases, the amount of exercise needed to produce intermittent claudication is reduced, and often the next noted symptom is a sense of coldness in the distal part of the extremity. In addition, cramping muscle pain eventually occurs at rest and during sleep, and this "rest pain" becomes especially disturbing as the disease progresses. However, the most distressing symptom of occlusive vascular disease is the neuritic pain of ischemia. This differs from intermittent claudication and rest pain in that it is constant, burning, and oppressing in quality.

ANCILLARY TEST (NONISOTOPIC)

Objective means of evaluating the extent of intravascular disease semiquantitatively include thermography; fluorescein circulation time, Doppler-effect ultrasound, and histamine wheal tests; venous occlusive and electrical strain plethysmography; oscillometry; indicator dilution and thermodilution techniques and arteriography. Most of these methods for evaluating blood flow are indirect, and despite refinements in the techniques and equipment, most are still subject to considerable technical error and variation.

To date only contrast arteriography has been widely accepted as essential in the evaluation of peripheral vascular disease. This method provides detailed anatomic information of the larger vessels, which is particularly important if reconstructive vascular surgery is being contemplated. Arteriography, however, is an anatomic study and does not necessarily yield data suitable for quantitation or the evaluation of the microcirculation in an extremity or the physiologic significance of the arteriographically demonstrable disease.

ISOTOPIC TESTS

Historical considerations

Since the time of William Harvey, major efforts have been made to comprehend and evaluate the physiologic aspects of blood flow. Long before the introduction of radioactive tracer, Fick, Stewart, Henriques, Hamilton, Kety, and others developed the theoretic foundation for the use of indicators to study blood flow. The introduction of radioactive tracers has certainly been influential in the adaption of many of these pioneering concepts to the clinical setting. Utilization of radioisotopes in the study of the circulation began in 1927 when Blumgart and Yens injected NaCl, activated by radium C, intravenously and measured circulation using a cloud chamber as a detector.[1]

Theoretic considerations

The method used by Blumgart and Yens was proposed by Fick in 1870 for the measurement of cardiac output; it can be applied to determining blood flow in an extremity. Assume that the rate of change in the amount of inert indicator Q in the circulation should equal the difference between inflow and outflow of the indicator from the region, that is:

$$\frac{dQ}{dt} = F(Ca - Cv)$$

where

F = Total rate of blood flow (ml/min)
Ca = Concentration of indicator in arterial blood
Cv = Concentration in venous blood

Therefore

$$F = \frac{dQ \div dt}{Ca - Cv}$$

By injecting radioactive tracer into the femoral artery, an external detector can be used to determine $\frac{dQ}{dt}$. Ca − Cv is obtained through the appropriate arterial and venous samples. The method is cumbersome and has not been widely applied clinically to the evaluation of total blood flow in an extremity.

The Stewart-Hamilton method originally proposed by Stewart in 1897 is based on the dilution of an indicator as measured, for example, in the femoral vein after injection into a systemic artery. Referring again to the equation $\frac{dQ}{dt} = F(Ca - Cv)$, if F is assumed to be constant, one can integrate from time 0 to time t, when all the indicator Q has left the area of interest.

From this one obtains the following:

$$Q = F - \int_0^t Cvdt$$

This can be rearranged as follows:

$$F = \frac{Q}{\int_0^t Cvdt}$$

Since $\int_0^t$ Cvdt is the area under the time-activity curve, this can be simplified as follows:

$$F = \frac{Q}{A}$$

where A is the total area under the venous time-concentration curve. To correct for recirculation, an exponential extrapolation of the descending part of the curve is carried out. However, as a means of measuring total blood flow to an extremity this too has not been widely accepted clinically.

The desirability of measuring total blood flow by using either of these two techniques, with or without utilization of radioisotopes, has been challenged. In many patients with peripheral vascular disease, the disease is located distally in the vascular bed of the extremities, and thus the total blood flow to the involved extremity may be normal.

Other techniques that have been proposed include use of intravenous ^{32}P with evaluation of time-activity curves over the soles of the feet, injection of intravenous ^{131}I-labeled serum albumin with subsequent determination of equilibration time, and determination of various circulation times, that is, arm to foot, using intravenous injections of ^{24}Na- and ^{131}I-labeled serum albumin. None of these have found clinical acceptance in the evaluation of peripheral vascular disease.

Present techniques

CLINICAL APPLICATIONS

The use of radioactive tracers in the diagnostic and prognostic evaluation of the patient with peripheral arterial disease has produced clinically useful information. Their application has provided a means by which the following can be accomplished:

1. The regional distribution of perfusion at the level of the microcirculation, under various stresses, can provide objective evidence of diffuse small vessel disease.
2. The determination of graft patency can be simply and repeatedly performed.
3. The relative perfusion of ischemic ulcers can be determined and utilized to prognosticate their healing potential.
4. The presence of arteriovenous shunts can be verified and quantified.
5. The skin perfusion pressure can be determined and used to assess the healing potential of surgical amputations.

TRACERS

Most present techniques for the study of the peripheral circulation involve the study of various circulation times, local tissue clearance rates, or regional distribution of blood flow.

The radioactive tracers presently used for the evaluation of peripheral arterial vascular disease have been quite varied and include the following: ^{99m}Tc-labeled albumin, ^{131}I- and ^{99m}Tc-labeled MAA, ^{99m}Tc- and ^{113m}In-labeled albumin microspheres, ^{99m}Tc-labeled red blood cells, ^{133}Xe, [^{99m}Tc] diethylenetriamine pentaacetic acid (DTPA), ^{113m}In-labeled transferrin, $^{99m}TcO_4$, ^{43}K, and ^{201}Tl (Table 20-1).

These tracers are most conveniently classified into those which are nonparticulate and those which are particulate in nature.

Table 20-1. Radiotracers used for study of arterial diseases

Tracer	Indication
Particulate	
^{131}I-labeled MAA and ^{99m}Tc	Determination of relative distribution of perfusion
^{99m}Tc-labeled albumin microspheres	
^{113m}In-labeled albumin microspheres	
Nonparticulate	
^{99m}Tc-labeled serum albumin	Visualization of blood pool
^{99m}Tc-labeled red blood cells	
^{113m}In-labeled transferrin	
TcO_4^- (immediately after injection)	
^{43}K	Determination of relative distribution of perfusion in muscle
^{201}Tl	
^{133}Xe (in solution)	Blood flow per weight of tissue
[^{99m}Tc] DTPA and ^{113m}In	

Of the nonparticulate radiotracers, there are those which are not freely diffusible and thus act as blood volume indicators. These tracers such as ^{99m}Tc-labeled serum albumin, ^{99m}Tc-labeled red blood cells, ^{113m}In-labeled transferrin, or ^{99m}Tc as pertechnetate, at least immediately after injection, can be used in the production of radionuclide angiograms. Other radiotracers such as ^{43}K and ^{201}Tl are analogs of naturally occurring intracellular ions and appear to be distributed in proportion to the perfusion, thus permitting an estimate of relative perfusion.

The nonparticulate radiotracers that are diffusible include gases such as ^{133}Xe or metal chelate solutions such as [^{99m}Tc] DTPA or [^{113m}In] DTPA. After deposition of these tracers in the region of interest, a qualitative estimate of blood flow in milligrams per minute per 100 g of tissue can be derived from the washout rate of the tracer from the tissue, in which the blood flow is proportional to the steepness of the slope of the curve. The flow rates are calculated by plotting on semilog paper the activity versus time data that have accumulated. Initially, the clearance curve is thus essentially monoexponential. The half-period ($t_{1/2}$) is derived from the curve, and the following equation is used to calculate blood flow in the area injected:

$$F = \frac{0.693}{t_{1/2}} \times 100 \times \lambda$$

where

F = ml/min/100 g
λ = Partition coefficient for tissue studied

The local clearance method is based on the use of tracers that diffuse so freely between the tissue and the capillary blood that diffusion equilibrium is practically maintained regardless of the blood flow rate. For these tracers the blood flow is the limiting factor in the removal from a locally injected depot. Such a flow-limited tracer is the radioactive inert gas ^{133}Xe. On the other hand, small ionized tracer molecules such as [^{99m}Tc] pertechnetate pass more slowly across the capillary membrane. This means that during maximal hyperemia, at least in skeletal muscle, it is predominantly the capillary diffusion capacity (the permeability) that determines the rate of removal by the blood. Such tracers are thus mainly diffusion limited during hyperemia, but at very low blood flow the small ions can be assumed to reach close to diffusion equilibrium between the tissue and blood, and hence in such measurements even these tracers are predominantly flow limited.

The particulate tracers, larger in diameter than a capillary, are trapped within the first capillary bed they encounter and, if adequately mixed, are distributed in proportion to the perfusion of that capillary bed. This retention in the microcirculation eliminates the need for mathematic integration of time-radioactivity curve because no significant recirculation occurs. These tracers include ^{131}I- or ^{99m}Tc-labeled MAA or ^{99m}Tc- or ^{113m}In-labeled albumin microspheres (preferably for arterial work because of their decreased fragility) (Chapter 3). Kits for labeling microspheres with ^{99m}Tc are available commercially, and numerous methods have been reported to label the spheres with ^{113m}In.[7] Remember, however, that the particulate tracers do not by themselves yield direct quantitative data on blood flow but permit quantification of the relative distribution of perfusion.

Use of nonparticulate radiotracers

DETERMINATION OF BLOOD FLOW AND SKIN PERFUSION PRESSURE. A wide range of techniques measuring various circulation times, equilibration rates, and other such parameters has been proposed.

Since the clearance of nondiffusible tracers is determined by factors other than just blood flow, freely diffusible inert gases such as ^{133}Xe and ^{85}Kr are utilized in an attempt to avoid this difficulty encountered with the clearance techniques. In 1965 Lassen, Linjerg, and others demonstrated a good correlation between maximum blood flow as determined by venous occlusive plethysomography and the clearance of diffusible tracers. Tonnesen[15] verified these observations, reporting good agreement in animal studies between results obtained by direct metering of blood flow and by ^{133}Xe clearance.

Attempts were made to increase the sensitivity and specificity of these procedures. One approach was to apply stress to the circulation. Using standardized walking tests it was established that the clearance rates in normal subjects were significantly increased with exercise (Fig. 20-1). However, an increased rate of clearance was also seen in patients with arterial vascular

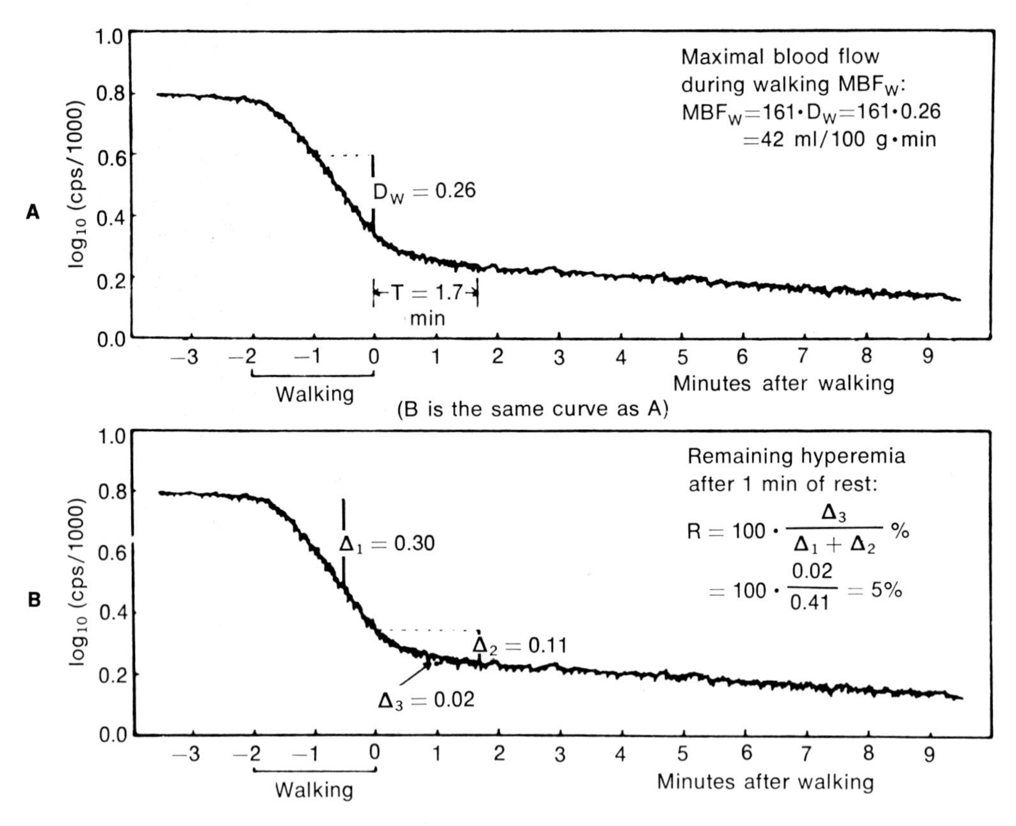

Fig. 20-1. ^{133}Xe clearance curves obtained by counting over right anterior tibial muscle in normal human subject. **A,** Calculation of maximal blood flow during exercise (walking). **B,** Calculation of remaining hyperemia after 1 min of rest. These are two of various parameters that have been studied with use of ^{133}Xe clearance curves during exercise. (From Alpert, J. S., Garcia de Rio, H., and Lassen, N. A.: Circulation **38:**849, 1966, by permission of the American Heart Association, Inc.)

disease, despite the patients' complaints of intermittent claudication during the exercise. This supported the hypothesis that patients with symptomatic arterial vascular disease can increase the blood flow to the symptomatic portion but the relative increase is not adequate to meet the increased demands. As will be discussed later in this chapter, this phenomenon is confirmed visually in the peripheral vascular perfusion scans using the particulate radiotracers. In normal individuals, ^{133}Xe clearance rates with and without reactive hyperemia were similar to those with and without exercise. The maximum clearance rate (peak reactive hyperemia) occurred within 1 min after release of the arterial occlusion. The maximum blood flow in milliliters per minute per 100 g and the time until maximum blood flow is reached after release of the arterial occlusion were thought to estimate the extent of involvement of the major arteries of the leg and, to some extent, the function of the collateral vessels. The patients with peripheral vascular disease had a delayed, diminished, and prolonged response to the reactive hyperemia.[5] With reactive hyperemia maximum blood flow in patients with peripheral vascular disease is approximately 17 ml/min/100 g, whereas in normal subjects under the same conditions it was approximately 50 ml/min/100 g. The figure widely accepted as the resting blood flow rate in the normal individual is 2 ml/min/100 g, whereas the flow rate during reactive hyperemia is considered to be 35 ml/min/100 g. In addition, to be considered normal, the maximum flow rate should be reached within 1 min after release of the occlusion pressure used to induce reactive hyperemia. Although the test has been reported to be quite sensitive (95% positive in patients with vascular disease between the heart and the site of measurement), false-negative results are encountered in patients with good collaterals and in those in which the lesion is close to the site of measurement. It should also be emphasized that one cannot rely purely on the calculated blood flow values with an evaluation of the overall configuration of the washout curve because the reproducibility of the flow values have about a 25% coefficient of variation.[5]

A useful technique for this approach utilizes an intramuscular injection of 0.1 ml of sterile saline solution containing 0.5 μCi of ^{133}Xe. This is made with a 27-gauge needle inserted approximately 1 to 2 cm deep and left in place for approximately 30 sec after completion of the injection. The appropriately collimated probe is placed 10 cm from the skin, and constant geometry must be maintained during the procedure. Precautions are necessary to avoid injecting gas because the gas-tissue partition coefficient for ^{133}Xe or ^{85}Kr is 10:1. The interpretation must, of course, take into account the cardiac status and metabolic state of the patient because these factors too are reflected in the clearance data.

An additional application of the diffusible gas–clearance approach is in the quantification of skin blood flow. Of particular interest was Serjsen's[10] finding of low flow values in the skin, particularly during and after exercise, in patients with peripheral vascular disease. This finding was qualitatively verified by the peripheral perfusion scanning with and without reactive hyperemia, as reported by Siegel and coworkers.[11] They found a relative decrease, during reactive hyperemia, of the nonmuscular-to-muscular blood flow rate in many of the patients with peripheral vascular disease.

Lassen and Holstein[5] and others have used the skin perfusion pressures as measured by ^{133}Xe-washout approach to predict the healing ability of amputation stumps. They use [^{131}I] iodoantipyrene or [^{125}I] iodoantipyrene as opposed to ^{133}Xe because of the significant diffusion of ^{133}Xe into the subcutaneous tissue, in which it is highly soluble. The iodoantipyrene (0.1 ml) is injected intradermally, and counterpressure is applied over the radioactive deposit with a blood pressure cuff. The skin perfusion pressure is that pressure at which perfusion stops (Fig. 20-2). In normal individuals this pressure approximates diastolic pressure.

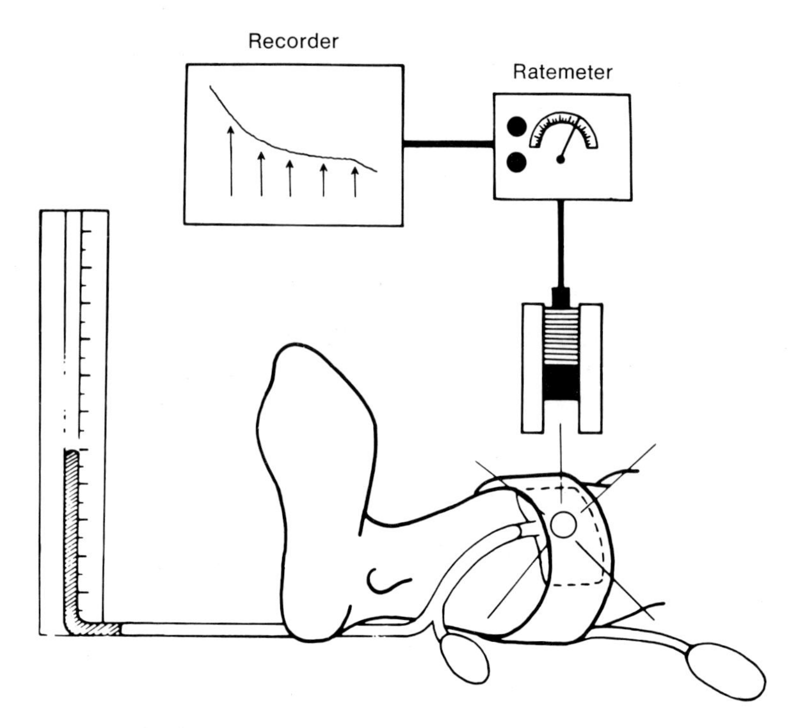

Fig. 20-2. Basic set-up for determining skin perfusion pressure from clearance of codo-antipyrene. Undertracing is displaying clots per unit of time on vertical axis versus pressure being applied (mm Hg) by inflatable cuff on horizontal axis. Perfusion pressure *(arrows)* is pressure at which $\Delta y/\Delta x$ approximates zero on release of pressure (last arrow on right); clearance resumes, as seen by increasing $\Delta x/\Delta y$. (From Lassen, N. A., and Holstein, P.: Surg. Clin. North Am. **54**:39, 1974.)

In a series of fifty-one postoperative leg amputees, results were as follows:

1. Five out of five stumps with skin pressures less than 20 mm Hg failed because of ischemic necrosis.
2. Three out of twelve with pressure in the range of 20 to 40 mm Hg failed.
3. None of the thirty-four who had pressures greater than 40 mm Hg failed.[5]

Preoperative assessment, however, is less reliable because of the changes in skin perfusion pressures that occur between the preoperative and postoperative period and thus the difficulty of predicting a postoperative pressure from a preoperative measurement.

PATENCY OF MAJOR ARTERIES AND GRAFTS. The utilization of intravenous radionuclide arteriography, using bolus injection of TcO_4^- and a scintillation camera has been proposed. Employing this approach, Dibos and associates[2] were not able to visualize a major vessel below the level of the popliteal vein, and thus for the visualization of subocclusive vascular disease, per se, it was somewhat limited. However, it appears as though the approach might be quite useful in verifying the patency of surgical bypass grafts without arterial catheterization. In addition, with the use of an analog to the digital computer and the production of transit time curves over the arterial trunk of interest, quantitative data regarding the patency of the bypasses could be obtained. Further work in this area is necessary.

DISTRIBUTION OF PERFUSION. Another approach that certainly warrants further investigation is the use of intravenously administered tracers such as ^{43}K or ^{201}Tl to visualize the distribution of perfusion among the muscle bundles of the leg[6] (Fig. 20-3). The distribution of these tracers to the muscles of the extremities is analogous to that seen in the cardiac musculature. However, it should be kept in mind that the tissue distribution of ^{43}K is related not only

Fig. 20-3. Perfusion scan of legs using ^{201}Tl. Note distribution of tracer in proportion to muscle mass in this patient with no known peripheral vascular disease.

to perfusion but also to the metabolic status of the tissue as well. How well this approach will permit an estimate of the relative skin perfusion, which may play an important role in amputation level selection, determine ulcer healing potential, and document the effect of arterial disease on the distribution of perfusion, remains to be demonstrated.

Use of particulate radiotracers

REGIONAL DISTRIBUTION OF BLOOD FLOW. Arteriography provides detailed anatomic information on the larger vessels, which is obviously a necessity in the preoperative evaluation of the patient with peripheral vascular disease. However, the physiologic significance of the intravascular disease and its ability to restrict perfusion to the regional capillary beds or the ability of the visualized collaterals to perfuse these beds is not directly evaluated by what is, essentially, a morphologic technique. With reconstructive vascular surgery becoming more commonplace, defining the physiologic significance of a lesion should prove useful in patient selection and operative approach. Moreover, many of these same patients are often burdened with ischemic ulcers, in which case the question of disfiguring surgery versus conservative therapy arises. An evaluation of the ability of the capillaries to perfuse the ulcer bed should, and did, prove useful as a prognosticator in these circumstances. The particle-distribution method for determining regional distribution of blood flow permits evaluation of the distribution of perfusion at the level of the microcirculation and the effect of intravascular disease on this distribution.

This method was introduced clinically in 1964 for the determination of regional pulmonary perfusion and has since been applied to the heart as well as to the extremities. The basic assumptions on which the particle-distribution method is based are as follows:

1. The particles do not alter the prevailing blood flow.
2. The particles are trapped and removed

from the circulation by the first capillary bed they encounter. This retention in the microcirculation eliminates the need for mathematic integration of the time-radioactivity curve because no significant recirculation occurs.

3. The particles have the same flow characteristics as the RBCs.
4. The particles are adequately mixed in the blood.

Only the fourth assumption has been recently challenged. To achieve "complete" mixing, a Reynolds number (Re) of at least 2000 must be obtained, which is believed to occur only in the aortic root. However, turbulent or nonlaminar flow may occur at values far less than 2000 Re[4]. Although the descending aorta and flow are inadequate for "complete" mixing, our laboratory's recent series of experiments using multiple sites and directions of injection of microspheres in the same patient revealed essentially identical distribution patterns.

As early as 1969, ^{131}I-labeled MAA was used for peripheral perfusion scanning, but subsequent reports have been limited. Recently, with the introduction of human serum albumin microspheres, which can be labeled with ^{99m}Tc or ^{131}In, the utilization of particulate radiopharmaceuticals in the evaluation of peripheral vascular disease has been rejuvenated. Remember, however, that the particulate tracers do not yield, by themselves, direct quantitative data of blood flow but only permit quantification of the relative distribution of perfusion.

My colleagues and I reported on a large group of normal individuals and patients with peripheral vascular disease studied by the particle-distribution method using injections of 5 to 10 mCi of ^{99m}Tc-labeled albumin microspheres injected either intra-aortically, with a translumbar approach, or through direct femoral artery catheterization with a Teflon-sheathed 19-gauge needle (Fig. 20-4). The patients were then scanned anteriorly and posteriorly with a dual 12.7 cm scanner using 1:5 minification. The distribution of radioactivity in the extremity was categorized into five distinct peripheral

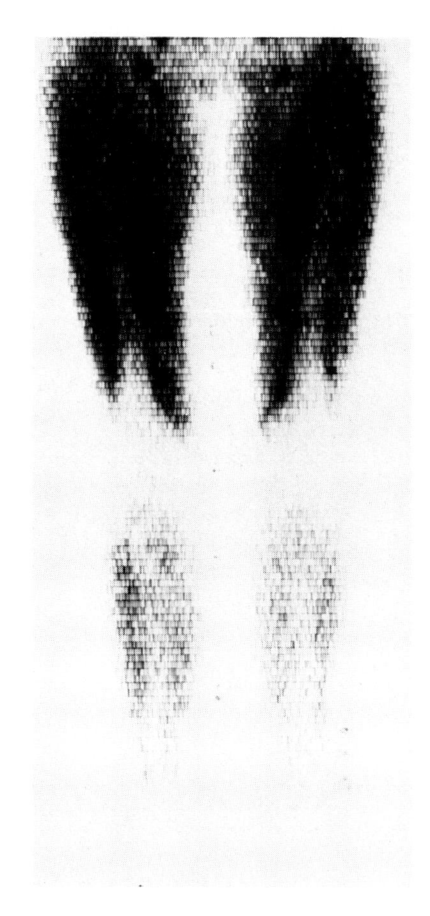

Fig. 20-4. Normal peripheral perfusion scan obtained after injections of ^{99m}Tc-labeled albumin microspheres into each common femoral artery. Tracer is distributed proportional to large muscle masses of thigh and calf. Increased radioactivity in musculature of thigh is related to greater muscle bulk in that area. Activity is determined in regions of lesser muscle mass, such as knees and ankles. (From Giangiana, F. A., Siegel, M. E., James, A. E., et al.: Radiology **108:**619, 1973.)

perfusion patterns.[9] More recently, using rest and stress scanning, a sixth pattern has been described. A technique to evaluate the peripheral vascular perfusion both at rest and under stress, simulating the conditions under which the patients are symptomatic, was reported.[11] Hashida[3] and others have implied that the evaluation of the peripheral vasculature with reactive hyperemia is indicative of the extremities' reaction to exercise. I utilized this and supplemented previous work with ^{131}I-labeled MAA by studying the same patient under multiphysiologic conditions. In addition, by using radioactive microspheres, I could also

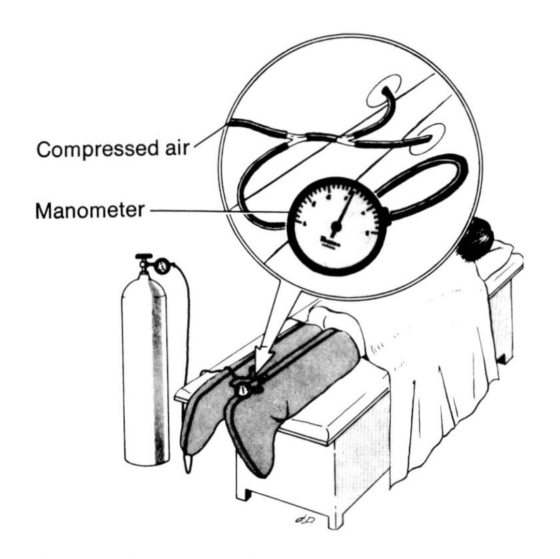

Fig. 20-5. Method used for producing arterial occlusion of lower extremities to evaluate perfusion during period of reactive hyperemia. Full-leg inflatable stockings, which are attached to each other, are connected to manometer and cylinder of compressed air. (From Siegel, M. E., et al.: Am. J. Roentgenol. Radium Ther. Nucl. Med. **118:**814, 1973.)

study other aspects of peripheral vascular disease, for example, A-V shunting.

These studies were performed as follows: After completion of the rest study as previously described, the patient's legs were then inserted into full-leg inflatable stockings, which were inflated for 5 min to 50 mm of Hg above the patient's systolic blood pressure (Fig. 20-5). After release of the pressure, 5 to 10 mCi of ^{113m}In microspheres was injected at the peak of the reactive hyperemia. The distributions of the two differently labeled microspheres are imaged separately, anteriorly and posteriorly, with 1:5 minification. The first pattern, seen in patients with no known peripheral vascular disease, demonstrated the activity to be distributed in proportion to muscle mass with a relative decrease in radioactivity in the nonmuscular regions, that is the knees and ankles (Fig. 20-6). The second pattern demonstrated the activity to be distributed

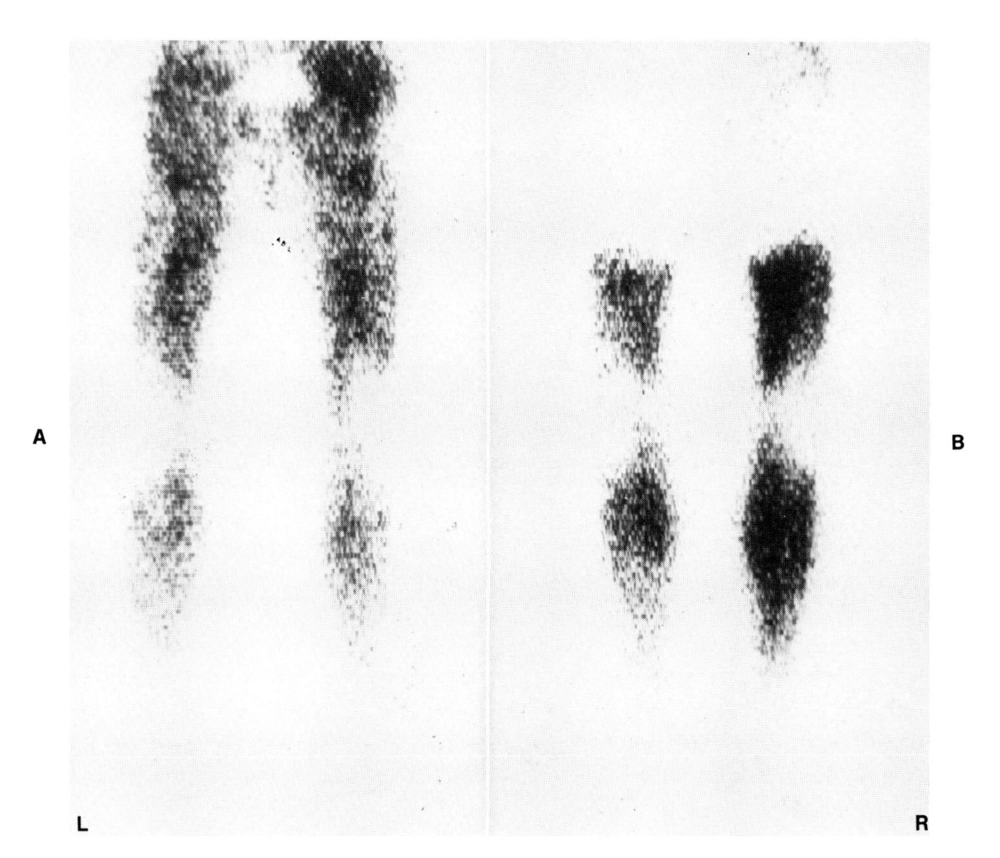

Fig. 20-6. Normal rest **(A)** and reactive hyperemia **(B)** perfusion scans with the distribution of perfusion in proportion to muscle mass.

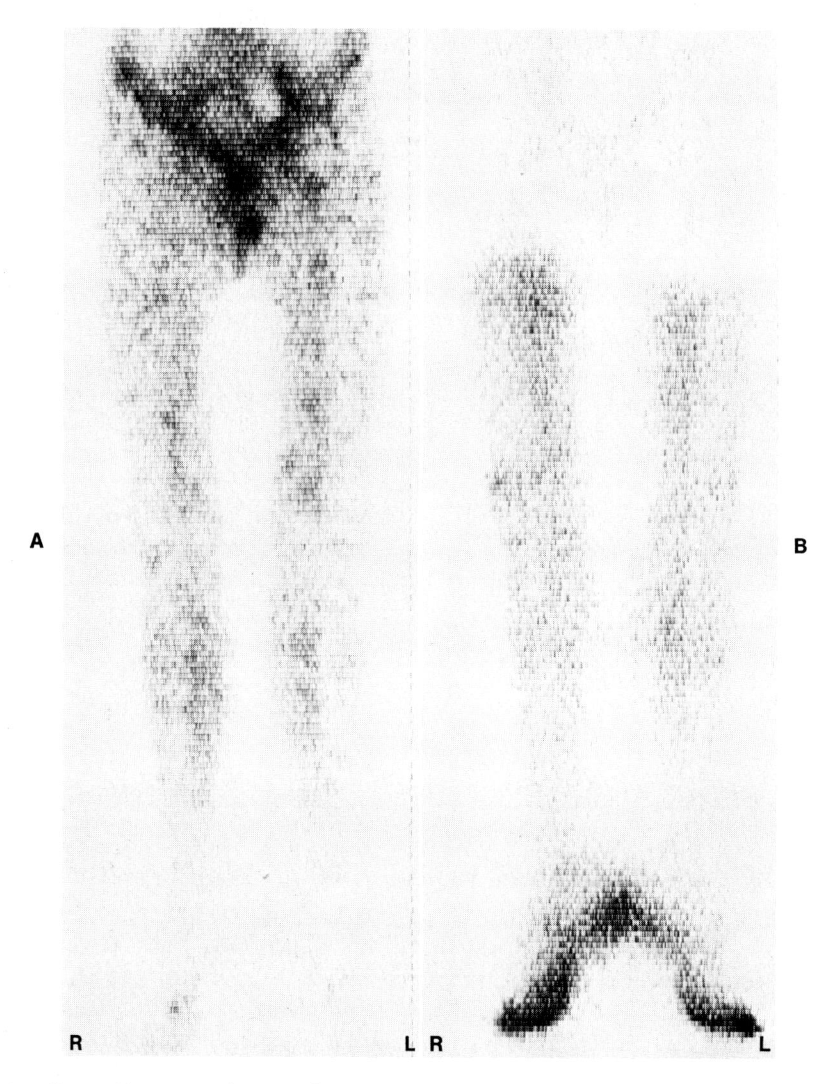

Fig. 20-7. Abnormal rest **(A)** and reactive hyperemia **(B)** perfusion scans demonstrating diffuse small-vessel disease pattern of perfusion proportional to skin surface rather than to muscle mass. This patient had claudication and normal arteriogram. (From Siegel, M. E., et al.: Am. J. Roentgenol. Radium Ther. Nucl. Med. **118:**814, 1974.)

more in proportion to skin surface than muscle mass with a more or less homogeneous distribution of activity (Fig. 20-7). This is frequently seen in patients with symptoms of ischemia, essentially normal arteriograms, and diabetes. It is thought these patients had diffuse small-vessel disease below the resolution of the arteriogram. The third pattern demonstrated an increase in the relative perfusion of the nonmuscular structures, that is, knees and ankles, while the major muscle masses remain easily identifiable (Fig. 20-8). This is seen in patients with known multifocal large-vessel disease. The fourth pattern demonstrated marked asymmetry in the distribution of radioactivity with a relative hypoperfusion of one or more of the muscle masses (Fig. 20-9). This pattern is seen in patients with focal large-vessel disease. The fifth pattern, seen in patients with ischemic ulcers of the distal extremity, demonstrated

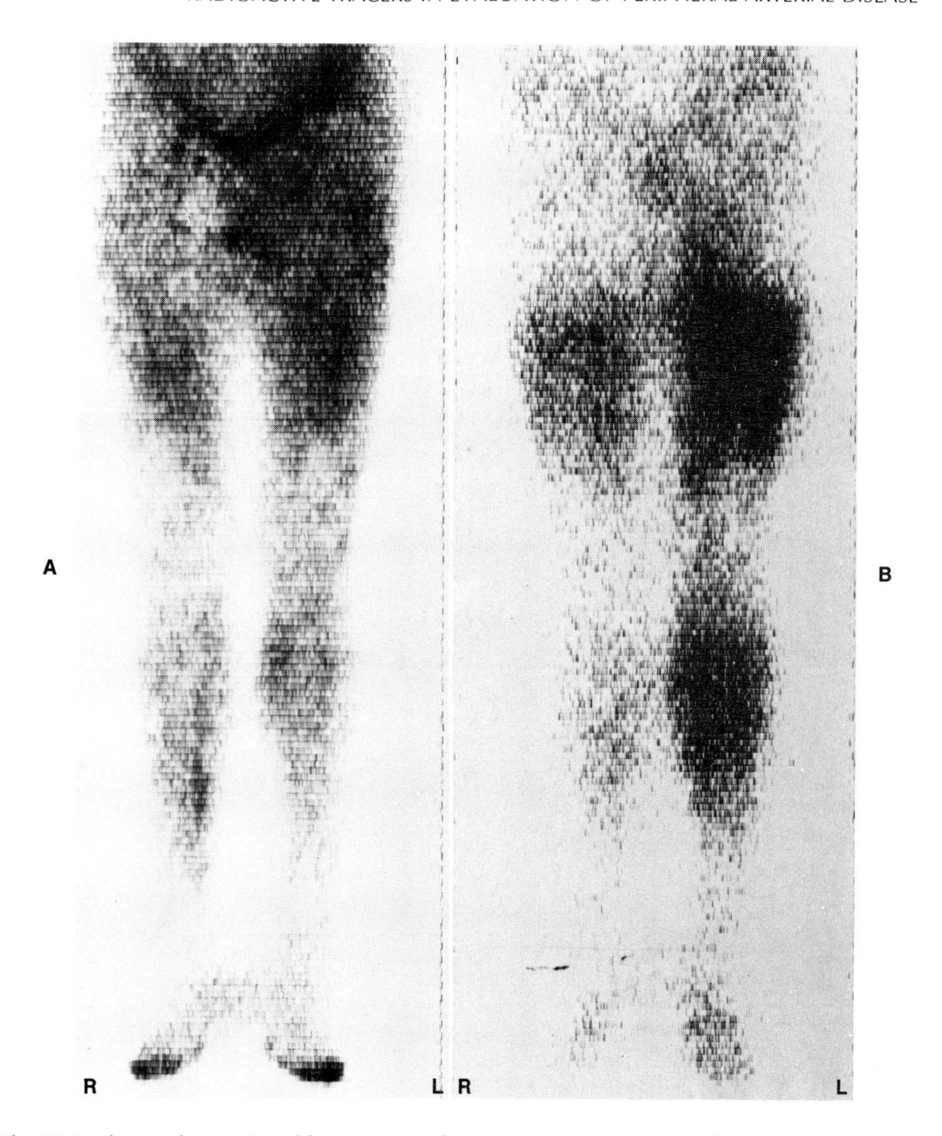

Fig. 20-8. Abnormal rest **(A)** and hyperemia perfusion **(B)** scans in patient with multifocal large-vessel disease, demonstrating relative increase in perfusion to nonmuscular structures, that is, knees and ankles, whereas major muscle bundles remain easily identifiable. (From Siegel, M. E., et al.: Am. J. Roentgenol. Radium Ther. Nucl. Med. **118:**814, 1974.)

increased activity in the region of the ulcer bed (Fig. 20-10). The sixth pattern demonstrated a relative increased activity in the osseous structures of the extremity (Fig. 20-11). This pattern was seen in patients with Paget's disease of fibrous dysplasia involving the lower extremities (Table 20-2). In recent unpublished animal studies, Giargiana and I have demonstrated that the change in blood flow during reactive hyperemia, at least in the major arterial trunks of the leg, appears to be quite similar in intensity and duration to that induced with radiographic contrast media. We have recently begun to do the rest and hyperemia studies in patients using this approach and have initially found the distribution pattern to be quite similar to that produced by the inflatable boot approach. The changes that occur due to the relative muscular perfusion when stressed estimate the pathophysiologic changes that occur when a patient ex-

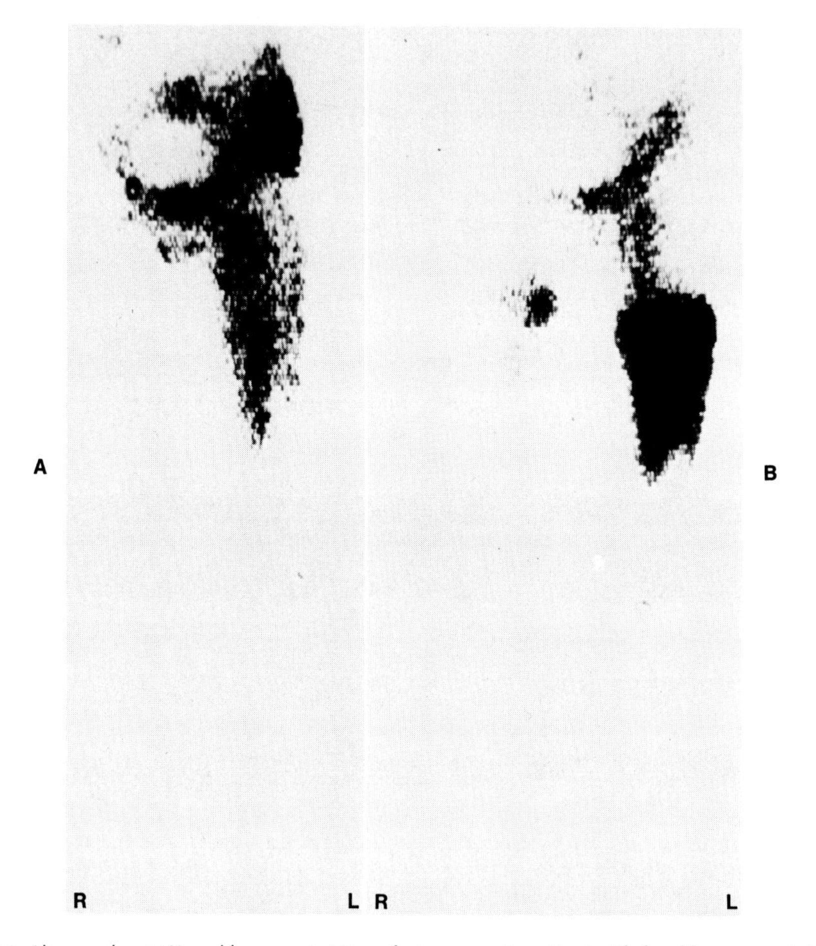

Fig. 20-9. Abnormal rest **(A)** and hyperemia **(B)** perfusion scans in patient with focal large-vessel disease, demonstrating perfusion in proportion to muscle mass, but with relative hypoperfusion of major muscle masses. Patient had severe right common iliac stenosis. (From Siegel, M. E., and Wagner, H. N.: Semin. Nucl. Med. **6:**253-278, 1976, by permission.)

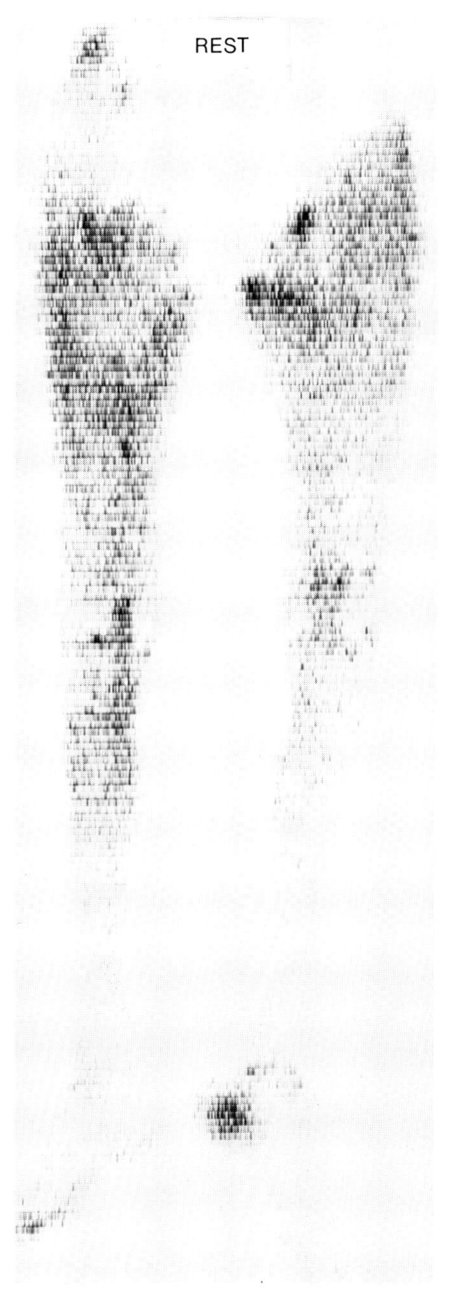

Fig. 20-10. Abnormal rest perfusion scan demonstrating relative hyperperfusion of left heel in identical location of patient's ischemic ulcer.

Table 20-2. Peripheral perfusion patterns

Disease state	Pattern description
Normal	Tracer distributed in proportion to muscle mass
Diffuse small-vessel disease	Tracer distributed in proportion to skin surface
Multifocal disease	Tracer distributed primarily in proportion to muscle mass yet relative increase in nonmuscular regions
Focal large-vessel disease	Tracer distributed in proportion to muscle mass yet relative decrease in one or more of the masses
Hyperperfused ischemic ulcer	Increase in tracer in ulcer bed compared to surrounding region
Hyperfused osseous disease	Increase in tracer in involved bone

periences symptoms such as claudication. It is thus thought that by evaluating the changes of relative distribution that occur in the extremities between the basal and hyperemic state, an estimation of the physiologically significant intravascular disease can be obtained. It seems logical that if the fixed resistance to blood flow in an extremity is peripherally located, surgically correcting only the proximal lesion may not alleviate the problem.

One prerequisite for a new diagnostic or prognostic procedure to be accepted is that it should offer previously obtainable information more easily or new, potentially useful clinical information about a disease process. My co-workers and I reported on

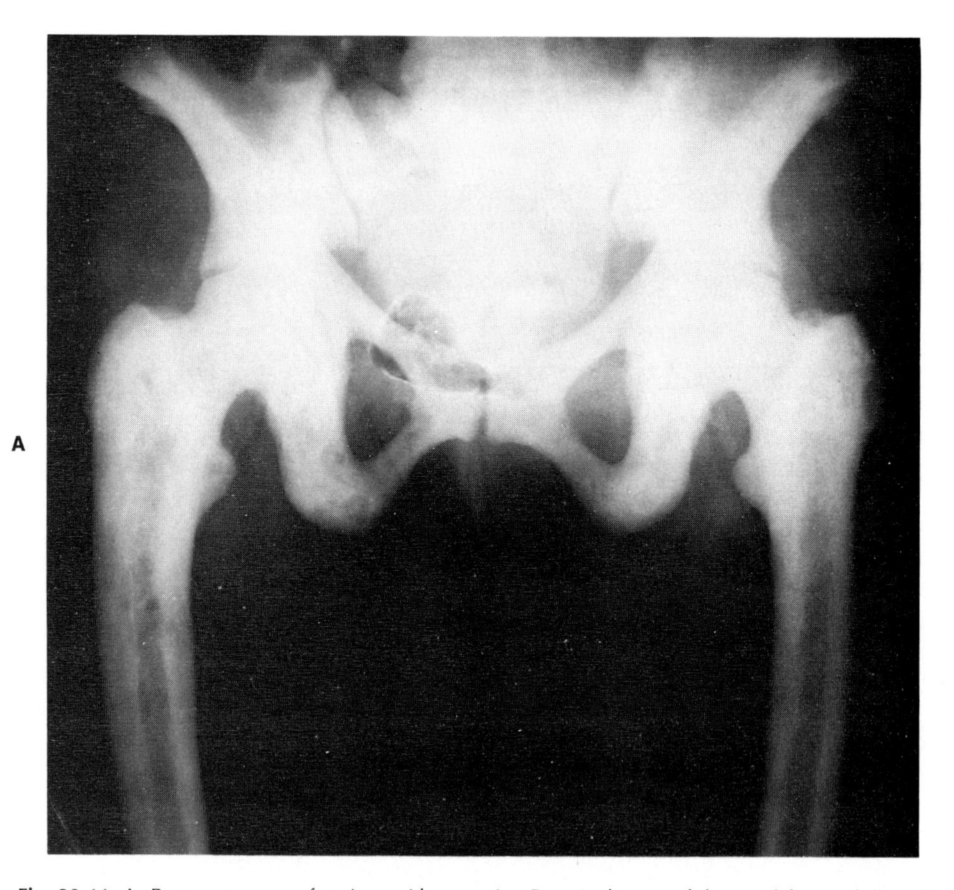

Fig. 20-11. A, Roentgenogram of patient with extensive Paget's disease of iliac and femurs bilaterally. Extent of involvement included entire femurs and proximal parts of both tibias. **B,** Perfusion scan in same patient performed by injection of microspheres into femoral artery. Relative hyperperfusion is seen in bones affected with Paget's disease. This is compatible with known increase in vascularity in bones involved with Paget's disease.

a study correlating the clinical, scan, and arteriographic findings in a group of patients with symptomatic peripheral vascular disease, which demonstrated that the scan does not necessarily offer the same information as obtained from the other two modalities and, indeed, may offer potentially useful physiologic information concerning the arterial disease.[12] There was, as expected, a poor correlation between the scans and the arteriograms, with 20% of the patients demonstrating more extensive disease on the scan than on the arteriogram and 40%, less. It was noted that the vast majority of the patients with more extensive disease on the scan were diabetics who had essentially normal arteriograms and scans, which suggested diffuse small-vessel disease. In fact, the scan is often the only objective evidence of these patients' diffuse arterial involvement.

There was agreement in 53% of the cases between the patient's symptoms and the arteriogram, whereas there was a positive

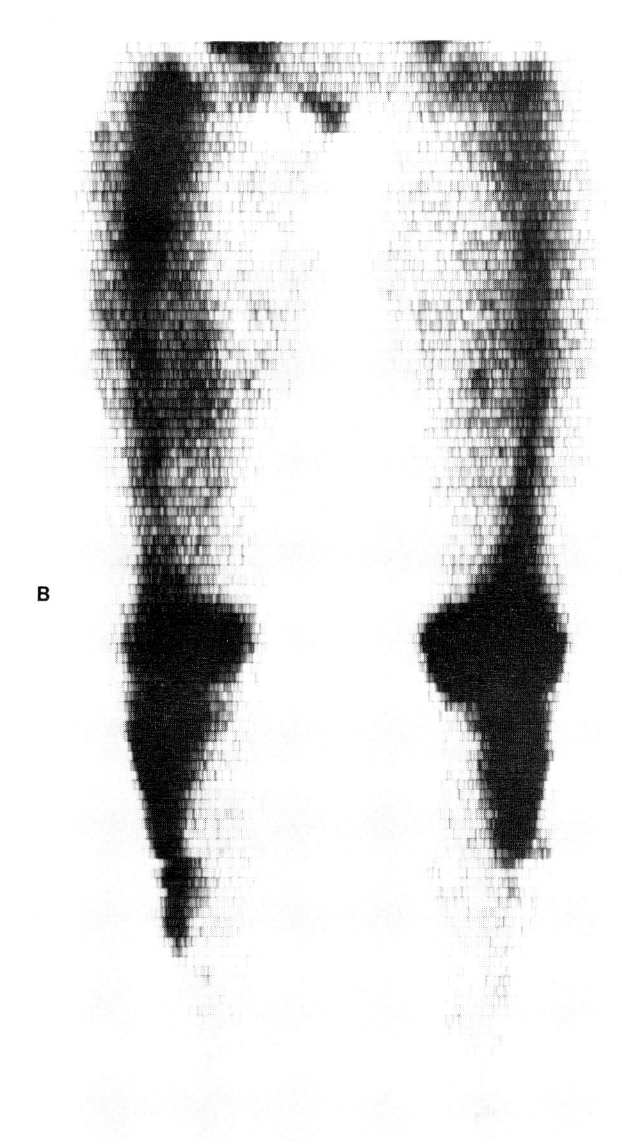

Fig. 20-11, cont'd. For legend see opposite page.

Table 20-3. Comparison of scans, patients' symptoms, and arteriograms

Categories	Number of patients	Percentage of total
Correlation of perfusion scans and arteriograms		
Rest scan		
Positive	11	37
Less extensive disease	15	50
More extensive disease	4	13
Reactive hyperemia scan		
Positive	12	40
Less extensive disease	12	40
More extensive disease	6	20
Correlation of patients' symptoms with arteriogram and scan findings		
Arteriograms and patients' symptoms		
Agreement	16	53
No agreement	14	47
Rest scans and patients' symptoms		
Agreement	18	60
No agreement	12	40
Reactive hyperemia scans and patients' symptoms		
Agreement	25	83
No agreement	5	17

correlation in 60% of the cases using the rest scan alone and in 83% using both the rest and the reactive hyperemia scans (Table 20-3). These findings support the concept that in comparing the two scans, one sees the distribution of perfusion at rest, when many patients are asymptomatic, and under stress, when the patient is exercising and symptomatic. It is this change in perfusion that indicates the physiologic significance of an arterial lesion and may be helpful in locating points of fixed resistance, which, if the perfusion past them is primarily to muscle, might be worth bypassing. The poor correlation between the arteriogram and the patient's symptoms does not degrade the arteriogram but affirms the hypothesis that the arteriogram is basically a morphologic study and does not necessarily yield data regarding the physiologic significance of the disease it detects.

The definitive answer to what are the predictive points to observe when viewing the scans is not yet available, but preliminary studies suggest utilization in the manner of the following example: A 45-year-old nondiabetic man with a history of intermittent claudication on the right and no rest pain was studied. No pulses were palpable below either femoral artery. The arteriogram revealed a right common iliac stenosis, bilateral profound stenosis, and bilateral superficial femoral artery stenosis. Trifurcated vessels were demonstrated on both sides. The rest scan revealed almost symmetric perfusion with good muscle delineation, although there was a slight relative hypoperfusion of the left leg compared to the right. The reactive hyperemia scan, however, demonstrated marked asymmetry of the perfusion with a relative hypoperfusion of the right calf with respect to the left (Fig. 20-12). This suggested that, although both legs were abnormal arteriographically, the peripheral vasculature on the left was much better able to respond to stress with the production of a needed relative muscle hyperemia. In addition, it suggested that there were fixed points of resistance at the level of the right superficial femoral lesion, even though it appeared arteriographically equal to that on the left. Because the perfusion distal to the lesion appeared to be distributed primarily to muscle in which it was needed, if surgical bypass could be performed there should be increased flow to the muscle groups. Bypass procedure was performed on the right only, and the patient is doing well.

Assessment of healing potential of ischemic ulcers. One evaluation of regional perfusion that has already proved to be prognostically significant is the evaluation of ischemic ulcers. It is often difficult to determine whether a lesion will heal with conservative management or a disfiguring operation is necessary. Previous guidelines were often inadequate, leading to prolonged hospitalization or premature surgery. The term ischemic ulcer is used to denote a skin lesion with tissue loss relative to arterial disease,

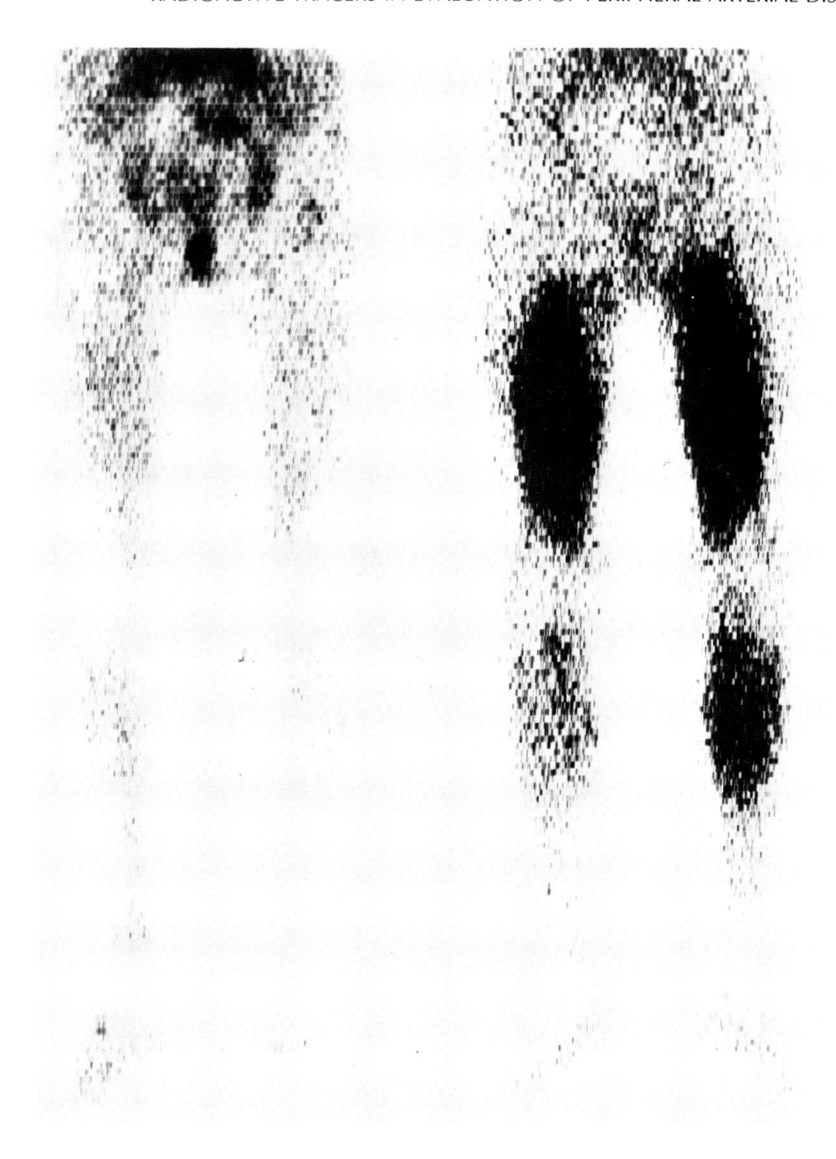

Fig. 20-12. Physiologic response of circulation to stress is well demonstrated. Although perfusion is primarily to muscle, fixed resistance proximal to right calf inhibits its response to stress, in comparison to left side.

rather than the perfusion state of the lesion. As mentioned earlier, an objective, reliable prognostic criterion for determining the healing potential of an ischemic ulcer is needed to prevent unnecessarily prolonged hospitalization or premature surgery.

Inherent in the healing process is the ability to develop an inflammatory response and hyperemia, which are requirements for the revitalization process. As previously described, after intra-arterial injections of degradable radioactive human serum albumin microspheres (15 to 30 μm in diameter), the relative blood flow in the extremity is proportional to the distribution of microspheres, which can be visualized and quantitated by scintillation scanning. In patients with arterial diseases, the arteriole may be maximally dilated secondary to ischemia, and no further response is possible in areas of injury or infection. Thus the ratio of activity per unit area of lesion to that of a neighboring area reflects the degree of ischemia on a local or microcirculatory level, which in turn may predict whether or not the lesion will heal with conservative measures.

I used this principle to investigate the relationship between the presence and degree of hyperemia in the ulcer bed and the ability of the ulcer to heal with conservative management. I also investigated the value of the arteriographic finding of patent trifurcation vessels and the presence of diabetes or palpable peripheral pulses in predicting which ulcers will heal.

Peripheral vascular perfusion scans were performed in a group of sixty patients who had, clinically, ischemic lesions of the lower extremity.[14] In 50% of these patients, the procedure was performed at the time of arteriography. From 5 to 10 mCi of ^{99m}Tc-labeled albumin microspheres (15 to 30 μm in diameter) were injected 15 min after completion of the arteriography via a translumbar catheter placed into the descending aorta. The catheter tip was located just below the level of the renal arteries. In the remaining 50% of the patients either arteriography was not requested or the patient was an outpatient. In these cases approximately 5 mCi of ^{99m}Tc-labeled albumin microspheres was injected via a Teflon-sheathed 19-gauge needle, which was placed in the femoral artery of the extremity in question. The injection was made upstream through this small catheter.

After injection of the particulate tracer, one is then able to visualize and, in a relative sense, quantify the distribution of perfusion in and around the region of interest. At the completion of the procedure, the extremity was scanned anteriorly and posteriorly with a dual-probe 12.7 cm rectilinear scanner, using 1:5 minification. The actual scan was not essential to this investigation but, as will be discussed later in this chapter, may be useful for optimizing the section of amputation level in the event that amputation is considered. The amount of radioactivity per unit area in the ulcer bed and in the adjacent areas and thus the relative hyperemia in these areas was determined by point counting.

Table 20-4. Presence of adequate relative hyperemia* and ulcer healing

	Hyperemia present	Hyperemia absent
Healed	27	4
Did not heal	3	26

*Adequate relative

$$\text{hyperemia} = \frac{\text{Activity/unit area over ulcer}}{\text{Activity/unit area over surrounding tissue}} = 3.5{:}1$$

Table 20-5. Clinical indicators of ulcer healing

Variable	Healed	Did not heal
Diabetes mellitus		
Present	13	15
Absent	18	14
Runoff		
Good*	6	8
Poor	9	9
Palpable peripheral pulses†		
Palpable	5	7
Not palpable	26	22

*At least two patent trifurcation vessels.
†Below femoral pulse.

Clinical information about each patient was recorded with particular reference to the presence of diabetes, peripheral pulse rates, and the status of the trifurcation vessels, if known. With the cooperation of surgical colleagues, the maximum, clinically feasible, conservative, nonamputative course of management, ranging from ten days to four months, was afforded each patient.

To evaluate relative perfusion and maximize the separation of those who healed from those who did not heal, a ratio of 3.5:1 was taken as the minimum hyperemic response adequate for healing. At or above this ratio, 90% (twenty-seven out of thirty) of the patients' ulcers healed. Conversely, of the thirty patients whose ulcers did not display at least this magnitude of hyperemia, only 10% (four out of thirty) went on to heal; the remainder underwent amputation (Table 20-4).

We also found, however, no consistent association between the presence or ab-

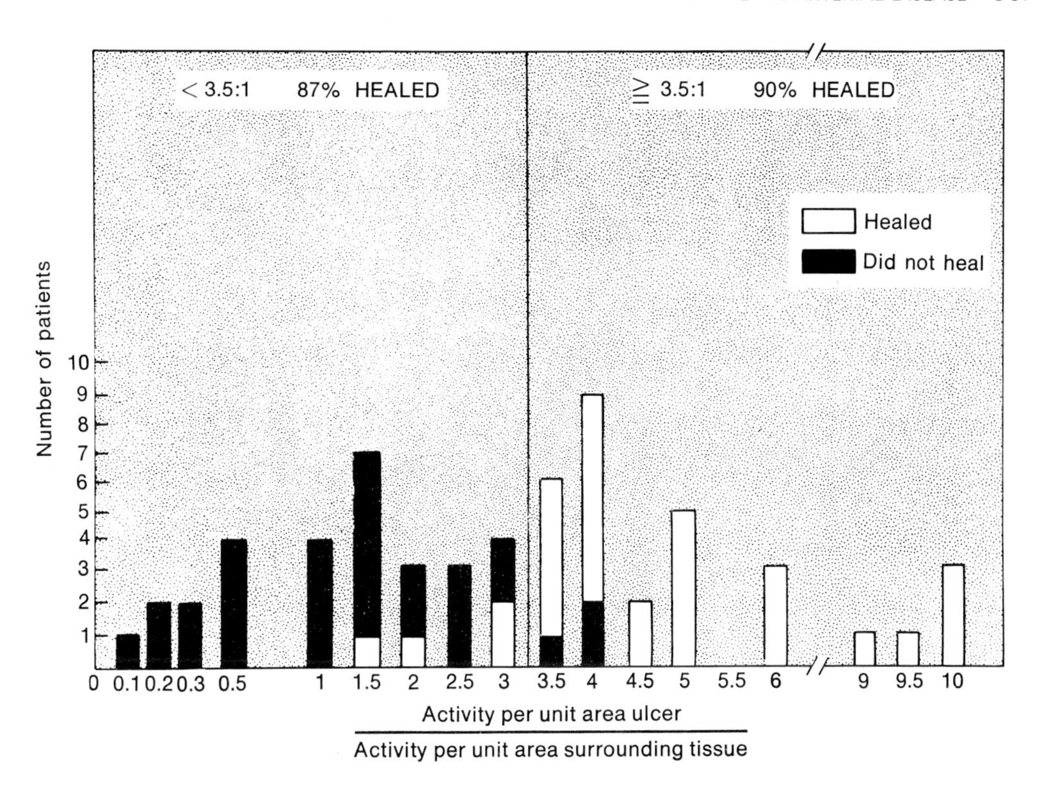

Fig. 20-13. Relationship between hyperemia of ulcer bed to adjacent tissue and ulcer healing.

sence of diabetes, peripheral pulse rates, or two-vessel runoff on the arteriogram and the ability of an ischemic lesion to heal with conservative management (Table 20-5). However, there was a striking association between the relative hyperemia of the ulcer bed to the adjacent tissue of at least 3.5:1 and the ability of the ischemic ulcer to heal.[14] With at least this degree of hyperemia in their ulcers, 90% of the patients' lesions healed. Conversely, 90% of the patients without this degree of hyperemia required an amputation because of nonhealing of their lesions (Fig. 20-13). These findings have proved to be of significant clinical value in the management of patients thought to have ischemic ulceration.

ARTERIOVENOUS SHUNT QUANTIFICATION, VERIFICATION, AND LOCALIZATION. Intra-arterially injected particles, larger in diameter than the capillaries, are entrapped in the microcirculation in proportion to the blood flow. Blood flow through an arteriovenous communication larger than the capillaries carries a proportional fraction of the injected particles past this capillary bed to the lungs, where the microcirculation removes the particles from the circulation. This fraction, which bypasses the initial capillary bed, depends on the fraction of blood flowing through the communication as well as the size of the particles relative to the communicating vessels. If the particles used are larger than the capillaries as well as any arteriovenous (A-V) communications present, the perfusion visualized in the extremity will represent total blood flow, and false-negative shunt determinations will be derived. When the particle size is widely variable, visualization of peripheral perfusion will represent something between nutritive and total blood flow, and, of course, shunting studies are meaningless under these conditions. If the particles used are large enough to be trapped by the normal microcirculation but small enough to pass through abnormal A-V communications, the visualized peripheral perfusion will represent nutritive flow,

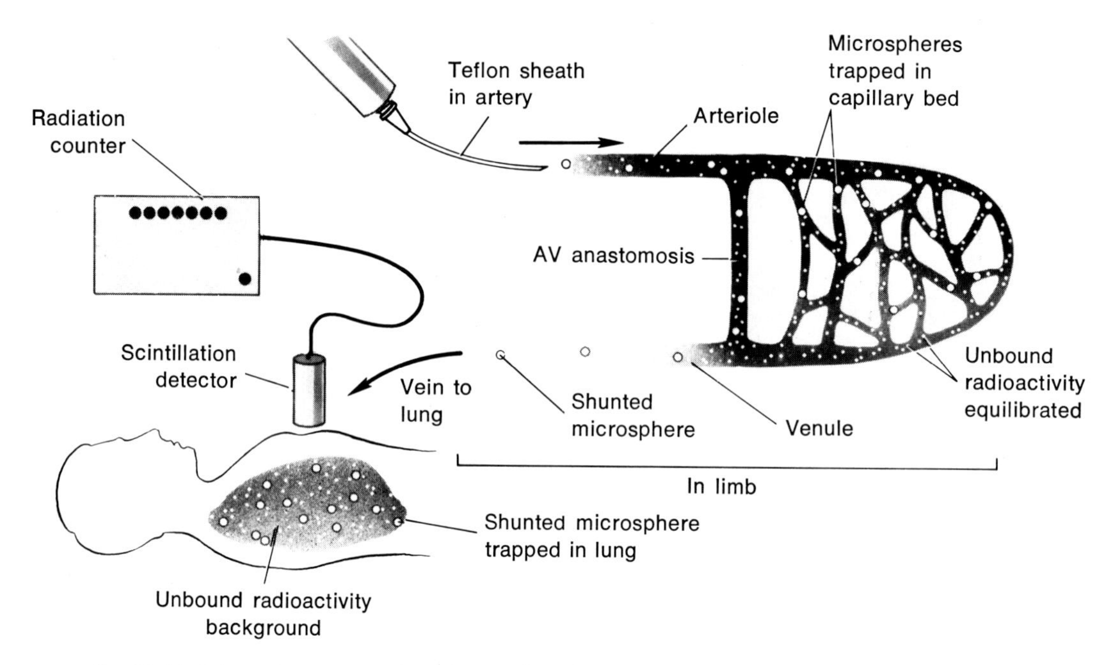

Fig. 20-14. Schematic representation of principles of A-V shunt detection by microsphere technique. (From Rhodes, B. A., Greyson, N. D., Hamilton, C. R., et al.: N. Engl. J. Med. **28:**686, 1972, reprinted, by permission.)

and the presence of A-V communications can be documented and quantitated. By counting over a representative section of lung while maintaining constant geometric relationships, the following data are collected:

1. A known quantity of supernate containing unbound activity is given intra-arterially to determine the amount of nonparticulate radioactivity that will be given in subsequent injection.
2. A known quantity of radioactively labeled particles is injected intra-arterially, and that activity in the lungs arising from trapped particles is monitored.
3. Finally, a known quantity of labeled particles is given intravenously, is trapped in the lung, and represents 100% shunting (Fig. 20-14).

Then by the use of the following equation, the percentage of the injected particle passing through the A-V communication is determined:

$$\text{Shunting}\,(\%) = \frac{C_i - C_{i-1} - C_1\left(\dfrac{V_i}{V_1}\right) \div A_i}{C_n - C_{n-1} - C_1\left(\dfrac{V_n}{V_1}\right) \div A_n} \times 100$$

where

C_1 = Counting rate after injection suspending solution (background)
C_i = Counting rate after ith arterial injection
C_n = Counting rate after last or venous injection
V_1 = Volume of suspending solution
V_i = Volume of ith arterial injection
V_n = Volume of last venous injection
A_i = Amount of ith arterial injection (mCi)
A_n = Amount of last or venous injection (mCi)

Patients with peripheral vascular disease often demonstrate, arteriographically, early venous opacification. The explanation of this finding had been that this was a manifestation of the A-V shunting that occurred in this group of patients. Early shunt studies performed with ^{131}I-labeled MAA supported the presence of this suspected shunting.

However, more recent studies using radioactive human serum albumin microspheres have challenged this finding. Using radioactive albumin microspheres, it has been shown that there are no significant A-V shunts in patients with peripheral vascular disease in the basal state, during reactive hyperemia, and immediately after contrast media injection. The early venous opacification is thought to be due to decreased resistance with resultant decreased transit time rather than actual A-V communications. The procedure has also been used clinically to document shunting in a patient with pulmonary osteoarthropathy and hypernephroma and to quantify the degree of shunting in patients with known A-V malformations.

This technique has already made significant contributions to understanding of the underlying pathophysiologic process in Paget's disease of the bone. Similar to patients with A-V fistula, patients with Paget's disease have an increased oxygen concentration in the venous blood and some demonstrate early venous opacification on angiograms. Because of this opacification and other cardiovascular findings, it has been generally considered that A-V communications are present in Paget's disease of the bone, even though A-V communications have never been demonstrated morphologically. A group of patients with Paget's disease involving the extremity were studied with anatomic, 15 to 30 μm in diameter ^{99m}Tc-labeled albumin microspheres, and it was concluded that anatomic A-V communications greater than 15 μm were not present in the involved bone.[8]

Amputation level selection. A procedure to amputate a lower extremity in the treatment of peripheral vascular disease should combine maximum potential for rehabilitation with minimum morbidity; it remains a difficult goal to achieve.

Not uncommonly, the surgeon must subject a patient to progressive levels of amputation and months of hospitalization before a viable, well-healed amputation stump can be obtained. A simplified approach taken by some is to select the level of amputation at the lowest level of skin viability as determined by testing for bleeding at the time of operation. This clinical test is very nonspecific and does not relate to the ability of the skin to heal or to the degree of muscle ischemia. Of the many ancillary aids available to the surgeon, angiography is the most widely accepted. It provides detailed information as to the gross vascular anatomy, including the presence of collateral circulation and an indication of the degree of atherosclerotic disease. However, it provides only indirect evidence of the actual perfusion of capillary beds, and most surgeons have become discouraged with this approach in determining the level of amputation. An example of how the microsphere approach is being applied, at least initially, can be illustrated by the following case (Fig. 20-15): The patient presented with a barely perceptible ischemic ulcer of the great toe of his left foot. The initial perfusion scan demonstrated no hyperemia, and thus the lesion had only 10% probability of healing with conservative management. A repeat study one month later confirmed the lack of hyperemia, and, in fact, the relative perfusion had decreased further. By this time the ulcer had enlarged and other toes were now involved. An arteriogram revealed diffuse disease with poor trifurcation of the vessels. There was debate among the surgeons as to whether to do an above knee (AK) or a below knee (BK) amputation. Because the counts per unit area at the level of a BK amputation were greater than 50% of the midthigh counts, I suggested that a BK would heal. (In a limited group of perfusion studies on normal patients, there has been up to a 60% decrease of perfusion to the calf in comparison to the thigh. This is most likely related to a large extent on the relatively greater muscle mass in the thigh in comparison to the calf.) The BK amputation was done, and it healed well.

Perhaps somewhat more promising at

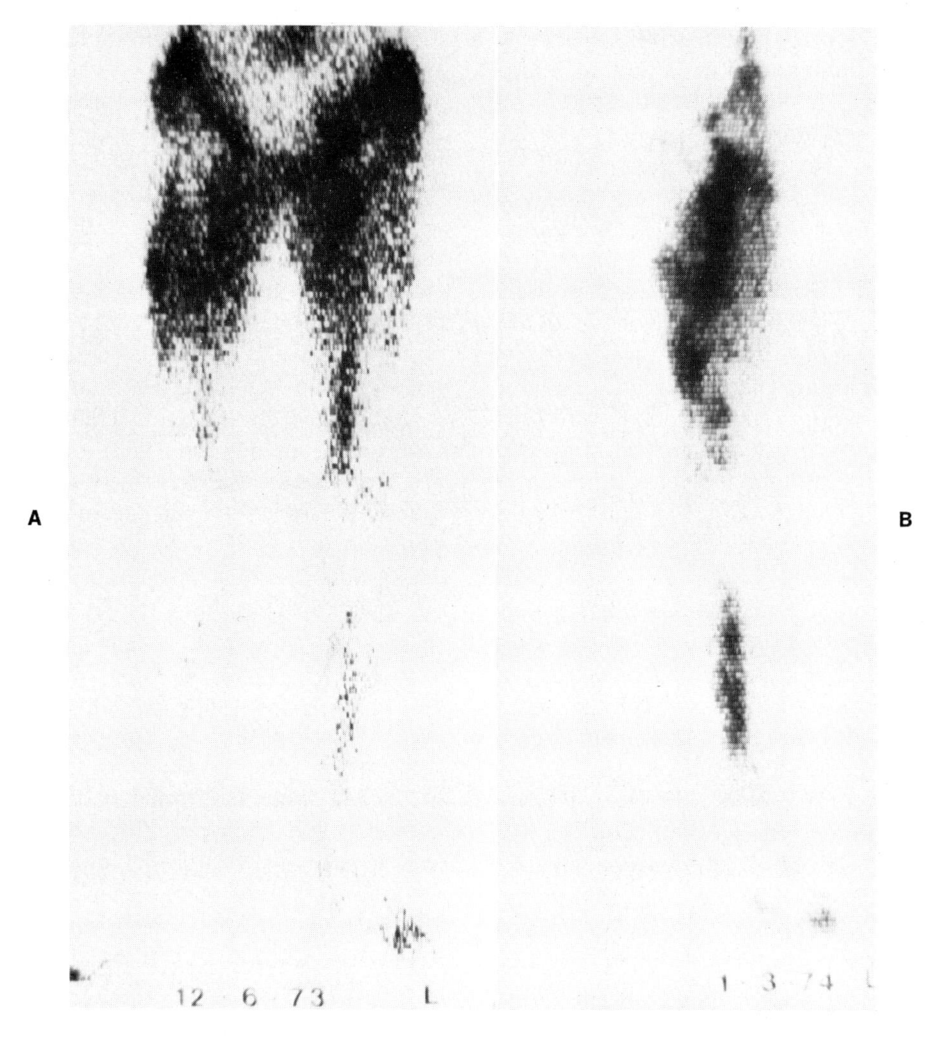

Fig. 20-15. A, Initial perfusion scan demonstrates relative hypoperfusion of region of left great toe. **B,** Repeat study done reveals increase in relative hypoperfusion in region of toes of left foot. (Increased tracer in left foot is in midmetatarsal region.) (From Siegel, M. E., and Wagner, H. N.: Semin. Nucl. Med. **6**:253-278, 1976, by permission.)

the present time is the use of the iodoantipyrene-determined skin perfusion pressures, as previously described, to determine the level at which healing will occur. Lassen and Holstein[5] consider a pressure less than 20 to 30 mm of Hg unsatisfactory for healing, and thus suggest consideration of amputation at more well-perfused proximal level.

REFERENCES

1. Blumgart, H. L., and Yens, O. C.: Studies on velocity of blood flow, J. Clin. Invest. **4**:1, 1927.
2. Dibos, P. E., Muhletaler, C. A., Natarjan, T. K., et al.: Intravenous radionuclide arteriography in peripheral occlusive arterial disease, Radiology **102**:181, 1972.
3. Hashida, Y.: Peripheral arteriography using reactive hyperemia, Jpn. Circ. J., **27**:349, 1963.
4. Krovetz, L., and Benson, R.: Mixing of dye and blood in the canine aorta, J. Appl. Physiol. **20**:922, 1950.
5. Lassen, N. A., and Holstein, P.: Use of radioisotopes in assessment of distal blood flow and distal blood pressure in arterial insufficiency, Surg. Clin. North Am. **54**:39, 1974.
6. Miyamoto, A. T., Mishkin, F. S., and Maxwell, T. M.: Non-invasive study of extremity perfusion by potassium-43 scanning, J. Nucl. Med. **15**:518, 1974.

7. Raban, P., Gregora, V., Sindelar, J. et al.: Two alternate techniques of labeling iron-free albumin microspheres with ^{99m}Tc and ^{113}In, J. Nucl. Med. **12:**616, 1971.
8. Rhodes, B. A., Greyson, N. D., Hamilton, C. R., et al.: Absence of anatomic arteriovenous shunts in Paget's disease of bone, N. Engl. J. Med. **287:** 686, 1972.
9. Rhodes, B. A., Greyson, N. D., Siegel, M. E., et al.: The distribution of radioactive microspheres in patients with peripheral vascular disease, Am. J. Roentgenol. Radium Ther. Nucl. Med. **118:**820, 1973.
10. Serjrsen, P.: The epidermal diffusion barrier to Xe-133 in man and studies of clearance of Xe-133 by sweat, J. Appl. Physiol. **24:**2111, 1968.
11. Siegel, M. E., Giargiana, F. A., Rhodes, B. A., et al.: Effect of reactive hyperemia on the distribution of radioactive microspheres in patients with peripheral vascular disease, Am. J. Roentgenol. Radium Ther. Nucl. Med. **118:**814, 1973.
12. Siegel, M. E., Giargiana, F. A., White, R. I., et al.: Peripheral vascular perfusion scanning: Correlation with the arteriogram and clinical assessment in the patient with peripheral vascular disease, Am. J. Roentgenol. Radium Ther. Nucl. Med. **122:**628, 1975.
13. Siegel, M. E., and Wagner, H. N., Jr.: Radioactive tracers in peripheral vascular disease. In Freeman, L. M. and Blaufox, M. D., editors: Seminars in nuclear medicine, New York, 1976, Grune & Stratton, Inc.
14. Siegel, M. E., Williams, G. M., Giargiana, F. A., and Wagner, H. N., Jr.: An objective useful criterion for determining the healing potential of an ischemic ulcer, J. Nucl. Med. **16:**933, 1975.
15. Tonnesen, K. H.: The blood flow through calf muscle during rhythmic contraction and in rest and in patients with occlusive arterial disease measured by Xe-133, Scand. J. Clin. Lab. Invest. **17:**433, 1965.

BIBLIOGRAPHY

Alpert, J. S., Garcia del Rio, H., and Lassen, N. A.: Diagnostic use of radioactive xenon clearance and a standardized walking test in obliterative arterial disease of the legs, Circulation **34:**849, 1966.

Alpert, J. S., Larsen, O. A., and Lassen, N. A.: Evaluation of arterial insufficiency of the legs—a comparison of arteriography and the Xe^{133} walking test, Cardiovasc. Res. **2:**161, 1968.

Alpert, J. S., Larsen, O. A., and Lassen, N. A.: Exercise and intermittent claudication. Blood flow in the calf muscle during walking studied by the Xe^{133} clearance method, Circulation **39:**353, 1968.

Andersen, A. M., and Ladefoged, J.: Partition coefficient of Xe^{133} between various tissues and blood in vivo, Scand. J. Clin. Lab. Invest. **19:**72, 1967.

Bardfeld, P. A., Wagner, H. N., Jr., Jones, E., et al.: The regional circulation in the extremities in normal man and in patients with peripheral arteriosclerosis, Abstracted, J. Nucl. Med. **8:**279, 1967.

Bell, G., and Harper, A.: Measurement of regional blood flow through the skin from the clearance of Kr^{85}, Nature (London) **202:**704, 1964.

Buchanan, J. W., Rhodes, B. A., and Wagner, H. N., Jr.: Labeling iron-free albumin microspheres with In^{113m}, J. Nucl. Med. **12:**616, 1971.

Christopher, F.: Peripheral vascula. In Davis, L., editor: Textbook of surgery, ed. 9, Philadelphia, 1968, W. B. Saunders Co., p. 1257.

Cuypers, Y., and Merchie, G.: Etude de la circulation sanguine peripherique a l'aide de serum albumine humaine marquee a I^{131}. La circulation an neveau du mollet, Cardiology **41:**166, 1962.

Cuypers, Y., and Merchie, G.: Etude de la circulation sanguine peripherique a l'aide de serum albumine humaine marquee a I^{131}. La circulation an neveau de pred, Acta Cardiol. **17:**117, 1962.

Dahn, I., Lassen, N. A., and Westling, H.: Blood flow in human muscles during external pressure or venous stasis, Scand. J. Clin. Lab. Invest. **99** (suppl.):160, 1967.

Dobson, E. L., and Warner, G. F.: Measurement of regional Na turnover rates and their application to the estimation of regional blood flow, Am. J. Physiol. **189:**269, 1957.

Fick, A.: Uber die Messung des Blutquantums in den Herzentrikeln, S. B. Phys. Med. Gest. Wurzb. **12:**16, 1870.

Gardner, T. J., Greyson, N. D., Rhodes, B. A., et al.: The clinical usefulness of leg scanning in arterial insufficiency using radioactive microspheres, Surg. Forum **13:**247, 1972.

Giargiana, F. A., Siegel, M. E., James, A. E., et al.: A preliminary report on the complimentary roles of arteriography and perfusion scanning in assessment of peripheral vascular disease, Radiology **108:**619, 1973.

Giargiana, F. A., White, R. I., Greyson, N. D., Rhodes, B. A., Siegel, M. E., Wagner, H. N., Jr., and James, A. E., Jr.: Absence of arteriovenous shunting in peripheral arterial disease. Presented at the annual meeting of the Radiological Society of North America, Chicago, 1972, Invest. Radiol. **9:**222, 1974.

Greyson, N. D., Rhodes, B. A., Williams, G. M., et al.: Radiometric detection of venous function and disease, Surg. Gynecol. Obstet. **137:**220, 1973.

Groom, A., Roberts, P. W., Roulands, S., et al.: Radioactive isotopes in studies on hemodynamics, Br. J. Radiol. **32:**641, 1959.

Henkins, R. E., Yao, J. S., Westerman, B. R., et al.: Radionuclide venography with albumin microspheres. Presented at the annual meeting of the Society of Nuclear Medicine, Miami Beach, Fla., June, 1973.

Herman, B., Dworecka, F., and Wisham, L.: Increase

in dermal blood flow after sympathectomy as measured by radioactive Na uptake, Vasc. Surg. **4**:161, 1970.

Hlavona, A., Lenhart, J., Prerovsky, I., et al.: Leg blood flow at rest during and after exercise in normal subjects and in patients with femoral artery occlusion, Clin. Sci. **29**:555, 1965.

Hobbs, J., and Flipov, A. M.: A new treatment for obtaining indicator dilution curves using a radioactive tracer, I^{125}, Br. Heart J. **26**:75, 1964.

Hosain, F., Pravoni, P., and Wagner, H. N., Jr.: Measurement of blood flow in skeletal muscle with chelated Yb-169 and electrical stimulation, J. Biol. Nucl. Med. **15**:21, 1971.

Hyman, C., and Lenthall, J.: Analysis of clearance of intra-arterially administered labels from skeletal muscle, Am. J. Physiol. **203**:1173, 1962.

Jones, E. L., Wagner, H. N., Jr., and Zuidema, G.: A new method for studying peripheral circulation in man, Arch. Surg. **91**:725, 1965.

Kety, S. S.: Measurement of regional circulation by the local clearance of radioactive sodium, Am. Heart J. **38**:321, 1949.

Kety, S. S.: The theory and applications of the exchange of inert gas at the lungs and tissues, Pharmacol. Rev. **3**:1, 1951.

Lassen, N. A.: Muscle blood flow in normal man and in patients with intermittent claudication evaluated by simultaneous Xe^{133} and Na^{24} clearance, J. Clin. Invest. **43**:1805, 1964.

Lassen, N. A., and Kamp, M.: Calf muscle blood flow during walking studied by the Xe^{133} method in normals and in patients with intermittent claudication, Scand. J. Clin. Lab. Invest. **17**:447, 1965.

Lassen, N. A., Lindbjerg, I. F., and Dahn, I.: Validity of the Xe^{133} method for measurement of muscle blood flow evaluated by simultaneous venous occlusion plethysmography. Observations in the calf of normal man and patients with occlusive vascular disease, Circ. Res. **16**:287, 1965.

Lassen, N. A., Lindbjerg, I. F., and Munch, O.: Measurement of blood flow through skeletal muscle by intramuscular injection of Xe^{133}, Lancet **1**:686, 1964.

Lindbjerg, I. F.: Measurement of muscle blood flow with Xe^{133} after histamine injection as a diagnostic method in peripheral arterial disease, Scand. J. Clin. Lab. Invest. **17**:371, 1965.

Lindbjerg, I. F.: Diagnostic application of Xe^{133} method in peripheral arterial disease, Scan. J. Clin. Lab. Invest. **17**:589, 1965.

Munck, O., Anderson, A. M., and Binder, C.: Clearance of 4-iodoantipyrine-125-I after subcutaneous injection in various regions, Scand. J. Clin. Lab. Invest. **19**(suppl. 99):39, 1967.

Natarajan, T. K., and Wagner, H. N., Jr.: A new image display and analysis system (IDA) for radionuclide imaging, Radiology **93**:823, 1969.

Pabst, H. W.: Strahlentherapie **102**:602, 1957.

Petersen, F., and Siggaard-Anderson, J.: Simultaneous venous occlusion plethysmography and Xe^{133} clearance in patients with peripheral vascular disease, Scand. J. Thorac. Cardiovasc. Surg. **3**:20, 1969.

Powell, M. R., and Anger, H. O.: Blood flow visualization with the scintillation camera, J. Nucl. Med. **7**:729, 1966.

Resisch, W., and Tango, F.: Peripheral circulation in health and disease, New York, 1957, Grune & Stratton, Inc.

Rhodes, B. A.: Blood flow through arteriovenous anastomoses. In Horst, W., editor: Frontiers of nuclear medicine, Berlin, 1971, Springer-Verlag.

Rhodes, B. A., Rutherford, R. B., Lopez-Majano, V., et al.: Arteriovenous shunt measurements in extremities, J. Nucl. Med. **13**:357, 1972.

Rhodes, B. A., Zolle, I., Buchanan, J. W., et al.: Radioactive albumin microspheres for studies of the pulmonary circulation, Radiology **92**:1453, 1969.

Rutherford, R. B., Reddy-Walker, C. M., et al.: A new quantitative method of assessing the functional status of the leg vein, Am. J. Surg. **122**:594, 1971.

Rutherford, R. B., Rhodes, B. A., and Wagner, H. N., Jr.: The distribution of extremity blood flow before and after vagectomy in a patient with hypertrophic pulmonary osteoarthropathy, Chest **56**:19, 1967.

Sapirstein, L. A., and Goodwin, R. S.: Measurement of blood flow in the human hand with radioactive potassium, J. Appl. Physiol. **13**:81, 1956.

Schaffner, F., Friedell, M. T., Pickett, W. J., and Hummon, D. F., Jr.: Radioactive isotopes in study of peripheral vascular disease; method of evaluation of various forms of treatment, Arch. Intern. Med. **83**:608, 1949.

Sheda, H., and O'Hara, I.: Study on peripheral circulation using ^{133}I and macroaggregated serum albumin, J. Exp. Med. **101**:311, 1970.

Shepherd, R. C., and Warren, R.: Studies on the blood flow through the lower limb in man by nitrous oxide technique, Clin. Sci. **20**:99, 1960.

Siegel, M. E., Giargiana, F. A., Rhodes, B. A. and Wagner, H. N., Jr.: Peripheral vascular perfusion scanning: a prognostic indicator of ulcer healing, abstracted, J. Nucl. Med. **15**:533, 1974.

Smith, B. C., and Quimby, E. A.: Use of radioactive sodium in studies of circulation in patients with peripheral vascular disease: preliminary report, Surg. Gynecol. Obstet. **79**:142, 1944.

Spence, R. J., Rhodes, B. A., and Wagner, H. N., Jr.: Regulation of arteriovenous anastomotic and capillary blood flow in the dog leg, Am. J. Physiol. **222**: 326, 1972.

Stewart, G. N.: Research on the circulation time in organs and on the influences which affect it. IV, The output of the heart, J. Physiol. **22**:159, 1897.

Strandness, O., Schultz, R., Summer, D., et al.: Ultrasonic flow detection—a useful technique in the evaluation of peripheral vascular disease, Am. J. Surg. **113**:311, 1967.

Wagner, H. N., Jr., Jones, E., Tow, D., et al.: A method for study of the peripheral circulation in man, J. Nucl. Med. **6**:150, 1965.

Wagner, H. N., Jr., Rhodes, B. A., Sasaki, Y., et al.:

Studies of the circulation with radioactive microspheres, Invest. Radiol. **4**:374, 1970.

Walder, D. N.: A technique for investigating the blood supply of muscle during exercise, Br. Med. J. **1**:255, 1958.

Warner, G. F., Dobson, E. L., Pace, N., et al.: Studies of human peripheral blood flow: effect of injection volume on the intramuscular radiosodium clearance rate, Circulation **8**:732, 1953.

Yeh, S. Y., and Peterson, R. E.: Soluability of krypton and xenon in blood, protein solutions and tissue homogenates, J. Appl. Physiol. **20**:1041, 1965.

Zankel, H. T., Clark, R. E., and Shipley, R. A.: Effect of physical modalities upon saphenous circulation time using a radioactive tracer, Arch. Phys. Med. Rehabil. **34**:360, 1953.

Zierler, K. L.: Equations for measuring blood flow by external monitoring of radioisotopes, Circ. Res. **16**: 309, 1965.

SECTION SIX

CLINICAL APPLICATIONS IN VITRO

21 □ Fundamentals of radioimmunoassay and other radioligand assays

Jan I. Thorell
Steven M. Larson

Radioimmunoassay (RIA) and related radioligand assay methods began a little over twenty years ago as a cumbersome research method used in a few specialized laboratories. Since that time there has been a phenomenal proliferation of these techniques to measure hormones and other substances present in minute quantities in biologic fluids. The applications for these methods are exceedingly diverse, and virtually every branch of medical research has been affected by these techniques. In particular, endocrinology has been greatly enriched by the new knowledge that has come as a direct result of RIA methods. In recognition of the importance of these contributions, Dr. Rosalyn Yalow, co-discoverer of RIA, was awarded the Nobel prize in 1977.

The revolution in the measurement of biologic substances that was brought about by RIA was primarily an effect of a dramatic increase in *sensitivity* compared to other chemical methods. In part, this sensitivity is due to the ready availability of methods for detection of radioactivity, which can be measured in very tiny quantities. For example, it is possible to measure ^{125}I in quantities as low as 10^{-12}M with good precision. Since the quantities of drugs and hormones in biologic fluids are usually much greater than this (10^{-8}M to 10^{-10}M), with the use of tracer ^{125}I labeling, detection of drugs and hormones at the lowest physiologic concentrations is now possible.

The other main feature of these methods is *specificity*. Most biologically active compounds are built up by the common elements carbon, oxygen, hydrogen, and a few others. The uniqueness of these compounds is conferred by the manner in which these common elements are put together. It is the architecture of the compounds that determines their function. A unique structure will give a unique external configuration. This is the basis for assay specificity. Drugs, hormones, and other small substances (ligands) will react with specific binding sites on a large protein molecule (binder). The binding site has a configuration that is complementary to a particular area on the surface of the ligand. The binding between binder and ligand can be viewed as the fitting of a key into a lock. The ability of a binder to recognize the external configuration is therefore the basis for identification of a particular compound in a biologic fluid. Binders may be antibodies, plasma-transporting proteins, or cellular receptors. Antibodies have been produced that have such high specificity that they can distinguish between two compounds that differ by as little as one atom, for example, thyroxine and triiodothyronine. The fit between binder and ligand also influences the strength of the receptor-ligand bond, a factor that helps determine assay sensitivity.

Radioimmunoassay and related techniques (radioligand assays) have been refined to the point that they are relatively easy to produce on a mass scale with a high degree of precision. These techniques lend themselves well to automation. In this regard we are beginning to see in the hospital laboratory the fruit of a decade or more of rapid commercial development.

BASIC PROCEDURE

Radioligand assays involve the following steps (Fig. 21-1):

1. The production of the reagents is often the most difficult step in developing new assays. For established assays, the rapidly increasing number of commercially available standardized reagents, particularly in the form of complete kits, have greatly facilitated this step.

2. The second step is the incubation of the binder (all proteins with specific binding properties are called binders) and the substance to be assayed (substances bound to specific binders are called ligands) together with a constant amount of that same substance labeled with a radionuclide (radioligand). Incubation periods range from minutes to several days, depending on the properties of the reagents of the specific

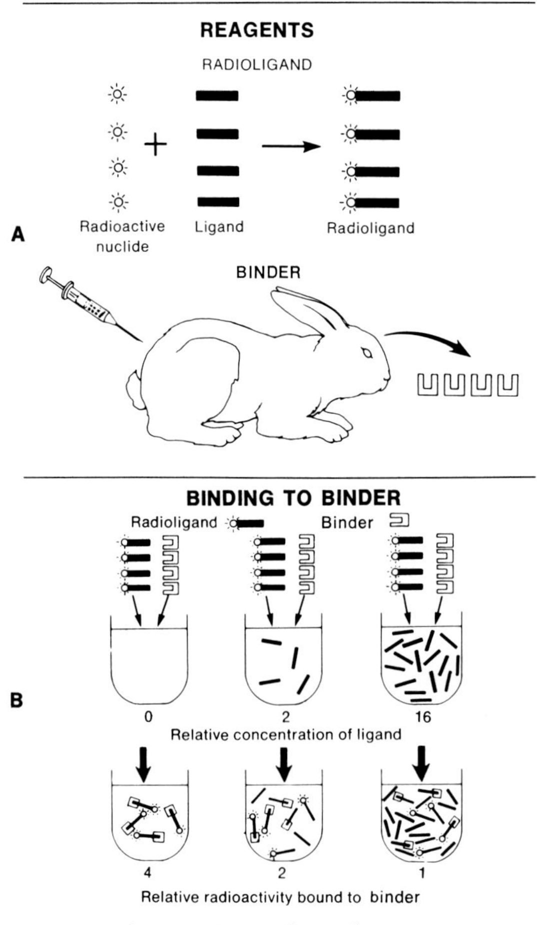

Fig. 21-1. See text for explanation.

assay. During the incubation period, both the unlabeled ligand and the radioligand bind with the binder. However, the amount of binder added is chosen so that its binding capacity approximately corresponds to the amount of the radioligand present. Therefore the ligand and the radioligand compete for the limited number of binding sites on the binder. The unlabeled ligand in the system dilutes the amount of radioligand being bound to the binder and thus reduces the amount of bound radioactivity.

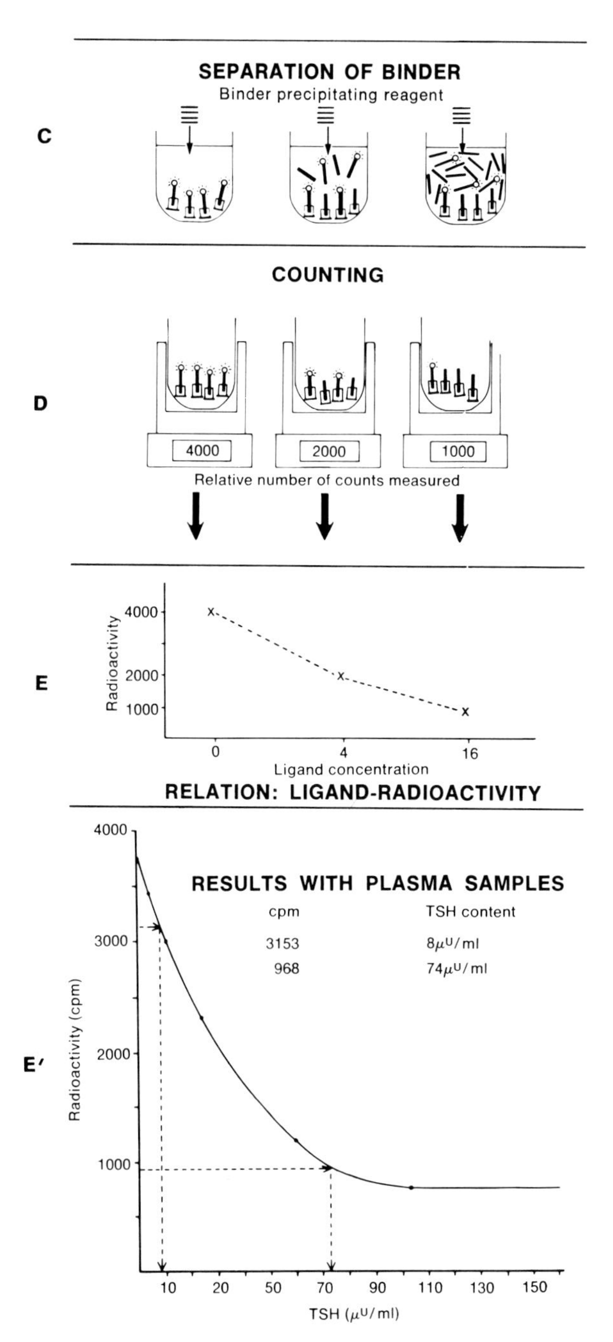

Fig. 21-1, cont'd. See text for explanation.

3. Isolation of the binder fraction occurs at the end of the incubation period. The amount of radioactivity bound to the binder is inversely proportional to the concentration of the substance to be assayed. This step is often referred to as the separation step.

4. Radioactivity bound to the isolated binder fraction is then measured.

5. Radioactivity measured is translated into an estimate of the concentration of the substance to be assayed. The counts bound to the binder in the unknown sample are compared with the amounts recorded from a series of standards made up with samples to which known amounts of the ligand have been added. The result of this standard series is most often plotted as a standard curve, which gives the relationship between the radioactivity measurement (y axis) and the amount of added ligand (x axis). E′ is an example of how the quantity of thyroid-stimulating hormone (TSH) in two samples was determined from a standard curve. In the first sample, 3153 cpm were measured; in the second sample, 968 cpm were found. From the standard curve, this activity corresponds to 8 and 74 μU, respectively.

CONSTITUENTS OF RADIOLIGAND METHOD

The development of a radioligand assay involves the following prerequisites: (1) it must be possible to label the ligand with a radionuclide, (2) it must be possible to produce a binder with a relatively high affinity for the ligand, and (3) methods for isolating the binder from the reaction mixture must be available.

Radioactive labeling

Some substances naturally contain elements that have useful radioactive isotopes (Table 21-1). For example, vitamin B_{12} contains stable cobalt, which may be substituted with the radioisotope ^{57}Co or ^{60}Co. In thyroxine, the native iodine may be exchanged for ^{125}I or ^{131}I. The chemical properties of the molecule are not altered by this type of substitution reaction. However, many substances do not contain elements that have nuclides which can be used to produce efficient radioligands for receptor assays. Unfortunately, the half-lives of radionuclides of the most common biologic elements (carbon, nitrogen, oxygen, and hydrogen) are either too short or too long. Radionuclides with very short half-lives are unsuitable because of transportation and storage problems. Radionuclides with very long half-lives, such as ^{14}C and ^{3}H, have low specific radioactivities that reduce the sensitivity of a radioligand assay to a degree that makes it impossible to measure some biologic substances.

Table 21-1. Principle radionuclides used in radioligand assay

Radionuclide	Half-life	Principle energy and radiation
^{125}I	60.2 days	28 keV (gamma)
^{131}I	8.15 days	364 keV (gamma)
^{14}C	5730 yr	158 keV (gamma)
^{3}H	12.3 yr	18 keV (gamma)
^{57}Co	267 days	122 keV (gamma)

Thus it is frequently necessary to introduce a radionuclide of an element that is not present in the native molecule, if this can be achieved without significant alteration of the configuration of the native molecule. The iodine isotopes ^{125}I and ^{131}I have been most useful in this regard, since it is possible to iodinate most proteins and polypeptides. Both these isotopes are gamma emitters with specific activities and decay rates high enough to allow good assay sensitivity. The half-lives of ^{125}I (sixty days) and of ^{131}I (eight days) are long enough to allow a practical shelf life for labeled reagents. Furthermore, the energy of their emitted radiation permits the use of crystal scintillation techniques for efficient and simple radioactivity measurements. Various methods have been described for the iodination of polypeptides with iodomonochloride, chloramine-T, the electrophoresis process, and lactoperoxidase.

Production of binder

Binders can be produced in several different ways. Antibodies are the most widely used of the binder types. Radioimmunoassays are, in general, restricted to relatively large molecules that are good immunogens. This limitation has been partially overcome by the coupling of small molecules to large proteins such as albumin and polylysine, in which the small molecule acts as a hapten. Radioligand assays have also been developed with certain plasma proteins, such as thyroxine-binding globulin and transcortin, as specific binding reagent. Receptors of tissue origin, either from cytoplasm or membranes, have been used; some examples are estrogen and gonadotropin assays with cytoplasmic receptors and insulin and growth hormone assays with membrane receptors. The properties of the various types of receptors are discussed in greater depth in Chapter 22.

Separation step

The isolation of the binder with its bound radioactivity has been accomplished in several ways. Conventional protein separation techniques, such as ammonium sulfate precipitation, electrophoresis, and gel chromatography, can be used for the separation of antibodies from nonbound proteins. Specific procedures have been devised for processing a large number of samples. The most commonly used is the so-called double-antibody technique for radioimmunoassay, in which the primary antibodies are precipitated with a second antibody system; an example of this procedure is precipitation of primary antibodies with goat antiserum against rabbit immunoglobulins, if the primary antibody was harvested from rabbits.

Other separation techniques that have been used are paper chromatography or a combination of chromatography and electrophoresis called chromatoelectrophoresis, ethanol precipitation, and adsorption by a solid phase such as charcoal, chalk, or silica. Some receptors of tissue origin are insoluble and can be isolated by simple centrifugation. Antibodies can be rendered insoluble by coupling with a solid phase like cellulose, sephadex, polyacrylamide, or polystyrene. Thus the inside of disposable polystyrene reagent tubes can be coated with antibodies so that the separation procedure is a simple matter of decanting the liquid from a tube.

TECHNICAL PERFORMANCE

One of the features of the radioligand assays that distinguishes them from most other ultrasensitive techniques is that these methods require only a few analytic steps and usually no primary extraction or purification procedures. Therefore technical performance consists primarily of repetitive pipeting. For this reason, a large number of samples can be conveniently analyzed by one technician, who often is assisted by equipment for automatized pipeting.

The other steps in the assay procedure can be performed on a large scale as well. Reliable counting equipment with a capacity for handling more than 1000 samples automatically has made the radioactivity measurement one of the easiest parts of the assay. Treatment of the large amount of counting data varies from simple manual reading of the values from the standard curve to the use of a computer that automatically determines the appropriate value. In addition, the computer may be used for correcting various errors in the assay and establishing confidence limits and quality-control data.

APPLICATION

RIA has been applied to an extremely broad range of biologic measurements (Table 21-2). In particular, the specific binding properties of antibodies allowed the initial determination of trace quantities of polypeptide hormones, such as insulin, glucagon, growth hormone, and gonadotropins. Later, with improvements in the production of specific antibodies, assays were developed for small polypeptides like adrenocorticotropic hormone (ACTH) and angiotensin and even for such small compounds as thyroxin. The specific binding properties of nonantibody proteins may also be used for radioligand assay (Table 21-3). The specific binding properties of plasma proteins have been exploited for measuring corticosteroids, thyroxine, and

Table 21-2. Representative radioimmunoassays (RIAs)

Ligand	Radioligand	Immunogen	Reporter(s)
ACTH	[^{125}I] ACTH	ACTH	Felber (1963)
Aldosterone	[1,2-^{3}H] aldosterone	Aldosterone-3-oxime-bovine serum albumin	Mayes et al. (1970)
	[^{3}H] aldosterone-γ-lactone	Aldosterone-γ-lactone-3-carboxymethyl oxime	Farmer et al. (1972)
Angiotensin	[^{125}I] angiotensin I	Angiotensin I	Haber (1969)
Arginine vasopressin (AVP)	[^{125}I] AVP	AVP–bovine serum albumin complex	Beardwell (1971)
Calcitonin	[^{131}I] calcitonin	Calcitonin	Deftos (1971)
Carcinoembryonic antigen (CEA)	[^{125}I] CEA	CEA	Gold and Freedman (1965)
Digitoxin	[^{125}I] 3-O-succinyl digitoxigenin tyrosine	Bovine albumin complex	Oliver et al. (1968)
Digoxin	[^{3}H] digoxin	Digoxin	Smith et al. (1969)
Estrogen	[^{3}H] estradiol	Estradiol	Abraham (1969) Mikhail et al. (1970)
α-Fetoprotein	[^{125}I] α-fetoprotein	α-Fetoprotein	Seppälä et al. (1972)
Follicle-stimulating hormone (FSH)	[^{125}I] FSH	FSH	Midgley (1967)
Gastrin	[^{125}I] gastrin	Gastrin	McGuigan (1967)
		Gastrin albumin complex	Yalow and Berson (1970)
Glucagon	[^{131}I] glucagon	Glucagon	Unger et al. (1959)
Growth hormone	[^{131}I] growth hormone	Growth hormone	Hunter and Greenwood (1962)
	[^{125}I] growth hormone	Growth hormone	Utiger et al. (1962)
Hepatitis B virus antigen	[^{125}I] Au antigen	Serum of Au carrier	Walsh et al. (1970)
		Au antigen	Overby et al. (1973)
Human chorionic gonadotropin (HCG)	[^{125}I] HCG	HCG	Midgley (1966)
Human placental lactogen (HPL)	[^{125}I] HPL	HPL	Jacobs et al. (1972)
IgE	[^{125}I] IgE	IgE	Wide et al. (1967)
Insulin	[^{131}I] insulin	Insulin	Yalow and Berson (1960)
	[^{125}I] insulin	Insulin	Morgan and Lazarow (1962) Hales and Randle (1963)
Parathyroid hormone	[^{131}I] parathyroid hormone	Parathyroid hormone	Berson et al. (1963)
Triiodothyronine (T_3)	[^{125}I] T_3	T_3	Chopra et al. (1971)
Thyroxine (T_4)	[^{125}I] T_4	T_4	Chopra et al. (1971)
Testosterone	[1,2-^{3}H] testosterone	Testosterone-3-oxime-bovine-serum albumin	Furuyama and Nugent, (1970)
Thyroid-stimulating hormone (TSH)	[^{125}I] TSH	TSH	Utiger (1965) Odell et al. (1965)

Table 21-3. Representative nonimmuno radioligand assays

Ligand	Radioligand	Binder	Reporter(s)
Competitive protein-binding assay			
Corticosteroids	[^{3}H] cortisol	CBG (corticosteroid binding globulin)	Nugent & Mayes (1966)
Cortisol	[^{3}H] cortisol	CBG	Murphy et al. (1963)
Cyanocobalamin	[^{57}Co] cyanocobalamin	Intrinsic factor	Barakat and Ekins (1961)
Estrogen	[^{3}H] estradiol	TBP (testosterone-binding protein)	Mayes and Nugent (1970)
Thyroxine (T_4)	[^{131}I] T_4	TBG (thyroxine-binding globulin)	Ekins (1960)
Testosterone	[1,2-^{3}H] testosterone	TBP	Mayes and Nugent (1968)
Cellular binding sites used as receptors*			
ACTH	[^{125}I] ACTH	Adrenal tumor membrane	Lefkowitz et al. (1970)
Estrogen	^{3}H	Uterine cytoplasm	Gorski et al. (1968)
Human chorionic gonadotropin (HCG)	[^{125}I] HCG	Membrane	Catt et al. (1972)
Insulin	[^{125}I] insulin	Solubilized binders	Ozaki and Kolant (1977)

*For a more complete review of this rapidly expanding field see Kahn, 1976.

sex hormones. Cell binding sites that are being used in radioligand assays include membrane receptors for insulin, ACTH, and catecholamines and cytoplasmic receptors for estrogen.

BIBLIOGRAPHY

Abraham, G. E.: Solid-phase radioimmunoassay of estradiol-17 beta, J. Clin. Endocrinol. Metab. **29:** 866, 1969.

Barakat, R. M., and Ekins, R. A.: Assay of vitamin B-12 in blood. A simple method, Lancet **2:**25, 1961.

Beardwell, C. G.: Radioimmunoassay of arginine vasopressin in human plasma, J. Clin. Endocrinol. Metab. **33:**254, 1971.

Berson, S. A., Yalow, R. S., Aurbach, G. D., and Potts, J. T., Jr.: Radioimmunoassay of bovine and human parathyroid hormone, Proc. Natl. Acad. Sci. U.S.A. **49:**613, 1963.

Berson, S. A., Yalow, R. S., Bauman, A., et al.: Insulin-I^{131} metabolism in human subjects, demonstration of insulin binding globulin in the circulation of insulin treated subjects, J. Clin. Invest. **35:**170, 1956.

Catt, K. J., Dufau, M. L., and Tsuruhara, T.: Radioligand-receptor assay of luteinizing hormone and chorionic gonadotrophin, J. Clin. Endocrinol. Metab. **34:**123, 1972.

Chopra, I. J., Nelson, J. C., Solomon, D. H., et al.: Production of antibodies specifically binding triiodothyronine and thyroxine, J. Clin. Endocrinol. Metab. **32:**299, 1971.

Chopra, I. J., Solomon, D. H., and Beall, G. H.: Radioimmunoassay for measurement of triiodothyronine in human serum, J. Clin. Invest. **50:**2033, 1971.

Deftos, L. I.: Immunoassay for human calcitonin 1, Metabolism **20:**1122, 1971.

Ekins, R. P.: The estimation of thyroxine in human plasma by an electrophoretic technique, Clin. Chim. Acta **5:**453, 1960.

Farmer, R. W., Roup, W. G., Jr., Pellizzari, E. D., et al.: A rapid aldosterone radioimmunoassay, J. Clin. Endocrinol. Metab. **34:**18, 1972.

Felber, J. P.: ACTH antibodies and their use for a radioimmunoassay for ACTH, Experientia **19:**227, 1963.

Furuyama, S., and Nugent, C. A.: A radioimmunoassay for plasma testosterone, Steroids **16:**415, 1970.

Gharib, H.: Radioimmunoassay for triiodothyronine, J. Clin. Endocrinol. Metab. **31:**709, 1970.

Gold, P., and Freedman, S. O.: Demonstration of tumor-specific antigens in human colon carcinomata by immunological tolerance and absorption techniques, J. Exp. Med. **121:**439, 1965.

Gorski, J., Toff, D., Shymala, G., Smith, D., and Notides, A.: Hormone receptors: studies on the interaction of estrogen with the uterus, Recent Prog. Horm. Res. **24:**45, 1968.

Haber, E.: Recent developments in pathophysiologic studies of the resin-angiotensin system, N. Engl. J. Med. **280:**148, 1969.

Hales, C. N., and Randle, P. J.: Immunoassay of insulin with insulin-antibody precipitate, Biochem. J. **88:** 137, 1963.

Hunter, W. M., and Greenwood, F. C.: Preparation of ^{131}I labeled human growth hormone of high specific activity, Nature (London) **194:**495, 1962.

Jacobs, L. S., Mariz, I. K., and Daughaday, W. H.: A

mixed heterologous radioimmunoassay for human prolactin, J. Clin. Endocrinol. Metab. **34**:484, 1972.

Kahn, C. R.: Membrane receptors for polypeptide hormones. In Korn, E. D., editor: Methods in membrane biology, vol. 3, New York, 1975, Plenum Publishing Co.

Lefkowitz, R. J., Roth, J., and Pastor, I.: Radioreceptor assay of adrenocorticotrophin hormone. New approach to assay of polypeptide hormone in plasma, Science **170**:633, 1970.

Mayes, D., Furuyama, S., Kem, D. C., et al.: A radioimmunoassay for plasma aldosterone, J. Clin. Endocrinol. Metab. **30**:682, 1970.

Mayes, D., and Nugent, C. A.: Determination of plasma testosterone by the use of competitive protein binding, J. Clin. Endocrinol. Metab. **28**:1169, 1968.

Mayes, D., and Nugent, C. A.: Plasma estradiol determined with a competitive protein binding method, Steroids **15**:389, 1970.

McGuigan, J. E.: Antibodies to the carboxylterminal tetrapeptide of gastrin, Gastroenterology **53**:697, 1967.

Midgley, A. R., Jr.: Radioimmunoassay: a method for human chorionic gonadotropin and human luteinizing hormone, Endocrinology **79**:10, 1966.

Midgley, A. R., Jr.: Radioimmunoassay for human follicle-stimulating hormone, J. Clin. Endocrinol. Metab. **27**:295, 1967.

Mikhail, G., Wu, C. H., Ferin, M., and Vande Wiele, R. L.: Radioimmunoassay of estrone and estradiol, Acta Endocrinol. **64**(suppl. 147): 347, 1970.

Morgan, C. R., and Lazarow, A.: Immunoassay of insulin using a two-antibody system, Proc. Soc. Exp. Biol. Med. **110**:29, 1962.

Murphy, B. E. P., Engelberg, W., and Pattee, C. J.: Simple method for the determination of plasma corticoids, J. Clin. Endocrinol. Metab. **23**:293, 1963.

Murphy, B. E. P., and Pattee, C. J.: Determination of thyroxine using the property of protein binding, J. Clin. Endocrinol. Metab. **24**:187, 1964.

Nishi, S.: Isolation and characterization of a human fetal α-globulin from the sera of fetuses and a hepatoma patient, Cancer Res. **30**:2507, 1970.

Nugent, C. A., and Mayes, D. M.: Plasma corticosteroids determined by the use of corticosteroid binding and dextran-coated charcoal, J. Clin. Endocrinol. Metab. **26**:1116, 1966.

Odell, W. D., Wilber, J. F., and Paul, W. E.: Radioimmunoassay of thyrotropin in human serum, Metabolism **14**:465, 1965.

Oliver, G. C., Jr., Parker, B. M., Brasfield, D. L., and Parker, C. W.: The measurement of digitoxin in human serum by radioimmunoassay, J. Clin. Invest. **47**:1035, 1968.

Overby, L. R., Miller, J. P., Smith, D. D., Decker, R. H., and Ling, C. N.: Radioimmunoassay of hepatitis B virus associated (Australia) antigen employing ^{125}I-antibody, Vox Sang. **24**(suppl.):102, 1973.

Ozaki, S., and Kolant, N.: A radioreceptor assay for serum insulin, J. Lab. Clin. Med. **90**:686, 1977.

Seppälä, M., Bagshaw, K. D., and Ruoslahti, E.: Radioimmunoassay of alpha-fetoprotein: a contribution to the diagnosis of choriocarcinoma and hydatidiform mole, Int. J. Cancer **10**:478, 1972.

Smith, T. W., Butler, V. P., Jr., and Haber, E.: Determination of therapeutic and toxic serum digoxin concentrations by radioimmunoassay, N. Engl. J. Med. **281**:1212, 1969.

Unger, R. H., Eisentraut, A. M., McCall, M. S., et al.: Glucagon antibodies and their use for immunoassay for glucagon, Proc. Soc. Exp. Biol. Med. **102**:621, 1959.

Utiger, R. D.: Radioimmunoassay of human plasma thyrotrophin, J. Clin. Invest. **44**:1277, 1965.

Utiger, R. D., Parker, M. L., and Daughaday, W. H.: Studies on human growth hormone. I. A radioimmunoassay for human growth hormone, J. Clin. Invest. **41**:254, 1962.

Walsh, J. H., Yalow, R., and Berson, S. A.: Detection of Australia antigen and antibody by means of radioimmunoassay techniques, J. Infect. Dis. **121**:550, 1970.

Wide, L., Bennich, H., and Johansson, S. G. O.: Diagnosis of allergy by an in-vivo test for allergen antibodies, Lancet **2**:1105, 1967.

Yalow, R. S., and Berson, S. A.: Immunoassay of endogenous plasma insulin in man, J. Clin. Invest. **39**:1157, 1960.

Yalow, R. S., and Berson, S. A.: Radioimmunoassay of gastrin, Gastroenterology **58**:1, 1970.

22 □ Role of the radioligand-binding assay in characterizing cellular receptors

Charles J. Homcy
Edgar Haber

Although the concept of specific cellular receptors that can combine with hormones or drugs was first proposed over a century ago,[25,26] the technology for defining and characterizing such sites at a molecular level has only been developed within recent years. This achievement is based solely on the present availability of radiolabeled ligands of sufficiently high specific activity to permit detection of high-affinity binding sites, which are typically present in extremely low concentrations. Over the past ten years, research into cytoplasmic and membrane-bound receptors has burgeoned.[12,14] Not only has such research quantitatively characterized receptor number and affinity, but it also has permitted definition of the molecular events underlying well-known physiologic phenomena such as desensitization.[36,37] Perhaps of ultimate importance, however, the ability to track cellular receptors will aid in eventually determining the mechanisms through which hormonal binding leads to a specific cellular biochemical event. Since Sutherland's[47] classic elucidation of catecholamine-induced adenylate cyclase stimulation, the membrane events leading to the generation of the "second messenger" (cyclic adenosine 5′-monophosphate) remain a central focus in membrane research.

REQUIREMENTS FOR A RECEPTOR

It is critical from the onset to formulate criteria which ensure that a particular binding interaction as observed in a radioligand-binding assay is reflecting a true receptor in the pharmacologic sense. Any given ligand will bind with varying degrees of specificity and sensitivity to a number of sites within the cell or its membrane. Some of these sites will be nonspecific, simply reflecting the physical characteristics of the ligand. For example, hydrophobic molecules would tend to dissolve within the lipid bilayer, whereas positively charged molecules might react with negatively charged carbohydrate moieties on the cell surface. Other interactions, however, may indeed be specific (defined as the fraction of total binding that can be displaced by unlabeled ligand at a reasonable concentration range) but not be due to an interaction with a true receptor. Obvious examples would include a hormone such as norepinephrine, which in target tissues undergoes transport, storage, or metabolism by enzymes such as catechol-*O*-methyltransferase and monoamine oxidase.[6,21,48] In attempting to characterize beta-receptor binding, some mechanism for differentiating these other "specific" binding sites must be devised. Similarly, nicotinic receptor binding of acetylcholine must be distinguished from its interaction with the intimately associated degrading enzyme acetylcholinesterase. Therefore, before attributing binding of a radioligand to the molecular counterpart of a pharmacologic receptor, certain criteria should first be met. Cuatrecasas[11] proposed the following as necessary for receptor binding:

1. Binding should be saturable and reversible with its kinetics correlating with the known biologic properties of the hormone.
2. Binding should only be demonstrable in tissues that show a biologic response to the ligand.
3. The affinity of the binding site should parallel biologic dose-response curves.
4. In vivo potency series of agonists, partial agonists, and antagonists should correlate with binding.
5. If the in vivo pharmacologic effects of the ligand demonstrate stereospecificity, so should the in vitro binding site.

Although the simple criterion of specific binding was once thought to be sufficient in defining receptor binding, recent reports clearly document specific, nonreceptor binding occurring as methodologic artifacts.[13] Such findings dictate careful evaluation of any radioligand binding assay to be certain that the observed binding reflects in every possible way the known biologic properties of the "receptor" in question.

BINDING ASSAY

Radioligand-binding assays simply involve the measurement of the binding of the radiolabeled ligand to a cell membrane or extract. As with any competitive protein-binding assay, this requires separation of free and bound ligand. Since one wishes to describe this interaction under equilibrium conditions, it is important that a rapid and efficient method of separation be developed that preserves equilibrium as much as possible. Again, as noted in Chapter 21, this latter point becomes critical for binding sites with rapid ligand dissociation rates. If nonequilibrium conditions are employed during the separation, it is imperative to validate the assay by demonstrating that in the time required for separation of bound from free ligand, an insignificant dissociation of the bound ligand from the receptor occurs. A variety of methods have been successfully employed and, in general, meet these criteria. In the investigation of membrane-bound receptors, several techniques are easily applicable and afford rapid separation of particulate-bound from free ligand.[14] The most extensively employed procedure involves rapid filtration of the particulate preparation on any of a variety of synthetic membranes. Centrifugation provides a rapid alternative means of separation. Equilibrium dialysis is an elegant technique, in which the assay is performed entirely under equilibrium conditions and is particularly useful for assaying relatively low-affinity binding sites (where the dissociation rate of the ligand-receptor complex may be so rapid as to preclude either filtration or centrifugation). The technique is cumbersome, however, and not easily applied to large numbers of samples.

It is also possible to assay receptors in soluble form, either cytoplasmic receptors such as for estrogens[39] or detergent-solubilized membrane receptors,[14] using much the same technique as described previously for competitive protein-binding assays. These include (1) adsorption of free ligands onto materials such as charcoal or talc, (2) polyethylene glycol precipitation of the ligand-receptor complex, (3) gel filtration and ion-exchange chromatography, and (4) equilibrium dialysis. There is an additional technique described by Stone and Metzger,[45] which is remarkable for its preservation of equilibrium as well as its simplicity. Ligand, receptor protein, and a porous bead such as Sephadex G-50, which will exclude receptor but include ligand, are incubated until equilibrium is reached. An aliquot of the supernatant not containing gel is then counted and compared with a control tube containing only ligand without receptor protein. The increase in the counts over the control is thus a measure of protein-bound ligand. This technique has been applied successfully in the characterization of a partially purified beta receptor.[51]

MODEL SYSTEM: THE BETA RECEPTOR

Ahlquist[2] first coined the terms alpha- and beta-adrenergic effects in 1948 to describe the two patterns of response that were observed in several different organs to a variety of catecholamine preparations. The pattern characterized by greatest sensitivity to isoproterenol and least to norepinephrine was a beta effect. As radiolabeled catecholamines became available, it was natural to attempt to identify binding sites in those tissues where beta-adrenergic effects were typically found.[9,27,45] However, it became apparent that tritiated catecholamine binding was not identifiable with the pharmacologic beta receptor.[15,50] First, the interaction was not completely reversible and probably reflected the tendency of the phenolic hydroxyl group of the catechol ring to oxidize and covalently bind nonspecifically to cellular proteins. Second, the concentration of a known beta receptor antagonist, propranolol, required to effect a 50% displacement of ^{3}H-labeled norepinephrine was approximately 10^{-4}M, which was three orders of magnitude greater than that necessary to inhibit adenylate cyclase stimulation. Third, the kinetics of binding and dissociation were inappropriately long when compared to the rapid onset and decline of catecholamine effects in vivo and in vitro. Fourth, the binding did not demonstrate stereospecificity with d and l isomers showing equivalent affinity. Significant progress was made only when radiolabeled beta antagonists were employed as the binding probe.[31] Although a high concentration of ^{3}H-labeled propranolol binding was still nonspecific in that it could not be displaced with high concentrations of unlabeled propranolol, the "specifically bound" fraction demonstrated all the properties that one could propose for the true beta receptor. The sites were of high affinity, saturable, and stereospecific with a dissociation constant identical to that calculated from cyclase inhibition. Typically in these experiments, a purified membrane preparation was incubated with increasing concentrations of a radiolabeled antagonist such as propranolol. Specific binding is that fraction that is displaced by a reasonably high con-

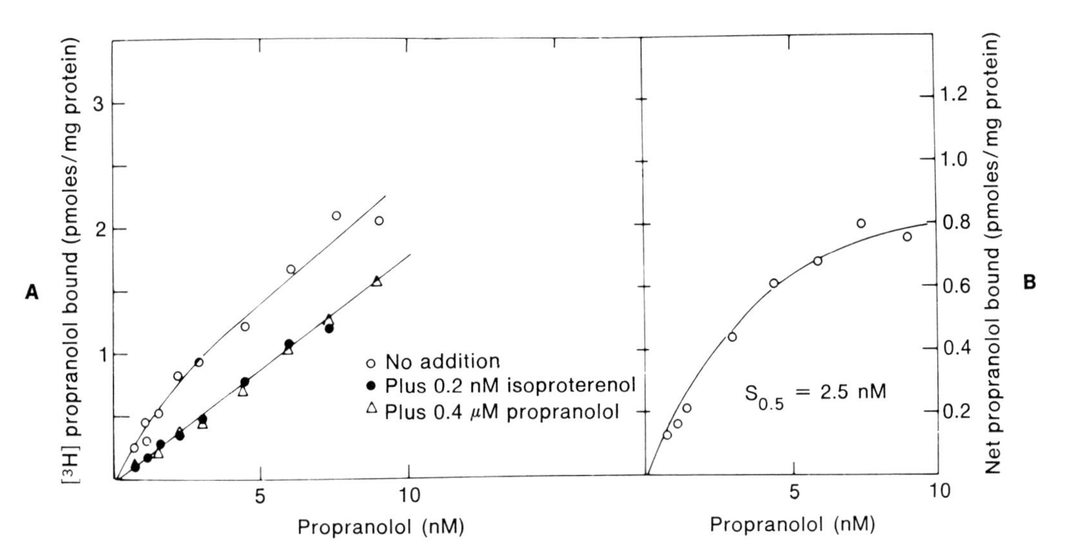

Fig. 22-1. Binding of [^{3}H] propranolol to turkey erythrocyte ghosts measured by ultracentrifugation. **A,** Binding in absence of competing ligands (○), in presence of excess nonradioactive propranolol (△), or isoproterenol (●). **B,** Binding of propranolol displaceable by isoproterenol (○). (From Levitzki, A.: Proc. Natl. Acad. Sci. **71**:2775, 1974.)

centration of unlabeled agonist or antagonist (Fig. 22-1, *A*). Saturation of available binding sites is then achieved and their concentration easily calculated (Fig. 22-1, *B*). Similarly, one can develop displacement curves wherein a constant amount of radiolabeled ligand is added and then the dissociation constant for a particular agonist or antagonist is calculated by determining the concentration required to effect a 50% displacement of bound ligand. The kinetics of binding and dissociation can be similarly investigated. The potency series for inhibition of binding by various catecholamine derivatives was that predicted for the beta receptor both from in vivo studies and cyclase activation (Fig. 22-2, *A* and *B*). Following these results, other radiolabeled probes of even greater specific activity were developed. ^{125}I-labeled hydroxybenzylpindolol could be separated from noniodinated starting material, resulting in a specific activity equivalent to the theoretic activity of ^{125}I (2200 Ci/mmol). Furthermore, a much higher fraction of total binding with either this label[5,35] or ^{3}H-labeled alprenolol (approximately 30 to 40 Ci/mmol) was specific in nature.

Recently, in vivo labeling of the beta receptor has been attempted.[10,46] Preliminary experiments have suggested that in the lung up to 50% of radiolabeled antagonist binding is inhibited by prior treatment with l-propranolol, which indicates that this represented a specific beta receptor interaction (Fig. 22-3).[21a,21b] Such techniques may eventually allow one to determine that a patient has received enough propranolol therapy to effect complete beta blockade by in vivo imaging of receptor occupancy with

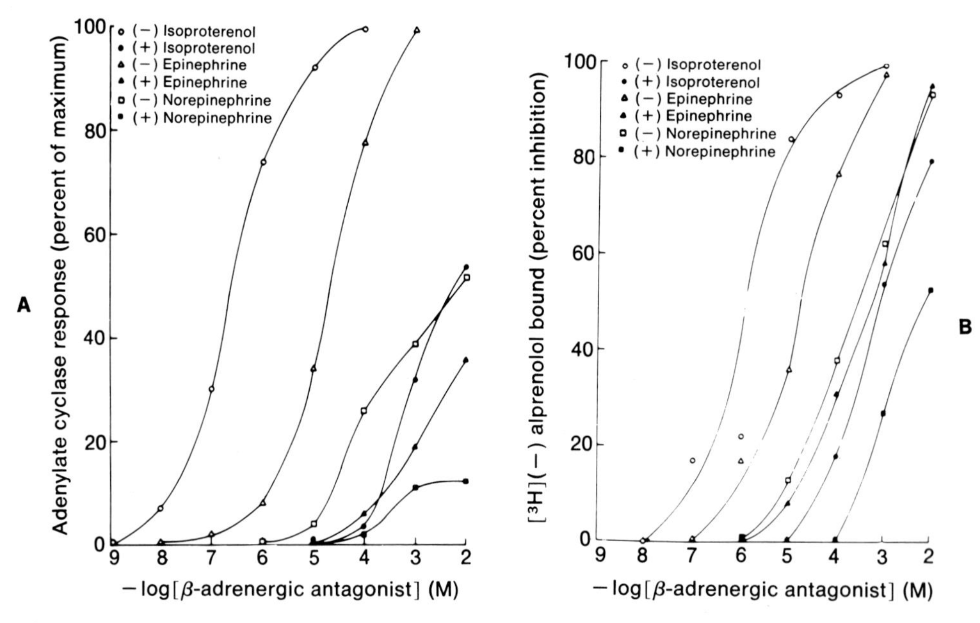

Fig. 22-2. A, Stimulation of frog erythrocyte membrane adenylate cyclase by (+) and (−) stereoisomers of isoproterenol, epinephrine, and norepinephrine. Maximum response refers to amount of cyclic AMP generated in presence of a maximum stimulatory concentration of (−) isoproterenol (0.1 mM). This was generally a fivefold stimulation of basal enzyme activity. Values shown are means of four determinations from two separate experiments. **B,** Inhibition of (−) [^{3}H] alprenolol binding to frog erythrocyte membranes by (+) and (−) stereoisomers of isoproterenol binding was determined in presence or absence of indicated concentrations of various agonists; *100% inhibition* of binding refers to complete blockade of specific binding, that is, inhibition of binding equivalent to that observed with 10 μM (−) alprenolol or (−) propranolol. Values shown are means of four determinations from two separate experiments. (From Mukheijep, C., Caron, M. C., Coverstone, M., and Lefkowitz, R. J.: J. Biol. Chem. **250:**4873, 1975.)

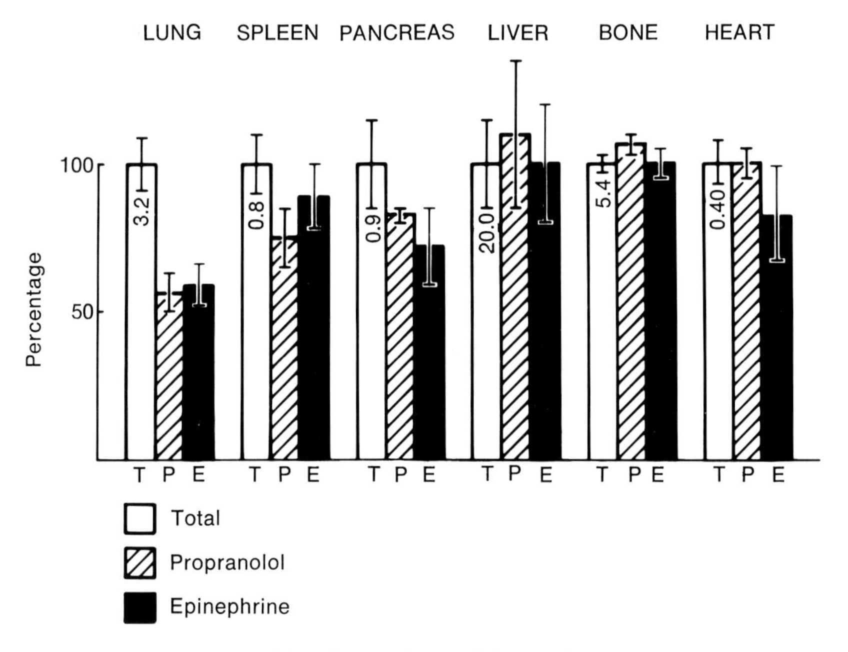

Fig. 22-3. Propranolol and epinephrine inhibition of [^{125}I] HBP binding.

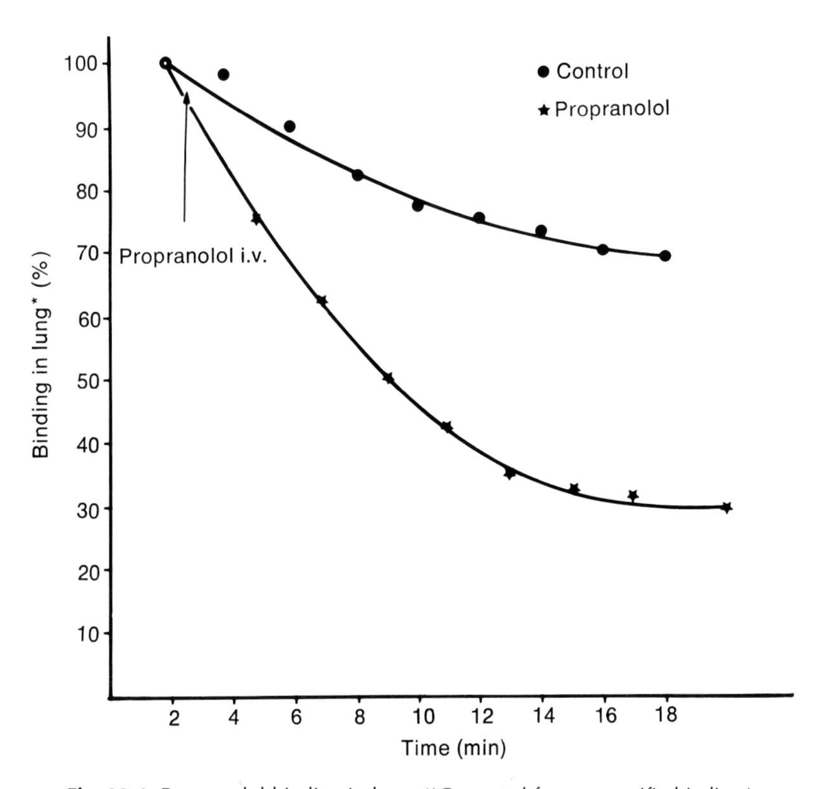

Fig. 22-4. Propranolol binding in lung. (*Corrected for nonspecific binding.)

radiolabeled antagonists. Finally, imaging with such agents may provide a useful technique for assessing beta-receptor function (number and affinity) in a variety of disease states. It has already been possible to visually and quantitatively detect [^{125}I] hydroxybenzylpindolol displacement from the lung of an anesthetized rabbit during intravenous propranolol administration. Reversal of binding was rapid and an in vivo displacement curve was generated. Such a method provides the potential for longitudinally assessing β-receptor occupancy and apparent affinity directly in the human (Fig. 22-4).

With the availability of radiolabeled antagonists of high specific activity, many basic questions concerning control of receptor function as well as its interaction with adenylate cyclase have also been approached directly. Previously, it had been shown that adenylate cyclase responsiveness to catecholamine stimulation was diminished with prolonged exposure to agonistic but not antagonistic beta-adrenergic agents.[20,36,41] Using the radioligand binding assay, an actual decrease in the receptor number but not in the affinity or total stimulatable cyclase was shown to be the underlying mechanism.[29,37] Recently, the relationship of adenylate cyclase to the beta receptor has been more thoroughly investigated. Following detergent solubilization of membranes possessing catecholamine-stimulatable adenylate cyclase, the ^{3}H-labeled alprenolol binding fraction in the resulting soluble fraction eluted in a separate fraction from adenylate cyclase activity during gel filtration.[32] Furthermore, the receptor also bound to an alprenolol-affinity resin with apparently complete resolution from adenylate cyclase.[47] These experiments strongly suggest that the beta receptor and adenylate cyclase are distinct molecular entities that are functionally linked in the membrane. Furthermore, an antibody to the partially purified beta receptor has recently been produced that will stereospecifically inhibit ^{3}H-labeled propranolol binding and isoproterenol-stimulatable adenylate cyclase but not basal or fluoride-stimulatable cyclase.[51] In addition, the antibody will not block glucagon-stimulatable cyclase in cardiac membrane, reflecting the distinct characters of the two receptors. Remarkably, the specificity of the cardiac beta receptor against which the immune sera was raised is such that the β_1 receptor from the liver of the same species of animal is not inhibited by the same antibody fraction, implicating actual structural differences as the basis of β_1- and β_2-adrenergic functional subclasses.

ROLE OF RECEPTOR ASSAY IN CLINICAL RESEARCH

It is logical that the next step in the application of the radioligand-binding assay will be in the realm of clinical research. With the technology to characterize cellular receptors both in terms of density and affinity, many investigators have begun to document alterations in receptor function in a variety of disease states as well as the following pharmacologic manipulation.

Diabetes was one of the first areas to be investigated extensively in the expectation that a primary derangement in insulin receptors was the underlying problem. Bennett and Cuatrecasas,[8] however, first reported that both the number and affinity of insulin-binding sites per cell were unchanged in rats made diabetic by steroid administration, starvation, or streptozotocin ablation of the pancreas as well as in naturally occurring strains with spontaneously decreased insulin responsiveness. Similar binding studies employing human fat cells obtained from the subcutaneous tissue of obese individuals who exhibited decreased insulin responsiveness have demonstrated that the total number of receptors per cell is not different from that of lean controls.[3] These results can be contrasted with those of Kahn and co-workers, who have extensively investigated the nature of insulin resistance in genetically obese (ob/ob) hyperglycemic mice.[24] A variety of tissues were examined as to the quantitative and qualitative differences between the insulin receptors of the ob/ob strains versus normal controls. Receptors from the two strains were identical in terms of receptor affinity, kinetics of binding, and characteristic site-to-site interactions, first noted for the insulin receptor by DeMeyts and colleagues.[17] Receptor density, however, was clearly decreased whether expressed per cell number, per cell volume, or per milligram of protein. In further experiments, this same group examined the factors controlling the insulin receptor population by examining alterations in receptor density when factors such as obesity and serum insulin concentrations were controlled. The results suggested that the increase in serum insulin concentration that occurs parallel to the development of obesity and the diabetic state was the factor determining receptor density.[44] It thus remains unclear whether cellular insulin receptors are actually decreased in states of insulin resistance. Even if this is the case, it may not represent a primary event but rather a response to increased circulating insulin levels. Such desensitization phenomena associated with exposure to high concentrations of agonistic hormones have been described with a variety of cellular receptors.[30,36,37]

Immunoglobulins directed against cellular receptors that functionally decrease the number of available binding sites has now been documented as a cause of hormone resistance. Recently, the etiology of the marked insulin resistance associated with the syndrome acanthosis nigricans has been clarified, using the radioligand receptor-binding assay.[19,23] These studies have revealed a marked decrease in insulin binding to its membrane receptor, which does not correlate with ambient insulin levels and thus does not appear to be a secondary phenomenon. Furthermore, in a large proportion of these patients, circulating antibodies to the receptor are present that effectively block ^{125}I-labeled insulin binding when added to normal cells. In those without this apparent immunologic disturbance, a primary defect in the receptor population has been postulated.

Recent investigations have also implicated the existence of antibodies directed against the acetylcholine receptor in the clinical syndrome of myasthenia gravis. There are now two lines of evidence supporting this hypothesis. Fambrough and colleagues[18] first demonstrated a reduction in the number of acetylcholine-binding sites in muscle biopsies from myasthenic patients, employing the ^{125}I derivative of the cholinergic antagonist, α-bungarotoxin. Subsequently, several groups have reported the presence of globulins in the sera of myasthenic patients, which inhibits bungarotoxin binding to a variety of acetylcholine receptor preparations.[1,4,38] The second kind of evidence is quite provocative in that an active myasthenic syndrome can be initiated in animals by immunization with heterologous purified acetylcholine receptor.[22,40] If blocking antibodies are the culprits responsible for abnormal neuromuscular excitation in the myasthenic syndrome, the mechanism leading to their production remains to be elucidated.

Finally, it is important to bear in mind that the simple demonstration of alterations in cellular receptors in any particular pathologic state does not necessarily lead to the conclusion that this represents the primary event. For example, exposure of receptors to high concentrations of agonistic hormones results in a decline in receptor concentration, probably reflecting the cellular mechanism underlying the physiologic phenomenon of desensitization. Therefore if a particular disease state is associated with an increased concentration of a circulating hormone, then one might predict that available binding sites for the hormone would decrease. Such findings have been documented for acetylcholine,[36] catecholamine,[37] insulin,[44] and growth hormone[30] receptors during prolonged exposure to high concentrations of these agents. Furthermore, in certain conditions, variations in receptor number and affinity may simply represent part of a generalized alteration in cellular metabolism. Thyroid hormone, for example, has been shown to increase the density of beta receptors in the hearts of rats chronically exposed to the hormone.[7,16,49] However, thyroid hormone also increases the actual number of protein copies of Na^+, K^+-ATPase per milligram of membrane protein.[33,34] It has also been demonstrated that thyroid hormone can regulate the number of muscarinic cholinergic receptors in the heart.[43] It is difficult, then, to determine which of these alterations, if any, is responsible for inducing the metabolic changes associated with thyroid excess. These examples illustrate the importance of critically evaluating any experimental protocol that suggests that an alteration in receptor number or affinity is the primary event leading to a pathophysiologic state.

The tool of radiolabeled ligand binding, critically applied, has opened the exciting world of direct receptor study. No longer is the drug or hormone receptor a hypothetical entity, but an identifiable molecule to be purified, characterized, and examined with respect to its interaction with its specific effector.

REFERENCES

1. Ahacanow, A., Abramsky, O., Tarrah-Hazdai, R., et al.: Humoral antibodies to acetylcholine receptor in patients with myasthenic gravis, Lancet **2**:340, 1975.
2. Ahlquist, R. P.: A study of the adrenotropic receptors, Am. J. Physiol. **153**:586, 1948.
3. Amatruda, J. M., Livingston, J. N., and Lockwood, D. H.: Insulin receptor: role in the resistance of human obesity to insulin, Science **188**:264, 1975.
4. Appel, S. H., Olmon, R. R., and Levy, N.: Acetylcholine receptor antibodies in myasthenia gravis, N. Engl. J. Med. **293**:760, 1975.
5. Aurbach, G. D., Fedak, S. A., Woodard, C. J., and Palmer, J. S.: β-Adrenergic receptor: stereospecific interaction of iodinated β-blocking agent with high affinity site, Science **186**:1223, 1974.
6. Axelrod, J., and Tomchick, R.: Enzymatic O-methylation of epinephrine and other catechols, J. Biol. Chem. **233**:702, 1958.
7. Banergie, S. P., and Kung, L. S.: β-adrenergic receptors in rat heart: effects of thyroidectomy, Eur. J. Pharmacol. **43**:207, 1977.
8. Bennett, G. V., and Cuatrecasas, P.: Insulin receptor of fat cells in insulin-resistant metabolic states, Science **176**:805, 1972.
9. Bilezikian, J. P., and Aurbach, G. D.: β-adrenergic receptor of the turkey erythrocyte, J. Biol. Chem. **248**:5575, 1973.
10. Bylund, D., Charness, M. E., and Snyder, S. H.: Beta-adrenergic receptor labeling in intact animals with ^{125}I-hydroxylbenzylpindolol, J. Pharmacol. Exp. Ther. **201**:644, 1977.
11. Cuatrecasas, P.: Commentary: insulin receptors, cell membranes, and hormone action, Biochem. Pharmacol. **23**:2353, 1974.
12. Cuatrecasas, P.: Membrane receptors, Ann. Rev. Biochem. **43**:169, 1974.
13. Cuatrecasas, P., and Hollenberg, M.D.: Binding of insulin and other hormones to non-receptor materials. Saturability, specificity, and apparent "negative cooperativity," Biochim. Biophys. Res. Comm. **62**:31, 1975.
14. Cuatrecasas, P., and Hollenberg, M.D.: Membrane receptors and hormone action, Adv. Protein Chem. **30**:251, 1976.
15. Cuatrecasas, P., Tell, P., Sica, V., Parikh, V., and Chang, K. J.: Noradrenalin binding and the search for catecholamine receptors, Nature **247**:92, 1974.
16. Curaldi, T., and Maunetti, G. V.: Thyroxine and propylthiouracil effects in vivo on alpha and beta adrenergic receptors in rat heart, Biochim. Biophys. Res. Comm. **74**:984, 1977.
17. DeMeyts, P., Roth, J., Neville, D. M., Jr., Gavin, J. R., III, and Lesniak, M. A.: Insulin interaction with its receptors: experimental evidence for negative cooperativity, Biochim. Biophys. Res. Comm. **55**:154, 1973.
18. Fambrough, D. M., Drachman, D. B., and Satyamurti, S.: Neuromuscular junction in myasthenic gravis: decreased acetylcholine receptors, Science **182**:293, 1973.
19. Flier, J. S., Kahn, C. R., Roth, J., and Bar, R. S.: Antibodies that impair insulin receptor binding in an unusual diabetic syndrome with severe insulin resistance, Science **190**:63, 1975.
20. Franklin, T. J., and Foster, S. J.: Hormone-induced desensitization of hormonal control of cyclic AMP levels in human diploid fibroblasts, Nature (New Biol.) **246**:146, 1973.
21. Haber, E., and Wrenn, S.: Problems in identification of the beta-adrenergic receptor, Physiol. Rev. **56**:317, 1976.

21a. Homcy, C. J., and Strauss, H. W.: Evaluation of ^{125}I-hydroxybenzylpindolol as an indicator of in vivo β-receptor occupancy, Circulation (suppl. II):158, 1978.

21b. Homcy, C. J., Strauss, H. W., and Kopiwoda, S.: Beta receptor occupancy: its assessment in the intact animal (submitted for publication).

22. Hullronn, E., Mattsoon, C. H., Stalberg, E., et al.: Neurophysiological signs of myasthenia in rabbits after receptor antibody development, J. Neurol. Sci. **24**:871, 1975.
23. Kahn, C. R., Flier, J. S., Bar, R. S., Archer, J. A., Gorden, P., Martin, M. M., and Roth, J.: The syndromes of insulin resistance and acanthosis nigricans: insulin-receptor disorders in man, N. Engl. J. Med. **294**:739, 1976.
24. Kahn, C. R., Neville, D. M., Jr., and Roth, J.:

Insulin-receptor interaction in the obese-hyperglycemic mouse: a model of insulin resistance, J. Biol. Chem. **248**:244, 1973.

25. Langley, J. N.: On the physiology of the salivary secretion. Part II, On the mutual antagonism of atropin and pilocarpin having especial reference to their relations in the submaxillary gland of the cat, J. Physiol. (London) **1**:339, 1878.
26. Langley, J. N.: On the reaction of cells and of nerve endings to certain poisons, chiefly as regards the reaction of striated muscle to nicotine and to curari, J. Physiol. (London) **33**:374, 1905.
27. Lefkowitz, R., and Haber, E.: A fraction of the ventricular myocardium that has the specificity of the cardiac β-adrenergic receptor, Proc. Natl. Acad. Sci. U.S.A. **68**:1773, 1971.
28. Lefkowitz, R. J., Mukherjie, C., Coverstone, M., and Caron, M. G.: Stereospecific [^{3}H] (-)-alprenolol binding sites, β-adrenergic receptors and adenylate cyclase, Biochim. Biophys. Res. Comm. **60**:703, 1974.
29. Lefkowitz, R. J., Mullikin, D., and Caron, M. G.: Regulation of β-adrenergic receptors by guanyl-5′-yl imidodiphosphate and other purine nucleotides, J. Biol. Chem. **251**:4686, 1976.
30. Lesniak, M. A., and Roth, J.: Regulation of receptor concentration by homologous hormone: effect of human growth hormone on its receptor in IM-9 lymphocytes, J. Biol. Chem. **251**:3730, 1976.
31. Levitzki, A., Sevilia, N., Atlas, D., and Steer, M. L.: The binding characteristics and number of β-adrenergic receptors on the turkey erythrocyte, Proc. Natl. Acad. Sci. U.S.A. **71**:2773, 1974.
32. Limbird, L. E., and Lefkowitz, R. J.: Resolution of β-adrenergic receptor binding and adenylate cyclase by gel exclusion chromatography, J. Biol. Chem. **252**:799, 1977.
33. Lo, C.-S., August, T. R., Liberman, U. A., and Edelman, I. S.: Dependence of renal (Na^+ + K^+)-adenosine triphosphatase activity on thyroid status, J. Biol. Chem. **251**:7826, 1976.
34. Lo, C.-S., and Edelman, I. S.: Effect of triiodothyronine on the synthesis and degradation of renal cortical (Na^+ + K^+)-adenosine triphosphatase, J. Biol. Chem. **251**:7834, 1976.
35. Maguire, M. E., Wiklund, R. A., Anderson, H. J., and Gilman, A. G.: Binding of [^{125}]-iodohydroxybenzylpindolol to putative β-adrenergic receptors of rat glionic cells and other cell clones, J. Biol. Chem. **251**:1221, 1976.
36. Meledi, R., and Potter, L. T.: Acetylcholine receptors in muscle fibers, Nature **233**:599, 1971.
37. Mickey, J., Tate, R., and Lefkowitz, R. J.: Subsensitivity of adenylate cyclase and decreased β-adrenergic receptor binding after chronic exposure to (-)-isoproterenol in vitro, J. Biol. Chem. **250**:5727, 1975.
38. Mittag, T., Kornfield, P., Tormay, A., and Woo, C.: Detection of anti-acetylcholine receptor factors in serum and thymus from patients with myasthenia gravis, N. Engl. J. Med. **294**:691, 1976.
39. O'Malley, B. W., and Hardman, J. G., editors: Methods in enzymology. Vol. 36, Hormone action, part A; steroid hormones, 1975, New York, Academic Press, Inc.
40. Patrick, J., and Lindstrom, J.: Autoimmune response to acetylcholine receptor, Science **180**:871, 1973.
41. Remold-O'Donnell, E.: Stimulation and desensitization of macrophage adenylate cyclase by prostaglandins and catecholamines, J. Biol. Chem. **249**:3615, 1974.
42. Schramm, M., Feinstein, H., Neum, E., Lang, M., and Lasser, M.: Epinephrine binding to the catecholamine receptor and activation of the adenylate cyclase in erythrocyte membranes, Proc. Natl. Acad. Sci. U.S.A. **69**:523, 1972.
43. Sharma, V. K., and Banerjee, S. P.: Muscarinic cholinergic receptors in rat heart: effects of thyroidectomy, J. Biol. Chem. **252**:7444, 1977.
44. Soll, A. H., Kahn, C. R., Neville, D. M., Jr., and Roth, J.: Insulin receptor deficiency in genetic and acquired obesity, J. Clin. Invest. **56**:769, 1975.
45. Stone, M. J., and Metzger, H.: Study of macromolecular interactions by equilibrium molecular sieving, J. Biol. Chem. **243**:5049, 1968.
46. Sutherland, E. W., and Rall, T. W.: Fractionation and characterization of a cyclic adenine ribonucleotide formed by tissue particles, J. Biol. Chem. **232**:1077, 1958.
47. Vauquelin, G., Geynet, P., Hanoune, J., and Strosberg, A. D.: Isolation of adenylate cyclase-free, β-adrenergic receptor from turkey erythrocyte membrane by affinity chromatography, Proc. Natl. Acad. Sci. U.S.A. **74**:3710, 1977.
48. Von Euler, U. S., and Zetterstrom, B.: The role of amino oxidase in the inactivation of catechol amines injected in man, Acta Physiol. Scand. **33**(suppl. 118):26, 1955.
49. Williams, L. T., Lefkowitz, R. J., Watanabe, A. M., Hathaway, D. R., and Besch, H. R., Jr.: Thyroid hormone regulation of β-adrenergic receptor number, J. Biol. Chem. **252**:2787, 1977.
50. Wolfe, B. B., Zivvolli, J. A., and Molinoff, P. B.: Binding of dl-[^{3}H] epinephrine to proteins of rat ventricular muscle: nonidentity with beta adrenergic receptors, Mol. Pharmacol. **10**:582, 1974.
51. Wrenn, S., and Haber, E.: An antibody specific for the propranolol binding site of cardiac muscle, J. Biol. Chem. (in press).

23 □ Radioimmunoassay of cardiac glycosides

Thomas W. Smith
Gregory D. Curfman

The narrow margin between therapeutic and toxic doses of cardiac glycosides and the resulting high incidence of digitalis intoxication in clinical practice has provided a potent stimulus for recent progress in the measurement of circulating concentrations of these drugs.[4] Radionuclide technology has played a major role in these advances, and both methodologic aspects and clinical applications will be considered in this chapter. References representative of the rapidly expanding literature on these subjects were selected for citation, and the reference list is not intended to be a comprehensive listing of all contributions to the field.

METHODS FOR MEASUREMENT OF SERUM OR PLASMA DIGITALIS GLYCOSIDE CONCENTRATIONS

From a historical perspective, the ability to measure clinically relevant cardiac glycoside concentrations in biologic fluids is a very recent development. Following is a list of the chemical methods of limited sensitivity and the methods of higher sensitivity applicable to measurement of quantities in the nanogram range:

I. Chemical and physicochemical methods of limited sensitivity (microgram to milligram range)
   - A. Alkaline picrate: digitoxin[125]
   - B. Alkaline dinitrobenzene: digoxin[125]
   - C. Xanthydrol: digitoxin and digoxin[66]
   - D. Mass spectroscopy[52]
   - E. Differential pulse polarography[70]

II. Micro methods suitable for measurement of nanogram quantities
   - A. Duck embryo bioassay[47,48]
   - B. Radionuclide labeling[11]
   - C. Physicochemical methods
     1. Double-radionuclide dilution derivative: digitoxin[84,86]
     2. Gas-liquid chromatography: digoxin,[93,126] digitoxin, lanatoside A, lanatoside C, desacetyllanatoside C, digitoxigenin, and digoxigenin[93]
     3. Fluorometry[18,94]
   - D. Na^+,K^+- ATPase inhibition
     1. Red blood cell rubidium uptake inhibition
        - a. ^{86}Rb radioactivity measurement: digitoxin* and digoxin†
        - b. Atomic absorption spectrophotometry of nonradioactive rubidium: digitoxin and lanatoside C[33,98]
     2. Microsomal Na^+,K^+-activated ATPase inhibition: digitoxin[20,23] and digoxin[21]
   - E. Competitive protein binding
     1. Radioimmunoassay: digoxin,‡ digitoxin,[96,112] deslanoside,[120] ouabain,[109,115] acetyl strophanthidin,[108] α-acetyl digoxin,[2] β-acetyl digoxin,[2,6] β-methyl digoxin,[2,5,67] α-propionyl

□ Parts of the work described were supported by United States Public Health Service grant no. HL 14325.

*See references 49, 82, 83, 102, and 103.
†See references 10, 14, 50, 82, 83, and 103.
‡See references 30, 45, 60, 79, 95, 113, 116, and 118.

digoxin, g-strophanthin, and proscillaridin[2]

2. ATPase enzymatic radionuclide displacement: digoxin and digitoxin,[19] ouabain, proscillaridin, and digoxigenin[7]

The development of methods for ^{3}H or ^{14}C labeling of cardiac glycosides in the 1950s, in addition to providing important information on pharmacokinetics of these drugs by direct administration to experimental animals and human volunteers, was an essential step in the development of nearly all methods subsequently devised.[39]

Lukas and Peterson developed an elegant but laborious double-radionuclide dilution derivative assay for digitoxin that has been useful in elucidating pharmacokinetics and metabolic pathways of this drug.[84,85] A gas-liquid chromatography approach for quantitation of digoxin was reported by Watson and Kalman.[126] Although technically demanding, this procedure also has high specificity and should be useful in studying metabolic patterns of digoxin.

A number of assay procedures are based on the ability of cardenolides to bind with and inhibit Na^{+},K^{+}–activated ATPase with resulting inhibition of transmembrane monovalent cation transport. These include the red blood cell ^{86}Rb uptake inhibition assay originally reported by Lowenstein and Corrill and since modified by a number of other workers.[82,83] An interesting variant of this technique in which atomic absorption spectrophotometry is used is the only reported method that does not depend on radionuclide measurement at some step of the procedure.[17] Direct inhibition of Na^{+},K^{+}-ATPase is the basis for the assay methods developed by Burnett and Conklin and coworkers.[8,20,21]

The methods in widest use at the present time are competitive protein-binding assays. Cardiac glycoside–specific antibodies can serve as specific binding proteins,[24,96,117] or Na^{+},K^{+}-ATPase can be used as the binding substance, as reported by Brooker and Jelliffe.[19] All these approaches to digitalis glycoside concentration measurements have been reviewed in greater detail elsewhere.[23,119]

Our own efforts have been directed toward the development and application of radioimmunoassay techniques. It has been shown that cardiac glycosides covalently coupled with carrier proteins elicit specific antibody responses in rabbits and sheep.[36,117] The high affinity and specificity of selected antisera for cardiac glycosides allow the measurement of subnanogram amounts of digoxin, digitoxin, deslanoside, ouabain, and acetyl strophanthidin.* The techniques involved are sufficiently simple to be applicable in the well-equipped clinical chemistry laboratory and are described in detail elsewhere.*

As shown for the case of digoxin in Fig. 23-1, an aliquot of serum or plasma containing unlabeled cardenolide, without prior extraction, is mixed with the tracer in a convenient buffer volume, and then a specific antibody is added. Labeled and unlabeled molecules compete for a limited number of antibody binding sites, and after an appropriate incubation period, dextran-coated charcoal is added to separate the antibody-bound tracer from the free tracer. The importance of careful control of the duration of charcoal exposure is illustrated subsequently. Antibody-bound tritiated glycoside remaining in the supernatant phase after centrifugation is added to the liquid scintillation medium and counted in a liquid scintillation spectrometer. In an alternative variation, ^{125}I-labeled tracer digoxin or digitoxin is used, which can be counted in a gamma well scintillation counter.

The use of ^{125}I-labeled compounds offers the advantage of not requiring a quench correction, as discussed subsequently.[31,42] Although the two most commonly used compounds (3-0-succinyl-digoxigenin-[^{125}I] tyrosine and [^{125}I] tyrosine methylester of digoxin) are structurally different from digoxin, they cross-react with digoxin-specific antibodies and can provide results

*See references 96, 108, 109, 112, and 115-118.

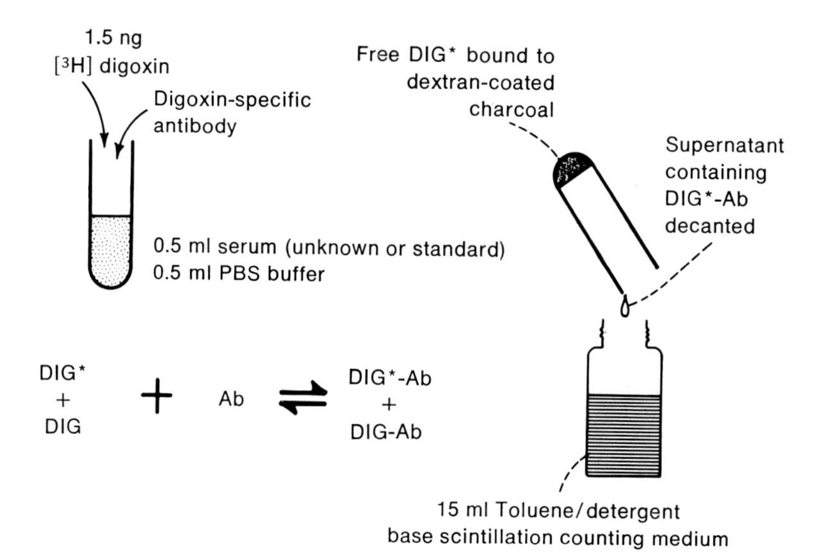

Fig. 23-1. Schematic outline of radioimmunoassay procedure for digoxin. (*DIG* = digoxin, *DIG** = ^{3}H-digoxin, and *Ab* = digoxin-specific antibody.)

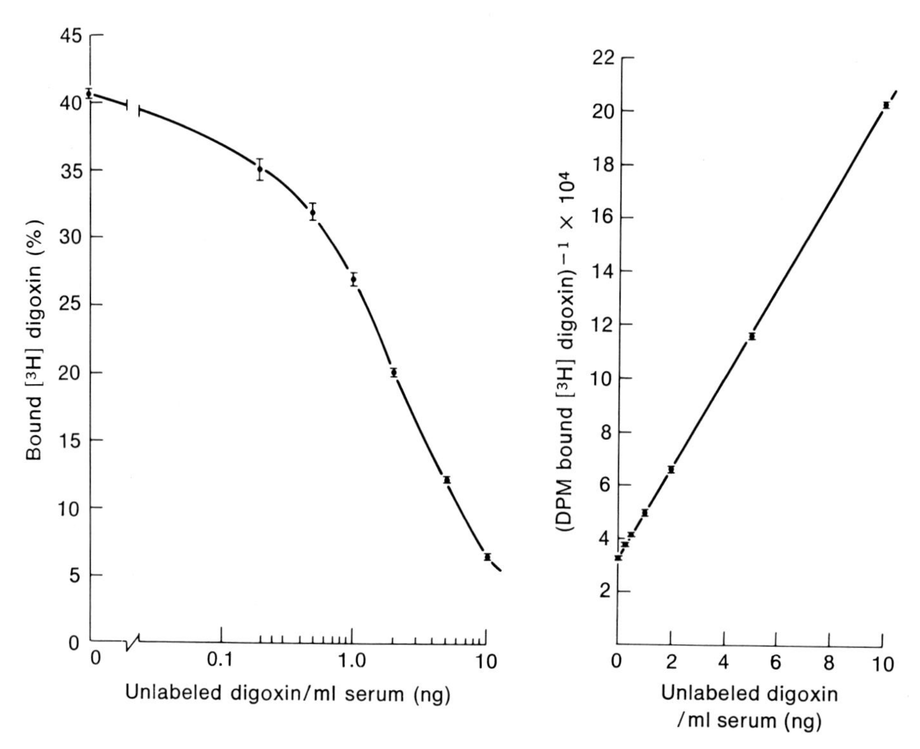

Fig. 23-2. Standard curves for radioimmunoassay of digoxin. *Left:* Data are plotted in semilogarithmic form. *Right:* Same data are plotted as reciprocal antibody-bound counts against linear scale of unlabeled digoxin concentration, yielding rectilinear plot.

that correlate closely with the ^{3}H-labeled digoxin method.[100,124]

Standard curves are constructed with known amounts of the cardiac glycoside to be measured. We have generally plotted these standard curves as illustrated in Fig. 23-2. The rectilinear reciprocal plot on the right is convenient for use with a computer program that facilitates processing of the data. A complete description and listing of this computer program has been provided in a previous edition of this book and an additional program has been published by Besch and Watanabe.[13]

It is necessary next to review certain technical aspects of cardiac glycoside radioimmunoassay procedures and to point out some of the important pitfalls. A sound understanding of the molecular interactions involved is necessary for the proper employment of these methods, and meticulous attention to details of quality control is necessary to ensure accurate results.

It is axiomatic that results in assays of this type can be no more accurate than the samples prepared for determining the standard curve. The crystalline compound to be used as standard must be tested for purity by thin-layer chromatography or some other sensitive analytic method. After the compound dries to constant weight under vacuum, standard solutions are prepared by gravimetric determination and serial dilution. Fresh standards should be made every few weeks. It is preferable to store standard solutions at 0° C or below in ethanol at such concentrations that serum or plasma standards can be prepared by addition of small volumes of the ethanol solution when each set of unknowns is run. Purity of tracer compounds must be monitored as well, since there is substantial variation from lot to lot from commercial suppliers.

The problem of quenching correction deserves special attention, since many clinical samples contain substantial amounts of strongly quenching substances such as hemoglobin and bile pigments.[29,58] The most straightforward approach is the use of internal standards.[116] External standardization is more convenient, but it requires critical application and frequent experimental validation of the relation between quenching of internal (^{3}H) radiation and that from the external standard (Chapter 1).

It is apparent that exogenous radioactivity present in the serum of patients after radionuclide scanning procedures can produce erroneously low results. Since these radionuclides are generally gamma emitters, their presence is easily detected by appropriate setting of an auxiliary channel of the scintillation spectrometer. When exogenous radioactivity is detected, it can be dealt with by extraction of the sample with water-immiscible organic solvents or by the method outlined by Butler.[19,22]

The major source of difficulty in cardiac glycoside radioimmunoassay procedures is the selection of antibody preparation. The thorough characterization of antisera prior to their use in radioimmunoassay procedures has frequently been overlooked and deserves special consideration. The average intrinsic affinity constant of an antibody population is important as a determinant of sensitivity and also in terms of antibody-hapten interaction stability.[130] The practical importance of the latter can be illustrated by a comparison of the properties of two antisera. There is evidence that the dissociation rate constant for the reversible interaction antibody + hapten $\leftrightarrows$ antibody − hapten complex is the major variable governing the equilibrium constant.[37] We have determined the dissociation kinetics for a number of cardiac glycoside–antibody combinations and have documented major differences in dissociation rate constants. The method used involves observation of the rate at which dextran-coated charcoal sequesters labeled cardenolide tracer molecules as they dissociate from the antibody complex. Fig. 23-3 shows the dissociation kinetics of the digoxin-specific antiserum designated as antiserum 1, used to construct the standard curves shown in Fig. 23-2. Fig. 23-3 also shows, in contrast, the dissociation kinetics of an antiserum, antiserum 2, of inferior quality, which has an average intrinsic association constant about

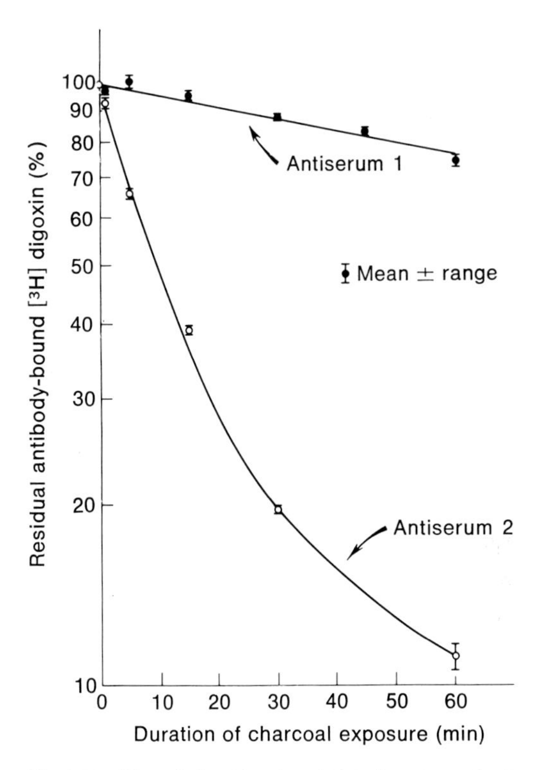

Fig. 23-3. Dissociation kinetics of [^{3}H] digoxin-antibody complexes. Plots were obtained by exposure of complex to dextran-coated charcoal for varying lengths of time. Antiserum 1, with high affinity for digoxin, shows slow dissociation. Antiserum 2 shows rapid dissociation kinetics, which could lead to substantial error in radioimmunoassay use.

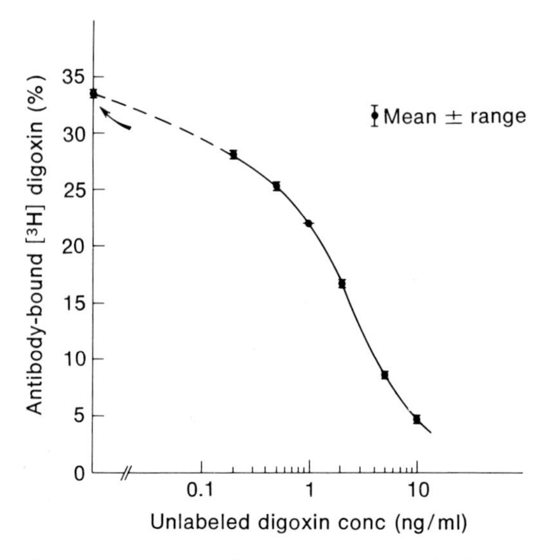

Fig. 23-4. Digoxin radioimmunoassay standard curve obtained with antiserum 2. Despite relatively low affinity, rapid dissociation kinetics, and low specificity of this antiserum, standard curve can be obtained that superficially appears to be satisfactory for analytic use.

fifteen times lower than that of antiserum 1. The complex formed by antiserum 2 dissociates quite rapidly compared with that of antiserum 1.

In practical terms, the use of an antiserum with rapid dissociation kinetics poses special problems that can result in significant error if not recognized. Sequential addition of charcoal to a series of standard and unknown samples results in lower supernatant count rates for samples exposed for longer durations. This presents no problem with antiserum 1, but special precautions are necessary to avoid this pitfall with antiserum 2. Meade and Kleist[88] and Kuno-Sakai and Sakai[76] have documented the existence of this problem of antibody-hapten dissociation with several commercially available digoxin radioimmunoassay kits.

Recently, alternative methods of separating antibody-bound digoxin from free digoxin that avoid charcoal-mediated disturbance of the equilibrium state have been proposed. These include solid-phase radioimmunoassays in which digoxin-specific antibody is conjugated to Sepharose, bromoacetylcellulose, or porous glass beads,[2,15,16,81] a gel equilibration technique,[51] and ammonium sulfate precipitation.[42]

Despite the problem of the relatively rapid dissociation kinetics, antiserum 2 provides a standard curve that superficially appears to be satisfactory for analytic use (Fig. 23-4). Still further hazards exist in the use of this antiserum, however, as illustrated in Figs. 23-5 and 23-6. The hapten inhibition experiment shown in Fig. 23-5 documents the high specificity of antiserum 1 for the homologous hapten digoxin.[117] In contrast, Fig. 23-6 shows a substantially lower degree of specificity for antiserum 2, which could result in falsely high values for unknowns because of displacement of tracer digoxin from antibody binding sites by nondigoxin compounds. Thus the demonstration of displacement of tritiated

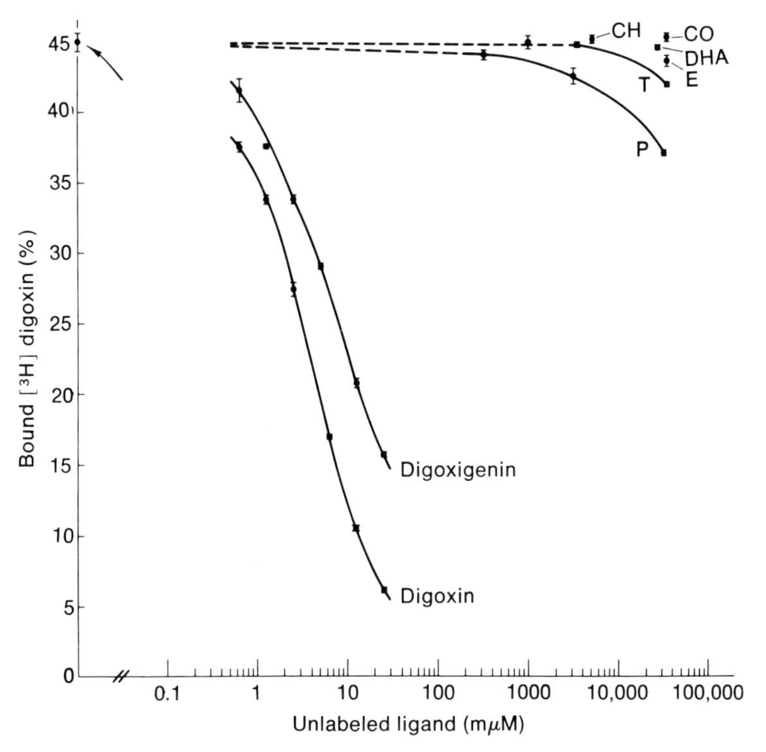

Fig. 23-5. Specificity of selected digoxin-specific antiserum. Only digoxin and digoxigenin compete effectively with ^{3}H-labeled digoxin for antibody binding sites. Cholesterol *(CH)*, cortisol *(CO)*, dehydroepiandrosterone *(DHA)*, 17-β-estradiol *(E)*, progesterone *(P)*, and testosterone *(T)* cause measurable displacement only when present in concentrations greater than 1000 times above those of digoxin. Arrow on vertical axis denotes binding in absence of competing ligand. Horizontal lines indicate ranges of duplicate determinations. (Reprinted with permission from Smith, T. W., Butler, V. P., Jr., and Haber, E.: Biochemistry **9:**331, 1970, copyright by the American Chemical Society.)

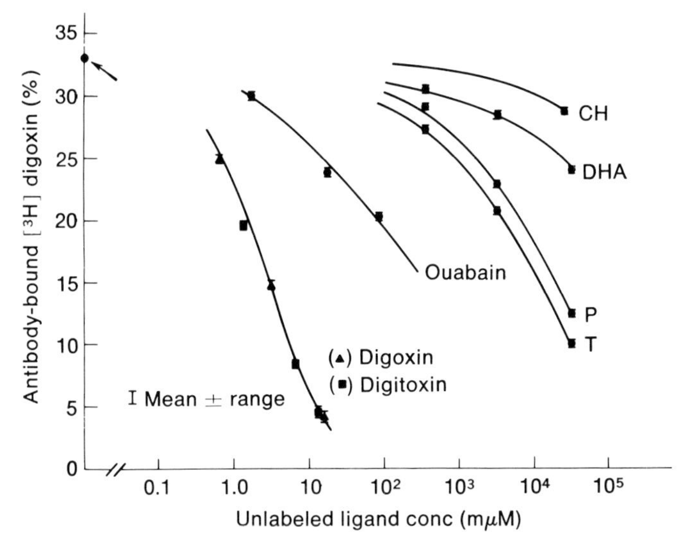

Fig. 23-6. Specificity of antiserum 2. Relatively poor specificity is evident; this results in potential for substantial error in radioimmunoassay. See Fig. 23-5 for explanation of abbreviations.

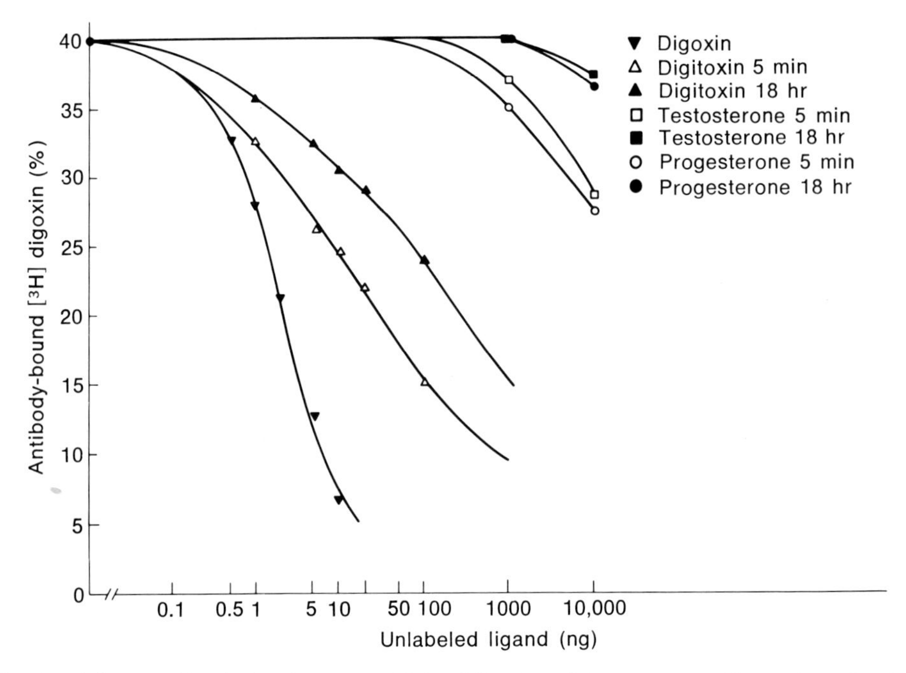

Fig. 23-7. Influence of incubation time on specificity of digoxin radioimmunoassay. Times refer to duration of incubation time prior to separation of antibody-bound [^{3}H] digoxin from free [^{3}H] digoxin with dextran-coated charcoal.

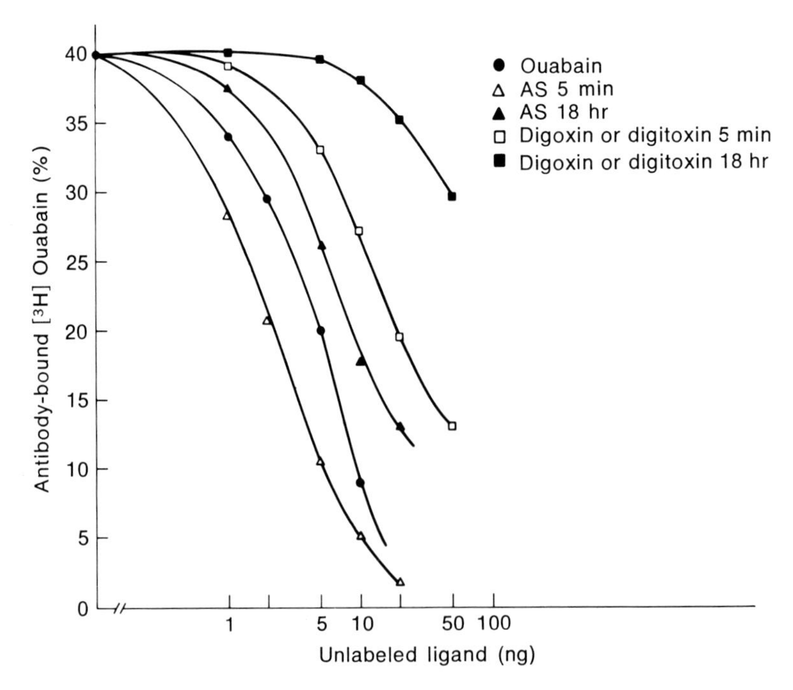

Fig. 23-8. Influence of incubation time on specificity of ouabain radioimmunoassay. Digoxin and digitoxin gave essentially identical results, as shown in curves denoted by open and closed squares. (*AS* = acetyl strophanthidin.)

digoxin from antibody binding sites by unlabeled digoxin, as shown in Fig. 23-4, does not constitute adequate evidence of suitability for use in radioimmunoassay. Since commercial suppliers of radioimmunoassay reagents often do not supply the purchaser with adequate documentation of the properties of their products, the user must assume responsibility for such testing. This point merits special attention because of the recent proliferation of commercial radioimmunoassay kits.[75] At the minimum, hapten inhibition studies for defining specificity and charcoal contact time studies for defining hapten-antibody dissociation kinetics are required.

We have documented an important dependence of the specificity of cardiac glycoside radioimmunoassay procedures on the duration of the incubation step. This relationship is a consequence of the fact that the rate of formation of the antibody-ligand complex is very rapid compared to the rate of dissociation of the complex. As shown in Fig. 23-7, ^{3}H-labeled digoxin is displaced from antibody combining sites of digoxin-specific antibody to a similar extent whether incubation continues for 5 min or 18 hr. In the case of digitoxin, however, substantially greater displacement is evident at 5 min than at 18 hr. Similarly, both testosterone and progesterone produce much greater displacement of ^{3}H-labeled digoxin from antibody binding sites at 5 min than at 18 hr.

Although the quantitative degree of radioimmunoassay specificity dependence on incubation time varies with the specific antiserum and ligands tested, it is a general phenomenon. Fig. 23-8 shows the results of an analogous experiment with ouabain-specific antibody. Displacement by the homologous hapten ouabain is similar at both 5 min and 18 hr, whereas both digoxin and digitoxin show considerably greater apparent cross-reaction after an incubation of 5 min than after one of 18 hr. This same phenomenon is even more strikingly demonstrated with acetyl strophanthidin. After a 5 min incubation, displacement of ^{3}H-labeled ouabain by acetyl strophanthidin is even greater than that by ouabain; this presumably reflects more rapid association kinetics of the aglycone. The degree of displacement by acetyl strophanthidin at 5 min remains significantly greater when compared on the basis of molar concentrations of the competing ligands. After an 18 hr incubation, however, the acetyl strophanthidin displacement curve has shifted to the right of that of ouabain, as would be expected if the dissociation rate of the ouabain-antibody complex were sufficiently slower than that of acetyl strophanthidin to overcome the somewhat slower association kinetics. In other words, in assessing antibody specificity, one must take into consideration not only the average equilibrium constants of the various antibody-ligand interactions but also the forward and reverse rate constants. The incubation time that will provide optimum specificity for a given antiserum can be easily determined by experiments of the type just described.

Our intent in discussing the foregoing considerations is not to convey the impression that cardiac glycoside radioimmunoassay procedures are unduly complex and demanding; they are not. Rather, our intent is to call attention to some relatively simple and straightforward principles that must be borne in mind to avoid erroneous results that will be useless or even misleading to the clinician.

RATIONALE FOR USE OF SERUM DIGITALIS GLYCOSIDE CONCENTRATION MEASUREMENTS

There are several lines of evidence that suggest a potentially useful relationship between serum cardiac glycoside concentrations and clinical or pharmacologic effects. First, both inotropic and toxic electrophysiologic effects of digitalis are known to be dose-related phenomena.[89] Since all studies to date have shown increasing serum or plasma cardiac glycoside concentrations with increasing dosage,[119] one would expect at least a statistical correlation between the circulating concentration and the clinical state, assuming, of course, that sufficient time has elapsed since the last previous dose to allow equilibration of the drug between intravascular and tissue compartments. A safe assumption would be that this takes about 6 hr. Recent clinical studies have provided additional support for a relationship between serum levels and inotropic effect assessed by systolic time intervals.[62]

Second, several investigators have documented a relatively constant ratio of digoxin concentration in serum to that in myocardium in both animals and human subjects after serum-tissue equilibration; this led to the prediction of the potential clinical usefulness of knowledge of serum concentrations.* This line of argument should not be pushed too far, however, since a constant myocardial-to-serum digoxin ratio has not been confirmed in all studies,[35] and total myocardial concentration has not been directly shown to parallel effective receptor concentration over the entire spectrum of serum levels. Thus total myocardial glycoside concentration cannot be assumed a priori to bear a direct relationship to effect.

Third, there is an impressive body of circumstantial evidence suggesting that Na^+, K^+-activated ATPase is involved in at least some of the actions of digitalis glycosides.† This membrane-bound enzyme system is affected by cardiac glycosides only when these agents are present at the outer cell surface in the case of the squid giant axon and the human erythrocyte.[25,61] This places at least one putative digitalis receptor close to the plasma compartment, promoting the translation of serum concentration to myocardial effect.

Finally, experimental work in a canine model documented a close relationship between serum digoxin concentration and cardiac electrophysiologic effect as judged by acetyl strophanthidin tolerance or response to low-energy endocardial electrical stimulation.[3] Experiments conducted in healthy dogs cannot, of course, be readily extrapolated to the patient with cardiac disease, in whom myriad factors affect myocardial sensitivity to digitalis glycosides.

*See references 26, 34, 53, 54, and 69.
†See references 1, 12, 78, 107, and 120.

CLINICAL STUDIES

The literature has expanded rapidly in the past few years with respect to clinical studies in which serum or plasma cardiac glycoside measurements are used.[119] Despite the multiplicity of techniques and laboratories involved, there is substantial agreement regarding digoxin and digitoxin levels in patients maintained on usual doses of these drugs. Table 23-1 summarizes studies involving more than 1000 patients. The mean digoxin concentration of groups of patients without evidence of toxic reaction was about 1.4 ng/ml. Increasing digoxin doses or decreasing renal function produced increases in mean serum levels,[119] as would be expected from our current understanding of digoxin pharmacokinetics. Data concerning the relation of serum digoxin concentration to toxic effect in the heart are also summarized in Table 23-1. Mean serum or plasma digoxin concentrations tended to be two to three times higher in patients with clinical evidence of toxic reaction to digoxin, and the difference was statistically significant in all studies except those of Fogelman and co-workers[46] and Howard and associates.[63] Nevertheless, overlap of levels between groups in toxic condition and those in nontoxic condition was observed in most series and was quite substantial in the prospective study by Beller and associates.[4]

Available studies correlating serum digitoxin concentrations with clinical state are summarized in Table 23-2. These data are qualitatively similar to those for digoxin, with mean levels in patients with clinical evidence of toxic reaction about two times higher than those without evidence of toxic reaction. Values in the groups of toxic condition and in those in nontoxic condition exceed usual digoxin concentrations by a factor of about 10 because of the relatively high affinity of serum albumin for digitoxin.[85] Digitalis leaf in usual maintenance

Table 23-1. Serum or plasma digoxin concentrations: patients with and without toxicity

Source	Method	Mean concentration		Statistical significance
		With toxicity	Without toxicity	
Beller et al.[4]	Radioimmunoassay	2.3	1.0	Yes
Bertler and Redfors[11]	^{86}Rb uptake	2.4	0.9	Yes
Bertler et al.[9]	^{86}Rb uptake	3.1	1.4	Yes
Brooker and Jelliffe[19]	Enzymatic displacement	3.1	1.4	Yes
Burnett and Conklin[21]	ATPase inhibition	5.7	1.2	Yes
Carruthers et al.[28]	Radioimmunoassay	2.76	1.21	
Chamberlain et al.[30]	Radioimmunoassay	3.1	1.4	Yes
Evered and Chapman[44]	Radioimmunoassay	3.36	1.38	Yes
Fogelman et al.[46]	Radioimmunoassay	1.7	1.4	No
Grahame-Smith and Everest[50]	^{86}Rb uptake	5.7	2.4	Yes
Hayes et al.[55]*	Radioimmunoassay			
Infants		4.4	2.8	Yes
Children		3.4	1.3	Yes
Hoeschen and Proveda[60]	Radioimmunoassay	2.8	0.8-1.3	Yes
Howard et al.[63]	Radioimmunoassay	0.91	0.97	No
Huffman et al.[65]	Radioimmunoassay	3.32	1.49	Yes
Iisalo, et al.	Radioimmunoassay	3.1	1.2	Yes
Johnston et al.[68]	Radioimmunoassay	2.2	1.1	Yes
Krasula et al.[74]*	Radioimmunoassay			
Infants		3.6	1.7	Yes
Children		2.9	1.1	Yes
Lader et al.[77]	Radioimmunoassay	2.2	1.1	Yes
McCredie et al.[87]*	Radioimmunoassay			
Infants		—	3.45	
Children		3.81	1.41	Yes
Morrison et al.[92]	Radioimmunoassay	3.35	0.76	Not stated
Oliver et al.[95]	Radioimmunoassay	3.0	1.6	Yes
Park et al.[97]	Radioimmunoassay	3.8	1.1	Yes
Ritzmann et al.[104]	^{86}Rb uptake	5.5 (median)	1.2 (median)	Yes
Scherrmann and Bourdon[106]	Radioimmunoassay	4.58	1.	Yes
Singh et al.[111]	Radioimmunoassay	4.79	2.91	Yes
Smith et al.[116]	Radioimmunoassay	3.3	1.3	Yes
Smith and Haber[118]	Radioimmunoassay	3.7	1.4	Yes
Weissel et al.[127]	Radioimmunoassay	2.97	1.38	Yes
Whiting et al.[129]	Radioimmunoassay	3.5	1.4	Yes
Zeegers et al.[131]	Radioimmunoassay	4.4	1.6	Yes

*Pediatric patients.

Table 23-2. Serum or plasma digitoxin concentrations: patients with and without toxicity

Source	Method	Mean concentration		Statistical significance
		With toxicity	Without toxicity	
Beller et al.[4]	Radioimmunoassay	34.0	25.0	Yes
Bentley et al.[8]	ATPase inhibition	39.0	23.0	Yes
Brooker and Jelliffe[19]	Enzymatic displacement	48.8	31.8	Not stated
Chiche et al.[32]	Radioimmunoassay	57.0	25.4	Yes
Dessaint[38]	Radioimmunoassay	96.0	26.8	Yes
Hillestad et al.[59]	^{86}Rb uptake	28.3	16.8	Yes
Lukas and Peterson[86]	Double isotope dilution derivative	43-67	20.0	Not stated
Morrison and Killip[90]	Radioimmunoassay	53.0	25.0 (0.1 mg/day) 44.0 (0.2 mg/day)	Yes
Peters et al.[99]	Radioimmunoassay	56.4	28.8	Yes
Rasmussen et al.[102]	^{86}Rb uptake	48.7	16.6	Not stated
Ritzmann et al.[103]	^{86}Rb uptake	37.0 (median)	20.5 (median)	Yes
Smith[112]	Radioimmunoassay	34.0	17.0	Yes

doses produces serum digitoxin concentrations that are quite similar to those in patients receiving digitoxin, whether measured by radioimmunoassay or by enzymatic displacement.[19,112] As in the case of digoxin, despite significant differences in mean serum digitoxin concentration between patients with toxic reaction and those without toxic reaction, significant overlap clearly does occur.

Thus it is apparent from the experience of a number of investigators that no level can be selected that unequivocally separates toxic and nontoxic cardiac glycoside concentrations in the usual clinical setting. Optimal doses of these drugs must still be based to a large extent on clinical response. The many factors that must be taken into account in the clinical evaluation of individual response to digitalis include serum potassium levels, calcium and magnesium concentrations, adequacy of tissue oxygenation, acid-base balance, renal function, thyroid status, autonomic nervous system tone, other drugs concurrently received, and type and extent of underlying heart disease. Our own experience suggests that ischemic heart disease is especially prevalent among patients who show evidence of toxic reaction at relatively low serum digoxin or digitoxin concentrations.[114] A number of other studies further document the important relationship between intrinsic cardiac disease and response to digitalis.*

Further experience is necessary to define the ultimate role of serum concentration measurements in the management of individual problem patients. Without question, however, these techniques are proving to be highly useful in the study of various problems in the clinical pharmacology of digitalis glycosides. The biologic availability of digoxin in various commercial preparations, in which there has been a recent flurry of interest, is representative of an area that has depended entirely on such measurements for its data.[80,110] A recent study has suggested that appropriate use of digoxin serum levels may significantly reduce the incidence of digoxin toxicity in hospitalized patients.[43] Effects of cardiopulmonary bypass on digoxin pharmacokinetics have been productively studied by radioimmunoassay techniques.[28,33,73,91] Gastrointestinal absorption has been investigated both in normal subjects and in pa-

*See references 4, 30, 44, 71, 119, and 121.

tients with malabsorption syndrome.[57,64,128] Digoxin pharmacokinetics in the pediatric age group have been studied* and transplacental passage of drug has been documented.[105] The clinical pharmacology of digitoxin has also been more completely outlined by studies involving radioimmunoassay procedures as well as other measurement techniques.† Recent reviews have provided a more exhaustive compilation of advances in these areas, made possible by serum or plasma concentration measurements.‡

*See references 55, 56, 74, 87, 105, and 123.
†See references 8, 101, 112, and 122.
‡See references 13, 23, 72, 114, and 126.

REFERENCES

1. Akera, T., Larsen, F. S., and Brody, T. M.: Correlation of cardiac sodium- and potassium-activated ATPase activity with ouabain-induced inotropic stimulation, J. Pharmacol. Exp. Ther. **173**:145, 1970.
2. Arndts, D.: A solid-phase radioimmunoassay for digoxin and its acylated derivatives, Naunyn-Schmiedeberg Arch. Pharmacol. **287**:309, 1975.
3. Barr, I., Smith, T. W., Klein, M. D., Hagemeijer, F., and Lown, B.: Correlation of the electrophysiologic action of digoxin with serum digoxin concentration, J. Pharmacol. Exp. Ther. **180**:710, 1972.
4. Beller, G. A., Smith, T. W., Abelmann, W. H., Haber, E., and Hood, W. B., Jr.: Digitalis intoxication: a prospective clinical study with serum level correlations, N. Engl. J. Med. **284**: 989, 1971.
5. Belz, G. G., and Kleeberg, U. R.: Plasma half-life of β-methyl digoxin following repetitive application in man, Klin. Wochenschr. **53**:491, 1975.
6. Belz, G. G., and Nubling, H.: Half-life in plasma following repetitive application of β-acetyl digoxin in man, Klin. Wochenschr. **53**:543, 1975.
7. Belz, G. G., and Pflederer, W.: Studies on a plasma cardiac glycoside assay based upon displacement of ^{3}H-ouabain from Na^+-K^+-ATPase, Basic Res. Cardiol. **70**:142, 1975.
8. Bentley, J. D., Burnett, G. H., Conklin, R. L., and Wasserburger, R. H.: Clinical application of serum digitoxin levels—a simplified plasma determination, Circulation **41**:67, 1970.
9. Bertler, A., Gustafson, A., Ohlin, P., Monti, M., and Redfors, A.: Digoxin intoxication and plasma glycoside levels. In Storstein, O., editor: Symposium on digitalis, Oslo, 1973, Gyldendal Norsk Forlag, p. 300.
10. Bertler, A., and Redfors, A.: An improved method of estimating digoxin in human plasma, Clin. Pharmacol. Ther. **11**:665, 1970.
11. Bertler, A., and Redfors, A.: Plasma levels of digoxin in relation to toxicity, Acta Pharmacol. Toxicol. (Kbh.) **29**(suppl. 3):281, 1971.
12. Besch, H. R., Jr., Allen, J. C., Glick, G., and Schwartz, A.: Correlation between the inotropic action of ouabain and its effects on subcellular enzyme systems from canine myocardium, J. Pharmacol. Exp. Ther. **171**:1, 1970.
13. Besch, H. R., Jr., and Watanabe, A. M.: Radioimmunoassay of digoxin and digitoxin, Clin. Chem. **21**:1815, 1975.
14. Binnion, P. F., Morgan, L. M., Stevenson, H. M., and Fletcher, E.: Plasma and myocardial digoxin concentrations in patients on oral therapy, Br. Heart J. **31**:636, 1969.
15. Boguslaski, R. C., and Denning, C. E.: A column radioimmunoassay method for the determination of digoxin, Biochem. Med. **14**:83, 1975.
16. Boguslaski, R. C., and Schwartz, C. L.: Column radioimmunoassay method for the determination of digitoxin, Anal. Chem. **47**:1583, 1975.
17. Bourdon, R., and Mercier, M.: Dosage des heterosides cardiotoniques dans les liquides biologiques par spectrophotometric d'absorption atomique, Ann. Biol. Clin. (Paris) **27**:651, 1969.
18. Britton, A. Z., and Njau, E.: The specific fluorometric determination of digoxin, Anal. Chim. Acta **76**:409, 1975.
19. Brooker, G., and Jelliffe, R. W.: Serum cardiac glycoside assay based upon displacement of ^{3}H-ouabain from Na-K ATPase, Circulation **45**:20, 1972.
20. Burnett, G. H., and Conklin, R. L.: The enzymatic assay of plasma digitoxin levels, J. Lab. Clin. Med. **71**:1040, 1968.
21. Burnett, G. H., and Conklin, R. L.: The enzymatic assay of plasma digoxin, J. Lab. Clin. Med. **78**:779, 1971.
22. Butler, V. P., Jr.: Digoxin radioimmunoassay, Lancet **1**:186, 1971.
23. Butler, V. P., Jr.: Assays of digitalis in the blood, Prog. Cardiovasc. Dis. **14**:571, 1972.
24. Butler, V. P., Jr., and Chen, J. P.: Digoxin-specific antibodies, Proc. Natl. Acad. Sci. U. S. A. **57**:71, 1967.
25. Caldwell, P. C., and Keynes, R. D.: The effect of ouabain on the efflux of sodium from a squid giant axon, J. Physiol. **148**:8P, 1959.
26. Carroll, P. R., Gelbart, A., O'Rourke, M. F., and Shortus, J.: Digoxin concentrations in the serum and myocardium of digitalized patients, Aust. N. Z. J. Med. **3**:400, 1973.
27. Carruthers, S. G., Cleland, J., Kelly, J. G., Lyons, S. M., and McDevitt, D. G.: Plasma and tissue digoxin concentrations in patients undergoing cardiopulmonary bypass, Br. Heart J. **37**: 313, 1975.

28. Carruthers, S. G., Kelly, J. G., and McDevitt, D. G.: Plasma digoxin concentrations in patients on admission to hospital, Br. Heart J. **36:**707, 1974.
29. Cerceo, E., and Cipriano, A. E.: Factors affecting the radioimmunoassay of digoxin, Clin. Chem. **18:**539, 1972.
30. Chamberlain, D. A., White, R. J., Howard, M. R., and Smith, T. W.: Plasma digoxin concentrations in patients with atrial fibrillation, Br. Med. J. **3:**429, 1970.
31. Chambers, R. E.: Digoxin radioimmunoassay: improved precision with iodinated tracer, Clin. Chim. Acta **57:**191, 1974.
32. Chiche, P., Baligadoo, S., Larvelle, P., and Borgard, J. P.: Intoxications digitaliques et déviations de l'activité thérapeutique de la digitaline, Coeur Med. Interne **15:**249, 1976.
33. Coltart, D. J., Chamberlain, D. A., Howard, M. R., Kettlewell, M. G., Mercer, J. L., and Smith, T. W.: The effect of cardiopulmonary bypass on plasma digoxin concentrations, Br. Heart J. **33:**334, 1971.
34. Coltart, D. J., Gullner, H. G., Billingham, M., Goldman, R. H., Stinson, E. B., Kalman, S. M., and Harrison, D. C.: Physiological distribution of digoxin in human heart, Br. Med. J. **4:**733, 1974.
35. Coltart, D. J., Howard, M., and Chamberlain, D.: Myocardial and skeletal muscle concentrations of digoxin in patients on long-term therapy, Br. Med. J. **2:**318, 1972.
36. Curd, J. G., Smith, T. W., Jaton, J. C., and Haber, E.: The isolation of digoxin-specific antibody and its use in reversal of the effects of digoxin, Proc. Natl. Acad. Sci. U. S. A. **68:**2401, 1971.
37. Day, L. A., Sturtevant, J. M., and Singer, S. J.: The kinetics of the reaction between antibodies to the 2,4 DNP group and specific haptens, Ann. N. Y. Acad. Sci. **103:**611, 1963.
38. Dessaint, J. P.: Dosage radio-immunologique des digitaliques (digitoxine et digoxine) dans le sang, Lille Med. **19:**156, 1974.
39. Doherty, J. E.: The clinical pharmacology of digitalis glycosides: a review, Am. J. Med. Sci. **255:** 382, 1968.
40. Doherty, J. E., and Perkins, W. H.: Tissue concentration and turnover of tritiated digoxin in dogs, Am. J. Cardiol. **17:**47, 1966.
41. Doherty, J. E., Perkins, W. H., and Flanigan, W. J.: The distribution and concentration of tritiated digoxin in human tissues, Ann. Intern. Med. **66:** 116, 1967.
42. Drewes, P. A., and Pileggi, V. J.: Faster and easier radioimmunoassay of digoxin, Clin. Chem. **20:**343, 1974.
43. Duhme, D. W., Greenblatt, D. J., and Koch-Weser, J.: Reduction of digoxin toxicity associated with measurement of serum levels, Ann. Int. Med. **80:**516, 1974.
44. Evered, D. C., and Chapman, C.: Plasma digoxin concentrations and digoxin toxicity in hospital patients, Br. Heart J. **33:**540, 1971.
45. Evered, D. C., Chapman, C., and Hayter, C. J.: Measurement of plasma digoxin concentrations by radioimmunoassay, Br. Med. J. **3:**427, 1970.
46. Fogelman, A. M., LaMont, J. T., Finkelstein, S., Rado, E., and Pearce, M. L.: Fallibility of plasma-digoxin in differentiating toxic from non-toxic patients, Lancet **2:**727, 1971.
47. Friedman, M., and Bine, R., Jr.: A study of the rate of disappearance of a digitalis glycoside (lanatoside C) from the blood of man, J. Clin. Invest. **28:**32, 1949.
48. Friedman, M., St. George, S., and Bine, R., Jr.: The behavior and fate of digitoxin in the experimental animal and man, Medicine (Baltimore) **33:**15, 1954.
49. Gjerdrum, K.: Determination of digitalis in blood, Acta Med. Scand. **187:**371, 1970.
50. Grahame-Smith, D. G., and Everest, M. S.: Measurement of digoxin in plasma and its use in diagnosis of digoxin intoxication, Br. Med. J. **1:**286, 1969.
51. Greenwood, H., Howard, M., and Landon, J.: A rapid, simple assay for digoxin, J. Clin. Pathol. **27:**490, 1974.
52. Greenwood, H., Snedden, W., Hayward, R. P., and Landon, J.: The measurement of urinary digoxin and dihydrodigoxin by radioimmunoassay and by mass spectroscopy, Clin. Chim. Acta **62:**213, 1975.
53. Güllner, H.-G., Stinson, E. B., Harrison, D. C., and Kalman, S. M.: Correlation of serum concentrations with heart concentrations of digoxin in human subjects, Circulation **50:**653, 1974.
54. Härtel, G., Kyllönen, K., Merikallio, E., Ojala, K., Manninen, V., and Reissell, P.: Human serum and myocardium digoxin, Clin. Pharm. Ther. **19:**153, 1976.
55. Hayes, C. J., Butler, V. P., and Gersony, W. M.: Serum digoxin studies in infants and children, Pediatrics **52:**561, 1973.
56. Hayes, C. J., Gersony, W. M., Smith, W. B., and Butler, V. P., Jr.: Serum digoxin studies in infants and children, Bull. N.Y. Acad. Sci. **47:**1226, 1971.
57. Heizer, W. D., Smith, T. W., and Goldfinger, S. E.: Absorption of digoxin in patients with malabsorption syndromes, N. Engl. J. Med. **285:**257, 1971.
58. Helman, E. Z., Spiehler V., and Holland, S.: Elimination of error caused by hemolysis and bilirubin-induced color quenching in clinical radioimmunoassays. Clin. Chem. **20:**1187, 1974.
59. Hillestad, L., Hansteen, V., Hatle, L., Storstein, L., and Storstein, O.: Digitalis intoxication. In Storstein, O., editor: Symposium on digitalis, Oslo, 1973, Gyldendal Norsk Forlag, p. 281.

60. Hoeschen, R. J., and Proveda, V.: Serum digoxin by radioimmunoassay, Can. Med. Assoc. J. **105:** 170, 1971.
61. Hoffman, J. F.: The red cell membrane and the transport of sodium and potassium, Am. J. Med. **41:**666, 1966.
62. Horschen, R. J., and Cuddy, T. E.: Dose-response relation between therapeutic levels of serum digoxin and systolic time intervals, Am. J. Cardiol. **35:**469, 1975.
63. Howard, D., Smith, C. I., Stewart, G., Vadas, M., Tiller, D. J., Hensley, W. J., and Richards, J. G.: A prospective survey of the incidence of cardiac intoxication with digitalis in patients being admitted to hospital and correlation with serum digoxin levels, Aust. N.Z. J. Med. **3:**279, 1973.
64. Huffman, D. H., and Azarnoff, D. L.: Absorption of orally given digoxin preparations, J.A.M.A. **222:**957, 1972.
65. Huffman, D. H., Crow, J. W., Pentikäinen, P., and Azarnoff, D. L.: Association between clinical cardiac status, laboratory parameters, and digoxin usage, Am. Heart J. **91:**28, 1976.
66. Jelliffe, R. W.: A chemical determination of urinary digitoxin and digoxin in man, J. Lab. Clin. Med. **67:**694, 1966.
67. Johnson, B. F., Bye, C. E., Jones, G. E., and Sabey, G. A.: The pharmacokinetics of beta-methyl digoxin compared with digoxin tablets and capsules, Eur. J. Clin. Pharmacol. **10:**231, 1976.
68. Johnston, C. I., Pinkus, N. B., and Down, M.: Plasma digoxin levels in digitalized and toxic patients, Med. J. Aust. **1:**863, 1972.
69. Jusko, W. J., and Weintraub, M.: Myocardial distribution of digoxin and renal function, Clin. Pharm. Ther. **16:**449, 1974.
70. Kadish, K. M., and Spiehler, V. R.: Differential pulse polarographic determination of digoxin and digitoxin, Anal. Chem. **47:**1714, 1975.
71. Klein, M. D., Lown, B., Barr, I., Hagemeijer, F., Garrison, H., and Axelrod, P.: Comparison of serum digoxin level measurement with acetyl strophanthidin tolerance testing, Circulation **49:** 1053, 1974.
72. Koch-Weser, J.: The serum level approach to individualization of drug dosage, Eur. J. Clin. Pharm. **9:**1, 1975.
73. Krasula, R. W., Hastreitor, A. R., Levitsky, S., Tanagi, R., and Soyka, L. F.: Serum, atrial, and urinary digoxin levels during cardiopulmonary bypass in children, Circulation **49:**1047, 1974.
74. Krasula, R. W., Tanagi, R., Hastreider, A. R., Levitsky, S., and Soyka, L. F.: Digoxin intoxication in infants and children, J. Pediatr. **84:**265, 1974.
75. Kubasik, N. P., Brody, B. B., and Barold, S. S.: Problems in measurement of serum digoxin by commercially available radioimmunoassay kits. Am. J. Cardiol. **36:**975, 1975.
76. Kuno-Sakai, H., and Sakai, H.: Effects on radioimmunoassay of digoxin of varying incubation periods for antigen-antibody reaction and varying periods of absorption by dextran-coated charcoal, Clin. Chem. **21:**227, 1975.
77. Lader, S., Bye, A., and Marsdin, P.: The measurement of plasma digoxin concentration: a comparison of two methods, Eur. J. Clin. Pharmacol. **5:**22, 1972.
78. Langer, G. A.: The intrinsic control of myocardial contraction—ionic factors, N. Engl. J. Med. **285:**1065, 1971.
79. Larbig, D., and Kochsiek, K.: Radioimmunochemische Bestimmung von Digoxin in menschlichen Serum, Klin. Wochenschr. **49:**1031, 1971.
80. Lindenbaum, J., Mellow, M. H.: Blackstone, M. O., and Butler, V. P., Jr.: Variation in biologic availability of digoxin from four preparations, N. Engl. J. Med. **285:**1344, 1971.
81. Line, W. F., Siegel, S. J., Kuon, A., Frank, C., and Ernest, R.: Solid-phase radioimmunoassay for digoxin, Clin. Chem. **19:**1361, 1973.
82. Lowenstein, J. M.: A method for measuring plasma levels of digitalis glycosides, Circulation **31:** 228, 1965.
83. Lowenstein, J. M., and Corrill, E. M.: An improved method for measuring plasma and tissue concentrations of digitalis glycosides, J. Lab. Clin. Med. **67:**1048, 1966.
84. Lukas, D. S.: Some aspects of the distribution and disposition of digitoxin in man, Ann. N. Y. Acad. Sci. **179:**338, 1971.
85. Lukas, D. S., and DeMartino, A. G.: Binding of digitoxin and some related cardenolides to human plasma proteins, J. Clin. Invest. **48:**1041, 1969.
86. Lukas, D. S., and Peterson, R. W.: Double isotope dilution derivative assay of digitoxin in plasma, urine and stool of patients maintained on the drug, J. Clin. Invest. **45:**782, 1966.
87. McCredie, R. M., Chia, B. L., and Knight, P. W.: Infant versus adult plasma digoxin levels, Aust. N.Z. J. Med. **4:**223, 1974.
88. Meade, R. C., and Kleist, T. J.: Improved radioimmunoassay of digoxin and other sterollike compounds using Somogyi precipitation, J. Lab. Clin. Med. **80:**748, 1972.
89. Moe, G. K., and Farah, A. E.: Digitalis and allied cardiac glycosides. In Goodman, L. S., and Gilman, A., editors: The pharmacological basis of therapeutics, ed. 3, New York, 1965, The Macmillan Co.
90. Morrison, J., and Killip, T.: Radioimmunoassay of digitoxin, Clin. Res. **14:**668, 1970.
91. Morrison, J., and Killip, T.: Serum digitalis and arrhythmia in patients undergoing cardiopulmonary bypass, Circulation **47:**341, 1973.
92. Morrison, J., Killip, T., and Stason, W. B.:

Serum digoxin levels in patients undergoing cardiopulmonary bypass, Circulation **42:**110, 1970.
93. Nachtmann, R., and Spitzy, H.: Rapid and sensitive high-resolution procedure for digitalis glycoside analysis by derivatization liquid chromatography, J. Chromatogr. **122:**293, 1976.
94. Naik, D. V., Groover, J. S., and Schulman, S. G.: Simplified fluorometric determination of digitalis alkaloids, Anal. Chim. Acta **74:**29, 1975.
95. Oliver, G. C., Parker, B. M., and Parker, C. W.: Radioimmunoassay for digoxin. Technic and clinical application, Am. J. Med. **51:**186, 1971.
96. Oliver, G. C., Jr., Parker, B. M., Brasfield, D. L., and Parker, C. W.: The measurement of digitoxin in human serum by radioimmunoassay, J. Clin. Invest. **47:**1035, 1968.
97. Park, H. M., Chen, I. W., Manitassas, G. T., Lowey, A., and Saenger, E. L.: Clinical evaluation of radioimmunoassay of digoxin, J. Nucl. Med. **14:**531, 1973.
98. Pebay-Peyroula, F., Gaultier, M., and Nicaise, A. M.: Assay of digitalis glycosides: its application in clinical toxicology, Clin. Toxicol. **4:**419, 1971.
99. Peters, U., Hausamen, T.-U., and Grosse-Brockhoff, F.: Serial tests of serum digitoxin levels during digitoxin treatment, Dtsch. Med. Wochenschr. **99:**1701, 1974.
100. Pippin, S. L., and Marcus, F. I.: Digoxin immunoassay with the use of [^{3}H] digoxin vs. [^{125}I] tyrosine methyl-ester of digoxin, Clin. Chem. **22:**286, 1976.
101. Rasmussen, K., Jervell, J., Storstein, L., and Gjerdrum, K.: Digitoxin kinetics in patients with impaired renal function, Clin. Pharmacol. Ther. **13:**6, 1972.
102. Rasmussen, K., Jervell, J., and Storstein, O.: Clinical use of a bio-assay of serum digitoxin activity, Eur. J. Clin. Pharmacol. **3:**236, 1971.
103. Ritzmann, L. W., Bangs, C. C., Coiner, D., Custis, J. M., and Walsh, J. R.: Serum glycoside levels in digitalis toxicity, Circulation **40:**111, 1969.
104. Ritzmann, L. W., Bangs, C. C., Coiner, D., Custis, J. M., and Walsh, J. R.: Serum glycoside levels by rubidium assay, Arch. Intern. Med. **132:**823, 1973.
105. Rogers, M. C., Willerson, J. T., Goldblatt, A., and Smith, T. W.: Serum digoxin concentrations in the human fetus, neonate, and infant, N. Engl. J. Med. **287:**1010, 1972.
106. Scherrmann, J. M., and Bourdon, R.: Dosage de la digoxine par la méthode radioimmunologique, Eur. J. Toxicol. **9:**133, 1976.
107. Schwartz, A.: Is the cell membrane NaK-ATPase enzyme system the pharmacologic receptor for digitalis? Circ. Res. **39:**2, 1976.
108. Selden, R., Klein, M. D., and Smith, T. W.: Plasma concentration and urinary excretion kinetics of acetyl strophanthidin, Circulation **47:**744, 1973.
109. Selden, R., and Smith, T. W.: Ouabain pharmacokinetics in dog and man: determination by radioimmunoassay, Circulation **45:**1176, 1972.
110. Shaw, T. R. D., Howard, M. R., and Hamer, J.: Variation in the biological availability of digoxin, Lancet **2:**303, 1972.
111. Singh, R. B., Rai, A. N., Srivastav, D. K., Somani, P. N., and Katiyar, B. C.: Radioimmunoassay of serum digoxin in relation to digoxin intoxication, Br. Heart J. **37:**619, 1975.
112. Smith, T. W.: Radioimmunoassay for serum digitoxin concentration: methodology and clinical experience, J. Pharmacol. Exp. Ther. **175:**352, 1970.
113. Smith, T. W.: The clinical use of serum cardiac glycoside concentration measurements, Am. Heart J. **82:**833, 1971.
114. Smith, T. W.: Contributions of quantitative assay techniques to the understanding of the clinical pharmacology of digitalis, Circulation **46:**188, 1972.
115. Smith, T. W.: Ouabain specific antibodies: immunochemical properties and reversal of Na^{+}, K^{+}-activated adenosine triphosphatase inhibition, J. Clin. Invest. **51:**1583, 1972.
116. Smith, T. W., Butler, B. P., Jr., and Haber, E.: Determination of therapeutic and toxic serum digoxin concentrations by radioimmunoassay, N. Engl. J. Med. **281:**1212, 1969.
117. Smith, T. W., Butler, V. P., Jr., and Haber, E.: Characterization of antibodies of high affinity and specificity for the digitalis glycoside digoxin, Biochemistry **9:**331, 1970.
118. Smith, T. W., and Haber, E.: Digoxin intoxication: the relationship of clinical presentation to serum digoxin concentration, J. Clin. Invest. **49:** 2377, 1970.
119. Smith, T. W., and Haber, E.: The current status of cardiac glycoside assay techniques. In Yu, P. N., and Goodwin, J. F., editors: Progress in cardiology, vol. II, Philadelphia, 1973, Lea & Febiger.
120. Smith, T. W., Wagner, H., Jr., Markis, J. E., and Young, M.: Studies on the localization of the cardiac glycoside receptor, J. Clin. Invest. **51:**1777, 1972.
121. Smith, T. W., and Willerson, J. T.: Suicidal and accidental digoxin ingestion: report of five cases with serum digoxin level correlations, Circulation **44:**29, 1971.
122. Solomon, H. M., and Abrams, W. B.: Interaction between digitoxin and other drugs in man, Am. Heart J. **83:**277, 1972.
123. Soyka, L. F.: Clinical pharmacology of digoxin, Pediatr. Clin. North Am. **19:**241, 1972.
124. Taubert, K., and Shapiro, W.: Serum digoxin levels using an ^{125}I-labeled antigen: validation of

the method and observations on cardiac patients, Am. Heart J. **89:**79, 1975.

125. The United States Pharmacopeia, revision 17, 1965, pp. 191-197.
126. Watson, E., and Kalman, S. M.: Assay of digoxin in plasma by gas chromatography, J. Chromatogr. **56:**209, 1971.
127. Weissel, M., Fritzsche, H., and Fuchs, G.: Salivary electrolytes and serum digoxin in the assessment of digitalis intoxication, Wein. Klin. Wochenschr. **88:**455, 1976.
128. White, R. J., Chamberlain, D. A., Howard, M., and Smith, T. W.: Plasma concentrations of digoxin after oral administration in the fasting and postprandial state, Br. Med. J. **1:**380, 1972.
129. Whiting, B., Sumner, D. J., Goldbert, A.: An assessment of digoxin radioimmunoassay, Scott. Med. J. **18:**69, 1973.
130. Yalow, R. S., and Berson, S. A. In Margoulies, M., editor: Protein and polypeptide hormones, Amsterdam, 1969, Excerpta Medica Foundation, pp. 36-44, 71-76.
131. Zeegers, J. J. W., Maas, A. H. J., Willebrands, A. F., Kruyswijik, H. H., and Jambroes, G.: The radioimmunoassay of plasma-digoxin, Clin. Chim. Acta **44:**109, 1973.

24 □ Radionuclide displacement assays

Charles J. Homcy
Edgar Haber

A great deal of the current knowledge concerning physiologic regulation by circulating hormones or by pharmacologic agents was built on a variety of ingeniously devised bioassays. The development of these often cumbersome and laborious methods was necessitated by the very low concentrations at which these agents manifest their effects; the detection of these concentrations is far beyond the capacity of conventional chemical methods of detection. Development of the double-radionuclide derivative method permitted the accurate measurement of certain substances, particularly steroid hormones, at very low concentrations.[9] This method is predicated on the availability of a reactive chemical group within a class of compounds with which a radionuclide label of high specific activity can be coupled. The compound to be measured is then selectively purified of all other labeled compounds. Losses during purification are controlled by the addition of a pure derivative compound labeled with another radionuclide. Generally, at least three chromatographic separations are required; this results in very low recovery.

A major advance in the measurement of compounds of physiologic interest occurred when Yalow and Berson[31] described a method for measuring plasma insulin concentrations in which antibodies with high affinity for this hormone were used. This principle has proved to be of great usefulness, and much of the appreciation of endocrinology and pharmacology today is based on the application of this principle to the measurement of a wide variety of hormones and drugs. More recently, binding proteins other than antibodies have been employed in an analogous fashion with gratifying success. In this chapter we will consider both antibodies and other proteins as reagents of unique specificity and affinity for their respective ligands and the relative usefulness of these several classes of compounds with respect to their application to radiodisplacement assays.

PRINCIPLES OF RADIODISPLACEMENT ASSAY

The concept of competitive protein-binding analysis relies on the ability of the substance (ligand) to be assayed to compete for binding sites of defined specificity on any of a variety of macromolecules. When a solution containing the binding protein and the labeled ligand is allowed to come to equilibrium, the labeled ligand distributes itself between the protein's binding site and the solution in a ratio that depends on the concentration of the substituents and the binding constant of the protein. The protein-bound and free labels can then be separated by a variety of techniques, and their relative radioactivities can be determined. If unlabeled ligand is then added to this mixture, it competes with the labeled for the protein's combining site. Simple considerations of mass action dictate that the concentration of unbound labeled ligand is then increased and that of bound labeled ligand is decreased.

If the interaction between macromolecule and ligand and the methods employed in the assay each fulfill certain requirements, then classical equilibrium analysis can be brought to bear to describe the relationship in simple quantitative terms. The conventional assay is calibrated by determination of the ratio of protein-bound radioactivity to free radioactivity (B/F ratio) against a range of unlabeled ligand concentrations. This calibration or standard curve makes possible the determination of unknowns.

Requirements for macromolecules with suitable binding sites

The critical factors that determine the utility of a particular macromolecule as a vehicle for developing an assay are both the specificity and the sensitivity of its binding. The specificity of an assay is solely determined by the characteristics of the binding protein. The macromolecule must bind with significant affinity only the substance to be measured and not related precursors, metabolic products, drugs, or hormones. The sensitivity of the assay depends not only on the characteristics of the macromolecular binding sites but also on the specific activity of the tracer to be measured.[1] The ability to label ligands with radionuclides of high specific activity, however, has provided such sensitivity of detection that, typically, the affinity of the binding protein for the ligand is the factor determining sensitivity. However, it is the property of binding site specificity, that usually remains the limiting factor in devising a particular assay.

These points can best be illustrated by considering the difficulties associated with developing antibodies of requisite specificity and sensitivity for use in a variety of radioimmunoassays. The radioimmunoassay is of special importance in that it now represents the most accessible technique for generating a competitive protein-binding assay. It appears that essentially all classes of compounds are antigenic in their own right, whereas smaller compounds, such as drugs, steroid hormones, lipid hormones, and certain catecholamine derivatives, can become effective antigens when covalently linked to carrier proteins as haptenic determinants. The typical immune response to any antigen is heterogeneous, with the production of antibodies of varying specificity and sensitivity. In general, then, one immunizes a large number of animals and selects the most suitable antibody. Certain guidelines for the preparation of antigen and subsequent immunization have been developed, which provide some rationale in optimizing the specificity and sensitivity of the immune response. The specific form of the antigen has been of value in directing antibody specificity. For example, Midgley and Niswender[23] pointed out that those antigenic determinants of the steroid molecule which are farthest away from the carrier protein are most likely to provide the major determinants of antibody binding. In terms of the immunization pattern, the best results are usually obtained by intracutaneous injection of antigen mixed with complete Freund's adjuvant into the toe pads or multiple skin sites at intervals of weeks to months with periodic testing of the antisera.

Only rather infrequently has it been impossible to develop an appropriate antibody for assay purposes. As mentioned previously, very rarely has the problem been affinity, with association constants as high as 10^{11} to 10^{12} having been observed. Certain peptides, such as glucagon and adrenocorticotropic hormone (ACTH), have been notably poor immunogens, and appropriate antibodies have only rarely been obtained.[28,32] The problem of binding specificity is often more difficult to overcome. Even a small protein has a number of antigenic sites; only some of them are relevant to that protein's biologic function. A biologically inactive precursor of that protein such as a prohormone may share common antigenic sites with the hormone to be measured, as may the hormone's inactive degradation products. The investigator is generally interested in the concentration of active hormone and not in the total concentration of hormone precursor and metabolic products. Consequently, the selection of an antibody directed primarily at the active site of the hormone is essential. However, this is not always possible. For example, it is rather easy to find antibodies in which there is little cross-reaction between angiotensin I (the prohormone) and angiotensin II (the active hormone).[13] To date, however, antibodies have been found that effectively differentiate angiotensin II and its degradation products, the hexapeptide and the heptapeptide. Consequently, it has not been possible to measure circulating angiotensin II concentrations without prior fractionation. This problem has been much more serious in the measurement of parathyroid hormone. It has only recently been recognized that the confusing and conflicting results reported concerning the metabolism of parathyroid hormone as well as the inability to differentiate between results obtained from some patients with parathyroid adenoma and normal subjects are related to problems of cross-reaction between the active hormone and several long-lived degradation products.[26] In another example, insulin determinations by radioimmunoassay include the simultaneous determination of proinsulin[10]; this has led to the observed lack of correlation between insulin bioassay and radioimmunoassay in certain disease states such as insulin-secreting adenoma. Radioimmunoassays have been difficult to devise for certain hormones that contain structures and sequences in common with others. For example, extensive cross-reaction between luteinizing hormone (LH) and human chorionic gonadotropin (HCG) was initially observed in many antisera. However, immunization with a unique peptide fragment from HCG has allowed a much higher degree of immune specificity to be achieved.[7] Recently an immunoassay with the ability to discriminate between these two proteins has been developed.[6]

Other classes of binding proteins useful in radiodisplacement assays have included physiologic carrier proteins, cytoplasmic binding proteins, and membrane-bound receptors. Among the transport proteins, intrinsic factor (which specifically binds vitamin B_{12}), thyroxine-binding globulin, and cortisol-binding globulin have been used extensively.[14,24,25] The highest affinity and consequently the greatest potential sensitivity is found among the transport proteins for vitamin B_{12}.[16] Transcobalamin 1 and 2 have affinities for vitamin B_{12} that are about 10^{14} mol^{-1}.

The binding affinity of some transport proteins such as transcobalamin exceeds that of any available antibodies. However, in general, the dissociation constant of a receptor or transport protein closely relates to the circulating concentration of the hormone. Consequently, it may be difficult to measure low concentrations of certain hormones through the use of assays based on their receptors without a prior concentrating procedure.

A more recently introduced class of binding proteins includes the receptor proteins of cells derived either in soluble form from the cytosol or in membrane-bound form. The former include the receptor proteins for estrogens, which have been successfully employed[18] in quantitative binding assays. Such a technique affords particular advantages in that one need not rely on the ligand behaving as an effective antigen, if specific receptor protein can easily be harvested. A relatively crude preparation of a target cell will often be of suitable sensitivity and specificity. The greatest single advantage of receptor proteins may be their ability to discriminate between substances with respect to the presence of structures requisite for biologic activity. Lefkowitz and coworkers[19] demonstrated that the binding of ACTH and several of its less active derivatives to a preparation of receptors derived from an adrenal tumor was directly proportional to their biologic activity in a bioassay that measured cyclic adenosine monophosphate (CAMP) production. Such an approach has been applied with some success to the measurement of ACTH concentrations in plasma.[30]

The relative efficacies of different competitive binding proteins is afforded by comparing the characteristics of the radioimmunoassay and the receptor-binding assay in the following instances. Antibodies described for use in a digoxin radioimmunoassay bind digitoxin with an affinity approximately fifty times less than that with which they bind digoxin.[27] On the other hand, a radionuclide displacement assay in which a crude preparation of Na^+, K^+-ATPase from the brain is used does not differentiate in the binding of digitoxin and digoxin.[3] This correlates with the equipotency of these drugs in the in vitro inhibition of this enzyme. In the differentiation of luteinizing hormone, thyroid-stimulating hormone, and follicle-stimulating hormone, considerably less cross-reaction is seen in a displacement assay in which subcellular fractions of rat testis are used as compared to most of the early radioimmunoassays.[4]

Antibodies are inherently stable molecules, which are not readily degraded by proteases normally found in plasma. Certain receptor preparations, on the other hand, are far less stable, having been derived from intracellular or membrane sites that are not generally in direct contact with serum proteases.

Requirements for labeled ligand

It is clear that for the labeled ligand to behave as an effective probe in the binding assay, at least two conditions must be met. First, the preparation must be chemically homogeneous. If the labeled ligand is contaminated with other substances, then the assay may not only measure the desired substance but also associated materials with similar binding characteristics. Second, it is critical that the labeling procedure alter the chemical properties of the ligand as little as possible. Although the probe and unlabeled ligand need not be identical in binding characteristics for the assay to be effective, it is the affinity of the binding protein for the labeled ligand that will be the limiting factor in determining the sensitivity of the assay. This last point is a particularly important consideration in the development of receptor binding assays. If biologic activity is lost in the process of labeling, the tracer is not likely to bind to the receptor and therefore cannot be used for assay purposes. This may produce a problem in the labeling of certain small peptides by iodination. Thus Lin and Goodfriend[20] found that monoiodoangiotensin retained only 25% of biologic activity, whereas diiodoangiotensin was virtually devoid of biologic activity. In general, antigenic sites are retained in iodinated peptides, but quite often biologic activity is lost. The loss of biologic activity is of no consequence in a radioimmunoassay but of course can be critical in a receptor-binding assay.

There are a variety of commonly used labeling procedures that in most instances can fulfill the latter requirements just discussed. First, a variety of ^{14}C- and ^{3}H-labeled drugs and hormones are commercially available and of sufficient quality to be used for radioimmunoassay without prior purification. Labeling with radioactive ^{125}I is now a widely used method for developing compounds with extremely high specific activity. In certain instances, separation of labeled from unlabeled ligand has permitted specific activities that are equivalent to the theoretical activity of the ^{125}I tracer (2200 Ci/mmol) to be achieved.[21]

There are three easily applicable methods for ^{125}I labeling of compounds of biologic interest. Each provides distinct advantages and obvious shortcomings. Enzymatic labeling as described by Marchalonis[22] is a gentle technique often preferred for peptides and proteins; it is gentle in that damage is less, compared with the conventional chloramine-T procedure. With empirical titration, it is possible to limit incorporation of iodine to 1 molecule/molecule of peptide. The ability to generate monoiodinated peptides such as glucagon can be of critical importance because such derivatives can be shown to possess a high degree of biologic activity.[12] Probably the most commonly employed procedure for ^{125}I labeling is that of Hunter and Greenwood,[17] which is based on the ability of chloramine-T to oxidize iodide to an intermediate substance capable of substituting onto the phenyl ring of tyrosine. Although it is a more effective technique than the Marchalonis procedure for small nonprotein ligands, damage to peptides often occurs as a result of oxidation of labile amino acids by chloramine-T. More recently, a significant advance in radioiodine labeling has been introduced by Bolten and Hunter.[2] Their procedure does not require the presence of a phenol moiety for ^{125}I labeling but rather depends on the nucleophilic character of 1- or 2-degree amines to enable them to form covalent bonds with a chemically activated ^{125}I-labeled tyrosyl derivative. It should be emphasized, however, that if the interaction of the ligand with the binding protein of the assay depends on the availability of the free amine, then the Bolten-Hunter technique cannot be utilized. This can, of course, be determined empirically.

In predicting the optimal specific activity to be sought in labeling a particular ligand, two guidelines should be considered. First, it is usually the affinity of the binding protein that determines the detection limits of the assay, and a specific activity directed at that limit should be the goal. Second, if a binding protein with high-affinity binding sites is available, then a probe with high specific activity is only necessary if the concentration of ligand to be measured is extremely small.

Finally, certain points concerning the purity of unlabeled ligand are noteworthy. Since unlabeled ligand is employed as the reference in generating a standard curve, it must be identical to the substance to be measured in its interaction with the binding protein. Although it was pointed out earlier that chemical identity between labeled probe and the substance to be measured was not necessary, it should be clear that such a requirement for unlabeled ligand must usually be met.

Requirements for the assay

In general, most competitive binding assays are performed under equilibrium conditions. This removes time of incubation as a variable in determining the B/F ratio for any particular ligand concentration. Furthermore, since the association rate for ligand binding to a macromolecule is usually only diffusion limited, the off rate becomes the important variable in determining the equilibrium constant. Short incubation times that do not achieve equilibrium, therefore, allow contaminating ligands with lower equilibrium constants to compete more effectively with the ligand to be measured, since all the ligands have similar on rates. This then may result in loss of assay specificity.

Equilibrium considerations consequently dictate the method that would be most appropriate in separating free from bound ligand. Ideally, such separations should be carried out under actual equilibrium conditions. In practice, however, only a few methods that have practical application are carried out at equilibrium. Equilibrium methods are essential for macromolecular binding sites with high rate constants for dissociation. Antibodies of high affinity have relatively low dissociation rate constants, allowing nonequilibrium methods to be used. The lack of equilibrium during separation is partially compensated for by employing methods that are rapid and can be performed in all tubes at a constant temperature.

In constructing a competitive-binding assay, the method of separation should be carefully evaluated in light of the considerations just discussed. Although classical methods of protein separation, including electrophoresis, gel filtration, and ion-exchange chromatography, could theoretically be utilized, these methods suffer from their lack of simplicity and speed, especially when applied to hundreds of samples. Actually, only two or three techniques have found wide application. Adsorption of free ligand to materials such as charcoal represents the simplest technique, provided the adsorbing material does not effectively compete for bound ligand.[8,15] These methods effect disequilibration at the time of separation. Significant errors can occur because of transfer of bound ligand from antibody to charcoal. Solid-phase binding assays with prior attachment of either antigen or antibody to a solid support have also been used with good success.[5,29] One method with particular application to the radioimmunoassay involves precipitation of the antigen (ligand)-antibody complex with antigammaglobulin antibody.[11] This second antibody is raised in a different species and generally is directed against determinants on the Fc fragment of the first antibody, thus not interfering with its binding site. In effect, the last two separations are carried out at equilibrium, since some of the free ligand–containing solution remains in contact with either the solid-phase antibody or the second antibody precipitate.

SUMMARY

It is possible to measure any substance in low concentration in physiologic fluids, provided one has a radionuclide-labeled form of high specific activity and a highly specific binding protein that has a dissociation constant equal to the lowest concentration of the substance to be measured. Both antibodies and naturally occurring transport or receptor proteins provide sources for the necessary reagents. Antibodies offer the advantages of stability, great variety, and potentially very high association constants. Receptor and transport proteins offer the advantages of ready availability and discrimination of ligands in relation to their biologic activity. In the development of a new assay, both kinds of binding proteins must be considered, since each may offer special advantages to the investigator.

REFERENCES

1. Berson, S. A., Yalow, R. S., Glick, S. M., and Roth, J.: Immunoassay of protein and peptide hormones, Metabolism **13:**1135, 1964.
2. Bolten, A. E., and Hunter, W. M.: The labelling of proteins to high specific and iodoactivities by conjugation to a ^{125}I-containing acylating agent. Application to the radioimmunoassay, Biochem. J. **133:**529, 1973.
3. Brooker, G., and Jelliffe, R. W.: Serum cardiac glycoside assay based upon displacement of ^{3}H-ouabain from Na-K ATPase, Circulation **45:**20, 1972.
4. Catt, K. J., Dufau, M. L., and Tsuruhara, T.: Radioligand receptor assay of luteinizing hormone and chorionic gonadotrophin, J. Clin. Endocrinol. Metab. **34:**123, 1972.
5. Catt, K. J., Niall, H. D., Tregear, G. W., and Burger, H. C.: Disc solid-phase radioimmunoassay of human luteinizing hormone, J. Clin. Endocrinol. Metab. **28:**121, 1968.
6. Chen, H.-C., Hodgen, G. C., Matsiura, S., Lin, L. J., Gross, E., Reichert, L. E., Birken, S., Canfield, R. E., and Ross, G.: Evidence for a gonadotropin from nonpregnant subjects that has physical, immunological, and biological similarities to human chorionic gonadotropin, Proc. Natl. Acad. Sci. U.S.A. **73:**2885, 1976.
7. Coyvet, J. P., Ross, G. T., Birken, S., and Canfield, R. E.: Absence of neutralizing effect of antisera to the unique structural region of human chorionic gonadotropin, J. Clin. Endocrinol. Metab. **39:**1155, 1974.
8. Daughaday, W. H., and Jacob, L. S.: Methods of separating antibody-bound from free antigen. In Odell, W. D., and Daughaday, W. H., editors: Principles of competitive protein-binding assays, Philadelphia, 1971, J. B. Lippincott Co., p. 303.
9. Fraser, R., and James, V. H. T.: Double isotope assay of aldosterone, corticosterone, and cortisol in human peripheral plasma, J. Endocrinol. **40:**59, 1968.
10. Gorden, P., and Roth, J.: Circulating insulins big and little, Arch. Intern. Med. **123:**237, 1969.
11. Grant, G. H., and Buttl, W. R.: Immunochemical methods in clinical chemistry, Adv. Clin. Chem. **13:**383, 1970.
12. Greenwood, F. C.: Radioiodination of peptide hormones: procedures and problems. In Odell, W. D., and Daughaday, W. H., editors: Principles of competitive protein-binding assays, Philadelphia, 1971, J. B. Lippincott Co., p. 288.
13. Haber, E., Koerner, T., Page, L. B., Kliman, B., and Purnode, A.: Application of a radioimmunoassay for angiotensin I to the physiologic measurement of plasma renin activity in normal human subjects, J. Clin. Endocrinol. Metab. **29:**1349, 1969.
14. Herbert, V.: Studies on the role of intrinsic factor in vitamin B_{12} absorption, transport and storage, Am. J. Clin. Nutr. **7:**433, 1959.
15. Herbert, V., Lank, S., Gottlieb, C. W., and Bleicher, S. J.: Coated charcoal immunoassay of insulin, J. Clin. Endocrinol. Metab. **25:**1375, 1965.
16. Hippe, E., Olesen, H., and Haber, E.: Some observations on vitamin B_{12} binding proteins. In Arnstein, H. R. V., and Wrighton, R. J., editors: The cobalamins: a Glaxo Symposium, Edinburgh, 1971, Churchill Livingstone.
17. Hunter, W. N., and Greenwood, F. C.: Preparation of iodine-131 labelled human growth hormone of high specific activity, Nature (London) **194:**495, 1962.
18. Korenman, S. G.: Relation between estrogen in-

hibitory activity and binding to cytosol of rabbit and human uterus, Endocrinology **87:**1119, 1970.

19. Lefkowitz, R. J., Roth, J., Pricer, W., and Pastan, E.: ACTH receptors in the adrenal: specific binding of ACTH-^{125}I and its relation to adenyl cyclase, Proc. Natl. Acad. Sci. U.S.A. **65:**745, 1970.
20. Lin, S. Y., and Goodfriend, T. L.: Angiotensin receptors, Am. J. Physiol. **218:**1319, 1970.
21. Maguire, M. E., Wiklund, R. A., Anderson, H. J., and Gilman, A. G.: Binding of [^{125}I] iodohydroxybenzylpindolol to putative β-adrenergic receptors of rat glioma cells and other cell clones, J. Biol. Chem. **251:**1221, 1976.
22. Marchalonis, T. J.: Enzymic method for the trace iodination of immunoglobulins and other proteins, Biochem. J. **113:**299, 1969.
23. Midgley, A. R., Jr., and Niswender, G. D.: Radioimmunoassay of steroids, Acta Endocrinol. **147:**320, 1970.
24. Murphy, B. E. P.: Protein binding and its application in the assay of nonantigenic hormones, thyroxine and steroids. In Hayes, R. L., Goswitz, F. A., and Murphy, B. E. P., editors: Radioisotopes in medicine: in vitro studies, Oak Ridge, Tenn., 1968, U.S. Atomic Energy Commission.
25. Murphy, B. E. P., and Pattee, C. J.: Determination of plasma corticoids by competitive protein-binding analysis using gel filtration, J. Clin. Endocrinol. Metab. **24:**919, 1964.
26. Segre, G. V., Habener, J. F., Powell, D., Tregear, G. W., and Potts, J. T.: Parathyroid hormone in human plasma. Immunochemical characterizations and biological implications, J. Clin. Invest. **51:**3163, 1972.
27. Smith, T. W., Butler, V. P., and Haber, E.: Characterization of antibodies of high affinity and specificity for the digitalis glycoside digoxin, Biochemistry **9:**331, 1970.
28. Unger, R. H.: Glucagon antibodies and an immunoassay for glucagon, J. Clin. Invest. **40:**1280, 1961.
29. Wide, L., and Porath, J.: Radioimmunoassay of proteins with the use of Sephadex-coupled antibodies, Biochim. Biophys. Acta **130:**257, 1966.
30. Wolfsen, A. R., McIntyre, H. B., and Odell, W. D.: Adrenocorticotropin measurement by competitive binding receptor assay, J. Clin. Endocrinol. Metab. **34:**684, 1972.
31. Yalow, R. S., and Berson, S. A.: Immunoassay of endogenous plasma insulin in men, J. Clin. Invest. **39:**1157, 1960.
32. Yalow, R. S., Glick, S., Roth, J., and Berson, S.: Radioimmunoassay of human plasma ACTH, J. Clin. Endocrinol. Metab. **24:**1219, 1964.

Index

L

M

N

O

P

Q

R

T

U

V

W

X

Y

Z